International Conference on Electromagnetic Fields, Signals and BioMedical Engineering (ICEMS-BIOMED 2022)

IOP Conference Series: Materials Science and Engineering
Volume 1254

Sibiu, Romania
18 - 20 May 2022

Editors:

Simona Miclaus
Paul Bechet
Mihai Munteanu
Laurentiu Dan Milici

ISBN: 978-1-7138-6154-6
ISSN: 1757-8981

Printed from e-media with permission by:

Curran Associates, Inc.
57 Morehouse Lane
Red Hook, NY 12571

Institute of Physics
Dirac House, Temple Back
Bristol BS1 6BE UK

Phone: 44 1 17 929 7481
Fax: 44 1 17 920 0979

techtracking@iop.org

Additional copies of this publication are available from:

Curran Associates, Inc.
57 Morehouse Lane
Red Hook, NY 12571 USA
Phone: 845-758-0400
Fax: 845-758-2633
Email: curran@proceedings.com
Web: www.proceedings.com

TABLE OF CONTENTS

IOP Conf. Series: Materials Science and Engineering 1254 (2022) 011001 doi:10.1088/1757-899X/1254/1/011001

PREFACE

INTERNATIONAL CONFERENCE ON ELECTROMAGNETIC FIELDS, SIGNALS AND BIOMEDICAL ENGINEERING (ICEMS-BIOMED)

1st Edition, 19-20 May 2022, Sibiu, Romania

The ICEMS-BIOMED conference (https://icems-biomed.emcsb.ro/) was conceived as an biennial itinerant one, which - at the first edition, was organized by a consortium of three prestigious academic institutions from Romania, which share the same values, goals and scientific concerns: Technical University of Cluj-Napoca, “Nicolae Bălcescu” Land Forces Academy of Sibiu and “Ştefan cel Mare” University of Suceava.

The first edition was hosted by the “Nicolae Bălcescu” Land Forces Academy of Sibiu, Romania, in May 2022, and the next editions will be organized in Cluj-Napoca, Romania, (2024) and Suceava, Romania (2026).

The conference was a hybrid one - on-site and on-line - thus offering opportunities to all interested participants to attend the presentations. A number of ten Invited Speakers provided to the audience comprehensive and interesting lectures of forty minutes each. Regular papers were presented at either oral or poster sessions.

The main objective of this scientific event was to create emulation in the academic world in approaching the following topics of great relevance and interest in the present times:

- Electromagnetic Fields
- Electromagnetic Compatibility
- Interactions between Electromagnetic Waves and Biological Systems
- Biomedical Engineering

In addition, the aims were to stimulate young researchers to disseminate their newest results and ideas among the specialists of the academic community and to network together researchers that may further apply for common research projects.

A total of sixty participants were present in the auditory, of which at least half were present on-site, researchers from fourteen countries being among the authors of the papers. A number of approximately fifteen young Bachelor-, Master- and Ph.D.- students were also present in the hall during the delivery of the presentations, belonging mainly to the host institution. The on-line presenters and participants had the possibility to follow each presentation in very good visual and audio conditions, offered by a high quality Zoom platform connection being able to address questions and comments, as well as the attendees that were physically present. The regular presentations were fifteen minutes long, with approximately five minutes of plenary questions and answers. Two poster sessions were organized during the coffee brakes where separate questions could be addressed.

The conference was made possible due to the professional technical and administrative involvement of the staff at the next levels, whom we are addressing special thanks for sustaining the event:

IOP Conf. Series: Materials Science and Engineering 1254 (2022) 011001 doi:10.1088/1757-899X/1254/1/011001

STEERING COMMITTEE:

- Professor Ghiță BÂRSAN, Gl. Bg. – Rector, "Nicolae Bălcescu" Land Forces Academy of Sibiu, Romania
- Professor Vasile ȚOPA – Rector, Technical University of Cluj-Napoca, Romania
- Professor Valentin POPA – Rector, "Ştefan cel Mare" University of Suceava, Romania
- Professor Radu MUNTEANU, Technical University of Cluj-Napoca, Romania
- Professor Adrian GRAUR, "Ştefan cel Mare" University of Suceava, Romania

CONFERENCE CO-CHAIRS:

- Simona Miclaus, "Nicolae Balcescu" Land Forces Academy, Sibiu, Romania
- Paul Bechet, "Nicolae Balcescu" Land Forces Academy, Sibiu, Romania
- Mihai Munteanu, Technical University of Cluj-Napoca, Romania
- Laurentiu Dan Milici, "Stefan cel Mare" University, Suceava, Romania
- Delia Arnaud-Cormos, University of Limoges, France
- Călin Munteanu, Technical University of Cluj-Napoca, Romania
- Radu Ciorap, University of Medicine and Pharmacy, Iasi, Romania

SCIENTIFIC AND REVIEWERS COMMITTEE:

- Adina Giurgiuman - Technical University of Cluj-Napoca, Romania
- Aldo De Sabata - Politehnica University of Timisoara, Romania
- Alexandru Lodin - Technical University of Cluj-Napoca, Romania
- Alexandru Sălceanu - "Gheorghe Asachi" Technical University, Iasi, Romania
- Anca Nicu - Technical University of Cluj-Napoca, Romania
- Andrei Chiciuc - Technical University of Moldova, Chisnau, Moldova
- Angela Lungu - Technical University of Cluj-Napoca, Romania
- Annamaria Sârbu - "Nicolae Bălcescu" Land Forces Academy, Sibiu, Romania
- Antonio Manuel da Cruz Serra - University of Lisbon, Portugal
- Antonio Sarolic - University of Split, Croatia
- Aurelian Rotaru - "Ștefan cel Mare" University, Suceava, Romania
- Călin Corciova - University of Medicine and Pharmacy, Iasi, Romania
- Călin Munteanu - Technical University of Cluj-Napoca, Romania
- Călin Vaida - Technical University of Cluj-Napoca, Romania
- Camelia Pîrghie - "Ștefan cel Mare" University, Suceava, Romania
- Cătălin Popa - Technical University of Cluj-Napoca, Romania
- Claudia Răcăşan-Păcurar - Technical University of Cluj-Napoca, Romania
- Cornel Balint - Politehnica University of Timisoara, Romania
- Corneliu Rusu - Technical University of Cluj-Napoca, Romania
- Dan Doru Micu - Technical University of Cluj-Napoca, Romania
- Dan Mândru - Technical University of Cluj-Napoca, Romania
- Dan Rafiroiu - Technical University of Cluj-Napoca, Romania
- Daniela Irimia - "Ștefan cel Mare" University, Suceava, Romania
- Delia Arnaud-Cormos - University of Limoges, France
- Dorin Petreuș - Technical University of Cluj-Napoca, Romania
- Dragana Popovic Renella - Senis AG, Switzerland
- Dragoș Vicoveanu - "Ștefan cel Mare" University, Suceava, Romania
- Dumitru Ciorbă - Technical University of Moldova, Chisinau, Moldova
- Emanuel Puschiță - Technical University of Cluj-Napoca, Romania

IOP Conf. Series: Materials Science and Engineering 1254 (2022) 011001 doi:10.1088/1757-899X/1254/1/011001

- Enrique Arribas – University of Castilla - La Mancha, Albacete, Spain
- Elma Zanaj – Technical University of Tirana, Albania
- Eugen Coca - "Ștefan cel Mare" University, Suceava, Romania
- Ilie Nucă - Technical University of Moldova, Chisinau, Moldova
- Jakub Svatos - Czech Technical University, Prague, Czech Republic
- Jan Saliga - Technical University of Kosice, Slovakia
- Jana Olivová - University of Defence, Brno, Czech Republic
- Janusz Mindykowski - Gdynia Maritime University, Poland
- Jiří Hozman - Czech Technical University, Prague, Czech Republic
- Lăcrimioara Grama - Technical University of Cluj-Napoca, Romania
- Lars Ole Fichte - Helmut Schmidt University, Hamburg, Germany
- Laura Dărăbant - Technical University of Cluj-Napoca, Romania
- Laurentiu Dan Milici - "Ștefan cel Mare" University, Suceava, Romania
- Levente Czumbil - Technical University of Cluj-Napoca, Romania
- Liliana Sachelarie - Apollonia University, Iasi, Romania
- Luca de Vito - University of Sannino, Italy
- Marian Poboroniuc - "Gheorghe Asachi" Technical University, Iasi, Romania
- Mariana Ciorap - "Gheorghe Asachi" Technical University, Iasi, Romania
- Mariana Milici - "Ștefan cel Mare" University, Suceava, Romania
- Marie Richterova - University of Defence, Brno, Czech Republic
- Marinel Temneanu - "Gheorghe Asachi" Technical University, Iasi, Romania
- Martin Rožánek - Czech Technical University, Prague, Czech Republic
- Mihaela Creţu - Technical University of Cluj-Napoca, Romani
- Mihai Munteanu - Technical University of Cluj-Napoca, Romania
- Octavian Baltag - University of Medicine and Pharmacy, Iasi, Romania
- Octavian Postolache - Institute of Telecommunications, Lisbon, Portugal
- Paul Bechet - "Nicolae Bălcescu" Land Forces Academy, Sibiu, Romania
- Petr Kudrna - Czech Technical University, Prague, Czech Republic
- Philippe Leveque - University of Limoges, France
- Platon Sovilj - University of Novi Sad, Serbia
- Radu Ciorap - University of Medicine and Pharmacy, Iasi, Romania
- Simona Miclăuș - "Nicolae Bălcescu" Land Forces Academy, Sibiu, Romania
- Simona Vlad - Technical University of Cluj-Napoca, Romania
- Theodore Laopoulos - Aristotle University of Thessaloniki, Greece
- Valeriu David - "Gheorghe Asachi" Technical University, Iasi, Romania
- Victor Gropa - Technical University of Moldova, Chisinau, Moldova

ORGANIZING COMMITTEE:

- Paul Bechet - "Nicolae Bălcescu" Land Forces Academy, Sibiu, Romania
- Robert Helbet - "Nicolae Bălcescu" Land Forces Academy, Sibiu, Romania
- Simona Miclăuș - "Nicolae Bălcescu" Land Forces Academy, Sibiu, Romania
- Mihai Munteanu - Technical University of Cluj-Napoca, Romania
- Dan Milici - "Ştefan cel Mare" University of Suceava, Romania
- Levente Czumbil - Technical University of Cluj-Napoca, Romania
- Mihaela Paval - "Ştefan cel Mare" University of Suceava, Romania

LOCAL ORGANIZING COMMITTEE OF "NICOLAE BALCESCU" LAND FORCES ACADEMY, SIBIU, ROMANIA:

- Ioan Virca
- Ioan-Dan Popa
- Iulia Maria Boboi
- Marius Bărbat
- Maria-Lucia Rusu
- Eta Ceclan
- Elena Pleșanu

SPONSORS ACKNOWLEDGEMENTS:

Rohde & Schwarz Romania

Signatories of the Preface,

EDITORS OF THE PROCEEDINGS VOLUME:

- Simona Miclaus, "Nicolae Balcescu" Land Forces Academy, Sibiu, Romania
- Paul Bechet, "Nicolae Balcescu" Land Forces Academy, Sibiu, Romania
- Mihai Munteanu, Technical University of Cluj-Napoca, Romania
- Laurentiu Dan Milici, "Stefan cel Mare" University, Suceava, Romania

Peer Review Statement

All papers published in this volume have been reviewed through processes administered by the Editors. Reviews were conducted by expert referees to the professional and scientific standards expected of a proceedings journal published by IOP Publishing.

1. Type of peer review: Double anonymous
2. Conference submission management system: Morressier
3. Number of submissions received: 51
4. Number of submissions sent for review: 48
5. Number of submissions accepted: 42
6. Acceptance Rate (Submissions Accepted / Submissions Received × 100): 82.3 %
7. Average number of reviews per paper: 2.08
8. Total number of reviewers involved: 34
9. Contact person for queries:
   Name: Simona Miclaus
   Affiliation: "Nicolae Balcescu" Land Forces Academy, Sibiu, Romania
   Email: miclaus@armyacademy.ro

IOP Conf. Series: Materials Science and Engineering 1254 (2022) 012001 doi:10.1088/1757-899X/1254/1/012001

5G Field level measurement for human exposure assessment: A lesson for 6G

M D Migliore

DIEI, University of Cassino and Southern Lazio, Via Di Biasio 43, 03043, IT

E-mail: mdmiglio@unicas.it

Abstract. Field level measurement for human exposure assessment is an important and sometimes underestimated problem. The aim of this work is to describe the problem of field level measurement for the assessment of human exposure from 2G to 5G, showing the growing difficulties caused by an increasing flexibility of cellular communication systems. If on the one hand this flexibility allowed higher performances, on the other hand it made the estimation of the electromagnetic field more and more complex. 5G field level measurement for human exposure assessment represents a formidable complex problem, currently under study to identify reliable and low-cost methods. Vendors have recently understood this problem and have included a control on the power density radiated by antennas in 5G systems, known as the "power lock" function. It is worth noting that for the first time a sophisticated power control is introduced in base stations (BS) for the human exposure target. This should be a lesson for the development of 6G, suggesting a safety-on-design approach to human exposure for the next generation of personal communication systems.

1. Introduction

Measurement of field level for Human Exposure Assessment is an important and sometimes an underestimated problem.

Random nature of communication flow and hence of the electromagnetic field encoding information has been object of large research in the past. The simplest solution is a time- average of the signal. However, the time required, that depends on the local regulation, and ranges from 6 minutes to 24 hours, as in Italy, makes this approach time consuming and still depending on the traffic condition at the time of the measurement.

To overcome the problem related to a direct average, Maximum Power Extrapolation (MPE) [1] techniques have been introduced. MPE techniques use a reference signal transmitted at maximum power to estimate the maximum field level assuming that all the resources of the communication system are assigned to a single user. The value obtained from MPE procedures must subsequently be multiplied by an appropriate correction factor that takes into account the stochastic nature of the communication in order to obtain a realistic value of the field level [2-7].

However, 5G represents a huge step forward in the use of space resources. This introduces new difficulties in the measurement of the field level. In this contribution the problems related to measuring the 5G field level are discussed, showing some solutions. Finally, a quick look at 6G is taken, showing the incredible challenges that the new approaches proposed for the next generation of communication systems pose in field level measurement for human exposure assessment.

2. "Cell-coverage related" field level and "On user" field level

Broadly speaking, one of the challenges of 5G field measurement regards the presence of two different 'field level'. a "cell-coverage related" field level and the "on user" field level in 5G.

Currently deployed 5G systems use *two* different sets of beams, the broadcast beams and the traffic beams (see Figure 1). These sets of beams are designed to meet different requirements, and consequently their radiation characteristics are different.

In order to discuss Broadcast beams, let us recall that, as any cellular communication system, 5G requites to send control signals for User Equipment (UE) access and to compare the strengths of signals from individual cells in 5G networks for handover. These signals must be *always on air* regardless the presence of UEs. For energy efficiency, 5G limits the "always on" signals as much as possible, packing all the fundamental information required for the access to the network in the SS-PBCH signal structure, that is highly concentrated in frequency, time and eventually space [8]. Consequently, in 5G there is only one "always on" NR signal, the Synchronization Signal / Physical Broadcast Channel (SS/PBCH), also called "SS block" (SSB), including the Primary Synchronization Signals (PSS), the Secondary Synchronization Signals (PSS), the Physical Broadcast Channel (PBCH) and the relative demodulation signal PBCH- DMRS. The SS blocks are grouped in block patterns called SS bursts. The structure of the SSB is very compact. SSB is mapped into 4 OFDM symbols in the time domain and 240 contiguous subcarriers in the frequency domain. Each SSB of the SS Burst is transmitted along a specific *broadcast beam* (Figure 1) belonging to a set of beams whose number is equal to the SSBs in the SS burst. The number of beams can change for only one to up to 4 or 8 in FR1 and 64 in FR2. The structure of the SSB is designed to make the transmission of critical information on the cell as robust as possible. For this reason, low cardinality modulations are used, such as BPSK and QPSK, which allow robust decoding even in presence of a relatively low Signal to Interference Noise Ratio (SINR).

Loosely speaking, SSBs are a directional version of synchronization signals that are transmitted with high periodicity.

SSBs are easily identifiable in the 5G frame using spectrum analyzers in zero-span mode and have been proposed by many authors as reference quantity for 5G MPE procedures [8-14]. Furthermore, since the correct reception of the control signals transmitted in SSB is fundamental for accessing the cell, 3GPP considers a specific parameter, called SS-RSRP [15], that directly gives the power associated to the Secondary Synchronization Signal (SSS) [15-16]. The SS-RSRP value can be read using a 5G phone with proper software, or by the numerous portable automatic measurement systems offered on the market. This makes the use of the SS-RSRP very suitable for extensive 5G field measurement [17-19].

Summarizing, SSB related quantities are easily measurable. They include the PSS, SSS, PBCH, DMRS-PBCH and SS-RSRP.

However, payload data transmission must meet different specifications regarding data rates and transmission reliability requirements. For example, in Enhanced Mobile Broadband applications (which is the main "usage scene" of currently deployed 5G systems) a higher transmission rate with less critical reliability is required. In order to increase the data rate, high cardinality modulations are used, f.i. 256 QAM. This requires a high SINR, that forces toward the use of beams having higher directivity compared to the broadcast beams. Accordingly, payload data and in general information associated to a specific user is sent along specific high gain beams, called *traffic beams* (see Figure 1). These beams are present "on the air" only in presence of a UE, and are 'user specific' (more details are in [16]).

Consequently, in case of absence of users connected to the gNB in the cell, the field in the measurement position is due only to the SSBs, and consequently is rather low, as shown in Figure 2. The presence of users increases the average level of the field. It is important to stress that the field reaches the maximum level only when traffic beams are pointed toward the measurement position. As an example, in Figure 4 the frame in case of measurement close to User A (see Figure 3) is shown. The REs of the PDSCH associated to the User A are plotted in red, while the ones associated to the User B are plotted in yellow, indicating different received power in the measurement positions due to the different directions of the traffic beams (more details in [8] and [16]).

The use of flexible beams is a huge technological step-forward in 5G. In order to meet the sophisticated requirements in terms of beams, Active Antennas Systems (AAS) have been introduced in modern 5G base stations. The AAS are currently used to obtain a set of beams pointing toward different directions, called 'grid of beams' from which the most suitable traffic beam is selected. However, AAS are able to implement also more sophisticated communication strategies as MU-MIMO.

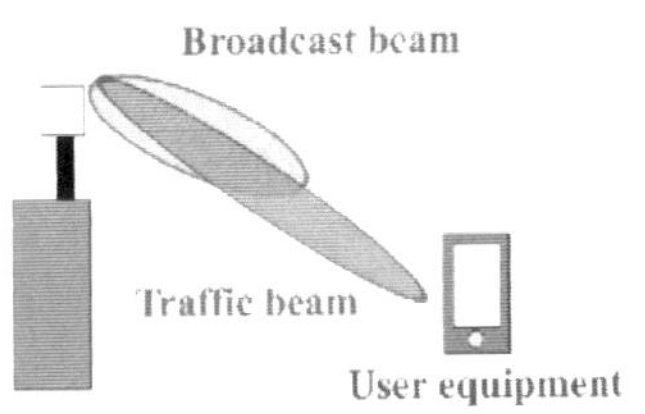

Figure 1. 5G uses two different kinds of grid of beams; the broad- cast beams have lower directivity and are used to transmit signals of interest to all the UEs, as SSP, SSS and PBCH; traffic beams have higher directivity and are steered toward the UE to send UE-specific information, as PDSCH.

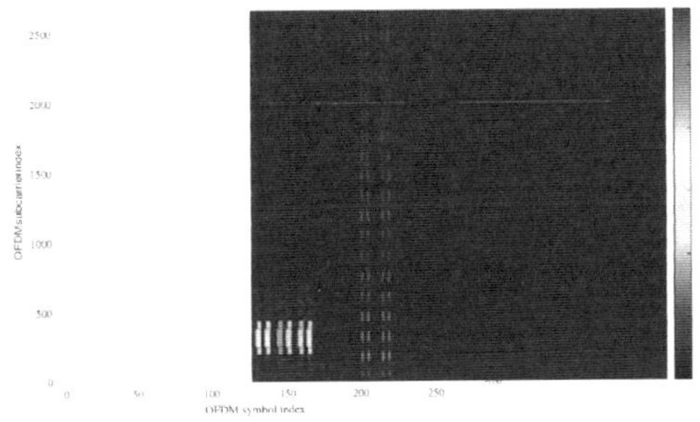

Figure 2. The figure (Power vs Symbol x Carrier map) shows the power level (in false colors, higher levels are in red) of the Resource Elements (REs) in a 5G frame without traffic data; the structures on the lower left corner of the figure are the SSBs; the third SSB is received with the highest power among the SSBs of the SS Block.

2. Maximum Power Extrapolation Techniques from 2G to 5G

The goal of the Maximum Power Extrapolation procedure is to estimate the maximum field level [V/m] that could be reached at the measurement point. We will call this quantity E^{max} [8]. This supposes an unrealistic condition in which a single user saturates all the resources of the communication system. As indicated in the Introduction, this quantity is used as a reference for estimating exposure to electromagnetic fields under realistic conditions using an appropriate scale factor.

Loosely speaking, the procedure requires the estimation of a 'reference' signal that is a cell specific signal always 'on air' and transmitted at maximum power. Instruments usually give the power level as it arrives at the instrument's input connector, let P be. From this value and the knowledge of the Antenna Factor and the attenuation of the cable connecting the antenna to the receiver it is possible to obtain the value of the field level.

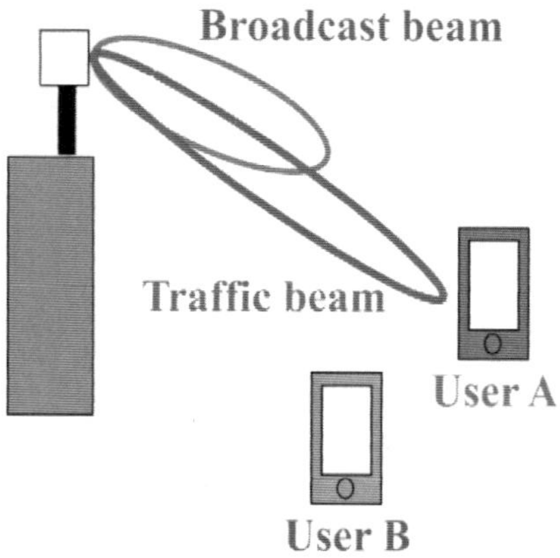

Figure 3. The traffic beam points toward the user that must be served (User A). The level of the field on the User A is high, while the field on the User B, that is not served, is low; when the User B is served the signal on the User A will be low.

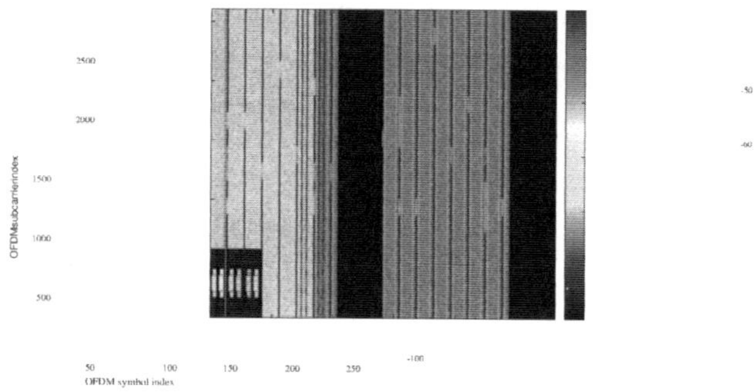

Figure 4. The figure (Power vs Symbol x Carrier map) shows the power level (in false colors, higher levels are in red) of the Resource Elements (RE) in a 5G frame in case of measurement position close to the User A. The REs of the PDSCH associated to the User A are plotted in red, while the ones associated to the User B are plotted in yellow, indicating different received power in the measurement positions due to the different directions of the traffic beams. For a deeper discussion see Ref. [8] and [16].

Finally, this value is multiplied by appropriate factors in order to obtain the maximum field level from the measured field level.
In 2G the signal used as reference is the Broadcast Control Channel (BCCH) whose power level can be measured using a standard spectrum analyzer. The MPE formula is:

$$E_{2G}^{max} = E_{BCCH}\sqrt{N} \tag{1}$$

wherein E_{BCCH} is the measured field level of the BCCH channel, and N is the number of channels of the GSM system.
In 3G the signal used as reference is the Common Pilot Channel (P-CPICH), whose power level measurement requires signal demodulation by signal analyzers. The MPE formula is:

$$E_{3G}^{max} = \frac{E_{CPICH}}{\sqrt{\rho CA}} \tag{2}$$

wherein E_{CPICH} is the Common Pilot Channel field level, and ρCA is the correction factor that takes into account the maximum number of channels that can be used.
In 4G the signal used as reference is the Reference Signal (RS), whose power level measurement requires the demodulation in of the signal by signal analyzers. The MPE formula is:

$$E_{4G}^{max} = E_{RS}\sqrt{N_{SC}} \tag{3}$$

wherein E_{RS} is the field level of a Resource Element of the Reference Signal, and N_{SC} is the number of subcarriers. It is useful to recall that if TDMA technique is used only a part of the Resource Elements are used in downlink connection; the fraction of the REs used in downlink is given by a parameter called F_{TDC}. This parameter is also used in 5G.
In 5G a similar approach can be used, and E^{max} can be estimated as:

$$E_{5G}^{max} = E_{SSB}\sqrt{N_{SC}}\sqrt{F_{TDC}}\sqrt{F_{beam}} \tag{4}$$

wherein the meaning of the parameters is the following: E_{SSB} is the field level of a Resource Elements of the reference signal, that in 5G is a SSB-related quantity, as the power of PSS, SSS, PBCH, PBCH-DMRS, or the value of SS-RSRP. The reference value can also be obtained by a direct measurement of the SSB using scalar analyzers [13]. All these quantities are associated to *broadcast* beams and are always "on air" (see Figure 2).

The factor F_{beam} takes into account the use of traffic beams to send payload data in presence of active users (see Figure 3).

The F_{beam} parameter has a key role in the MPE of 5G signals, since E_{5G}^{max} is obtained in presence of users. Note also that the effect of traffic beam cannot be measured without the presence of an active user placed along the measurement position.

The high flexibility of antennas used in 5G represents a huge challenge for F_{beam} estimation. Unfortunately, information on the configuration of the field radiated by Active Antenna Systems (AAS) provided by the vendors is limited, and not always available. In particular, the AAS radiation characteristics are summarized by the so call 'Envelope Radiation Patterns' [20], defined as *a non-physical radiation pattern obtained by taking, for each direction in azimuth and elevation, the maximum of the absolute, not peak normalized to its own peak, radiation pattern among the radiation patterns that the AAS can generate for a given operating condition(deployment/coverage scenario).*

In real applications, the field at the observation point is given by the interference of many contributions caused by reflection, refraction and diffraction phenomena, to which the direct path is added if the propagation is in LOS (Line of Sight) condition. Multiple reflections cause the presence of fast and slow fading and a significant decrease of the field level in NLOS (No Line of Sight) conditions compared to propagation in free space. In practice, the enormous flexibility in the choice of beams and the fact that they are provided in free space conditions makes an accurate estimation of the value of the factor F_{beam} at an observation point from the knowledge of the radiation pattern of the envelope of the Broadcast and Traffic beams a very cumbersome task.

Consequently, solutions based on 'on-air' measurements to estimate the F_{beam} value by means of numerical simulations, even if more complex and expensive, are of great interest.

A first possible solution is to perform measurements in a period of time, observing traffic data by 'waterfall' reconstruction [13]. In case of users requiring connection in the same direction of the measurement points it is possible to measure the field radiated by traffic beams, and hence the F_{beam} factor.

A different solution, proposed in [8], requires a UE able to force the traffic toward the measurement position. This allows to measure both the broadcast beam level and the traffic beam level, obtaining the value of F_{beam}. A video explaining this method is available in [21].

It should be emphasized that forcing traffic to the measurement point requires actively interacting with the gNB. This also allows access to the rich 5G UE-specific signaling structure (as PDSCH-DMRS, NZ CSI-RS, PDSCH and CSI-RSRP), opening up further interesting scenarios that require neither a cell reference signal nor F_{beam} estimate [16]. Experimental validations of some of the new ideas proposed in the paper [16] has been recently carried out and are reported in [22].

Developing accurate and hopefully low-cost MPEs for 5G is an active research field, filled with new ideas from many research groups.

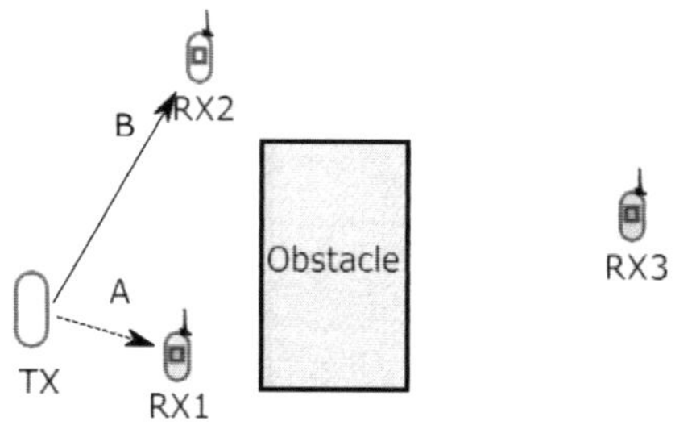

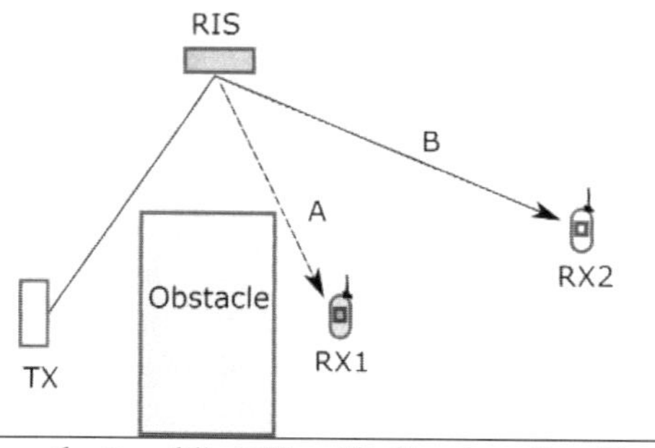

Figure 5. In 5G AASs (Active Antenna Systems) adapt the beams to the need of the communication system, allowing to maximize the communication rate on the user to be served; in the figure TX is the base station, RX1, RX2 and RX3 are two users; at mmW/THz frequency range the AAS allows a LOS connection; consequently RX3 cannot be served.	**Figure 6.** In 6G RISs (Reflecting Intelligent Surfaces) will adapt the environment to the need of the communication system, allowing to reflect the signal toward a specific user; in the figure TX is the base station, RX1 and RX2 are two users; the RIS allows a NLOS (Non Line of Sight) connection varying the angle of reflection according to the user to be served.

3. What should we expect from 6G?

5G introduced many advances compared to 4G at the physical layer level. Among them, the use of millimeter wave (mmW) band and advances space/time communication strategies represent the most critical novelties for field measurement. Millimeter wave requires quite sophisticated measurement devices, increasing the cost of the field estimation procedure. Instead, the use of complex space/time communication strategies, as SU-MIMO and MU-MIMO, as well as the introduction of Active Antennas that can radiate highly flexible patterns, represents a huge challenge in the definition of the measurement procedure. These antennas are able to take advantage of the electromagnetic environment distributing the radiated power density in an optimal way. To do this, they must dynamically adapt their patterns according to the variation of the electromagnetic environment, making the field value in a position depending on the specific scenario (including the positions of the users to be served).

Put simply, in 5G, smart antennas allow you to adapt the communication system to the electromagnetic environment.

6G is still in its infancy, but at the physical layer level the main idea is to use the mmW/THz (millimeter wave/THz) frequency range to achieve higher data rates [23-24]. However, the use of mmW/THz is limited to Line of Sight (LOS) links. In order to extend the coverage, the use of repeaters has been proposed. In particular, besides active repeaters, the research is currently focused on the massive use of low-cost passive repeaters whose reflection angle can be electronically controlled. These structures are known as RIS (Reflecting Intelligent Surfaces) and are based on the so called 'anomalous reflection', i,e. they are able to give an angle of reflection on the surface different from the incident angle. The 'anomalous reflection' is obtained by introducing a linear phase shift on the reflecting surface using small electronically controlled scattering elements placed on the surface.

RIS and smart active repeaters allow to build a reconfigurable electromagnetic environment able to optimize the communication channel with great advantage in terms of data rate and reliability of the communication.

Put simply, in 6G, smart active and passive repeaters will allow you to adapt the electromagnetic environment to the communication system.

This introduces a further variable in the estimation of the level of the electromagnetic field. Until 5G the random nature of the environment was well described by a "fast fading" related to the local variation of the scenario and a "slow fading", related to the large-scale variation of the scenarios, such as the urban structures of the center and of the uptown. The extensive use of controllable reflectors introduces a third 'fading' related to the intelligent variation of the environment, which is dynamically optimized according to the specific user to be served. Depending on the specific implementation, this 'RIS fading'could have the 'worse'characteristics of the two classic fading, i.e. it could be quite large since it involves a profound variation of the electromagnetic scenarios, as in slow fading', and fast enough since the optimization is based on the user served at a given time. Furthermore, the 'RIS fading' would affect both the 'user specific signals' and the 'broadcast signals', making cumbersome or eventually impossible to use broadcast signals as references, as still happen in 5G.

The 'SIR fading' will make field level estimation even more complex than 5G, and it will probably require a deep critical analysis of the approaches followed up to now for the field estimation. Possible solution is to include in the standard mechanisms that help the measurements of the field level and possibly allow also an intrinsic safety of the telecommunications system towards the radiated field level. In this perspective, most field measurements (which are likely to be complex and expensive in 6G) could be limited to confirming the proper functioning of safety mechanisms.

It is worth noting that this is a path that has already been taken to some degree in 5G.

Among the many recent innovations introduced in 5G, one has passed over in silence: the introduction of the so-called 'power lock' (PL) functionality. Basically, the PL is an automatic tool to monitor the power radiated by the antenna in specific sectors associated with specific groups of beams. Radiated power monitoring is a feature originally developed to improve the energy efficiency of the BS. However, it can also be used to monitor the EIRP in specific angular sectors covered by the antenna beams. This possibility is used in the PL to limit the maximum value by ensuring that the average power transmitted in the reference time interval does not exceed the permitted threshold set by local regulations.

The introduction of the PL functionality represents an important step forward toward human exposure control. With PL feature the BS has an *active* role in assuring that the exposure limits are respected, introducing for first time a sophisticated power control in BS having as target the limitation of the human exposure.

The PL functionality was introduced 'retrospectively' in 5G, i.e. after the installation of the first BS, as a response to the problems related to the control of field levels on the population

As discussed above, the complexity of 6G systems will make active control of radiated power from antennas increasingly important, suggesting that PL-like solutions should be considered in the development of the 6G standard.

4. Conclusions

The flexibility of 5G at the physical layer level makes the field level measurement for human exposure assessment a formidable complex problem. Many solutions are currently under study by many groups of research to identify reliable and low-cost methods. Vendors have recently understood this problem and have included a control on the power density radiated by antennas in 5G systems, known as the "power lock" function. It is worth noting that for the first time a sophisticated power control is introduced in BS for the human exposure target.

The experience of 5G suggests considering the problems related to human exposure assessment from the beginning, introducing a *safety-on-design approach* to exposure to electromagnetic fields from the very beginning of the development of the 6G standard.

5. Acknowledgments

The author thanks Dr. Daniele Franci (Arpa Lazio), Dr. Settimio Pavoncello (Arpa Lazio), Dr. Enrico Grillo (Arpa Lazio), Dr.ssa Sara Adda (Arpa Piemonte), Dr. Riccardo Suman (Vodafone Italia SpA), Dr. Stefano D'Elia (Vodafone Italia SpA), Dr. Roberto Cosentino (Rohde Schwarz company) and Dr. Massimo Bauco (Rohde Schwarz company) to provide the experimental data and for their invaluable help in analyzing the measured data.

This work was supported in part by the Ministry of Instruction, University and Research under Grant 'Dipartimenti di Eccellenza (2018-2022)' and Grant 2017SAKZ78 (PRIN Project MIRABILIS).

References

[1] International Electromechanical Commission 2017 IEC 62232:2017 Determination of RF field strength, power density and SAR in the vicinity of radiocommunication base stations for the purpose of evaluating human exposure

[2] Baracca P, Weber A, Wild T and Grangeat C 2018 A statistical approach for RF exposure compliance boundary assessment in massive MIMO systems *WSA 2018 22nd International ITG Workshop on Smart Antennas* VDE pp 1–6

[3] Thors B, Furusk¨ar A, Colombi D and T¨ornevik C 2017 Time-averaged realistic maximum power levels for the assessment of radio frequency exposure for 5G radio base stations using massive mimo *IEEE Access* vol **5** pp 19711–19719

[4] Pinchera D, Migliore M D and Schettino F 2020 Compliance boundaries in 5G communication systems: a statistical approach *IEEE Access* **1**(1) pp 620–628

[5] Colombi D, Joshi P, Xu B, Ghasemifard F, Narasaraju V and T¨ornevik C 2020 Analysis of the actual power and EMF exposure from base stations in a commercial 5G network *Applied Sciences* **10**(15) p 5280

[6] Migliore M D and Schettino F 2020 Power reduction estimation of 5G active antenna systems for human exposure assessment in realistic scenarios *IEEE Access* vol **8** pp 220095–220107

[7] Xu B, Sanjurjo D A, Colombi D and T¨ornevik C 2021 A Monte Carlo analysis of actual maximum exposure from a 5G millimeter-wave base station antenna for EMF compliance assessments *Frontiers in Public Health* vol **9**

[8] Adda S, Aureli T, D'Elia S, Franci D, Grillo E, Migliore M D, Pavoncello S, Schettino F and Suman R 2020 A theoretical and experimental investigation on the measurement of the electromagnetic field level radiated by 5G base stations *IEEE Access* vol **8** pp 101448–101463

[9] Keller H 2019 On the assessment of human exposure to electromagnetic fields transmitted by 5GNR base stations *Health physics* **117**(5) pp 541–545

[10] Swiss Federal Institute of Metrology 2020 Technical report: Measurement method for 5G NR base stations up to 6 GHz

[11] Franci D, Coltellacci S, Grillo E, Pavoncello S, Aureli T, Cintoli R and Migliore M D 2020 Experimental procedure for fifth generation (5G) electromagnetic field (EMF) measurement and maximum power extrapolation for human exposure assessment *Environments* **7**(3) p 22

[12] Franci D, Coltellacci S, Grillo E, Pavoncello S, Aureli T, Cintoli R and Migliore M D 2020 An experimental investigation on the impact of duplexing and beamforming techniques in field measurements of 5G signals *Electronics* **9**(2) p 223

[13] Aerts S, Verloock L, Van Den Bossche M, Colombi D, Martens L, T¨ornevik C and Joseph W 2019 In-situ measurement methodology for the assessment of 5G NR massive MIMO base station exposure at sub-6 GZz frequencies *IEEE Access* vol **7** pp 184658–184667

[14] Wali S Q, Sali A, Allami J K and Osman A F 2022 RF-EMF exposure measurement for 5G over Mm-wave base station with MIMO antenna *IEEE Access*

[15] 3GPP 2019 Base station (BS) radio transmission and reception (Release 15) 3 GPP technical specification group radio access network; nr.TS 38.104, v15.5.0 May 2019

[16] Migliore M D, Franci D, Pavoncello S, Grillo E, Aureli T, Adda S, Suman R, D'Elia S and Schettino F 2021 A new paradigm in 5G maximum power extrapolation for human exposure assessment: Forcing gNB traffic toward the measurement equipment *IEEE Access* vol **9** pp 101946–101958

[17] Lee A-K, Jeon S-Band Choi H-D 2021 EMF levels in 5G new radio environment in Seoul Korea *IEEE Access* vol. **9** pp 19716–19722

[18] Chiaraviglio L, Lodovisi C, Franci D, Pavoncello S, Aureli T, Blefari-Melazzi N and Alouini M-S 2021 Massive measurements of 5G exposure in a town: Methodology and results *IEEE Open Journal of the Communications Society* **2** pp 2029-2048

IOP Conf. Series: Materials Science and Engineering 1254 (2022) 012001 doi:10.1088/1757-899X/1254/1/012001

[19] Brahim S *et al* 2021 Exposure of the South Korean population to the 5th generation (5G) of mobile phone networks (3.4 to 3.8 GHz) and its respective contribution among other radio frequencies in the country *URSI GASS 2021 Rom.* IURSI pp 1–4
[20] N. Alliance 2020 *Recommendation on base station active antenna system standards*
[21] https://sites.google.com/unicas.it/electromagnetic-information/home accessed on 2021-03-01
[22] Adda S, Aureli T, Bastonero S, D'Elia S, Franci D, Grillo E, Migliore M D, Pasquino N, Pavoncello S,Schettino F *et al* 2022 Methodology based on vector and scalar measurement of traffic channel power levels to assess maximum exposure to electromagnetic radiation generated by 5G NR systems *IEEE Access* **10**
[23] Migliore M D 2020 Big, bigger, biggest. What should we expect from 6G antennas? *2020 XXXIIIrd General Assembly and Scientific Symposium of the International Union of Radio Science*. IEEE pp 1–4
[24] Migliore M2021 A quick look to 6G antenna concepts as seen from the deep physical layer level *2021 XXXIVth General Assembly and Scientific Symposium of the International Union of Radio Science (URSI GASS)* IEEE pp 1–3

RF-EMF Measurement for 5G over C-band and mm-Wave frequency band: Exposure assessment and Procedures

S Qahtan Wali

WiPNET Research Centre, Department of Computer and Communication Systems Engineering, Faculty of Engineering, Universiti Putra Malaysia (UPM), Serdang, Selangor 43400, Malaysia

Abstract. The 5G network is intended to accommodate a significant quantity of mobile data traffic as well as a great number of wireless connections. It improves cost, power consumption, and offers ultra-low latency and ultra-high dependability to enable new services in a variety of sectors. However, the general public is concerned about the possible health dangers linked with 5G equipment's Radio Frequency (RF) radiation, and numerous localities are actively lobbying to prevent 5G implementation. This aims to demonstrate the cause of increasing the amount of RF-EMF exposure, and the international standards of RF-EMF accepted limits. In addition, it aims to illustrate some measurement procedures to conduct RF-EMF measurement for different researchers and two other procedures and the result of the maximum exposure emitted by a 5G base station operating on 3.5 GHz and the mm-Wave frequency band by the author of this paper.

The 5G technologies standard, among others, is being supported and implemented globally [1]. The fifth-generation (5G) new radio (NR) systems have been deployed on different frequency bands since 2019 in commercial areas, like North America, Europe, Oceania, and portions of Asia [2]. The 5G network is expected to deliver faster throughputs, better bandwidth, more capacity, and extraultra-low latency [3]. The 5G network system is anticipated to be more reliable, adaptable, and dependable than other mobile generations. The 5G mobile network also provides many other services, like the ecology of the telecommunication network, which delivers energy-efficient services to the healthcare, agricultural, and smart city projects [4]. To achieve the goals of the 5G mobile communication wireless network, newly developed technologies including the use of different frequency ranges like C-band and millimeter-wave (mmWave), massive (MIMO) antennas, and small cells should be deployed [5].

On the one hand, Millimeter-wave(mm-Wave) has been known as an important technology for 5G wireless communications [6]. Usually, mm-Wave refers to frequency bands from 30 GHz to 300 GHz[7], but often 10 GHz to 30 GHz band are known as mm-Wave because of sharing certain propagation characteristics, and it is wavelength ranging located between 1mm and 100mm [8-9]. Mm-Wave provides a much larger bandwidth, higher throughput, faster data rate, and capacity compared with, 3G, 4G, and 5G C-band frequency ranges [9-12]. However, C-band is now the only core spectrum available for fifth-generation around the world. It is also known as the (Mid-band) or sub-6GHz band, which offers the best technological and economic benefits for long-term 5G implementation success. C-band operates at frequencies ranging from 3.4GHz to 4.2GHz. High transmission bandwidth, wide-area coverage, throughput constancy, cost-efficiency, boost successful peak, and cheaper end-user device costs are all advantages of C-Band. Furthermore, C-band has fewer attenuations to rain and oxygen compared to the other bands. C-band waves can send and receive information for 5G smartphones and equipment over longer distances at a slower speed, and C-band electromagnetic waves penetrate buildings easier than mmWave band to provide interior services [13]. Sub-6GHz offers a coverage advantage over the mmWave bands, which has been verified by experiments because of the bigger surface area of antenna elements and higher diffraction efficiency.

On the other hand, as these technologies become more widely adopted, there are growing concerns regarding the health and safety implications of exposure to radiofrequency electromagnetic fields (RF-EMF) radiated by 5G BSs [14-15]. EMF exposure is a source of concern for many people since it cannot be prevented. RF-EMF exposure has been linked to some health problems [16], including cancer, skin damage, ocular abnormalities, glucose metabolism, and oxygen effects. Even though some diseases have been only identified in animals, the argument over the potential health effects of Rf radiation isa critical subject. Three reasons lead to rising in the amount of RF-EMF exposure. The fifth-generation system requires a greater signal power at the receiver side to properly send energy where it is wanted to offer a higher data rate [17]. As a result, the amount of EMF produced will rise, and it will come into contact with consumers. Second, more base stations and small cells will be implemented in the 5G system [18]. These base stations, which will be situated near the customers, will provide connectivity for a small area. As a result, users will have a greater possibility of being exposed to EMF [19]. Third, in the 5G system, narrower beams (SSB) are used. Multiple antennas are employed in the 5G base stations to boost the antenna gain[20]. Humans are more susceptible to electromagnetic fields when electromagnetic energy is concentrated in a smaller area [21].

Depending on years of scientific research, the International Commission on Non-Ionizing Radiation Protection (ICNIRP) and the Institute of Electrical and Electronics Engineers (IEEE) have produced guidelines for reducing human exposure to electromagnetic fields (EMF) [22-23]. Internationally respected agencies such as the World Health Organization(WHO), the US Federal Communications Commission (FCC) [24], and the International Telecommunications Union (ITU) [25], have concluded that EMF exposure below the limit recommended by ICNIRP guidelines does not appear to have any known consequence to health and is safe.

In the last two decades, RF-EMF radiation has become a criticaltopic for the telecommunication industry, so several studies have investigated previous and current wireless network generations such as 2G, 3G, 4G, and 5G new radio. Here are some measurement procedures and results for some research:

Researchers in [26] monitored and studied the amount of RF-EMF exposure that is generated from 5G BSs by measuring the transmission power for 25 base stations in Telstra, Australia. Massive MIMO antennas were used at the BSs to use beamforming and optimize signal strength at the smartphone. They continued to measure 24 h over a week to know the actual transmitted power and EMF exposure. The data was collected through the network which is time-averaged transmitted power and beam scheduling for spatial distribution calculation of transmitted power. The study was carried out on a commercial 18 live network located in urban dense areas operating in new radio band n78 (3300-3800 MHz) with a channel bandwidth ranging between 40 and 80 MHz between the sites. The network vendor (Ericsson) network management tool was utilized for the data collection instead of measurement conduction. Their method was based on total cell time-averaged transmitted power through the whole 24 hours of 6 minutes. All of the base stations were found to operate a maximum average power of 22.5-25%. majority of average power were found within 0-2.5% of maximum. The study clearly shows the difference between the maximum and actual average power transmitted. The theoretical maximum radiation was found to be smaller than the maximum time-averaged power for each beam path.

The study's author [25] established and tested a new systematic approach using spectrum analyzer equipment to measure and calculate in-situ the time-averaged simultaneous emission and theoretical maximum exposure from the 5G NR station. Their approach includes procedures such as identifying synchronization signal block (SSB), which is the only 5G NR always-on element that is sent regularly and at constant power, determining the power density carried by its resources, and extrapolating to the theoretical E_{max} exposure amount. The approach was tested in the line-of-sight 5G base station that was operating on 3.5 GHz. The base station was built to function with a fixed beam indefinitely to validate the methods in a controlled environment. One user equipment (UE) device was used for five tests which are 100% downlink, 100% uplink, voice call, video call, and video streaming). The greatest amount of theoretical exposure from the 5G base station was found to be 5.537 V/m. The results indicated that the amount of exposure that is generated from the base station was well below the ICNIRP standard.

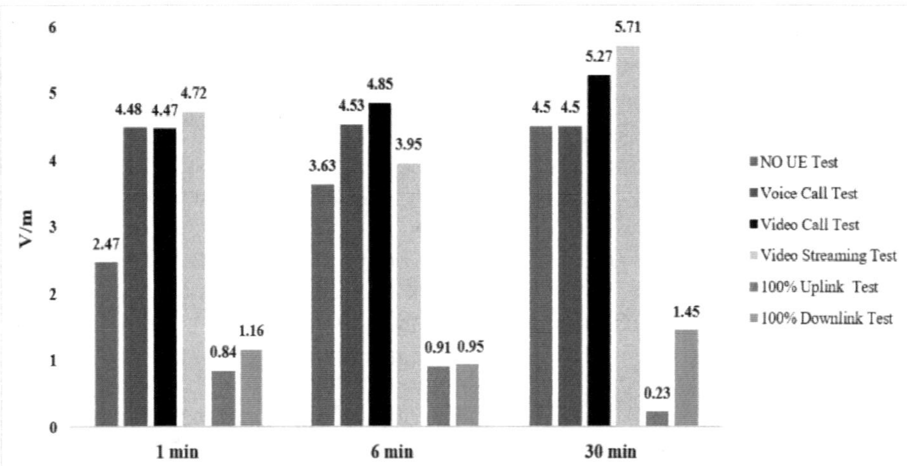

Figure 1. Maximum RF-EMF exposure generated from a 5G mm-Wave base station

In our study [27], the RF-EMF exposure for a 5G base station with a MIMO antenna that operates on the mm-Wave frequency band in Malaysia was conducted to investigate the maximum and average exposure. The study was based on the code selective method. The BS was operating on 29.5 GHz and had only one sector (PCI) that had four channels. Six different tests were taken except one of the test UE devices has not been used in the other 5 tests the UE device was utilized. The measurement was conducted at 1min, 6mins, and 30 mines. The location of the measurement was selected based on getting maximum RSRP power. The distance between Tx and Rx was 22m, the height of the Tx base station was 25 meters and Rx was 1.5m. The maximum exposure from the base station was found to be 5.71 V/m, and the average exposure was 2.02V/m (Figure 1).The Rohde & Schwarz 5G site testing solution (R&S 5G STS) was selected for the measurement. The (R&S 5G STS) consists of the R&S®TSMA6 scanner, R&S R&S®ROMES4 software, R&S®TSME30DC downconverter, and UX241 GPS (TSME-ZA4), and 26 - 40 GHz Vertically Polarized Omnidirectional Antenna fitted with a K type Connector (Figure 2).

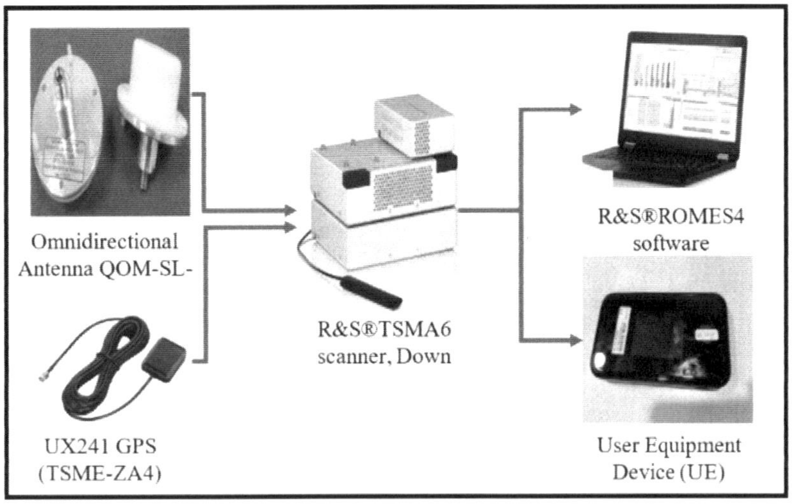

Figure 2. Measurement Equipment.

References:

[1] Mandl P, Pezzei P, and Leitgeb E 2020 Comparison of Radiation Exposure between DVBT2, WLAN, 5G and other Sources with Respect to Law and Regulation Issues *2020 International Conference on Broadband Communications for Next Generation Networks and Multimedia Applications (CoBCom)* IEEE pp 1-5

[2] M. Series 2015 IMT Vision–Framework and overall objectives of the future development of IMT for 2020 and beyond *Recommendation ITU* vol. 2083-0

[3] Niknejad A M, Thyagarajan S, Alon E, Wang Y and Hull C 2015 A circuit designer's guide to 5G mm-wave *2015 IEEE Custom Integrated Circuits Conference (CICC)* IEEE pp 1-8

[4] Kumar T A, Akash R and Kumar S A 2019 Role of 5G Networks: Issues, Challenges and Applications *International Journal of Engineering and Advanced Technology (IJEAT)* **8**(6) pp 3172-3178

[5] Bogale T E and Le L B 2016 Massive MIMO and mmWave for 5G wireless HetNet: Potential benefits and challenges *IEEE Vehicular Technology Magazine* **11**(1) pp 64-75

[6] Kyösti P, Lehtomäki J, Medbo J and Latva-aho M 2017 Map-based channel model for evaluation of 5G wireless communication systems *IEEE Transactions on Antennas and Propagation* **65**(12) pp 6491-6504

[7] Rangan S, Rappaport T S and Erkip E 2014 Millimeter-wave cellular wireless networks: Potentials and challenges *Proceedings of the IEEE* **102**(3) pp 366-385

[8] Huang J, Wang C-X, Feng R, Sun J, Zhang W and Yang Y 2017 Multi-frequency mmWave massive MIMO channel measurements and characterization for 5G wireless communication systems *IEEE Journal on Selected Areas in Communications* **35**(7) pp 1591-1605

[9] Lee J-H, Choi J-S, Lee J-L and Kim S-C 2017 Permittivity effect of building materials on 28 GHz mmWave channel using 3D ray tracing simulation *GLOBECOM 2017-2017 IEEE Global Communications Conference* pp 1-6

[10] Mahmoud K R and Montaser A M 2018 Synthesis of multi-polarised upside conical frustum array antenna for 5G mm-Wave base station at 28/38 GHz *IET Microwaves, Antennas & Propagation* **12**(9) pp 1559-1569

[11] Gao X, Edfors O, Rusek F and Tufvesson F 2015 Massive MIMO performance evaluation based on measured propagation data *IEEE Transactions on Wireless Communications* **14**(7) pp 3899-3911

[12] Ghosh A 2016 The 5G mmWave radio revolution *Microw. J* **59**(9) pp 22-36

[13] Halvarsson B, Simonsson A, Elgcrona A, Chana R, Machado P and Asplund H 2018 5G NR testbed 3.5 GHz coverage results *2018 IEEE 87th Vehicular Technology Conference (VTC Spring)* pp 1-5
[14] Adda S *et al 2020* A theoretical and experimental investigation on the measurement of the electromagnetic field level radiated by 5G base stations *IEEE Access* **8** pp 101448-101463
[15] Velghe M, Aerts S, Martens L, Joseph W and Thielens A 2021 Protocol for personal RF-EMF exposure measurement studies in 5th generation telecommunication networks *Environmental Health* **20**(1) pp 1-10
[16] Chiaraviglio L, Elzanaty A and Alouini M-S 2021 Health risks associated with 5G exposure: A view from the communications engineering perspective *IEEE Open Journal of the Communications Society* **2** pp 2131-2179
[17] Nasim I and Kim S 2019 Mitigation of human EMF exposure in downlink of 5G *Annals of Telecommunications* **74**(1) pp 45-52
[18] Chiaraviglio L, Bianchi G, Blefari-Melazzi N and Fiore M 2020 Will the Proliferation of 5G Base Stations Increase the Radio-Frequency "Pollution"? *2020 IEEE 91st Vehicular Technology Conference (VTC2020-Spring)* IEEE pp 1-7
[19] Nasim I 2019 Analysis of Human EMF Exposure in 5G Cellular Systems
[20] Kim S, Visotsky E, Moorut P, Bechta K, Ghosh A and Dietrich C 2017 Coexistence of 5G with the incumbents in the 28 and 70 GHz bands *IEEE Journal on Selected Areas in Communications* **35**(6) pp 1254-1268
[21] Moon J-H 2020 Health effects of electromagnetic fields on children *Clinical and experimental pediatrics* **63**(11) p 422
[22] Guidelines for limiting exposure to time-varying electric, magnetic, and electromagnetic fields (up to 300 GHz). International Commission on Non-Ionizing Radiation Protection. Available from: https://pubmed.ncbi.nlm.nih.gov/9525427/
[23] I. o. Electrical and E. Engineers 2006 *IEEE standard for safety levels with respect to human exposure to radio frequency electromagnetic fields, 3 kHz to 300 GHz*. (IEEE: New York)
[24] Fields R E 1997 Evaluating compliance with FCC guidelines for human exposure to radiofrequency electromagnetic fields *OET bulletin* **65**(10)
[25] Aerts S *et al 2019* In-situ measurement methodology for the assessment of 5G NR massive MIMO base station exposure at sub-6 GHz frequencies *IEEE Access* **7** pp 184658-184667
[26] Colombi D, Joshi P, Xu B, Ghasemifard F, Narasaraju V and Törnevik C 2020 Analysis of the actual power and EMF exposure from base stations in a commercial 5G network *Applied Sciences* **10**(15) p 5280
[27] Wali S Q, Sali A, Allami J K and Osman A F 2022 RF-EMF Exposure Measurement for 5G over mm-Wave Base Station with MIMO Antenna *IEEE Access*

IOP Conf. Series: Materials Science and Engineering 1254 (2022) 012003 doi:10.1088/1757-899X/1254/1/012003

Recent trends for 5G in vitro exposure systems dosimetry and setup

D Arnaud-Cormos[1,2], R Orlacchio[1] and P Leveque[1]

[1]Limoges University, CNRS, XLIM, UMR 7252 F-87000 Limoges, France
[2] University Institute of France, 75005, Paris, France
E-mail: delia.arnaud-cormos@unilim.fr

Abstract. 5G wireless communications networks are being massively studied and developed. This new technology introduces new operating frequencies and employs smart adaptative antennas that allow focusing electromagnetic waves in specific directions. Bioelectromagnetic investigations to assess exposure and interactions with living entities of these new 5G signals are carried out. This contribution aims to present current trends on exposure systems designs for bioelectromagnetic studies with associated computational and experimental dosimetry. The extension to 5G signals of classical exposure systems such as transverse electromagnetic cells (TEM cells) and new innovative designs are discussed.

1. Introduction

Wireless communications technologies such as mobile phones networks have been in progress for almost 50 years with the fifth generation of mobile communications (5G) being intensively developed and deployed in recent years [1]. Compared to the fourth generation, 5G proposes performance breakthrough in terms of data speed, which is multiplied by 10, ultra-low latency, which is divided by 10 and also more communication reliability, increased availability, massive network capacity, etc. In addition to enhanced mobile communication performance, this technology aims diverse uses such as massive internet of things (IoT), smart cities, autonomous and connected vehicles, or virtual reality applications [2-3]. 5G will employ smart adaptative antennas based on "Massive MIMO" (Multiple Input Multiple Output) composed of multiple transmitters integrated in a single device which will focus the electromagnetic waves over a given area, user, or endpoint. This focus, called beamforming, allows to increase the network capacity and a more efficient bandwidths allocation for communications and energy management [4-5].

With the 5G rollout, questions were raised concerning the exposure, interaction, and possible health effects on living entities of specific 5G generated electromagnetic waves. 5G will use new frequency bands from around 700 MHz (694–790 MHz) to 2.1 GHz, 3.5 GHz (3.4–3.8 GHz) and later the 26 GHz band (24.25–27.5 GHz) [6]. As the frequency increases, the penetration of electromagnetic waves in tissues and organs decreases. Indeed, 5G higher bands waves will mainly be present within a few millimeters of a tissue i.e., about 3 to 6 times less compared to 3G or 4G frequencies for which the specific absorption rate (SAR) is typically attenuated by a ratio of 100 over 2–3 cm. As a consequence of the weaker penetration, studies of effects and potential risks will be more focused on targets such as human skin, eyes, or nerves terminations [7].

The 700 MHz–2.1 GHz band includes frequencies of existing generations of mobile communication technology 2G, 3G and 4G. Although research continues, particularly on cancer, altered brain function, or fertility, numerous studies were carried with no established evidence of health effects associated with exposure to sources of electromagnetic fields in this band. On the contrary, existing literature concerning effects is rather limited in the 3.5 GHz frequency band and quite scarce at 26 GHz. Therefore, it is important to continue bioelectromagnetic investigations [6-8] to produce new data and knowledge as recommended by expert organizations [6] (ANSES, WHO, ICNIRP, etc). In Europe, a certain number of research projects founded by the European community or national organizations are currently underway with expected results in the next two to five years.

For bioelectromagnetic experiments, several types of exposure systems were developed over the past decades for both *in vivo* and *in vitro* studies [9]. Based on the electromagnetic wave coupling, the exposure systems can be classified as propagating TEM cells [10], [11], resonant waveguides [12], or radiating antennas [13]. Moreover, the systems can be integrated in specific containers such as incubators or chambers which ensure typical cellular maintenance conditions such as temperature, humidity, and CO_2.

2. 5G exposure systems and dosimetry

2.1. 3.5 GHz and prospective 26 GHz exposure systems

Characteristic dosimetry and associated exposure systems based on classical TEM cells are currently studied and proposed for the 5G middle band at 3.5 GHz. Indeed, TEM cells known for their wide frequency bandwidth and capacity to deliver homogenous fields were implemented for studies [10, 11] mainly up to 2.6 GHz (4G-LTE).

A small TEM cell accommodating a multi-electrode array (MEA) [14] is characterized and adapted for 5G 3.5 GHz electromagnetic fields (EMF) exposure for electrophysiological measurements of neuronal activity. Due to the proximity of phones and brain, the central nervous system is still considered one of the possible targets of 5G signals at 3.5 GHz. This specific dosimetry allows investigations of firing and bursting rates of neonatal rat neuronal cultures for the study of 5G exposures of neuronal electrical activity. A 3.5 GHz adapted TEM cell containing a Teflon chamber is also designed for using Bioluminescent Resonant Energy Transfer (BRET) measurements [15]. Examples of cells such as primary non-excitable Schwann cells, skin fibroblasts and keratinocytes, can be placed on glass coverslips into a white opaque measurement chamber containing saline solution. The chamber is designed to allow BRET measurements at molecular level for the monitoring of the activation of specific stress response pathways while simultaneously exposing cells to RF fields.

Moreover, we recently designed a new, innovative setup based on mode-stirred reverberation chamber (RC) implemented in a cell culture incubator. This system allows homogeneous EMF delivery to adherent cells in different tissue culture plates (e.g. 6 and 96- wells) or exposure conditions (SAR values, exposures up to several days, etc.) while simultaneously providing a controlled atmosphere (37°C, 5% CO_2, and 95% humidity).

Furthermore, with the frequency increase particularly for 5G at 26 GHz, propagating or resonant systems can become limited for exposures of typical biological holders. Indeed, the systems size diminishing with the frequency, biological holders such as flasks, 6-well, 96-well plates or Petri dishes can no longer be physically inserted in some classical systems such TEM cells. Therefore, for small volumes exposures, designs based on coplanar waveguides or, for classical holders, radiating antennas such as horn, printed microstrip patch or dipoles are currently explored.

2.2. Computational and experimental dosimetry of 5G EMF

In our studies, exposure systems are dosimetrically characterized in terms of SAR distribution and temperature within exposed samples. SAR levels are numerically evaluated using full-wave 3D electromagnetic simulation software based on Finite Difference Time Domain (FDTD) method [13, 15, 16]. SAR are also assessed through experimental measurements of temperature inside the exposed samples with optic fiber temperature probes (Luxtron One, Lumasense Technologies, CA, USA). These probes typically allow acquisitions every 250 milliseconds within a 1 mm^3 volume. At microscopic level of a few tens of micrometers, such as for biological cells or electrodes of an MEA, fluorescence measurements using the temperature dependent dye Rhodamine B, are investigated for assessing temperature [17].

Our studies on TEM cells evidence that for a normalized incident power of 1 W, SAR levels and distributions depend on the signal operating frequency. Indeed, for the same exposure configuration, e.g., TEM cell with an MEA containing neuronal networks within 1 mL of culture media, significantly higher SAR levels and less homogeneity are obtained when applying 5G-3.5 GHz signals compared to 3G-2.1 GHz for instance.

3. Conclusion

With the intensive development of 5G mobile communications, particularly within new frequency bandwidths i.e., 3.5 GHz, investigations on exposure assessment and effects on living entities are carried out. In this contribution, recent trends in terms of 5G in vitro exposure designs and associate dosimetry were presented. Computational and experimental dosimetry evidence for instance that TEM cell-based designs are compatible with 5G exposures at 3.5 GHz providing high SAR values and efficiency. New incubator-based reverberating chambers allow simultaneous exposures of large sample volumes under various conditions i.e., SAR levels, exposure durations, yielding powerful statistical results. Future investigations will include developments on 26 GHz exposure systems.

References

[1] Navarro-Ortiz J, Romero-Diaz P, Sendra S, Ameigeiras P, Ramos-Munoz J J and Lopez-Soler J M 2020 A Survey on 5G Usage Scenarios and Traffic Models *IEEE Communications Surveys Tutorials* **22** (2) pp 905–929 doi: 10.1109/COMST.2020.2971781

[2] Kumar S, Dixit A S, Malekar R R, Raut H Dand Shevada L K 2020 Fifth Generation Antennas: A Comprehensive Review of Design and Performance Enhancement Techniques *IEEE Access* **8**pp 163568-163593 doi: 10.1109/ACCESS.2020.3020952

[3] Shafique K, Khawaja B A, Sabir F, Qazi S and Mustaqim M 2020 Internet of Things (IoT) for Next-Generation Smart Systems: A Review of Current Challenges, Future Trends and Prospects for Emerging 5G-IoT Scenarios *IEEE Access* **8** pp 23022–23040 doi: 10.1109/ACCESS.2020.2970118

[4] 5G Networks, '5G beamforming, beam steering and beam switching with massive MIMO' *5G Networks* 2020 available from: https://www.5g-networks.net/5g-technology/5g-beamforming-beam-steering-and-beam-switching-with-massive-mimo/ (accessed Mar. 30, 2022)

[5] Suyama S, Okuyama T, Nonaka N and Asai T 2022 Recent Studies on Massive MIMO Technologies for 5G Evolution and 6G *2022 IEEE Radio and Wireless Symposium (RWS)* pp 90–93 doi: 10.1109/RWS53089.2022.9719949

[6] Agence nationale de sécurité sanitaire de l'alimentation, de l'environnement et du travail (ANSES) 2021*Expositions aux champs électromagnétiques liées au déploiement de la technologie de communication « 5G » et effets sanitaires éventuels associés.*

[7] Arnaud-Cormos D. and Leveque P 5G et santé: que sait-on? *The Conversation* available from: http://theconversation.com/5g-et-sante-que-sait-on-147295

[8] Di Ciaula A 2018 Towards 5G communication systems: Are there health implications? *International Journal of Hygiene and Environmental Health* **221**(30) pp 367-375 doi: 10.1016/j.ijheh.2018.01.011

[9] Hansen J W *et al.* 2020 A systematic review of in vitro and in vivo radio frequency exposure methods *IEEE Reviews in Biomedical Engineering* **13** pp 340351,.

[10] Iftode C and Miclaus S 2012 Design and Validation of a TEM Cell Used for Radiofrequency Dosimetric Studies *Progress In Electromagnetics Research* **132** pp 369-388 doi: 10.2528/PIER12081306

[11] Soueid M et al. 2014 Electromagnetic Analysis of an Aperture Modified TEM Cell Including an Ito Layer for Real-Time Observation of Biological Cells Exposed to Microwaves *Progress In Electromagnetics Research* **149** pp193-204 doi: 10.2528/PIER14053108

[12] Schuderer J, Spat D, Samaras T, Oesch W and Kuster N 2004 In vitro exposure systems for RF exposures at 900 MHz *IEEE Transactions on Microwave Theory and Techniques* **52**(8) pp 2067–2075doi: 10.1109/TMTT.2004.832010
[13] Collin A et al. 2016 In vivo setup characterization for pulsed electromagnetic field exposure at 3 GHz' *Phys Med Biol* **61**(16) pp 5925-5941 doi: 10.1088/0031-9155/61/16/5925
[14] Nefzi A et al. 2022 Dosimetry of Microelectrodes Array Chips for Electrophysiological Studies Under Simultaneous Radio Frequency Exposures *IEEE Transactions on Microwave Theory and Techniques* **70**(3) pp 1871-1881 doi: 10.1109/TMTT.2021.3136296.
[15] Poque E et al 2020 Effects of radiofrequency fields on RAS and ERK kinases activity in live cells using the bioluminescence resonance energy transfer technique *Int J Radiat Biol* **96**(6) pp 836-843 doi: 10.1080/09553002.2020.1730016
[16] Leveque P, Reineix A and Jecko B 1992 Modelling of dielectric losses in microstrip patch antennas: application of FDTD method *Electronics Letters* **28**(6) pp 539–541
[17] Nefzi A, Carr L, Dalmay C, Pothier A, Leveque P and Arnaud-Cormos D 2019 Microdosimetry using Rhodamine B within macro- and microsystems for radiofrequency signals exposures of biological samples *IEEE Transactions on Microwave Theory and Techniques* pp 19

Recent activity on electromagnetic exposure tests with cell culture samples

L O Fichte

Helmut Schmidt University, Hamburg, Germany

Abstract. The paper is aimed at reporting the recent findings of studies documenting cell cultures exposure to electromagnetic fields of various origin. Both thermal and non-thermal effects were analysed, highlighting the role of the specific absorption rate as a metric offering a common framework for comparison among different results. The main problems related to electromagnetic field exposure of cell cultures were identified as reporting accurate field magnitude and ensuring suitable environmental conditions for cell cultures in order to guarantee the reproducibility of the experiments.

1. Introduction

The use of mobile communication and social media has become an integral part of today's society. The steadily increasing volume of data leads to changes in the information transmission, which is achieved less by a higher dose of electromagnetic irradiation than by a changing the signal structure and its transmission on other, more suitable frequencies. Under various buzzwords like "intelligent household appliances", "internet of things or "industry 4.0", wireless communication is being integrated into other areas of everyday life and offers a new, more convenient alternative to use data. A large number of devices use electromagnetic waves at different frequencies, especially in the main frequency bands between 800 MHz and 5 GHz. The type and distribution of everyday individual radiation exposure is volatile in terms of intensity, directions of propagation, polarization plane(s), frequencies and waveforms. At present, only the thermal effect of electromagnetic fields is characterized by its SAR value [1]. To measure non-thermal effects, the latter must be substantiated in a reproducible way, followed by an identification of their mechanisms.

2. Past investigations

There are various indications of a non-thermal effect of electromagnetic fields: The Swiss Federal Office for the Environment (Schweizer Bundesamt für Umwelt, abbr. BAFU) lists epidemiological studies as well as laboratory experiments: statistically significant relationships between sleep disorders and a radio short-wave transmitter and physiological changes in cell cultures [2]. Also on behalf of the BAFU, investigations below the limit values for various health and physiological effects, animal and cell experimental studies as well as molecular and biophysical interaction mechanisms were evaluated: Hug et al. estimate the evidence of non-thermal effects of electromagnetic fields with regard to the examined endpoints as predominantly insufficient, but sufficient as brain waves are regarded. They conclude that in certain unfavorable situations the safety factors would be overestimated, which is why a better understanding of the biophysical and biological mechanisms is necessary for low doses and for long-term exposure [3].

The German Radiation Protection Commission (Strahlenschutzkommission, abbr. SSK) reports insufficient evidence for the occurrence of an increased cancer risk due to high-frequency electromagnetic fields (e.g. radio) and microwaves (e.g. mobile phone applications). The evidence of non-thermal biological effects from terahertz radiation could not be finally assessed due to the lack of data [4].

The influence of electromagnetic fields on biological samples, interesting for the planned exposure of cell cultures, has already been examined several times using different endpoints [5-10]. However, adverse effects on the human organism beyond thermal effects are not documented [5-8]. In studies on rats, it was found that age and stage of development have an influence on the sensitivity of the heart tissue to electromagnetic fields [11]. With GSM exposure (Global System for Mobile Communications or Groupe Spécial Mobile) of hematopoietic stem cells, the only documented effect was a decrease in DNA damage when compared to non-exposed cells [9]. Another source of extensive publications on this subject is the so-called "EMF portal", which is maintained by the Research Center for Electromagnetic Environmental Impact at RWTH Aachen University [12].

The thermal dose of electromagnetic radiation (e.g. administered in medical applications) is usually defined by the average power per mass unit, the so-called SAR (specific absorption rate), which the DUT receives from an external electromagnetic field. In order to be able to compare different exposure situations, in particular electromagnetic exposure on the one hand and other types of thermal exposure on the other, the SAR of a sample exposed with a measurement system must be determined as precisely as possible. Ensuring small sample sizes and using a suitable analytical model, the SAR exposure can be determined at any point in the sample volume from measured temperature profiles. Depending on the sample under consideration, the power flux balance used for the thermodynamic system under considerationcan become too complex to infer the SAR value using an analytically manageable model. In this case, even a numerical simulation of the SAR exposure is possible.

3. Dosimetry

If non-thermal effects can be identified, the SAR probably does not offer an appropriate dose term, such as for ionizing radiation. Nevertheless, the SAR offers a framework for comparison, especially since there is extensive experience as of which SAR exposure can be expected to have thermal effects.

4. Problems with exposure environment

One common way to investigate the effect of electromagnetic fields is to expose individual cell cultures to an electromagnetic field of known magnitude, because this approach EMF-induced biological effects can generally be reproduced. To exclude other influencing factors in addition to the electromagnetic radiation, suitable environmental conditions (temperature, air humidity, pH value) must be ensured for the cell cultures. Cell cultivation is generally carried out using so-called incubators to multiply the cells. At the same time, it must be ensured that, in addition to the defined and intended reproducible electromagnetic field, no other external fields can affect the cell cultures.

This talk will give a comprehensive overview of how to design an experiment to handle the problems mentioned above, and how to avert known biases that effect both experiments and scientists. Additionally, the results of some recent investigations will be presented and discussed.

References

[1] www.icnirp.com accessed on May 1st, 2022.
[2] Bundesamt für Umwelt, Schweiz; „Gesundheitliche Auswirkungen von Hochfrequenz-Strahlung“; https://www.bafu.admin.ch/content/bafu/de/home/themen/elektrosmog/fachinformationen/auswirkungen-elektrosmog/gesundheitliche-auswirkungen-von-hochfrequenz-strahlung.html,

[3] Hug K, Ackermann P, Dürrenberger G, Kuster N, Mevissen M, Schär P and Röösli M 2014 *Beurteilung der Evidenz für biologische Effekte schwacher Hochfrequenzstrahlung* im Auftrag des Bundesamt für Umwelt/Schweiz, Basel
[4] Strahlenschutzkommission, Deutschland 2011 Vergleichende Bewertung der Evidenz von Krebsrisikendurch elektromagnetische Felder und Strahlungen – Stellungnahme der Strahlenschutzkommissionmit wissenschaftlicher Begründung **248** *Sitzung der Strahlenschutzkommission* Bonn
[5] Roux D, Girard S, Paladian F, Bonnet P, Lalléchère S, Gendraud M, Davies E and Vian A 2011 Human keratinocytes in culture exhibit no response when exposed to short duration, low amplitude, high frequency (900 MHz) electromagnetic fields in a reverberation chamber *Bioelectromagnetics* **32** 4, pp 302-311 Wiley Online Library
[6] Schrader T, Münter K, Kleine Ostmann T and Schmid E 2008 Spindle disturbances in human hamster hybrid (AL) cells induced by mobile communication frequency range signals *Bioelectromagnetics* **29** 8 pp 626-639 Wiley Online Library
[7] Kleine-Ostmann T, Jastrow C, Baaske K, Heinen B, Schwerdtfeger M, Karst U, Hintzsche H, Stopper H, Koch M and Schrader T 2014 Field exposure and dosimetry in the THz frequency range *IEEE Transactions on Terahertz Science and Technology* **4**(1) pp12-25 IEEE
[8] Hintzsche H, Jastrow C, Kleine-Ostmann T, Schrader T and Stopper H 2012 900 MHz radiation does not induce micronucleus formation in different cell types Mutagenesis, **27**(4) pp 477-483 (Oxford Univ Press)
[9] Gläser K, Rohland M, Kleine-Ostmann T, Schrader T, Stopper H and Hintzsche H 2016 Effect of Radiofrequency Radiation on Human Hematopoietic Stem Cells *RADIATION RESEARCH* **186** pp 455–465
[10] Morgan L L, Kesari S and Davis D L 2014 Why children absorb more microwave radiation than adults: the consequences *J Microsc Ultrastruct* pp 197–204

Challenges and progress in the application of physiological models for clinical decision support in cardiovascular medicine

D R Hose[1,2], PV Lawford[1,2], I Halliday[1,2], D Rafiroiu[3], A Lungu[3]

[1]University of Sheffield, United Kingdom
[2]Insigneo Institute for In Silico Medicine
[3]Technical University of Cluj Napoca, Romania

E-mail: d.r.hose@sheffield.ac.uk

Abstract. The IUPS Physiome and the European Commission's Virtual Physiological Human (VPH), funded under the Framework 7 programme, are just two of the major initiatives that have supported the development of physiological models for clinical decision support. There is significantprogress in the formalisation of the concept of the Digital Twin for human physiology and in processes for model personalisation, verification and validation. The VPH Institute and the Avicenna Alliance continue to promote best practice in this area. One of the most important challenges is to achieve the right level of complexity in the model. The most comprehensive models seek to capture all that is known about the physiological system, including detailed anatomy, organ interactions, control functions, and physical, chemical and biological processes. These models can be of enormous value in understanding human physiology, but can also be terribly difficult to personalise to support the clinical management of an individual. There is inevitable conflict between model complexity and the pragmatic limitation of data collection in the clinical pathway. In this presentation we make the case for a 3-axis digital twin, including recognition of the individual physiological envelope, and introduce three applications in cardiovascular medicine.

1. Introduction

Physiological models of the cardiovascular system generally seek to represent the distribution of the fundamental physical quantities of pressure, volume and flow. Lumped parameter (0D) compartment models [1] describe grossly, but adequately for many clinical purposes, the temporal distribution of blood volume and pressure in the multiple components of the cardiovascular system. Essentially the model is a system of ordinary differential equations. They range in complexity from just a few variables up to hundreds or even thousands. They might include physiological control and regulation processes. They might represent just parts of the system, with appropriate boundary conditions, or they might represent the whole closed loop system with representation of the pumping heart. One dimensional models [2] add a representation of the distribution of pressure and flow along the vessels, and are thus able to capture wave effects including transmission times and reflections. Three dimensional models add detailed representation of anatomical compartments (e.g. chambers of the heart, vessels including curvature and local lesions) and support detailed assessment of quantities such as wall shear stress (important for many biological responses) that are unavailable from the lower dimensional models. All of these models have value for particular clinical applications. The challenge always is interpretation, especially in the context of what the computed quantities actually mean in terms of the effect on the life of the individual whose clinical management is under consideration. There is increasing recognition [3] that system models might be of the greatest value in mapping clinical observations, together now with consumer-level personal monitoring devices such as watches and phones, to physiological models that can operate predictively to assess the likely response to candidate interventions. Wouters et al [4] review the challenges of introducing modelling into a cardiovascular clinical pathway, including issues of model complexity, parameter sensitivity and identifiability, variation and uncertainty, and verification and validation.

2. The Digital Twin and the Physiological Envelope

A digital twin is a virtual representation that serves as the real-time digital counterpart of a physical object or process [5].Digital twins are widely used in engineering, with applications across the spectrum including the continuous monitoring and modelling of the health of jet engines [6]. The digital twin of a human being offers huge potential for monitoring the health of the individual and for representing and/or predicting changes as conditions evolve, whether as the result of inaction or as the result of some form of intervention. A human being is, however, a very complex organism and formalisation of methods for the representation and interpretation of the data embodied in the digital twin is a major challenge. Lamata et al [7] review progress and challenges in the development of a cardiovascular Digital Twin for medical application.

2.1. A Three-axis Digital Twin

We propose the idea of a three-axis digital twin as a means to capture and to organise important characteristics to support diagnostic and prognostic analysis of health.The digital twin requires a definition of the data that represents the individual. Every observation or measurement on the individual produces data. We label each of these observations and measurements, and call this label a 'concept'. Every concept has a magnitude in some sense. It might be binary (e.g. diabetic/non-diabetic), it might be an integer that represents a class (e.g. NYHA) or it might be a real number on a scale (e.g. systolic blood pressure). It is important that these concepts are consistently recognised and interpreted irrespective of how they are measured, although the means of measurement itself might have significant impact on the reported value, and indeed there might be a subjectivity in the observation (for example a clinical judgement of a diagnosis). Nevertheless each data entry is a quantified observation of a concept at some particular time and in some particular state. Especially for cardiovascular application we need to know under what physiological conditions a measurement was made. Was the individual at rest, or in a state of physical activity? We also wish to monitor how these measurements change over periods of weeks, months or years of observation.For each concept this can be represented on a three axis graph, illustrated in figure 1, with axes of state, time and numerical value.The units of the vertical axis, the concept, are the units of measurement.

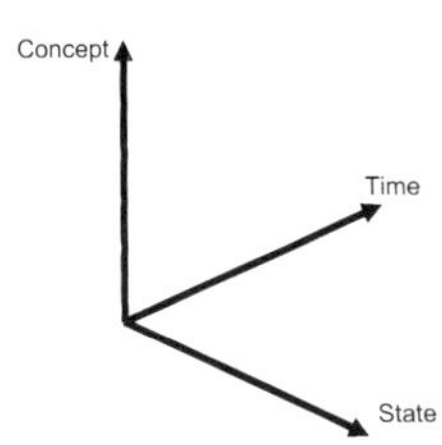

Figure 1. Three-Axis Digital Twin.

The second axis, the state, describes the physiological state in which the measurements are made. We need a consistent and quantitative way to describe the state. In the coarsest representation we might recognise only two states, perhaps 'rest' and 'stress', but we would prefer to have a more continuous representation, perhaps quantified by the metabolic demand in that state (MET), or by some surrogate (e.g. power to weight ratio (PWR) or heart rate). Some personalised normalisation similar to that performed in derivation of the Personal Activity Intelligence (PAI) measure [8] might be very appropriate for this axis.The third axis is time, where here we are looking at longer time scales, perhaps over multiple clinical visits or over weeks, months and years as the health (or morbidity) of the individual evolves.

This representation lends itself to the description of the diagnostic and prognostic operation of a physiological model, and to the understanding of the predictive power of the digital twin. Suppose that systolic and diastolic pressure are measured by a doctor using a sphygmomanometer. Based on this one observation of two concepts a diagnosis might be made, and one might receive medication. Each is just a single point on our graph, the measurement has been made in one state (probably approximating a rest state, 'white coat syndrome' aside) and at one point in time. Diagnosis might be improved if the measurements could be made in multiple physiological states, including when the individual is under some form of physiological stress. Then, at a single point in time, we have multiple measurements in the concept/state plane. Learning and modelling operations will be performed on multiple concepts, in this example systolic and diastolic blood pressure (figure 2) each represented in this three-axis system.

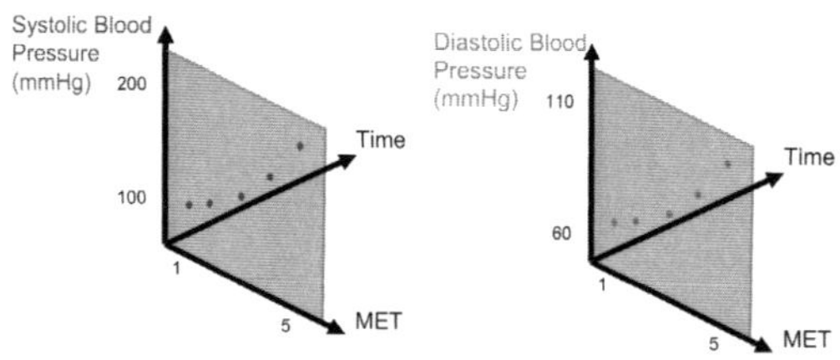

Figure 2. Systolic and diastolic blood pressure measured in multiple states at a given time: population of one plane of the 3-axis digital twin.

If we were to make these measurements at intervals over weeks, months or years we might develop a clear picture of how the cardiovascular health is changing. Each orange plane, or collection of orange planes, will contribute to the diagnosis. The trajectory will be important in any planning of intervention.These simple diagrams also illustrate the operation of the digital twin in a predictive mode, figure 3. The green planes in this diagram represent how the health of the patient will evolve in the future. Based on all observations, in multiple states and at multiple times, we use a model to predict the evolution at future times, with or without intervention. The model might be based on learning and observation of trajectories and outcomes in a cohort of similar individuals, or it might be based on the mechanistic modelling of physiology. Both have their place, and our modelling approach is based on the synergistic operation of artificial intelligence with physics-based physiological modelling.

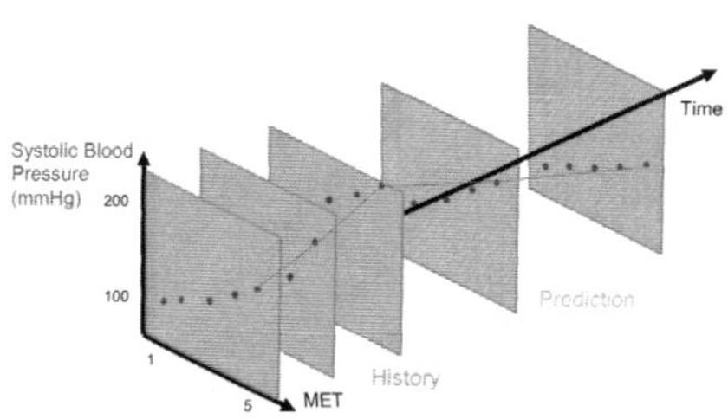

Figure 3. 3-axis digital twin illustrating capture of data associated with one concept (systolic blood pressure) at multiple times (orange planes), followed by predictions for the future as a result of a proposed intervention (green planes).

The 'concept' can be further refined by recognition that each concept is not necessarily single-valued at each observation. It might be a vector. For example blood pressure might be measured during the day and at night, or over a 24 hour period, or during performance of a six-minute walk test. These do not represent an evolution over the time axis in our three-axis model, but rather they are a complementary collection of observations of the concept at our current measurement interval, so the concept is now represented as a vector rather than as a scalar. This will be important for the structure of the databases but does not change philosophically the fundamentals of the representation of the digital twin. Vector representations will also arise to describe intra-cardiac-cycle variations of parameters as the heart beats.

We would suggest that it is very important that we routinely recognise, record and represent the second and third axes of this digital twin if we are to exploit its potential in clinical and personal health management.

2.2. *The Physiological Envelope*

A key feature of the 3-axis representation is the formal recognition of the physiological status in the measurement state. This naturally leads on to the question of how multiple states combine to influence the evolution of health and disease. It is well understood that exercise regimes that feature short, high intensity bursts are not the same as those in which activity is sustained at a lower level for longer periods of time. A model should recognise this distinction and make predictions accordingly. In a disease application one might wish to predict the risk of rupture of an aneurysm based on measurements of size, pressure, stress, strain, wall thickness, wall composition and other factors. It is very likely that, if the aneurysm were to rupture, this would occur under a stress condition, but do the long-term, consistent, low amplitude states contribute to the evolution towards a rupture condition? In engineering mechanics we have methods to combine fatigue damage over multiple states to make a prediction, but currently we have no similar methods for arterial structures.

The range of physiological conditions that might be experienced by a sedentary individual going about normal daily activities will be very different from those of an athlete in regular training. To represent this we might borrow an idea from aircraft design. A flight envelope describes the operational conditions during a typical flight (take-off, cruise, landing), and is not the same for short-haul and long-haul routes. Similarly we might construct a physiological envelope representing the activity excursions of an individual as they live their lives, and use this as the basis of our modelling operations. This envelope might be characterised by a series of measurements such as those collected by accelerometers or watches, or might be based on softer data for example from questionnaires or even from population averages.

3. Applications of Physiological Models in Cardiovascular Medicine

In this section we examine briefly just three applications of physiological models in cardiovascular medicine each of which has particular features that highlight different aspects of physiological modelling strategy and interpretation. The selected applications are: i) fractional flow reserve in treatment planning for coronary artery disease, ii) construction of a model-based decision support system for heart valve interventions and iii) diagnosis of pulmonary hypertension.

3.1. *Fractional Flow Reserve*

Fractional flow reserve (FFR) is perhaps the best example of the adoption of a physiological model to support decisions over intervention for cardiovascular disease (in this case stenotic lesions in the coronary arteries). It has been proven [9] that outcomes of interventions that are guided by the measurement of FFR are more successful than those that are based on anatomical measurements of the diameter of the stenosis. At the core of FFR is a very simple model of two resistors in series. The first resistance is that presented by the stenosis and the second is that of all of the coronary vasculature and microvasculature distal to the stenosis. If the pressure at the inlet (pa) and at a point just downstream of the stenosis (pd) are measured then the ratio, pd/pa, is the FFR. Based on this measurement, using the simple two-element model, a prediction can be made of the relative increase of flow that might be achieved if the lesion is removed by an intervention. There is no simpler or more powerful indicator of the benefits of using physiological data (in this case pressure data) to inform clinical intervention. Despite the proven value of FFR, for many reasons the measurement is often not made in the clinical workflow. A model-based virtual equivalent, vFFR, can be estimated based on image data and simple boundary conditions. Accuracy and challenges [10] of vFFR will be discussed in the presentation.

3.2. Heart Valve Interventions

EurValve [www.eurvalve.eu], a project funded under the Horizon 2020 programme, developed a decision support system (DSS) for intervention for heart valve disease. One of the elements of the DSS was the measurement and/or prediction of the left ventricular pressure-volume loop and cardiac energetics, pre- and post-intervention,in individuals suffering from heart valve disease. The analysis protocol, figure 4, included the personalisation, based on sparse physiological measurements, of a simple model of the left side of the heart and the coronary circulation coupled to a three-dimensional model of the diseased valve based on medical image data. Typical results are illustrated in figure 5.

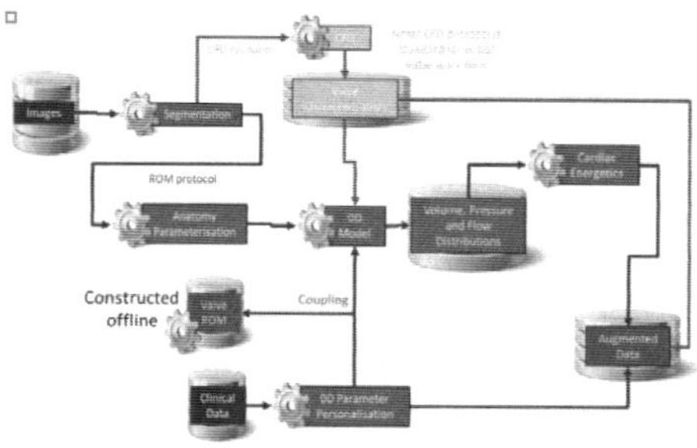

Figure 4. Analysis protocol for model-based assessment of cardiac energetics in valve disease cohort.

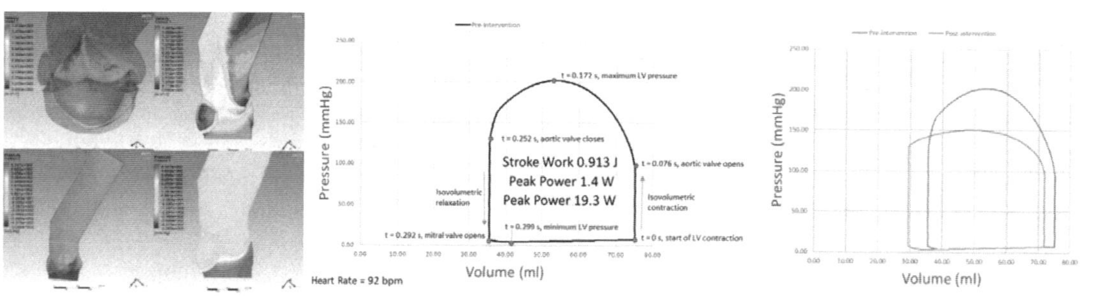

Figure 5. Valve haemodynamics, model-based PV loop and predictions for one case.

In our presentation we will:

i) outline the modelling protocol, focusingon the challenges of personalisation of the model parameters from sparse clinical data, on the characterisation of the physiological envelope for the individuals recruited to the study, and on the methods for extrapolation from rest to exercise states.
ii) present representative results from the model and compare with observations in the patient cohort [11].

3.3. Diagnosis of Pulmonary Hypertension

Pulmonary hypertension is often diagnosed late in the disease process, often with severe consequences for the patient. Unlike hypertension of the systemic circulation, there is no simple sphygmomanometer that can directly measure pulmonary arterial pressure. Gold standard diagnosis requires invasive catheterisation. This also makes it difficult to monitor, in a timely manner, changes associated with treatments. One of the characteristics of a one-dimensional physiological model is that it is able to represent and recognise waves in the arterial system. On the basis that pulmonary hypertension arises from pathological changes in the pulmonary arterial tree, Lungu et al [12] hypothesised that the disease state might be characterised by a mismatch in the impedance of vessels leading to a detectable increase in the power of reflected waves relative to forward-travelling waves, and developed an image-based process to determine the reflected power ratio. Combining these novel physiological data with other clinical measures she was able to demonstrate [13] improved correlation between non-invasive diagnosis of pulmonary hypertension and the gold-standard invasive process. Very recently, this has been further strengthened bynon-invasive estimation of mean pulmonary artery pressure, using machine learning based regression models [14].

Decision support for the management of pulmonary hypertension lends itself well to the personalised modelling protocols described previously for the valve application. A major difference is that for this application it is essential to add the pulmonary circulation and the right side of the heart into the underpinning lumped parameter model. This adds complexity and increases the challenges of model personalisation but, with the important addition of regulation processes into the model, makes it possible to represent the physiological changes under multiple external load conditions including the transition between states. Figure 6 shows a typical result of the transition between rest and the state of physical activity, for a normal individual. These left and right ventricular PV loops were obtained with a 0D model of the controlled cardiorespiratory system, which is under development, aiming for its extension to various valvular, coronary, and pulmonary pathological states.

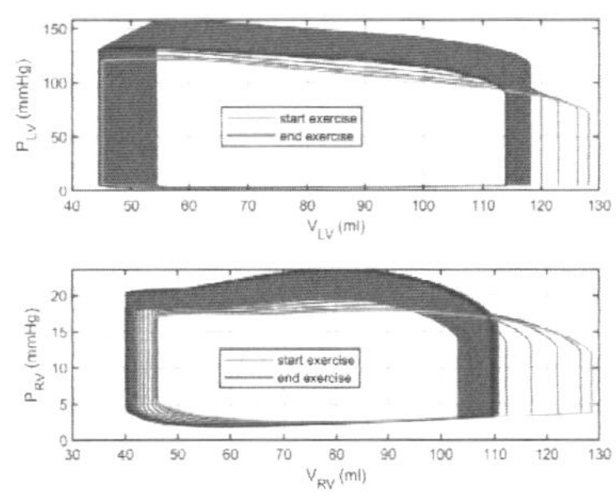

Figure 6. Model-based PV loops showing the transition between rest and a 50Watt physical activity state of a normal individual.

A major positive of this application from the perspective of modelling and machine learning development and validation is the wealth of physiological data that is available on pulmonary hypertension cohorts, in comparison with the very sparse data from other disease groups.We believe that model-based decision support for the management of treatment of pulmonary hypertension patients will, in the future, mirror the success of coronary FFR in improving clinical outcomes.

4. Conclusions

Personalised physiological models, underpinned by a digital twin, can support decision making in cardiovascular medicine, but many challenges remain in data collection, data representation, sensitivity analysis, parameter identifiability, variation and uncertainty representation, verification and validation.

Acknowledgements

The Eurvalve project has received funding from the European Union's Horizon 2020 research and innovation programme under grant agreement No 689617.

References

[1] Shi Y, Lawford P and Hose R 2011 Review of zero-D and 1-D models of blood flow in the cardiovascular system *Biomed Eng Online* **10** doi: 10.1186/ 1475-925X-10-33

[2] van de Vosse F N and Stergiopulos N 2011 Pulse wave propagation in the arterial tree. Annu Rev Fluid Mech **43** pp 467–99 doi: 10.1146/ annurev- fluid- 122109- 160730

[3] Hose D R, Lawford P V, Huberts W, Hellevik L R, Omholt S and van de Vosse F N 2019 Cardiovascular models for personalised medicine: where now and where next? *Medical Engineering and Physics* **72** pp 34-48

[4] Huberts W, Heinen S G H, Zonnebeld N, van den Heuvel D A F, de Vries J P P M, Tordoir J H M, Hose D R, Delhaas T and van de Vosse F N 2018 What is needed to make cardiovascular models suitable for clinical decision support? A viewpoint paper *J Comp Sci* **24** pp 68-84 doi: 10.1016/j.jocs.2017.07.006

[5] https://en.wikipedia.org/wiki/Digital_twin

[6] https://www.cio.com/article/188765/rolls-royce-turns-to-digital-twins-to-improve-jet-engine-efficiency.html

[7] Corral-Acero J et al 2020 The 'Digital Twin' to enable the vision of precision cardiology. *Eur Heart J.* **41**(48) pp 4556-4564 doi: 10.1093/euheartj/ehaa159

[8] Nauman J, Nes B M, Zisko N, Revdal A, Myers J, Kaminsky L A, Wisløff U 2019 Personal Activity Intelligence (PAI): A new standard in activity tracking for obtaining a healthy cardiorespiratory fitness level and low cardiovascular risk. Progress in Cardiovascular Diseases. **62**(2) pp 179-185 ISSN 0033-062

[9] Tonino PAL et al 2009 Fractional Flow Reserve versus Angiography for Giuding Percutaneous Coronary Interventions *N Engl J Med.* **360** pp 213-224 doi: 10.1056/NEJMoa/0807611

[10] Morris P D, Van De Vosse F N, Lawford P V, Hose D R, Gunn J P 2015 Virtual (Computed) fractional flow reserve current challenges and limitations *JACC Cardio- vasc Interv* **8** pp 1009–1017 doi: 10.1016/j.jcin.2015.04.006

[11] Archer G 2020 *Computational Modelling in the Management of Patients with Aortic Valve Stenosis* PhD Thesis (University of Sheffield) https://etheses.whiterose.ac.uk

[12] Lungu A, Wild J M, Capener D, Kiely D G, Swift A J and Hose D R 2014 MRI model-based non-invasive differential diagnosis in pulmonary hypertension *J Biomech* **47**(12) pp 2941-2947 doi: 10.1016/j.jbiomech.2014.07.024

[13] Lungu A, Swift A J, Capener D, Kiely D, Hose R and Wild J M 2016 Diagnosis of pulmonary hypertension from magnetic resonance imaging–based computational models and decision tree analysis *Pulmonary Circulation* **6**(2) pp 181-190 doi: 10.1086/686020

[14] Grzeszczyk M K, Satlawa T, Lungu A, Swift A, Narracott A, Hose R, Trzcinski T and Sitek A 2022 Noninvasive Estimation of Mean Pulmonary Artery Pressure Using MRI, Computer Models, and Machine Learning *International Conference on Computational Science* (London)

IOP Conf. Series: Materials Science and Engineering 1254 (2022) 012006 doi:10.1088/1757-899X/1254/1/012006

Smart Sensing and AI for Physical Therapy in IoT Era

O Postolache

Instituto de Telecomunicações and ISCTE- Instituto Universitário de Lisboa, Av. Rovisco Pais, No.1, 1049-001, Lisbon, Portugal

opostolache@lx.it.pt

Abstract. It is well known that medical spending increase with disability status. Per capita spending for people with five or more limitations in activities of daily living (ADLs) is nearly five times the amount incurred by those with limitations in only one instrumental activities of daily living (IADLs). Physical therapy is the way to improve the motor capabilities however it takes a lot of time, it requires physiotherapists services, is often painful and the outcome are evaluated in subjective way. New technologies including smart sensors were adopted in healthcare including wearable solutions for cardiac and respiratory activity monitoring and successfully are contributing to reduce the costs of services. In the case of motor activity and particularly in physical rehabilitation the developments are still reduced the physical therapy services are using as hardware mechanical equipment without sensing, embedded processing and internet connectivity that significatively reduce the possibility to measure and evaluate the physical training outcomes in objective way. In this paper the disruptive solutions for physical therapy are presented that are based on hot technologies such as smart sensors, IoT, virtual reality (VR), mixed reality (MR), and artificial intelligence (AI). Applied AI may conduct to develop models, classifiers (gait classification) and short term or medium term prediction of physical therapy outcomes. Highly motivation of the patients under physical rehabilitation can be increased promoting serious game characterized by VR and MR scenarios.

1. Introduction

The Physiotherapy treatment modalities and techniques are effective at facilitating muscle activation and strength, assisting with joint mobility and soft-tissue problems, modulating pain and inflammation. Treatment can involve joint mobilization, strength and stability exercises and stretching, manual soft tissue therapies (including massage, myofascial, trigger point, positional releases), hydrotherapy, electrotherapy, etc. With progress in methods of therapy and increasing demand on reducing costs in national health care, evidence-based physiotherapy has become a demanding endeavour with pressures from both insurers and employers

Recent work reported in the literature on tele-rehabilitation have been underscore the need for unobtrusive, robust to noise, reliable, mobile smart sensors for motor activity assessment in rehabilitation. Several solutions on smart equipment for physical rehabilitation monitoring are reported in [1-2] as so as the remote physical rehabilitation based on remote sensing and serious game [3-4]. The extraction of complementary information such as heart rate, SpO2, respiratory rate, heart rate variability during the physical training sessions in real, virtual, or mixed scenario can be used to extract useful data to assure more safety and to adapt the physical training plan to the patient cardiac and respiratory condition [5].

Regarding the remote sensing in the last years several commercial solution were adopted to serve as part of physical rehabilitation frameworks. Since 2011 Microsoft Kinect technology appear one of the main remote sensing solution adopted as part of tele-rehabilitation. The latest version of this type of remote sensing device, Azure Kinect dk that works together Azure Kinect Sensor SDK that provides low-level sensor access for Azure Kinect DK hardware sensors and device configuration, the main and applied functionalities being the access of human body joints Cartesian coordinates. Other types of remote sensing device that were used are LeapMotion that was used in different serious game developed by the group and reported in different publications [6].

An objective evaluation of functional status based on accurate measurement systems that could extract the information in non-invasive and unobtrusive way represents contributes to validate the applied physiotherapy techniques for individual cases together the cardiac and respiratory status that can assure optimal conditions for physical rehabilitation effort. Appropriate choice of the applied techniques might conduct to reduced recovery period without risks related undesired cardiac or other type of events. To increase the motivation on performing the recommended exercises and the quality of execution according with the physiotherapy recommendation as so as automatic generation alerts based on cardiac and respiratory condition continuous monitoring represent an important issue. Different solutions including serious game and 3D virtual reality are designed to increase the patient adherence proven their effectiveness to increase the patient engagement on proposed recovery program [7]

A way to increase the motivation for physiotherapy training is the usage of serious game adapted to the user rehabilitation needs that provide at the same time feedback to the users about health status condition (e.g. SpO2, HR, HRV). Ma and Bechkoum developed a serious-game based therapy to stimulate physical exercises practice by stroke patients with upper limb motor disorders [6] and in our group several serious game frameworks were developed, many of them being designed for upper limb rehabilitation [9]. In virtual scenarios the system allows patients to interact in real-time with virtual objects to practice specific motor skills through diverse modalities. On another hand, an interesting Android-based application is DroidGlove, a serious game for wrist rehabilitation that runs on small portable devices [8] . Motivational interaction with virtual reality scenario are provided also by extended systems reported by Postolache et al [9] where smart gloves are used to navigate and interact with virtual objects in a virtual scene considering semi-immersivity conditions.

A serious game for balance rehabilitation is presented by Betker et al [10] in which a video game-based tool is controlled by a center–of-pressure (CoP) for the maintenance of balance in a short-sitting position for people with chronic spinal cord and traumatic brain injuries. Similar work was reported in [11] where the data from a force platform is used to adapt the game scenarios and virtual object distribution that will be checked to avoid unbalance of the user and accidents. The gait recognition and classification is an important task as part of diagnosis and evaluation. Different classifiers for gait were developed to be applied to sensor data (shimmer IMU) disposed on the level lower limb. Several results are reported by Leite et al [12]. Latest development in the deep learning algorithms were created good opportunity to be used also in the physical rehabilitation for objective evaluation. Thus, smart sensor-based rehabilitation exercise recognition system was reported by Zhang et [13] al that are developed a deep learning framework for the mentioned task. Other technologies such as IoT are also part of the newest developments that assure the physical therapy services digitalization as part of general services digitalization direction.

2. Smart Sensors and IoT for healthcare services

The convergence of healthcare, instrumentation and measurement technologies will transform healthcare as we know it, improving quality of healthcare services, reducing inefficiencies, curbing costs and improving quality of life. Two technologies are frequently considered as part of tele-medicine solutions particularly for remote physical therapy implementations. Thus, can be mentioned the smart sensors as part of IoT ecosystem. The smart sensors and also smart actuators are characterized by sensing or actuating, embedded processing and communication capabilities and are the main component of smart equipment that are reported in the field of smart physical therapy. An example of smart walker prototype designed for gait rehabilitation after stroke events is reported in [14]. The general architecture and the hardware components are presented in Figure1.a.

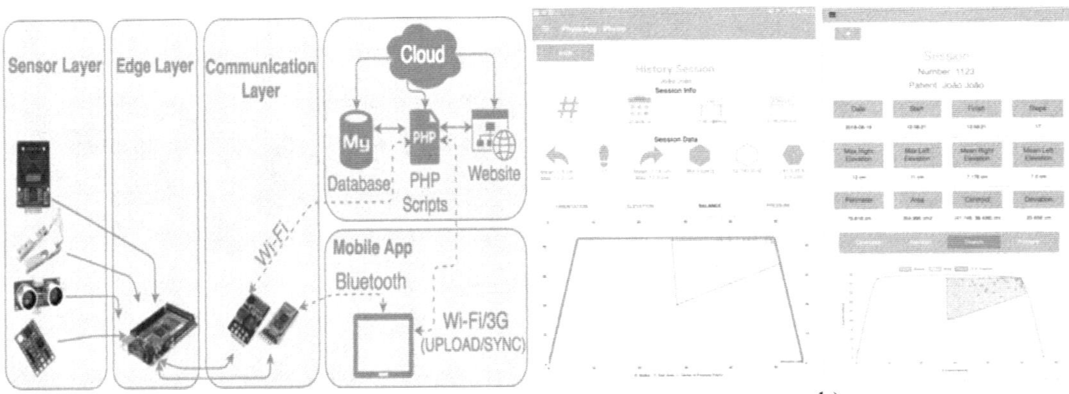

a) b)

Figure 1. Smart walker implementation a) IoT layers b) mobile and web APP for gait training based onsmart walker.

As it can be observed in Figure.1 the implemented smart walker prototype is characterized by several layers from sensor layer to cloud and mobile computing layer all of this layers materializing and IoT implementation for gait rehabilitation equipment. Can be underlined load cells as the main components of sensor layer associated with the measurement of floor reaction force during the walker usage but also the IMU usage for walker acceleration and direction of walker motion. The signals coming from the sensors is acquired by Arduino mega based platform pre-processed and transmitted using short range communication protocols expressed by Wi-Fi and Bluetooth. The Wi-Fi communication capabilities provides direct Internet connectivity while the Bluetooth is used for real time display of the data using a mobile APP (Figure 1.b). Additionally a web APP was also developed to make easily the data visualization through a common browser. The presented APPs are able to display the CoP (Center of Pressure) during the aided gait training highlighting the right user balance. The data can be also stored from different walkers and analysed using data science techniques to extract objective information about the evolution of user walking capabilities. IoT ecosystem will be discussed underlining a set of new sensing technologies and the integration of this on the smart sensors. History and future of IoT ecosystem are discussed starting from RFID and providing interesting approach about 5G for new implementation for physical therapy. Related IoT protocols such MQTT and IoT platforms such The Thing Network for healthcare will be presented including strength and drawbacks.

This brief example of the group of smart equipment for physical therapy as IoT implementation offer new and vision of exciting possibilities for more robust, reliable, flexible and low-cost healthcare systems and patient care strategies based on data. These may provide value-added information and functionalities for patients, particularly for those with neuro-motor impairments. It has great importance in developed countries in the context of population ageing.

3. Wearable and Remote Sensing for VR theragames

Theragames represents the serios games associated with physical rehabilitation that are characterized by virtual reality and mixed reality scenarios to be used for upper limb and lower limb rehabilitation for a higher user motivation. The interaction between user and VR scenario can be materialized using wearable smart sensors [15] or remote sensing materialized by commercial natural user interfaces such as Kinect or Leap Motion. Example of upper limb theragame is reported in [4] the implementation combines VR but also wearable devices (Figure 2) for upper limb rehabilitation.

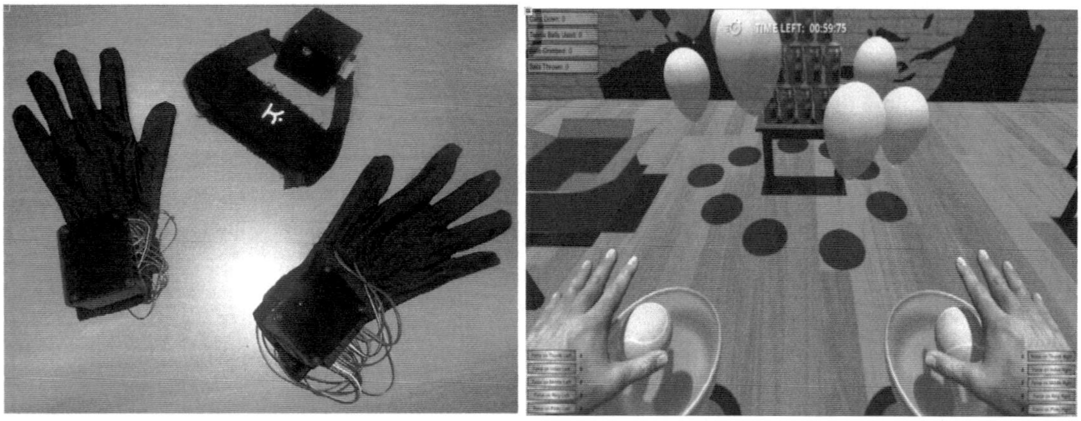

a) b)

Figure 2. Theragame and Wearable Interface a) smart gloves and head band game interfaces; b) cans down virtual reality serious game for upper limb rehabilitation.

As part of these interactive environments, 3D image sensors for natural user interaction with rehabilitation scenarios and remote sensing of user movement, as well as thermographic camera for remote evaluation of muscle activity will be presented.

4. AI for physical therapy

Artificial intelligence (AI) solutions applied for physical therapy data provided by the smart sensors associated with interactive environments as so as training support equipment (e.g. smart walkers, crutches, force platforms) are considered. Example of applied AI algorithms for gait classification, balance control subsystems, physical rehabilitation outcome prediction using classical AI algorithms but also the latest deep-learning algorithms will be presented. Will be discussed the importance of AI on diagnosis and evaluation but also on training plan evaluation for higher motivation and reduced rehabilitation periods.

5. Conclusion

Smart Physiotherapy based on smart sensors IoT and theragames supported by virtual reality or mixed reality assures an increased quality of services, reduced rehabilitation period. The smart sensors are producing big data that can be used for physical rehabilitation models and for predictionof the physicalrehabilitation outcome. Personalized serious games with natural user interfaces data storage and analysis associated with smart physiotherapy IoT compatible increase the patient motivation and physical therapyeffectiveness. AI andbig data Analysis can be used to improve the QoS for physical therapy services.

Acknowledgments

This work was funded by Instituto de Telecomunicações and FCT/MCTES through national funds and when applicable co-funded EU funds under the project UIDB/50008/2020.

References

[1] Postolache O, Dias Pereira J M, Viegas V, Pedro L, Girão P S, Oliveira R and Postolache G 2015 Smart Walker Solutions for Physical Rehabilitation *IEEE Instrumentation and Measurement Magazine* **18**(5) pp 21-30

[2] Frango F, Postolache O and Yang Y 2018 Smart Object for Physical Rehabilitation Assessment, *International Conf. and Exposition on Electrical and Power Engineering - EPE* **1** pp 1-6 (Iasi, Romania)

[3] Monge J and Postolache O 2018 Augmented Reality and Smart Sensors for Physical Rehabilitation *International Conf. and Exposition on Electrical and Power Engineering - EPE* **1** pp 1-6 (Iasi, Romania)

[4] Postolache O, Lourenço F L, Dias Pereira J M and Girão P M 2017 Serious Game for Physical Rehabilitation: Measuring the Effectiveness of Virtual and Real Training Environments *IEEE International Instrumentation and Measurement Technology Conference* **1** pp 1-6 I2MTC (Torino, Italy)

[5] Lohse K R, Hilderman C G E, Cheung K L, Tatla S and Van der Loos H F M 2014 *Virtual reality therapy for adults post-stroke: a systematic review and meta-analysis exploring virtual environments and commercial games in therapy* **9**(3) doi: 10.1371/journal.pone.0093318

[6] Ma M and Bechkoum K 2008 Serious games for movement therapy after stroke *IEEE International Conference on Systems, Man and Cybernetics*

[7] Ferreira D, Oliveira R and Postolache O 2017 Physical Rehabilitation based on Kinect Serious Games *International Conf. on Sensing Technology – ICST* **1** pp 1-6 (Sydney, Australia)

[8] Deponti D, Maggiorini D and Palazzi C E 2009 DroidGlove: An android-based application for wrist rehabilitation *International Conference on Ultra Modern Telecommunications & Workshops*

[9] Alexandre R, Postolache O and Girão P M 2019 Physical Rehabilitation based on Smart Wearable and Virtual Reality Serious Game *IEEE International Instrumentation and Measurement Technology Conference* **1** pp 1-6 (Auckland, New Zealand)

[10] Betker A L, Desai A, Nett C, Kapadia N and Szturm T 2007 Game-based Exercises for Dynamic Short-Sitting Balance Rehabilitation of People With Chronic Spinal Cord and Traumatic Brain Injuries *Physical Therapy* **87**(10) pp 1389–1398

[11] Postolache O, Girão P M, Dias Pereira J M and Postolache G 2016 Postural Balance Analysis using Force Platform for K-Theragame users *IEEE International Symp. on Medical Measurements and Applications - MeMeA* **1** pp 1-6 (Benevento, Italy)

[12] Leite P L, Postolache O, Dias Pereira J M and Postolache G 2019 Gait Rehabilitation Monitor *IMEKO Joint IMEKO TC1-TC7-TC13 Symposium* **1** pp 1-6 (St. Petersburg, Russia)

[13] Zhang W, Su C and He C 2020 Rehabilitation Exercise Recognition and Evaluation Based on Smart Sensors With Deep Learning Framework *IEEE Access* **8** pp 77561–77571

[14] Nave C and Postolache O 2018 Smart Walker based IoT Physical Rehabilitation System *IEEE International Symposium on Sensing and Intrumentation in IoT Era (ISSI)* **1** pp 1-6 (Shanghai, China)

[15] Postolache O, Hemanth J H, Alexandre R, Gupta D G, Geman O and Khanna A 2020 Remote Monitoring of Physical Rehabilitation of Stroke Patients using IoT and Virtual Reality *IEEE Journal on Selected Areas in Communications* **39**(2) pp 562-573

Wi-Fi inside an airplane

Enrique Arribas[1], Isabel Escobar[1], Antonio Martinez-Plaza[2], Carlos Adelantado[3], and Raquel Ramirez-Vazquez[1]*

[1]University of Castilla-La Mancha, Applied Physics Department, Faculty of Computer Science Engineering, Avda. de España s/n, University campus, 02071 Albacete, Spain
[2]University of Castilla-La Mancha, Mathematics Department, School of Industrial Engineering, Albacete, Spain
[3]University of Córdoba, Analytical Chemistry Department, Faculty of Science, 14071 Córdoba, Spain

*** *Corresponding author:*** *raquel.ramirez@uclm.es*

Abstract.

In this work, we present results for measurements of personal exposure to Radiofrequency electromagnetic Fields (RF-EMF) from 2.4 GHz Wi-Fi and 5.8 GHz Wi-Fi frequency bands while travelling by airplane. The flight was taken by a researcher of this work, it departed from Barajas Adolfo Suarez Airport in Madrid (Spain) and arrived in Benito Juárez International Airport (Mexico City – Mexico), on the 20th of June 2021. The measurements were carried out using two EME SPY 140 personal exposimeters. Results indicate that exposure levels change at any moment. Values are variable when the researcher is inside the airplane prior to takeoff or landing, when the airplane is airborne and when the Wi-Fi is in use or not. The average value registered during the whole flight was 123 $\mu W/m^2$ with a 95 percentile of 273 $\mu W/m^2$ in 2.4 GHz Wi-Fi frequency band, and the average value was 87.1 $\mu W/m^2$ with a 95 percentile of 184 $\mu W/m^2$ in 5.8 GHz Wi-Fi frequency band.

Keywords: Personal exposure, Radiofrequency Electromagnetic Fields, airplane, Wi-Fi connection, intensity of electromagnetic wave.

1. Introduction

We are living in an invisible ocean of electromagnetic waves. The implementation of new methodologies to measure intensity of an electromagnetic wave allows us to know personal exposure levels to Radiofrequency Electromagnetic Fields (RF-EMF), to which the population is subject in different microenvironments.

Different study methods, types of measuring equipment and means to carry them have been evaluated [1–19]. For instance, the personal exposimeter has been carried in a bag next to the researcher's body [10,11,20], in a plastic basket on a bicycle [21,22], in a cardboard tube held above the researcher's head [23,24], on the top of a car or different ExpoM-RF devices have been mounted inside a box in a small car [25], or on a drone [26]. All these methods allow us to know RF-EMF intensity levels taken at ground level in an urban environment, but they do not allow us to know RF-EMF intensity levels to which a passenger is exposed when travelling by air at 800 km/h at a height of 10000 meters. An airplane is an indoor means of transport, in which there is Wi-Fi connection that is necessary for communication between the airplane and the control tower, and this service is available to passengers.

Most studies about measuring personal exposure to RF-EMF have been carried out with the help of volunteers in different microenvironments in Europe [1–15,27–33], and a first study in Mexico [11].

Implementing new methodologies to carry out this kind of measurements allows us to reach microenvironments that are inapproachable by a researcher working at ground level in urban areas. One of these microenvironments would be inside an airborne airplane.

The increasing quality value brought in by Wi-Fi technology (Wireless-Fidelity) makes means of transport more advantageous in enclosed and controlled environments such as an airplane [34], train, bus or car [35]. Airplanes have always been a vehicle in which the use of electronic devices is prohibited during the flight to avoid radio interferences among the crew and not to disable frequency bands used for communication [36]. Nowadays, airplanes need a connection band for the Wi-Fi so that pilots can communicate with the control tower and offer services to passengers.

Fixed mobile phone towers establishing high speed Internet connectivity cannot be practically installed and connected everywhere because of the presence of oceans and seas in between countries, and the solution is to utilize space for connecting two points worldwide, such as airplanes, which use satellite connectivity (Figure 1).

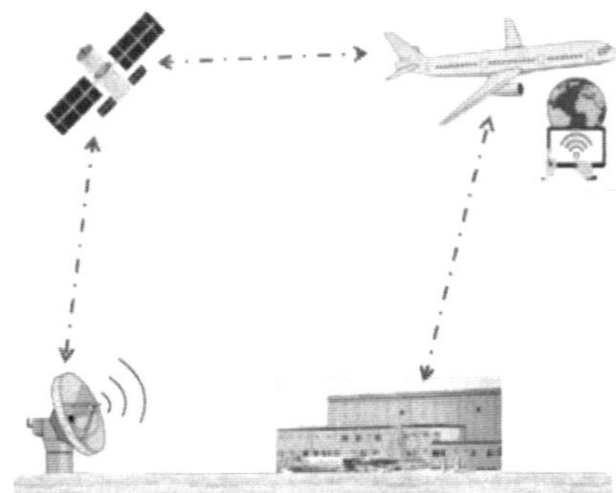

Figure 1. Wi-Fi connection method used in an airplane (Wi-Fi satellite).

Connectivity relationships can easily be distorted by commercially available Wi-Fi interfering devices. The possibility of communication between the operator and the aircraft expands to more radiofrequencies available, and therefore to technologies boosting their immunity [37]. Wi-Fi connectivity in airplanes is possible above 10000 meters, and the device must be set to flight mode. Most airplanes, particularly the Iberia airplane in which the study is carried out, provide a satellite-communication service with a global Wi-Fi coverage. The speed of Wi-Fi connection will depend on the device in use and the number of users connected at a certain moment.

Our research group has been interested in 2G Wi-F and 5G Wi-Fi (and future ones) from the moment that Pall published an article about Wi-Fi being an important threat to human health [38], which has received some comments from a scientific point of view [39,40]. When we refer to 2G we mean to 2.4 GHz Wi-Fi, and when we refer to 5G, we mean to 5.8 GHz. For this reason, we decided to measure Wi-Fi inside an airplane. The aim of this study was to measure exposure to RF-EMF from 2G Wi-Fi and 5G Wi-Fi frequency bands on a regular flight from Barajas Adolfo Suarez Airport in Madrid (Spain) to Benito Juárez International Airport in Mexico City (Mexico). The flight lasted 11 hours and it was a direct one.

2. Methodology

2.1. Exposimeters

Measurements of personal exposure to RF-EMF were carried out with two EME SPY 140 Satimo personal exposimeters (Figure 2). The main features of this device type are that 14 frequency bands between 88 MHz and 5.8 GHz can be measured, the maximum sensitivity level is 0.05 V/m and the device can distinguish contributions from each emitting frequency band and the total field [41]. The device is also able to register up to 12540 measurements in time periods between 4 and 255 seconds [41].

Figure 2. EME SPY 140 personal exposimeter measuring inside the airplane.

2.2. Measurements

Measurements were carried out by an author for this work. The events took place inside an airplane on a flight that departed from Barajas Adolfo Suarez Airport in Madrid (Spain) and arrived in Benito Juárez International Airport (Mexico City, Mexico), on the 20th of June 2021 (Figure 3). The flight was direct and lasted approximately 11 hours.

The researcher carried out two personal exposimeters, each one on her own shoulder strap, hanging on the waist, on each side of his body and with no metallic artefacts to avoid interferences with the exposimeters. The mobile phone was always far from the exposimeters. Nevertheless, whenever a Wi-Fi connection was established, the event and the time were recorded in a personal diary.

Measurements were conducted at the moment that the researcher entered the airplane, when the airplane took off and started to flight (while being in flight mode), during the whole flight from Madrid to Mexico City (Figure 3), including the moments when Wi-Fi was used for send a WhatsApp messages, search for information on the Internet, downloading and reading documents, downloading and watching photos and even a short videos send by another group member (the sender was in Spain). To carry out the whole procedure, everything was planned, and the researcher who performed measurements notified the other member of the research group when the flight departed, and she connected to a Wi-Fi band.

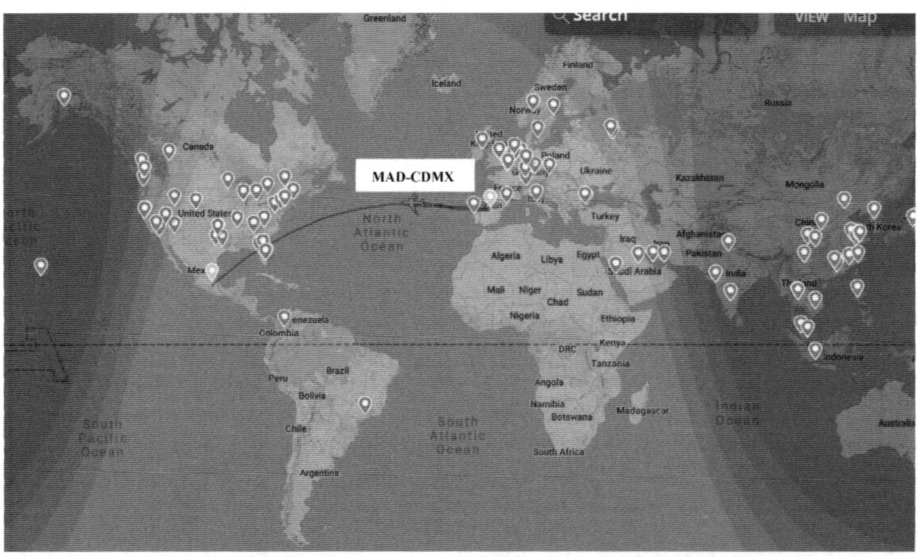

Figure 3. Flight from Madrid (Spain) to CDMX (México).

During the flight the researcher paid for the Wi-Fi connection, to make use of the Internet and identify differences between measurements. The contract for the Wi-Fi service took place at three different moments, each contract was 1 hour long, and the pay was the highest one (9 €), this being selected as it is the one that allows to chat, search for information on the internet, reading documents, and download and watch photos and even short videos.

When the researcher used the Wi-Fi connection, she contacted another group member by WhatsApp. This member sent her information files, pictures, and short videos. To synchronize time for each activity, the researcher who was travelling sent WhatsApp messages to the researcher in Spain at the moment of boarding the airplane and whenever she had Wi-fi service.

The researcher had a diary in which she recorded the exact time and the activity she was doing at the moment of boarding the airplane, while she was inside the airplane waiting for take off (origin country) and while she was waiting to disembark (destination country); besides, at the moment of taking off and landing (while mobile devices were in flight mode) and flight duration.

2.3. Software

The statistical analysis was conducted using the personal exposimeter own software, EME SPY Analysis V3. Microsoft Excel and Statistical Package for the Social Sciences (SPSS) were used for data processing and descriptive analysis. Google Earth Pro, Iberia App and Flightradar 24 App were of use for flight tracking. Results for measurements are presented in $\mu W/m^2$ with three significant figures [42].

2.4. Measurement procedure

Measurements were carried out with two personal exposimeters (EME SPY 140) inside two cloth bags at waist height, each one at each side of the researcher (left and right). The exposimeters were configurated to record measurements every 10 seconds.

Measurements were averaged over 6 minutes intervals for local exposure and 30 minutes intervals for a whole body exposure [43]. Before statistical analysis, all data collected were revised to identify mistakes and verify measurements with personal diary registers, correcting any mistake identified. We recorded values below the detection limit (called 'nondetects') from the exposimeter in the two frequency bands, namely 2G Wi-Fi and 5G Wi-Fi. Nondetects data percentage ranged from 80 to 85%. Values recorded as nondetects received a special treatment, and they were replaced by the detection limit divided by two [44].

3. Results

After carrying out statistical analysis in detail for all results, we have identified that intensity for electromagnetic wave changes at different moments. The same value is not obtained inside an airplane when making use of Wi-Fi (waiting for take off or disembark) or during the flight (using Wi-Fi service or not).

When the flight departed and arrived, a high value was also registered (Figure 4). We have investigated the reason for this event with the airline in whose airplanes measurements were performed, although we have been told not to share this information because of privacy policies. We were not able to obtain information about Wi-Fi functioning inside the airplane. Therefore, it can be deduced that values recorded both at the beginning and in the end correspond to connection and monitoring that the pilot and the crew maintain with the control tower.

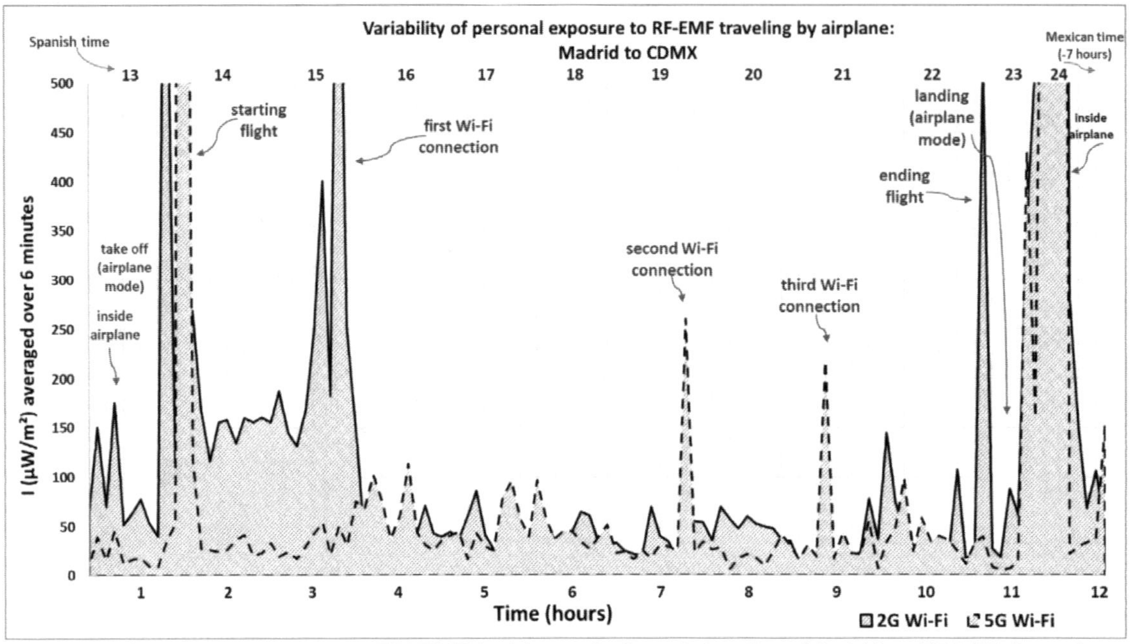

Figure 4. Personal exposure to RF-EMF inside an airplane flying from Madrid to CDMX. Intensity of electromagnetic wave in µW/m^2. Averaged over 6 minutes.

Another peak or a high value was recorded during the use of the Wi-Fi on behalf of the researcher, who paid for the Wi-Fi and used it to search for information on the Internet, download and watch information files, pictures and short videos sent by other group researcher. This event occurred at three different moments (as seen in Figure 4).

In Figure 4, we can see average intensity level over 6 minutes for local exposure [43], an arrow indicating the events registered in the personal diary, highlighting the higher peaks. Y axis shows average intensity value recorded every 6 minutes, both from 2G Wi-Fi and 5G Wi-Fi, and the lower X axis shows time in hours, starting at zero from the time the researcher entered the airplane in the country of origin (Spain) until the time she landed and disembarked in the country of destination (Mexico), while the upper X axis shows the real time at which the measurement was performed. The personal exposimeters were synchronized to the time zone for the origin country. It must be pointed out that the team did not change the local time as the flight traversed different time zones, as happens with a smart watch, then we bore in mind the 7-hour difference when the journey finished.

The highest value recorded during the whole flight was 1830 µW/m^2 in the 2G Wi-Fi frequency band and 1870 µW/m^2 in the 5G Wi-Fi frequency band, while the lowest value was 12.9 µW/m^2 in the 2G Wi-Fi and 4.91 µW/m^2 in the 5G Wi-Fi band. The frequency band with the highest activity was 2G Wi-Fi band compared to 5G Wi-Fi band.

When the flight departed, a high value was 832 µW/m^2 recorded in the 2G Wi-Fi frequency band and a higher value was 1870 µW/m^2 in the 5G Wi-Fi frequency band, due to the connection to the land antennas at the airport. When the flight arrived, a high value was 1830 µW/m^2 recorded in the 2G Wi-Fi frequency band and a higher value was 1680 µW/m^2 in the 5G Wi-Fi frequency band (Figure 4).

During the whole flight, there were three moments at which high values were recorded, due to the Internet connection that the researcher contracted and used. Recorded values were: 831 µW/m^2 in 2G Wi-Fi and 102 µW/m^2 in 5G Wi-Fi (first Wi-Fi connection); 69.7 µW/m^2 in 2G Wi-Fi and 261 µW/m^2 in 5G Wi-Fi (second Wi-Fi connection); and 144 µW/m^2 in 2G Wi-Fi and 217 µW/m^2 in 5G Wi-Fi (third Wi-Fi connection).

In Figure 5 we show variations for average intensity levels registered during the flight over 30 minutes for a whole body exposure [43], placing an highlighting on peaks or high values. Y axis shows average intensity value every 30 minutes for a whole body exposure from 2G Wi-Fi band and 5G Wi-Fi band, and the lower X axis shows time in hours, starting at zero from the time the researcher entered the airplane until the time she disembarked, while the upper X axis shows real time at which the measurement was conducted. High values were recorded when the flight departed (417 $\mu W/m^2$ in 5G Wi-Fi band), when the flight arrived (1050 $\mu W/m^2$ in 2G Wi-Fi band), and at the moments that Wi-Fi was accessed and used inside the airplane: 295 $\mu W/m^2$ in 2G Wi-Fi band (first Wi-Fi connection), 74.1 $\mu W/m^2$ in 5G Wi-Fi band (second Wi-Fi connection), and 65.9 $\mu W/m^2$ in 5G Wi-Fi band (third Wi-Fi connection).

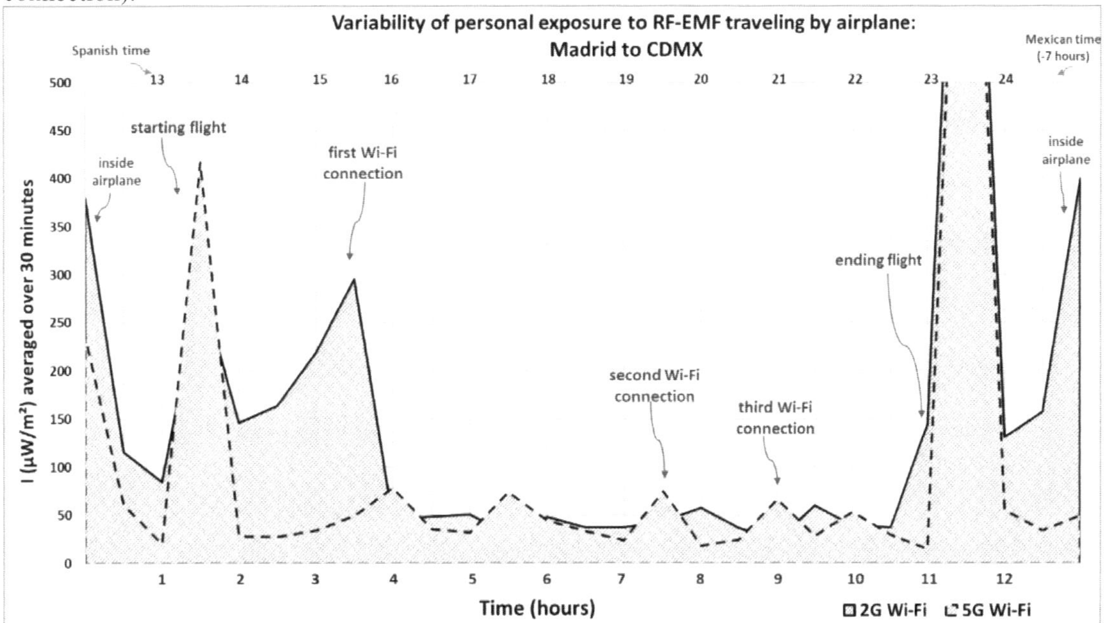

Figure 5. Personal exposure to RF-EMF inside an airplane flying from Madrid to CDMX. Intensity of electromagnetic wave in $\mu W/m^2$. Averaged over 30 minutes.

In Table 1 and Figure 6 we can see the average and 95 percentile of personal exposure levels to RF-EMF recorded at different times during the flight.

Table 1. Average personal exposure to RF-EMF flying inside an airplane: from Madrid (Spain) to CDMX (Mexico). Intensity of electromagnetic wave in $\mu W/m^2$ with three significant figures [42].

Measured moment	2G Wi-Fi ($\mu W/m^2$)		5G Wi-Fi ($\mu W/m^2$)	
	Average	P95	Average	P95
inside airplane (Madrid)	127	740	23.9	111
take off (airplane mode)	61.1	164	20.6	92.2
flight	123	273	87.1	184
landing (airplane mode)	107	487	31.8	188
inside airplane (Mexico)	46.8	205	57.1	297

The Y axis on the left in Figure 6 indicates the moment or the microenvironment in which the researcher recorded the measurement, the X axis on the right indicates the real time at which the measurement was recorded. The X axis indicates average intensity from 2G Wi-Fi and 5G Wi-Fi bands in $\mu W/m^2$.

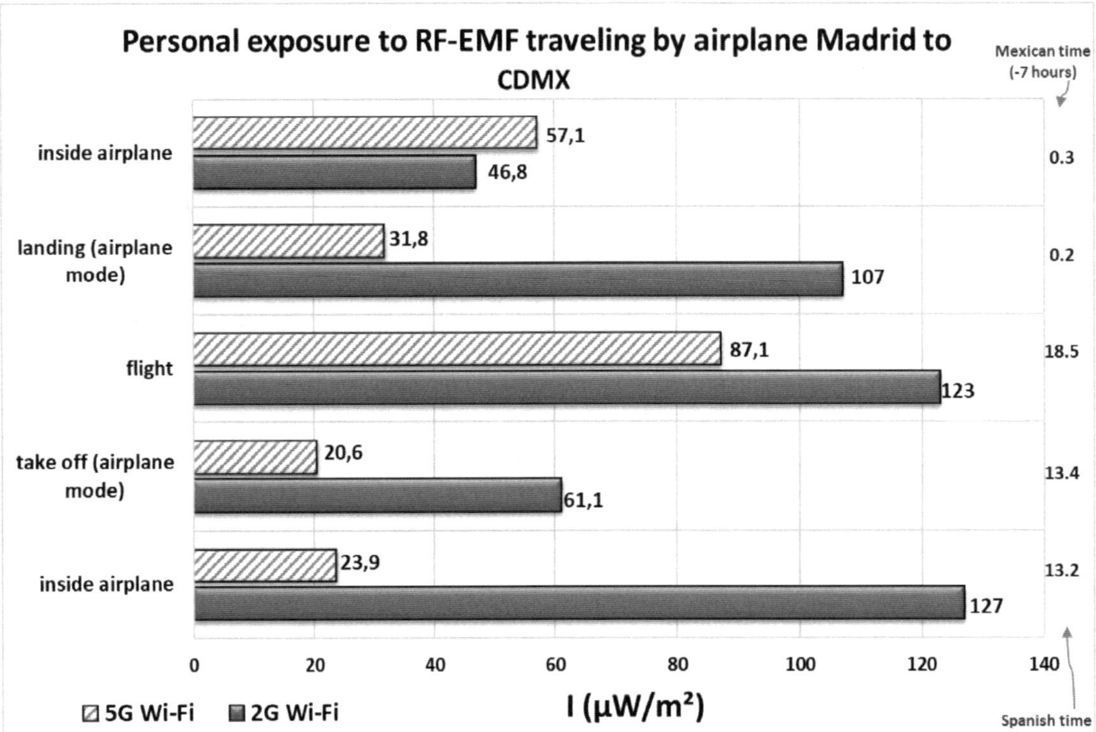

Figure 6. Average personal exposure to RF-EMF at different moments of the flight. Intensity of electromagnetic wave in μW/m^2 with three significant figures [42].

4. Discussion

To the best of our knowledge, there is no study in which personal exposure to RF-EMF inside an airplane during a flight has been measured. Our aim in the short term is to replicate this study in similar conditions so that results can be compared and discussed.

Personal exposimeters were set and synchronized to the time zone for the origin country; the time was not changed from one time zone to the next and to the destination country on the exposimeter, therefore, we have indicated in the figures (upper X axis) the time difference for the researcher when she arrived in the destination country, Mexico. There was a 7-hour difference.

The flight was long and exhausting the researcher was careful with the personal exposimeters for the whole duration of the flight.

No significant differences were recorded in the measurements of each exposimeter. To carry out the analysis of the results, we have calculated the average of both exposimeter, to consider the exposure on the left and right side of the researcher.

During the flight, high values were recorded affecting measurements; for example, the average value of 1870 μW/m^2 recorded in the 5G Wi-Fi band at the beginning of the flight. This value was not discarded. We concluded that it was due to the connection of the Wi-Fi antenna of the airplane to the ground antennas and the control tower, connections being made by the crew and the pilot in order to carry out the necessary controls.

The frequency band with the highest activity was 2G Wi-Fi.

One of the main limitations was the access to confidential information on behalf of the airline. Regarding measuring devices, one difficulty was the recording of nondetects values, which were processed before we carried out the statistical analysis, as indicated in methodology section.

5. Conclusion

The average value registered during the flight was 123 $\mu W/m^2$ with a 95 percentile of 273 $\mu W/m^2$ in the 2G Wi-Fi frequency band: and an average of 87.1 $\mu W/m^2$ with a 95 percentile of 184 $\mu W/m^2$ in the 5G Wi-Fi frequency band.

The highest values were registered at the beginning of the flight and during the three different moments in which the Wi-Fi connection was used. As explained in methodology and result sections, the payment of a Wi-Fi service had been planned and it was used for sending a WhatsApp message, search for information on the Internet, exchanging information files, sending, and receiving pictures and short videos (from 15 to 16 hours, from 19 to 20 hours and from 21 to 22 hours).

Size for the file and the video was small so that they both could be downloaded and viewed. These activities were recorded in the personal diary and coincided with the peaks and high values observed in the graphics. We therefore concluded that the airplane router manages and responds to a connectivity requirement, if there is one; that is to say, it works as a normal router on demand.

The 95 Percentile recorded inside the airplane during the flight corresponds to a value of 273 $\mu W/m^2$ in the 2G Wi-Fi band and 184 $\mu W/m^2$ in the 5G Wi-Fi band. The average value of 273 $\mu W/m^2$ is 46 dB below the maximum allowed level (10 W/m^2) for population personal exposure, established by international regulations (ICNIRP, 2020).

References

[1] Bhatt C R, Thielens A, Billah B, Redmayne M, Abramson M J, Sim M R, Vermeulen R, Martens L, Joseph W and Benke G 2016 Assessment of personal exposure from radiofrequency-electromagnetic fields in Australia and Belgium using on-body calibrated exposimeters *Environ. Res.* **151** 547–63

[2] Birks L E, Struchen B, Eeftens M, van Wel L, Huss A, Gajsek P, Kheifets L, Gallastegi M, Dalmau-Bueno A, Estarlich M, Fernandez M F, Meder I K, Ferrero A, Jimenez-Zabala A, Torrent M, Vrijkotte T G M, Cardis E, Olsen J, Valic B, Vermeulen R, Vrijheid M, Roeoesli M and Guxens M 2018 Spatial and temporal variability of personal environmental exposure to radio frequency electromagnetic fields in children in Europe *Environ. Int.* **117** 204–14

[3] Bogers R P, van Gils A, Clahsen S C S, Vercruijsse W, van Kamp I, Baliatsas C, Rosmalen J G M and Bolte J F B 2018 Individual variation in temporal relationships between exposure to radiofrequency electromagnetic fields and non-specific physical symptoms: A new approach in studying 'electrosensitivity' *Environment International* **121** 297–307

[4] Bolte J F B, Pruppers M J M, Kamer J, Van der Zande G, Schipper C M A, Fleurke S R, Kluwer T, Van Kamp I and Kromhout J 2008 The Dutch Exposimeter Study: Developing an Activity Exposure Matrix *Epidemiology* **19** S78–9

[5] Choi J, Hwang J-H, Lim H, Joo H, Yang H-S, Lee Y-H, Eeftens M, Struchen B, Roosli M, Lee A-K, Choi H-D, Kwon J H and Ha M 2018 Assessment of radiofrequency electromagnetic field exposure from personal measurements considering the body shadowing effect in Korean children and parents *Sci. Total Environ.* **627** 1544–51

[6] Eeftens M, Struchen B, Ellen Birks L, Cardis E, Estarlich M, Fernandez M F, Gajsek P, Gallastegi M, Huss A, Kheifets L, Meder I K, Olsen J, Torrent M, Trcek T, Valic B, Vermeulen R, Vrijheied M, van Wel L, Guxens M and Roosli M 2018 Personal exposure to radio-frequency electromagnetic fields in Europe: Is there a generation gap? *Environ. Int.* **121** 216–26

[7] Frei P, Mohler E, Neubauer G, Theis G, Bürgi A, Fröhlich J, Braun-Fahrländer C, Bolte J, Egger M and Röösli M 2009 Temporal and spatial variability of personal exposure to radio frequency electromagnetic fields *Environ. Res.* **109** 779–85

[8] Gallastegi M, Guxens M, Jiménez-Zabala A, Calvente I, Fernández M, Birks L, Struchen B, Vrijheid M, Estarlich M, Fernández M F, Torrent M, Ballester F, Aurrekoetxea J J, Ibarluzea J, Guerra D, González J, Röösli M and Santa-Marina L 2016 Characterisation of exposure to non-ionising electromagnetic fields in the Spanish INMA birth cohort: study protocol *BMC Public Health* **16** 167

[9] Massardier-Pilonchery A, Nerriere E, Croidieu S, Ndagijimana F, Gaudaire F, Martinsons C, Noe N and Hours M 2019 Assessment of Personal Occupational Exposure to Radiofrequency Electromagnetic Fields in Libraries and Media Libraries, Using Calibrated On-Body Exposimeters *Int. J. Environ. Res. Public Health* **16** 2087

[10] Ramirez-Vazquez R, Gonzalez-Rubio J, Arribas E and Najera A 2019 Characterisation of personal exposure to environmental radiofrequency electromagnetic fields in Albacete (Spain) and assessment of risk perception *Environ. Res.* **172** 109–16

[11] Ramirez-Vazquez R, Gonzalez-Rubio J, Escobar I, Suarez Rodriguez C D P and Arribas E 2021 Personal Exposure Assessment to Wi-Fi Radiofrequency Electromagnetic Fields in Mexican Microenvironments. *International Journal of Environmental Research and Public Health* **18** 1–20

[12] Roser K, Schoeni A, Struchen B, Zahner M, Eeftens M, Frohlich J and Roosli M 2017 Personal radiofrequency electromagnetic field exposure measurements in Swiss adolescents *Environ. Int.* **99** 303–14

[13] Thielens A, Bockstael A, Declerck S, Aminzadeh R, Aerts S, Botteldooren D, Martens L and Joseph W 2019 Mobile phones: A trade-off between speech intelligibility and exposure to noise levels and to radio-frequency electromagnetic fields *Environmental Research*

[14] Zeleke B M, Brzozek C, Bhatt C R, Abramson M J, Croft R J, Freudenstein F, Wiedemann P and Benke G 2018 Personal Exposure to Radio Frequency Electromagnetic Fields among Australian Adults *Int. J. Environ. Res. Public Health* **15** 2234

[15] Zeleke B M, Bhatt C R, Brzozek C, Abramson M J, Freudenstein F, Croft R J, Wiedemann P and Benke G 2019 Radiofrequency electromagnetic field exposure and risk perception: A pilot experimental study *Environ. Res.* **170** 493–9

[16] Frei P, Mohler E, Buergi A, Froehlich J, Neubauer G, Braun-Fahrlaender C and Roosli M 2010 Classification of personal exposure to radio frequency electromagnetic fields (RF-EMF) for epidemiological research: Evaluation of different exposure assessment methods *Environment International* **36** 714–20

[17] Röösli M, Frei P, Bolte J, Neubauer G, Cardis E, Feychting M, Gajsek P, Heinrich S, Joseph W, Mann S, Martens L, Mohler E, Parslow R C, Poulsen A H, Radon K, Schuez J, Thuroczy G, Viel J-F and Vrijheid M 2010 Conduct of a personal radiofrequency electromagnetic field measurement study: proposed study protocol *Environ. Health* **9** 23–37

[18] Jalilian H, Eeftens M, Ziaei M and Röösli M 2019 Public exposure to radiofrequency electromagnetic fields in everyday microenvironments: An updated systematic review for Europe *Environ. Res.* **176** 108517

[19] Dongus S, Jalilian H, Schurmann D and Röösli M 2021 Health effects of WiFi radiation: a review based on systematic quality evaluation. Published online. *Crit. Rev. Environ. Sci. Technol.*

[20] Ramirez-Vazquez R, Gonzalez-Rubio J, Arribas E and Najera A 2019 Personal RF-EMF exposure from mobile phone base stations during temporary events *Environ. Res.* **175** 266–73

[21] Gonzalez-Rubio J, Arribas E, Ramirez-Vazquez R and Najera A 2017 Radiofrequency electromagnetic fields and some cancers of unknown etiology: An ecological study *Sci. Total Environ.* **599** 834–43

[22] Gonzalez-Rubio J, Najera A and Arribas E 2016 Comprehensive personal RF-EMF exposure map and its potential use in epidemiological studies *Environ. Res.* **149** 105–12

[23] Ramirez-Vazquez R, Arabasi S, Al-Taani H, Sbeih S, Gonzalez-Rubio J, Escobar I and Arribas E 2020 Georeferencing of Personal Exposure to Radiofrequency Electromagnetic Fields from Wi-Fi in a University Area *International Journal of Environmental Research and Public Health* **17** 1898

[24] Ramirez-Vazquez R, Escobar I, Thielens A and Arribas E 2020 Measurements and Analysis of Personal Exposure to Radiofrequency Electromagnetic Fields at Outdoor and Indoor School Buildings: A Case Study at a Spanish School *IEEE Access* **8** 195692–702

[25] Eeftens M, Dongus S, Burgler A and Roosli M 2020 A real-world quality assessment study in six ExpoM-RF measurement devices *Environ. Res.* **182** 109049

[26] Joseph W, Aerts S, Vandenbossche M, Thielens A and Martens L 2016 Drone based measurement system for radiofrequency exposure assessment: Drone Based RF Exposure Measurements *Bioelectromagnetics* **37** 195–9

[27] Akkam Y, A. Al-Taani A, Ayasreh S, Almutairi A and Akkam N 2020 Correlation of Blood Oxidative Stress Parameters to Indoor Radiofrequency Radiation: A Cross Sectional Study in Jordan *Int. J. Environ. Res. Public Health* **17** 4673

[28] Joseph W, Verloock L, Goeminne F, Vermeeren G and Martens L 2012 Assessment of RF Exposures from Emerging Wireless Communication Technologies in Different Environments *Health Phys.* **102** 161–72

[29] Kuehnlein A, Heumann C, Thomas S, Heinrich S and Radon K 2009 Personal Exposure to Mobile Communication Networks and Well-Being in Children-A Statistical Analysis Based on a Functional Approach *Bioelectromagnetics* **30** 261–9

[30] Roser K, Schoeni A, Buergi A and Röösli M 2015 Development of an RF-EMF Exposure Surrogate for Epidemiologic Research *Int. J. Environ. Res. Public Health* **12** 5634–56

[31] Thomas S, Kühnlein A, Heinrich S, Praml G, Nowak D, von Kries R and Radon K 2008 Personal exposure to mobile phone frequencies and well-being in adults: A cross-sectional study based on dosimetry *Bioelectromagnetics* **29** 463–70
[32] Thuroczy G, Molnar F, Janossy G, Nagy N, Kubinyi G, Bakos J and Szabo J 2008 Personal RF exposimetry in urban area *Ann. Telecommun.-Ann Telecommun.* **63** 87–96
[33] Viel J-F, Cardis E, Moissonnier M, de Seze R and Hours M 2009 Radiofrequency exposure in the French general population: Band, time, location and activity variability *Environ. Int.* **35** 1150–4
[34] Guersu M, Vilgelm M, Zoppi S and Kellerer W 2016 Reliable co-existence of 802.15.4e TSCH-based WSN and Wi-Fi in an Aircraft Cabin *2016 Ieee International Conference on Communications Workshops (ICC)* (New York: Ieee) pp 663–8
[35] Trifiletti E G 2008 High Speed Train and New Technological and Functional Innovations *TeMA* **1** 101–8
[36] Justavino Castillo R A, Gruenheid R, Bauch G, Wolff F and von der Heide S 2018 Communication Analysis Between an Airborne Mobile User and a Terrestrial Mobile Network *IEEE Trans. Veh. Technol.* **67** 3457–65
[37] Sierzputowski R, Polak R, Wojtyra D and Laskowski D 2020 Advanced protection methods of unmanned aircraft vehicle against attack by radio techniques *Radioelectronic Systems Conference 2019* vol 11442, ed P Kaniewski and J Matuszewski (Bellingham: Spie-Int Soc Optical Engineering) p UNSP 114420Y
[38] Pall M L 2018 Wi-Fi is an important threat to human health *Environ. Res.* **164** 405–16
[39] Foster K R and Moulder J E 2019 Response to Pall, "Wi-Fi is an important threat to human health" *Environ. Res.* **168** 445–7
[40] Arribas E, Ramirez-Vazquez R and Escobar I 2018 Comment on "Wi-Fi is an important threat to human health" *Environ. Res.* **167** 639–639
[41] MVG 2017 EME SPY 200 | RF-Safety. URL http://www.mvg-world.com/es/products/field_product_family/rf-safety-3/eme-spy-200.
[42] Ramirez-Vazquez R, Escobar I, Franco T and Arribas E 2022 Physical units to report intensity of electromagnetic wave *Environ. Res.* **204** 112341
[43] ICNIRP 2020 International Commission on Non-Ionizing Radiation Protection. Guidelines for Limiting Exposure to Electromagnetic Fields (100 kHz to 300 GHz) *Health Phys.* **118** 483–524
[44] Najera A, Ramirez-Vazquez R, Arribas E and Gonzalez-Rubio J 2020 Comparison of statistic methods for censored personal exposure to RF-EMF data. *Environmental monitoring and assessment* **192** 77

Design, implementation and preliminary testing of an automated system for the evaluation of ionospheric channels

E Șorecău[1], M Șorecău[1], N Craiu[1], A Sârbu[2] and P Bechet[2]

[1] Department of Electrotechnics and Measurements, Technical University of Cluj Napoca, 400114 Cluj-Napoca, Romania
[2] Department of Communications, IT and Cyber Defence, "Nicolae Bălcescu" Land Forces Academy, 550170 Sibiu, Romania

E-mail: emil.sorecau@gmail.com

Abstract. Radio communications in the High Frequency (HF) range are used to transmit information over long distances between one or more correspondents. Techniques and methodsto improve the capacity and reliability of HF communications become key objectives as they are generally limited to narrow bandwidths and low transfer rates. This work aims to implement an automated system for real-time measurement of ionospheric radio channel parameters under Near Vertical Incidence Skywave (NVIS) propagation conditions using Software Define Radio (SDR) platforms. Preliminary testing of the system was carried out both under laboratory conditions using signal generators and high-performance spectral vector analyzers and under real propagation conditions by installation at the established sites. The results showed that the system is capable of performing real-time measurements of signal power, noise power and calculating signal-to-noise ratio for bandwidths up to 24 kHz.

1. Introduction

The ionosphere acts as a mirror for high frequency signals (3-30 MHz) and can be used as a medium for long-distance communication between radio equipment. Ionospheric propagation takes advantage of the signal mirroring benefits being employed particularly for emergency services, military applications and sensor interconnection.

The performance and quality of radio communications depend significantly on the type and behavior of the ionospheric channel. Therefore, it is necessary to study the availability, usable frequencies and capacity of ionospheric channels for different bandwidths, power levels and times of day and night [1-2].

The data transmission capacity on any channel or medium is defined by Shannon as follows [3]:

$$C=B*\log_2(1+S/N) \quad (1)$$

where: C is the channel capacity in bits per second (or maximum data rate), B is the bandwidth available for data transmission expressed in Hz, S is the received signal strength, and N is the total channel noise power over the entire B bandwidth. Therefore, the higher the signal-to-noise ratio (SNR) and channel bandwidth, the higher the data transmission rate is.

Noise in the HF range has mainly atmospheric noise components (the distribution and average power values do not change significantly over time) and man-made noise components, which include intentional and unintentional emissions from different sources [4-5]. Recently, several studies have been reported on the characteristics of ionospheric channels. These have led to the development of efficient measurement systems with higher sensitivity and real-time analysis capabilities, as well as customised procedures to measure the effects of phenomena contributing to the total noise level in the HF range [6-8].

The SNR parameter is used in most cases to improve radio channel characteristics and is a very important decision factor in adaptive communications, such as dynamic reconfiguration of cognitive radio stations, adaptive modulation and coding and adaptive power allocation [9]. Calculating, estimating or measuring this important parameter has been done in different ways, such as studying the amplitudes of waveforms [10] or examining the difference between the received signal and the transmitted signal [11]. The current interest in HF communications is to study channel parameters at higher bandwidths (3 kHz, 6 kHz, 12 kHz, 24 kHz) and to propose methods and techniques to increase the ability of systems to transmit data at higher rate [12-14].

To this extent, the use of SDR platforms became a prominent solution in the implementation of systems for measuring ionospheric channel parameters. SDRs have the advantage of flexibility to implement different applications on a single platform and provide high processing power. A variety of newly developed methods, algorithms and protocols can be easily implemented and verified in a similar way as they would be realized in a system built on conventional radio platforms [1].

An important measurement campaign has been carried out in the Antarctica Project in which a long-distance radio link was established between Spain and Antarctica. Channel parameters in the whole HF range, propagation time and Doppler frequency shift were included in the set of measurements. The SNR was calculated for 24h by comparing the received power measured during transmission with the noise power measured during the waiting periods between transmissions [15].

Another project worth mentioning is the Wideband HF Ionospheric Sounder for Propagation Environment Research (WHISPER) run by the Communications Research Centre in Ottawa, which aims to demonstrate the feasibility and benefits of wideband HF communications. The project objectives were to increase the data throughput of the HF channel, not only by increasing spectral efficiency in a standard 3 kHz bandwidth, but more importantly by using a wider bandwidth [1].

This work is in line with these research directions and aims at the design and preliminary testing of an ionospheric channel assessment system operating under NVIS propagation conditions, based on SDR platforms controlled by predefined software applications. The aim is to measure and analyze in real time parameters such as received signal strength, noise power and SNR calculation, using selectable frequencies across the HF spectrum and variable bandwidths up to 24 kHz. Data will be saved in phase (I) and quadrature (Q) format as well as in comma-separated values text format using local hard drives.

The paper is further organized as follows: section II presents in detail the communication system and explains the methods used to measure the parameters, section III presents the results of the system calibration under laboratory conditions and preliminary tests under real conditions and section IV presents the final conclusions of the paper.

2. Materials and Methods

In order to monitor rapid variations in signal level, an SDR platform with real-time measurement capability was used. We chose an SDR platform from National Instruments - Universal Software Radio Peripheral (NI USRP), controlled by an application developed in the GNU Radio environment.

The emission system configuration was achieved using a laptop running the Linux operating system (Ubuntu 20.04) connected to a NI USRP 2932 platform, with an LFRX model 0-30 MHz radiofrequency receiver board, a Rohde & Schwarz BBA 150 power amplifier and a Diamond W330 wideband dipole antenna mounted in a-inverted "V". The used antenna is optimal for NVIS communications and was elevated 12 m above the ground on a sectional mast (Figure 1).

The receiving system consists of the same laptop-USRP configuration, connected to a multiband HF T2FD (3-30MHz) antenna also mounted in a inverted "V". For the receiver, we have used a 9:1

IOP Conf. Series: Materials Science and Engineering 1254 (2022) 012008 doi:10.1088/1757-899X/1254/1/012008

balun and non-inductive end resistors to obtain a Voltage Standing Wave Ratio below 2 across the HF range. Figure 2 shows the components of the receiving system.

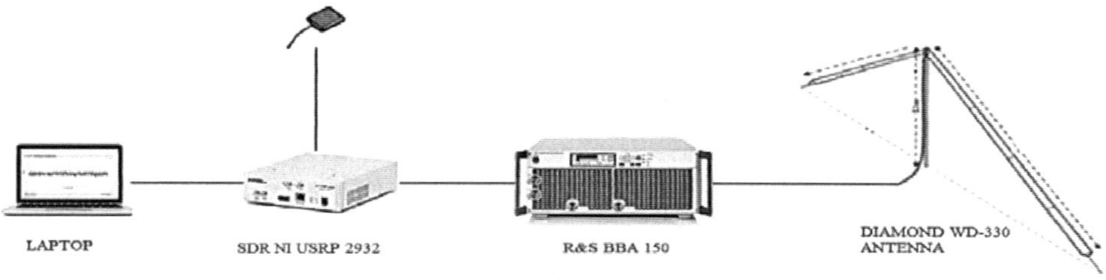

Figure 1. Emission system components.

Precise timing, channel estimation methods and a very accurate external frequency reference must be used for proper system operation. To this end, both the transmitting and receiving systems were installed using external Global Positioning System units connected to the SDR and implemented in Python script.

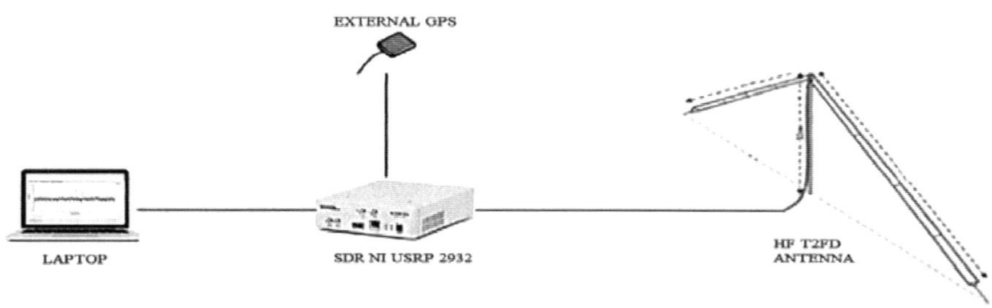

Figure 2. Components of the receiving system.

The frequency range of interest was achieved by setting the appropriate sampling frequency of the SDR platforms. From this point of view, the GNU Radio software application offered the possibility to monitor a wider range of frequencies. A sample rate of1MHz was first set and then by decimation (reduction of the sampling frequency) the monitored bandwidth was reduced to 100 kHz. To measure the two channel powers (for the signal and noise), two processing streams were enabled: one for measuring signal power and another for measuring noise power. The two resulting values were compared (subtracted) to display the SNR (Figure 3).

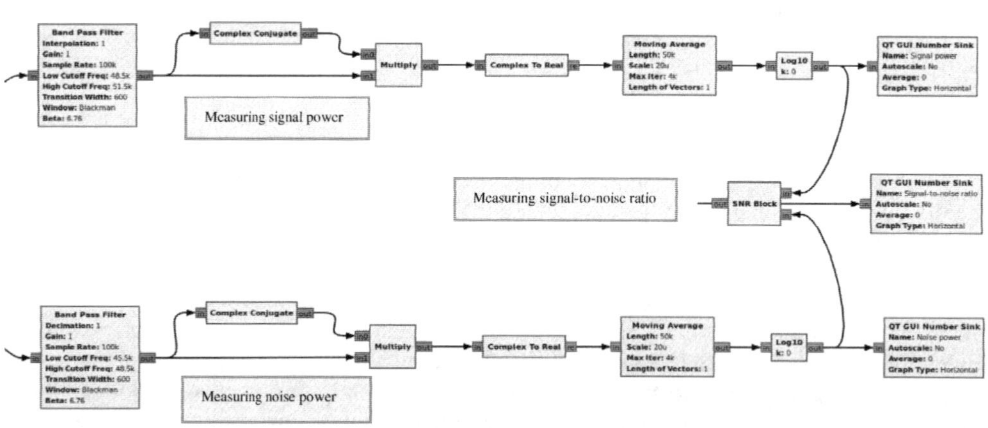

Figure 3. Software application developed in GNU Radio environment.

IOP Conf. Series: Materials Science and Engineering 1254 (2022) 012008 doi:10.1088/1757-899X/1254/1/012008

Control of the channel bandwidth was achieved by means of two bandpass filters configured to change their characteristics according to the selected central frequency and bandwidth: 3 kHz, 6 kHz, 12 kHz and 24 kHz. The bandpass filter used for measuring the noise power is offset by a bandwidth from the filter pass band that measures the signal channel power. This configuration allows us to display SNR values in real time. The evaluation of channel power (signal or noise) was performed in the time domain. For this purpose, the signal is multiplied by its conjugate, then averaged and logarithmed to express the power in dBm.

In order to save the sample values containing the signal strength and noise as well as the SNR value, a new block has been set up in GNU Radio specifically implemented with these capabilities. Since the GNU Radio application has only a block that saves in I and Q format, a Python script was programmed to save the actual samples. This block has been named "DATA SAVE" and saves the samples in a CSV format file (Figure 4).

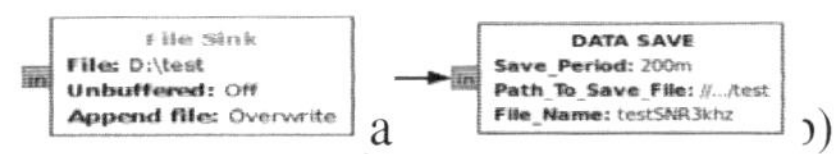

Figure 4. Data storage blocks a) IQ and b) CSV.

The design and software implementation of the transmitter and receiver systems was done to enable the following capabilities: selection of user defined time-delimited frequency sets, setting of spectrum analysis parameters (frequency, sweep time, frequency resolution, frequency range), channel bandwidth control and setting the start and stop times of the recordings. Figure 5 shows the data flow through the receiving system.

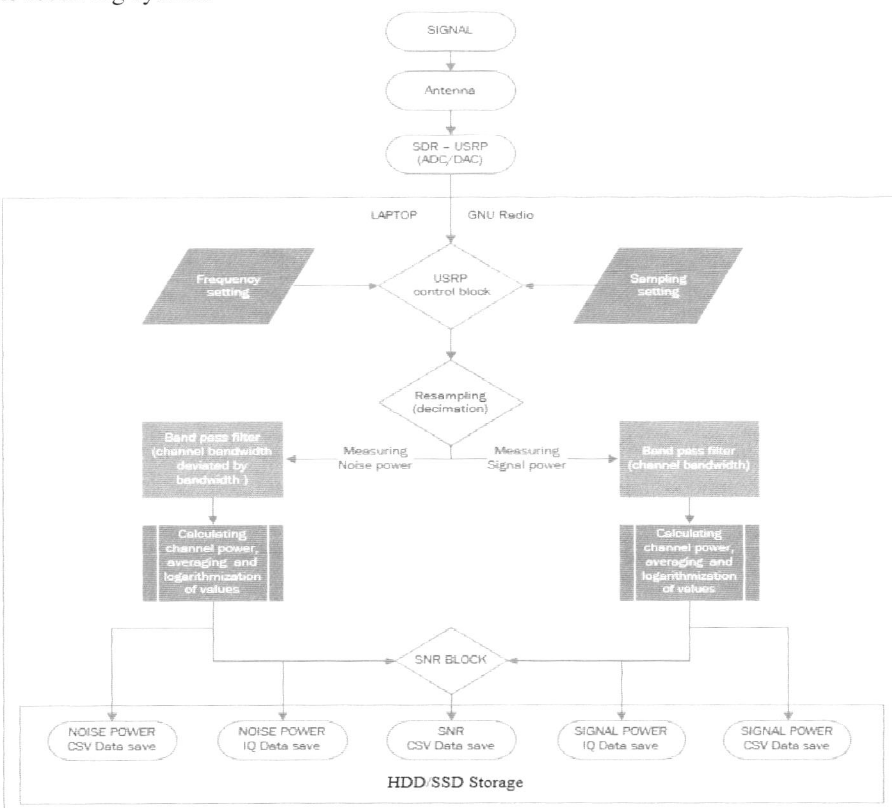

Figure 5. Data flow through the receiving system.

For high flexibility in making measurements, the Python scripts generated by GNU Radio were edited after making the desired flowgraph and implemented in graphical applications using the tkinter package ("Tk interface") (Figure 6).

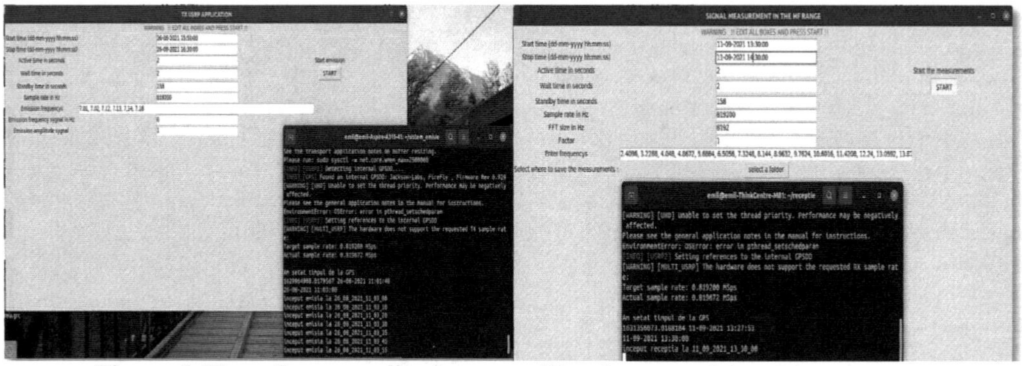

Figure 6. The software application controlling the transmit/receive system.

3. Results and Discussions

The evaluation of the system was performed in three steps: (1) calibration of transmission and reception using signal generators and analyzers, (2) back-to-back testing of the system under laboratory conditions and (3) verification under real operating conditions in NVIS propagation scenarios.

In the first phase a signal generator was connected via coaxial cable to an RX channel of the USRP device. An R&S FSV spectrum analyzer was used as the equipment providing the amplitude reference level. The amplitude and frequency response of the SDR system was evaluated. For this purpose, the frequency of the generated signal was varied in the range 3-30MHz with a step of 1MHz and the amplitude was varied from -90 dBm to -30 dBm with a step of 5dB. Figure 7 shows the test stand under laboratory conditions.

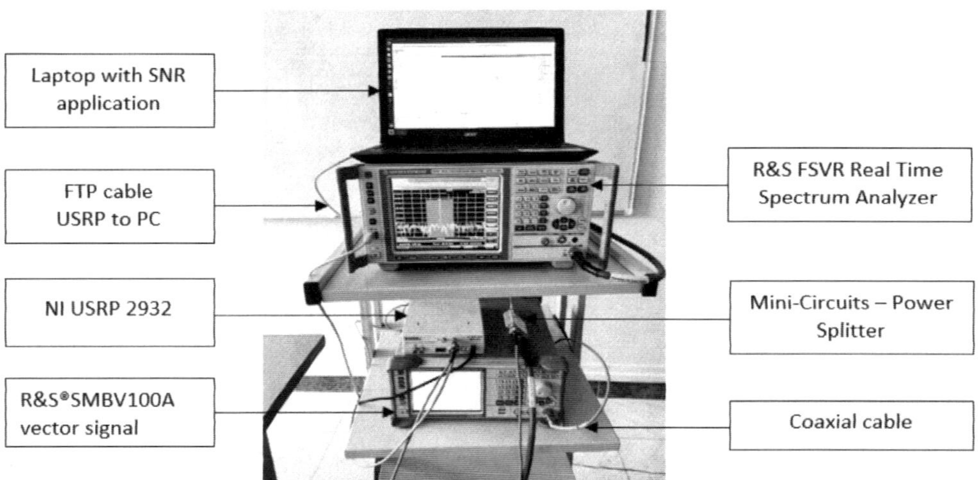

Figure 7. Test stand under laboratory conditions.

Calibration of the power level was performed over a 3kHz bandwidth, generating a sinusoidal signal with different power levels. As can be seen, after calibrating the parameters, the channel power level measured with the application made in GNU Radio is similar to that measured with the spectrum analyzer.

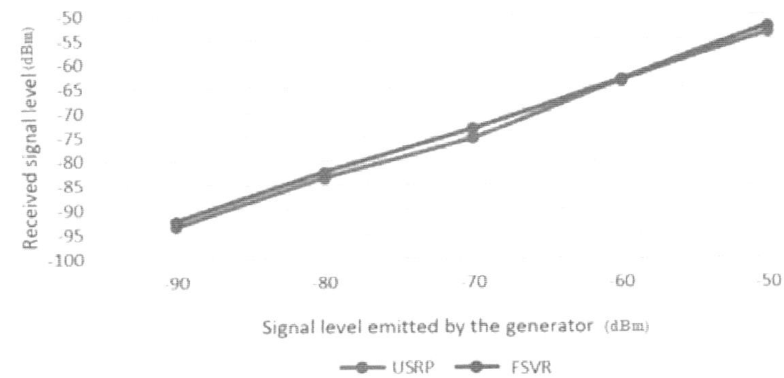

Figure 8. Comparative (FSVR-USRP) power measurement in a 3kHz bandwidth ionospheric channel.

Back-to-back testing was done by connecting the USRP of the transmitting system directly to the USRP of the receiving system using a coaxial cable. A list of 30 frequencies in the HF range was transmitted for 2 h. In this way, synchronization between the two systems was calibrated and linearity in amplitude was also achieved at reception.

The radio link under NVIS conditions for real system testing was made between the transmitting point installed in Sibiu (coordinates: Latitude: 45.7828, Longitude: 24.1452) and the receiving point installed in Sîngeorz-Băi (coordinates: Latitude: 47.3826, Longitude: 24.6559), which is 180 km away in a straight line (Figure 9).

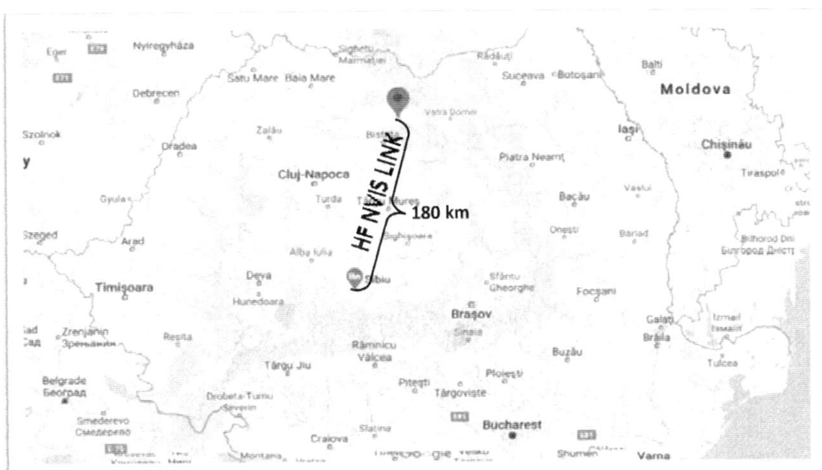

Figure 9. Locations of transceiver systems.

To configure and supervise the correct operation of the system, we chose to remotely control the receiving system from the location of the transmitter, using a "remote control" application available for Linux operating systems. The transmission was performed using 2 HF frequencies (6.7 MHz and 7.3 MHz), by generating a continuous sinusoidal baseband signal, amplified to 60W power. The transmit-receive duration for each frequency was set to 2 seconds, and the inter-frequency interval was set to 4 seconds. The list of frequencies, both transmit and receive, was repeated every 15 minutes for 6 hours.

Figure 10 shows the received power levels over 6hours for the frequency of 6.7MHz and 7.3 MHz at a bandwidth of 3 kHz. It can be seen that the signal starts to decrease from 18:00 pm for the frequency of 6.7 MHz and 18:45 pm for the frequency of 7.3MHz. They also reach the noise threshold (-95 dBm) after 20:30 pm for the frequency of 6.7 MHz and 19:30 pm for the frequency of 7.3 MHz.

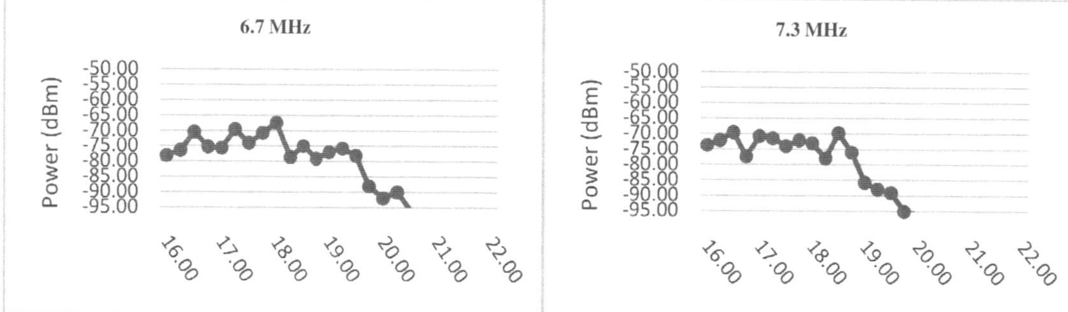

Figure 10. Variation of signal strength at reception for 6.7 MHz and 7.3 MHz carrier frequencies at 3 kHz channel bandwidth.

Using the application made in GNU Radio, the SNR value was calculated at the same times for a bandwidth of 3 kHz. The results reflect variations in SNR due to changes in signal strength and noise. If we consider an SNR level of at least 5 dB, we can state that in the time interval in which the measurement was made, for an ionospheric channel of 3 kHz bandwidth, the channel availability is only between 16:00 pm and 18:00 pm for the 6.7 MHz frequency and between 16:00 pm and 18:45 pm for the 7.3 MHz frequency.

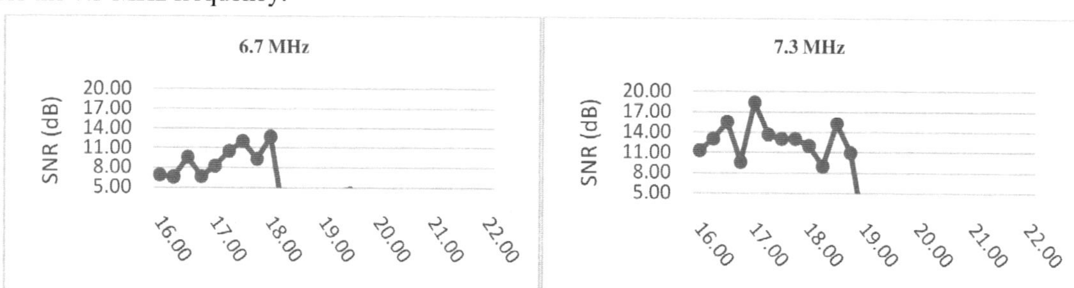

Figure 11. Variation of signal-to-noise ratio at reception for 6.7 MHz and 7.3 MHz carrier frequencies at a channel bandwidth of 3 kHz.

Figure 12 shows the variation of the measured signal-to-noise ratio for a continuous emission for 10 minutes with a carrier frequency of 5.7 MHz and using three different bandwidths. As expected, the SNR decreases with increasing channel bandwidth. If we analyze the ionospheric channel availability as a function of the SNR value we can see that for a minimum SNR value of 5 dB the availability is about 95% for an ionospheric channel bandwidth of 3 kHz and drops considerably to less than 50% for an ionospheric channel bandwidth of 24 kHz. This underlines the fact that increasing the channel bandwidth in order to achieve a higher data transmission rate must be considered very carefully from the perspective of SNR degradation.

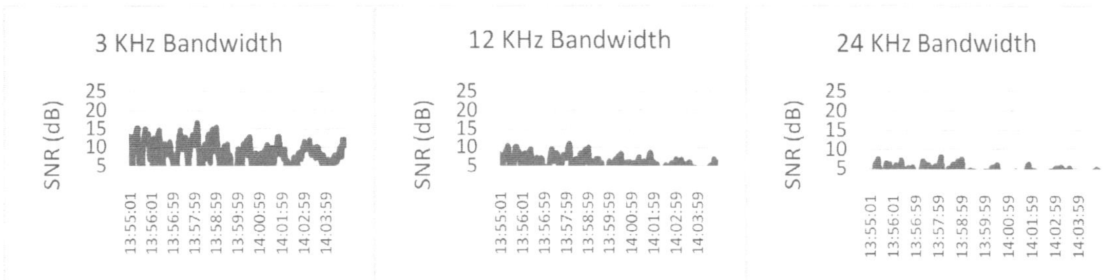

Figure 12. Variation of signal-to-noise ratio at reception for the carrier frequency of 5.7 MHz and bandwidths of 3 kHz, 12 kHz and 24 kHz.

In order to establish a radio link in the HF range that maintains the highest possible SNR, it is necessary to constantly and automatically adapt the transmit/receive parameters to the propagation conditions and channel availability. Therefore, parameters such as operating frequency, channel bandwidth, signal strength and channel noise level are very important to improve data through put in HF radio communications. From this perspective, an agile system is needed to allow real-time evaluation of these parameters.

4. Conclusions

This paper presents the design, implementation and testing of an ionospheric channel monitoring and evaluation system. The system is based on an SDR NI USRP 2932 platform that allows the implementation of transmit and receive channels. The system has been tested both under laboratory and real-world conditions by installing it under NVIS propagation conditions. The SDR platform control and data processing have been performed using a software application implemented in GNU Radio scripts. Calibration and testing of the system were performed using a signal generator and a spectral vector analyzer with real-time analysis capabilities.

Software applications have been developed for the acquisition and processing of radio signals. Samples associated with a frequency band were processed to estimate the signal-to-noise ratio. Real world testing showed that the system is flexible and easy to configure according to the measurement objectives. It can be noted that the system has real-time monitoring capabilities of the main parameters for ionospheric channels (channel power, noise power and signal-to-noise ratio) both in the short term (for the purpose of acquiring fast signal variations) and in the long term (for the purpose of statistical analysis of the received signal).

Future research directions consist in the measurement and statistical analysis of ionospheric parameters in order to experimentally assess the state of the ionospheric channel in measurement campaigns as a function of: operating frequency, temporal characteristics, emission power and channel parameters (bandwidth and noise level).

5. References

[1] Hervás M, Bergadà P and Alsina-Pagès R M 2020 Ionospheric Narrowband and Wideband HF Soundings for Communications Purposes: A Review Sensors **20** 2486

[2] Vilella C, Miralles D, Altadill D, Acosta F, Solé J G, Torta J M and Pijoan J L 2009 Vertical and oblique ionospheric soundings over a very long multihop HF radio link from polar to midlatitudes: Results and relationships *Radio Science* **44**(2) pp 1-11

[3] Sorokina M and Turitsyn SK 2013 Shannon capacity of nonlinear regenerative channels *ArXiv* abs/1305.1537

[4] Recommendation ITU-R SM.2155 2009 Man-made noise measurements in the HF range (Geneva)

[5] Recommendation ITU-R SM.1753-2 2012 Methods for measurements of radio noise (Geneva)

[6] Craiu N, Șorecău E, Stanca R, Miclăuș S, Bechet P and Bouleanu I 2020 HF Noise Measurements Based on Software Defined Radio Dual Receiver and Two Orthogonal Inverted Vee Antennas, *International Conference and Exposition on Electrical And Power Engineering (EPE)* (Iasi, Romania) pp 434-437

[7] Ruuben T, Meister M-A, Lossmann E, Berdnikova J and Madar U 2012 SNR estimation in HF communications channel *6th European Conference on Antennas and Propagation (EUCAP)* (Prague, Czech Republic) pp 2013-2015

[8] Walden M C 2012 Comparison of propagation predictions and measurements for midlatitude HF near-vertical incidence sky wave links at 5 MHz *Radio Science* **47**(4) pp 1-9

[9] Giesbrecht J E 2015 Estimation of SNR for modulation recognition in the presence of HF noise *International Conference and Workshop on Computing and Communication (IEMCON)* (Vancouver, Canada) pp 1-6

[10] Ijaz A, Awoseyila A B and Evans B G 2009 Improved SNR estimation for BPSK and QPSK signals *Electronics Letters* **45**(16) pp. 858-862

[11] Arslan H and Reddy S 2003 Noise power and SNR estimation for OFDM based wireless communication systems *Proc. of 3rd IASTED Int. Conf. on Wireless Optical Commun. (WOC)* (Canada)

[12] Bergada P, Alsina-Pages R M, Pijoan J L, Salvador M, Badia J R and Graells S 2014 Digital transmission techniques for a longhaul HF links: DSSS versus OFDM *Radio Science* **49** pp 1-13

[13] Antoniu S, Christofi L, Green P R and Gott G F 2006 Highrate data transmission in the mid-latitude NVIS HF channel *IEE Proceedings – Communications* **153**(2) pp 272-278

[14] Bechet P, Miclăuș S, Miclăuș A and Balint C 2016 Experimental analysis of noise level and channels availability for high frequency OFDM data transmission in NVIS propagation conditions *International Symposium on Electromagnetic Compatibility - EMC EUROPE* (Wroclaw, Poland) pp 844-849

[15] Ads A G, Bergadà P, Vilella C, Regué J R, Pijoan J L, Bardajà R and Mauricio J 2013 A comprehensive sounding of the ionospheric HF radio link from Antarctica to Spain *Radio Science* **48**(1) pp 1-12

IOP Conf. Series: Materials Science and Engineering 1254 (2022) 012009 doi:10.1088/1757-899X/1254/1/012009

Electromagnetic spectrum monitoring of LTE channels based on SDR portable sensor: preliminary analysis

M Șorecău[1], E Șorecău[1], A Sârbu[2], R Helbet[2] and P Bechet[2]

[1]Department of Electrotechnics and Measurements, Technical University of Cluj Napoca, 400114 Cluj-Napoca, Romania
[2]Department of Communications, IT and Cyber Defence, "Nicolae Bălcescu" Land Forces Academy, 550170 Sibiu, Romania

E-mail:mirela.doicin@gmail.com

Abstract. This paper presents preliminary research for the implementation and testing of a Software Defined Radio (SDR) sensor with electromagnetic spectrum monitoring capabilities for measuring signals specific to the Long-TermEvolution (LTE) mobile communications standard. The proposed sensor is developed on the ADALM Pluto SDR platform - and is designed to acquire a bandwidth of at least 20MHz. To determine the spatial extent of the measurements, the sensor also includes a GPS (Global Positioning System) module. The hardware control of the sensor is performed both in stationary variant using a laptop and in mobile mode using a low-cost module - Raspberry Pi. Present findings suggest that the designed SDR solution represents a reliable and low-cost alternative for real-time spectrum monitoring of wideband signals like the case of 4G-specific signals. The cost of the whole implemented monitoring system is about EUR 330.

1. Introduction

As the complexity of the spectrum increases, new challenges arise in its management. The secure and efficient management of this scarce resource must be based on a complete understanding of its usage. The characteristics of new generation communication signals that present variations in all four dimensions (time, amplitude, frequency and space) call for a new paradigm of performing spectrum monitoring activity using systems with real-time monitoring capabilities. Monitoring systems that meet this condition and that can perform real-time monitoring are vector signal analyzers with real-time analysis capabilities. This solution also comes with certain disadvantages, such as the high cost of the equipment and the lack of scalability.

The problem of real-time monitoring of wideband channels specific to 4th (4G) and 5th (5G) generation mobile communication standards, brings difficulties in creating a flexible system designed to monitor these channels with increasingly variable bandwidths. We observe bandwidths from 1.4MHz to 20MHz in case of 4G, and in the context of new generations of 5G standards or IEEE 802.11ax standard - the bandwidth increases up to 160MHz [1].

The 4G standard, the successor to the 2G and 3G standards, developed by 3GPP increases network capacity and provides high data transfer speeds for end users. The advantages of 4G are obvious and unprecedented. The quality of services and data transfer rates are the most important performances that stand out among all [2]. An important feature specific to the LTE standard is the Orthogonal Frequency Division Multiplexing (OFDM) technology, which uses and distributes the high-speed data stream over multiple orthogonal carriers, rather than a single wideband carrier. For the LTE standard used in the current 4G networks the minimum transmit unit is a resource block with a duration of 0.5ms occupying 180 kHz comprised of 12 sub-carriers 15kHz wide [3-4]. This technique brings many system advantages including spectral efficiency, reduced inter-symbol interference by addition of a cyclic prefix [5], and reduced delay propagation.

The activity of monitoring the electromagnetic spectrum is becoming more and more challenging in the context of monitoring increasingly variable signals with high bandwidths, and new solutions have been proposed to monitor the spectrum using sensors. Considering this need for real-time monitoring of the electromagnetic spectrum, highlighted by the International Telecommunication Union (ITU) [6], it becomes equally important to develop supporting applications and implement a database for storing data provided by monitoring stations.

Since 2013, the US government agencies “National Telecommunications and Information Administration” (NTIA) and “National Institute of Standards and Technology” (NIST) have been concerned with this issue and have developed a real-time spectrum monitoring solution by sharing spectrum data with all interested entities via the public Internet [7].

In recent years, there has been an evolution of SDR platforms used as solutions for electromagnetic spectrum monitoring activity, and a study on the implementation of a spectrum monitoring station using Software Defined Radio (SDR) technology is provided by the ElectroSense project [8]. This global initiative proposes the use of low-cost sensors as primary spectrum sensing devices to collect and analyse data, which are then made available to the general public.

Other initiatives to build a flexible software-defined sensor architecture are exemplified in [9-10]. The solution aims at a large-scale use of the architecture and includes sensors built from inexpensive Commercial off-the-self (COTS) hardware components such as RTL-SDR USB dongles and single board computers - Raspberry Pi. The main parameters of RTL-SDR device are [11]: the maximum sampling frequency is 3.2 MS/s, ADC resolution 8 bits, and frequency range is 25 MHz -1.75 GHz. The feasibility study of using RTL-SDR USB dongles as low-cost real-time spectrum monitoring solutions is also proved in [12].

An approach towards the same direction of using SDR platforms for real-time monitoring belongs to the authors of this paper in the study [13], who propose a low-cost, portable and flexible system for monitoring signals specific to the Terrestrial Trunked Radio (TETRA) communication standard. The solution demonstrated the capability of the SDR monitoring station to accurately measure the levels of signals emitted in the TETRA standard. The original SDR solution was improved and the research results are presented in [14]. The measurement capabilities of TETRA channels were analyzed using three SDR platforms: ADALM Pluto, Hack RF One and NI USRP 2930. The results obtained with the NI USP 2930 SDR platform were found to be comparable to those provided by the R&S FSV13 reference analyzer.

The present work is in line with these concerns and aims to identify a solution to the problem of real-time spectrum monitoring by implementing a portable system that meets certain performance objectives: the sensor must be COTS, provide scalability and offer supporting applications. Supporting applications solve one of the major problems of an efficient portable monitoring system, namely the processing of significant amounts of data provided by the sensor, especially under real-time conditions. Accordingly, we propose a COTS-type portable sensor, flexible, low form factor, scalable, for which we implement supporting applications such as Channel Power - by calculating channel power in the time domain and supporting application on IQ data collection for further processing.

In remainder of the paper is organized as follows: Section II details the implemented measurement system and the proposed method of measuring the channels specific to the LTE standard; Section III presents the results of the system calibration under laboratory conditions and preliminary field testing, and finally the conclusions of the study are presented in Section IV.

2. Materials and Methods

Spectrum monitoring for narrowband channels has been studied extensively, but monitoring of wideband channels is a recent research topic. In terms of electromagnetic spectrum monitoring, the 4G standard features scalable carrier bandwidths (from 1.4 MHz to 20 MHz), uses the MIMO technique, supports both frequency division duplexing (FDD) and time division duplexing (TDD) [15]. In the frequency domain, the LTE reception power measurement is the integration over a given bandwidth, which is the channel bandwidth. It is important to set the measured bandwidth according to the channel bandwidth configuration, which is illustrated in Table 1.

Table 1. Channel bandwidth configuration for LTE.

	The measurement bandwidths defined for channel power measurements					
Bandwidth	1.4 MHz	3 MHz	5 MHz	10 MHz	15 MHz	20 MHz
Measurement Bandwidth (in practice)	1.08 MHz	2.7 MHz	4.5 MHz	9 MHz	13.5 MHz	18 MHz

Our sensor design to measure the power of LTE channels includes: an ADALM Pluto SDR platform connected to the OmniLOG 70600 antenna (Frequency range: 680MHz - 6GHz, Omni-directional, Dimensions (L/W/D): 173 x 62 x 9 mm) and the software application in GNU Radio environment running on a Raspberry Pi minicomputer. Data display is achieved using a 7-inch touch display connected directly to the Raspberry Pi. The sensor has been designed to ensure its mobility and portability for various monitoring activities. The SDR platform, ADALM Pluto, corresponds to the system implementation by targeted features: instantaneous acquisition frequency band up to 20MHz, frequency range between 325 MHz and 3.8 GHz, 12-bit analogue-to-digital converter precision resolution, low acquisition cost - about $200, small size and very good precision spectrum monitoring capability. The system design was focused on meeting the specific performance characteristics of a monitoring sensor [9]: small form factor, low cost, wide bandwidth, COTS hardware components, flexibility. The component parts of the sensor are shown in Figure 1.

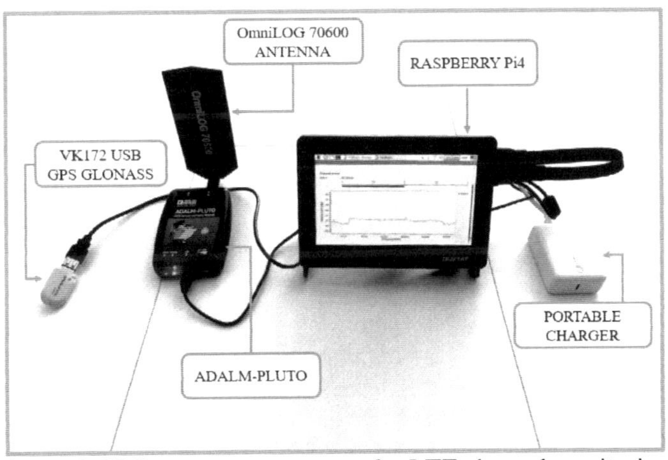

Figure 1. SDR sensor components for LTE channel monitoring.

In order to measure the channel power specific to the 4G standard, an application has been implemented in the GNU Radio programming environment (it is a free & open-source environment). The application has been built to solve the following operations: spectrum display; possibility to select different channel bandwidths (3 MHz, 5 MHz, 10 MHz, 15 MHz and 20 MHz) for which we want to measure power; calculation and display of measured channel power for the configured channels; saving of IQ samples. The carrier frequencies were chosen according to [16] in the frequency bands 791-821MHz, 925-960MHz, 1805-1880MHz, 2620-2650MHz, for downlink communications and frequency division duplexing (FDD). Although the OFDM access technique is specific to LTE channel design, for downlink communications the orthogonal frequency division multiple access (OFDMA) technique is used, and for uplink communications a pre-codified version of the same modulation technique - the single-carrier frequency division multiple access (SC-FDMA) technique - is used [5,17].

To correlate channel power values with the locations from which the measurements were taken, a low-cost GPS/GLONASS receiver was used, which was integrated into the Python code and whose coordinates, associated with the channel power values, can be displayed on a map of the area of interest. By adding the GPS module, the aim was to synchronize its measurements with the values provided by the monitoring sensor, so that the measurement data would be accurate and up-to-date in the coverage area.

The implemented application contains both blocks included in the GNU Radio library and custom-made processing blocks specifically designed to meet the needs of the monitoring system. Figure 2 shows the SDR platform control block, the module for signal level calibration (Multiply Const), the Band Pass Filter block for filtering the signal according to the bandwidth of the channels. The saving of IQ samples was implemented through the File Sink block for further analysis.

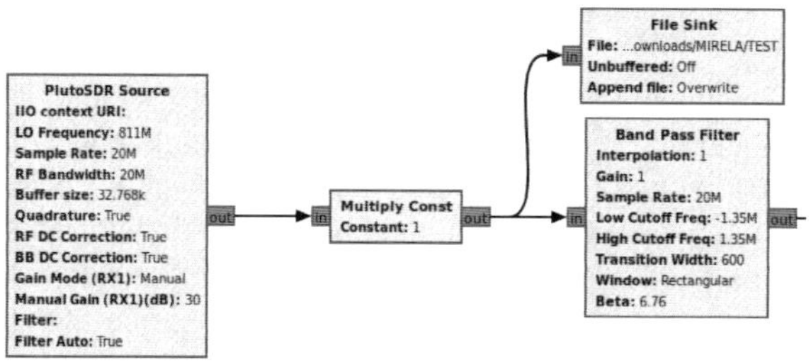

Figure 2. SDR platform control and signal filtering.

Figure 3 shows the calculation of channel power after filtering. This is performed by multiplying the signal by its conjugate followed by converting data from complex to float and performing a moving average on the signal samples. The values were converted to log scale by the Log10 block and displayed by the QT GUI Number sink block. The "Data sink" block was specially created in Python and imported into GNU Radio to save the channel power value, frequency, date, time and geographical coordinates of the measurements in an excel file.

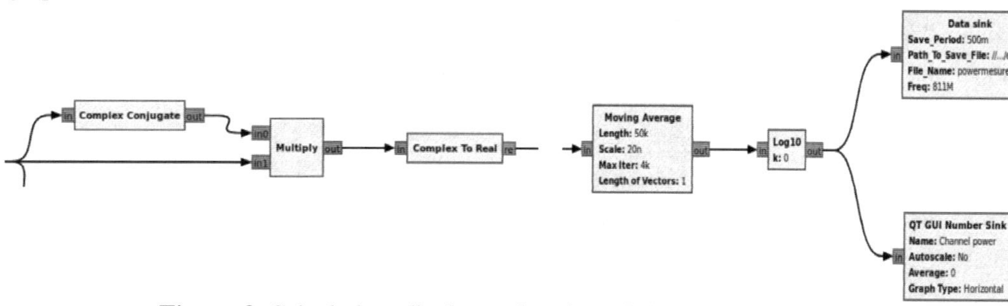

Figure 3. Calculation, display and saving of channel power values.

The sensor test was completed in two phases. In the first step, under laboratory conditions, the sensor was calibrated and its performance was evaluated. The following parameters were used for performance evaluation floor and channel noise power in the frequency range specific to the LTE standard, sensor sensitivity, amplitude dynamic range. The measurement stand under laboratory conditions is shown in Figure 4 and includes: the SMBV 100A signal generator, Rohde & Schwarz FSV 3013 vector spectrum analyzer with real-time analysis capabilities, radio frequency splitter, SDR platform. To calibrate the system, 50 ohm impedances were connected to the radio frequency inputs of the spectrum analyzer and the ADALM Pluto SDR platform. An LTE-specific test signal over a bandwidth of 10MHz and a channel power of -75dBm was used to determine the sensitivity, dynamic range analysis and frequency response of the SDR sensor. The test signal was provided by the SMBV 100A signal generator.

Sensor testing in real conditions was performed by positioning the sensor in a car and measuring the 4G signal channel power in motion on a travel route. The speed of the car highlighted the synchronization between the GPS module readings and the SDR sensor readings. For this purpose, the output data file includes: the coordinates of the measurement point and the technical characteristics of the signal (central frequency and channel power). The capability of the sensor to measure low-level signals was investigated.

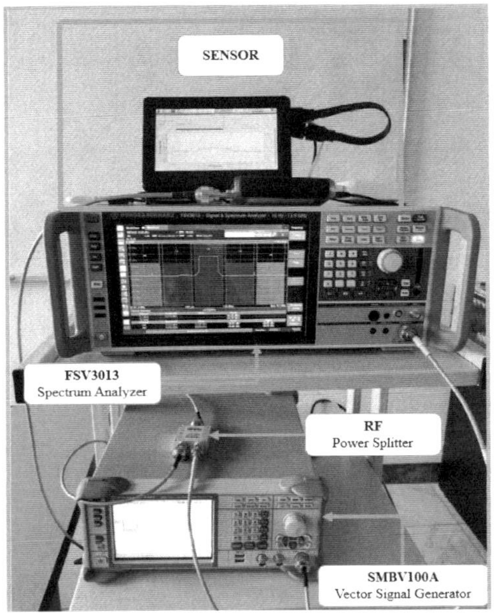

Figure 4. Experimental set-up for laboratory testing of the SDR sensor.

3. Results and Discussions

Laboratory evaluation and system calibration was performed using a test signal generated by the SMBV 100A signal generator. The test signal is the EUTRA/LTE digital standard for which the following characteristics were set: central frequency 811 MHz - LTE band 20; channel bandwidth: 10 MHz; FFT size - 1024; FDD - Frequency Division Duplex. Calibration was performed on the receive channel of the ADALM Pluto SDR platform by performing comparative measurements with the Rohde & Schwarz FSV3013 real-time capable vector spectrum analyzer (frequency range: 10Hz - 13.6GHz, bandwidth: up to 200MHz, DANL: -151dBm). Comparative measurements were performed for an LTE channel power of -75 dBm.

Channel noise power was analyzed in the laboratory by connecting a 50 ohms impedance to both the RF input of the SDR platform and the reference analyzer. The noise in the channel is influenced by the receiver noise factor and the channel bandwidth. The measurement results are shown in Table 2 for both the ADALM Pluto SDR platform and the R&S FSV3013 spectrum analyzer.

Noise in the channel will limit the minimum power level that can be measured. It can be seen that there is a difference of at least 12 dB between the SDR platform and the reference spectrum analyzer. This means that in terms of the ability to measure low-level signals, the SDR platform will have at least 12 dB lower sensitivity.

Table 2. Noise power for LTE channels of different bandwidths.

Channel Bandwidth [MHz]	3	5	10	15	20
Channel Power Noise measured with R&S FSV 3013 [dBm]	-101	-99	-96	-94	-93
Channel Power Noise measured with ADALM Pluto [dBm]	-88,5	-86,5	-83,5	-82	-80

The amplitude response of the SDR sensor was tested for a dynamic range of the input signal level between -90 dBm and -55 dBm. Comparative measurement results between the SDR platform and the reference spectrum analyzer are shown in Table 3. It can be seen that the SDR platform for the dynamic range of channel power between -82dBm and -55dBm has a good response.

Table 3. Amplitude response for an LTE channel: power variable between -90dBm and -55dBm; bandwidth: 10MHz; central frequency - 811MHz.

Output power given by signal generator SMBV 100A [dBm]	Channel power measured with R&S FSV3013 [dBm]	Channel power measured with SDR platform – ADALM Pluto [dBm]
-90	-89.5	-
-85	-85.3	-
-82	-82.6	-82.8
-80	-80.5	-79.4
-75	-75.7	-75.7
-70	-70.7	-71.1
-65	-65.7	-66.3
-60	-60.7	-61.4
-55	-55.7	-56.4

Under real environmental conditions, the sensor was set to take measurements under moving conditions, inside a car, on a set route. The GNU Radio application offers the possibility to select the time interval in which data acquisition can be done. Thus, for an average speed of 72km/h, the system allows data to be acquired at one second intervals with the measuring point every 20 m, which is acceptable from the perspective of a realistic evaluation. The software application allows, depending on the objectives of the measurements, to set the time interval between which data acquisition takes place. The data is saved using the "data sink" block, after which with the help of a script added to the application made in GNU Radio it is possible to display the channel power levels on a map using the coordinates associated with each value (Figure 5). The display can be done in real time at the moment of the measurement or latter by accessing the recorded data.

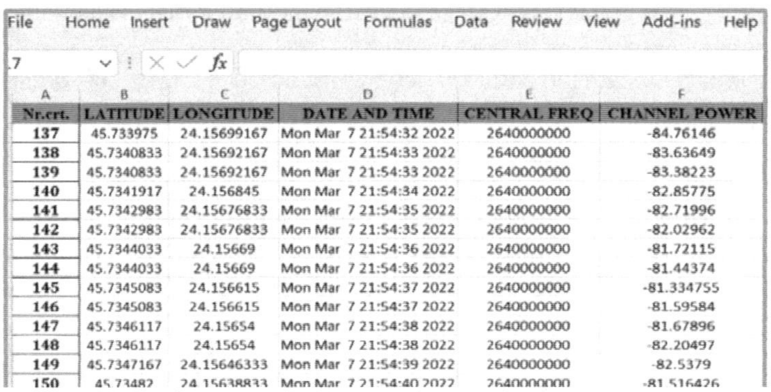

Nr.crt.	LATITUDE	LONGITUDE	DATE AND TIME	CENTRAL FREQ	CHANNEL POWER
137	45.733975	24.15699167	Mon Mar 7 21:54:32 2022	2640000000	-84.76146
138	45.7340833	24.15692167	Mon Mar 7 21:54:33 2022	2640000000	-83.63649
139	45.7340833	24.15692167	Mon Mar 7 21:54:33 2022	2640000000	-83.38223
140	45.7341917	24.156845	Mon Mar 7 21:54:34 2022	2640000000	-82.85775
141	45.7342983	24.15676833	Mon Mar 7 21:54:35 2022	2640000000	-82.71996
142	45.7342983	24.15676833	Mon Mar 7 21:54:35 2022	2640000000	-82.02962
143	45.7344033	24.15669	Mon Mar 7 21:54:36 2022	2640000000	-81.72115
144	45.7344033	24.15669	Mon Mar 7 21:54:36 2022	2640000000	-81.44374
145	45.7345083	24.156615	Mon Mar 7 21:54:37 2022	2640000000	-81.334755
146	45.7345083	24.156615	Mon Mar 7 21:54:37 2022	2640000000	-81.59584
147	45.7346117	24.15654	Mon Mar 7 21:54:38 2022	2640000000	-81.67896
148	45.7346117	24.15654	Mon Mar 7 21:54:38 2022	2640000000	-82.20497
149	45.7347167	24.15646333	Mon Mar 7 21:54:39 2022	2640000000	-82.5379
150	45.73482	24.15638833	Mon Mar 7 21:54:40 2022	2640000000	-81.516426

Figure 5. Saving data in xlsx format.

The monitoring system was field tested for the acquisition of an LTE channel provided by a base station with 20 MHz bandwidth. This scenario was chosen because it corresponds to the worst case where the noise power in the channel is the highest. The travel route was chosen such that we have channel power levels at the limit of the SDR sensor monitoring capability (-80dBm). The average vehicle speed was 72 km/h. Data readings were taken at one second intervals, which corresponds to a distance between measurement points of 20 m. Figure 6 shows the variation of channel power along the travel route. It can be seen that in the first part of the route, the sensor is not able to measure the channel power correctly, being below the sensitivity threshold. In the second part of the route the

channel power values are high enough to be accurately detected by the SDR sensor. This test demonstrates that the SDR
sensor has limitations in its ability to detect very low signals. To improve the sensitivity of the sensor, SDR platforms with a better resolution of the analogue-to-digital converter should be considered.

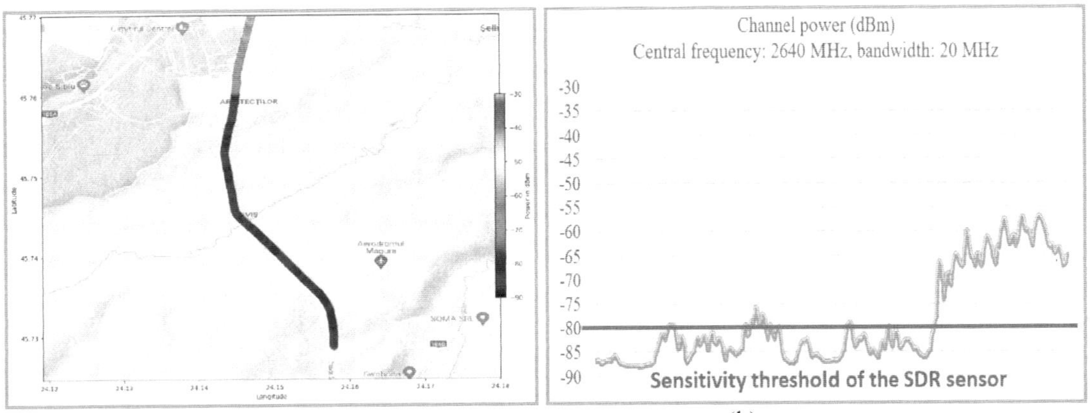

(a) (b)

Figure 6. Display channel power on frequency 2640 MHz, bandwidth 20 MHz: (a) map display; (b) graphical representation.

4. Conclusions

The present workaimed the implementation and preliminary testing of a sensor for real-time electromagnetic spectrum monitoring activity for wideband signals - LTE. The sensor was designed and implemented with components that have a low acquisition cost. From this perspective the sensor represents a viable and scalable alternative solution for replacing high-performance spectrum monitoring systems. Regarding data processing the free & open-source GNU Radio environment was used. Initial calibration and testing of the sensorwas performed under laboratory conditions. The sensor has also been tested in the real environment by measuring LTE signals received from base stations.

Preliminary test results demonstrate the capability and reliability of the system to monitor in real time the channel power of the LTE standard and to synchronize and collect the data provided by the GPS module and the sensor. Measurements performed by implementing the channel power calculation support, highlighted the system's ability to monitor 4G level signals with approximately 12 dB lower sensitivity compared to the reference spectrum analyzer. But the advantage is that it has been demonstrated that wideband channel monitoring with acceptable performance can be achieved using a flexible, low form factor, scalable and low-cost (about 300EUR) SDR sensor. A supporting application for collecting IQ samples for further analysis was also implemented at the sensor level. Future applications of the sensor will follow its use in a sensor network for storing IQ data. This will solve the problem of acquiring a large amount of data at the sensor level and using limited storage space at the sensor level.

Future research directions will include improving the SDR sensor for electromagnetic spectrum monitoring, integrating the sensor into a network and controlling it remotely, developing other supporting applications at the sensor level, testing low-cost sensors for real-time electromagnetic spectrum monitoring for wide bandwidth channels (at least 100 MHz).

Acknowledgment

This work was supported by a grant of the Romanian Ministry of Education and Research, CNCS - UEFISCDI, project number PN-III-P1-1.1-PD-2019-0500, within PNCDI III.

References

[1] Afaqui M S, Garcia-Villegas E and Lopez-Aguilera E 2017 IEEE 802.11ax: Challenges and Requirements for Future High Efficiency WiFi *IEEE Wireless Communications* **24**(3) pp 130-137

[2] QingXia D, Gang D and HaiYan C 2015 Research on Key Technology of TD-LTE Standard 4G Mobile Communications Network *The Open Cybernetics & Systemics Journal* **9**(1) pp 1436-1442

[3] Schaich F and Wild T 2015 Subcarrier Spacing - a Neglected Degree of Freedom? *IEEE 16th International Workshop on Signal Processing Advances in Wireless Communications (SPAWC)*(Stockholm, Sweden) pp 56-60

[4] Al-Hilfi H M 2020 Analysis of LTE Physical Channels Overhead *TELKOMNIKA Telecommunication, Computing, Electronics and Control* **18** pp 2800-2806

[5] Farhood A D, Agarwal N, Nitin N and Naji M K 2014 Performance Analysis of OFDMA in LTE *International Journal of Current Engineering and Technology* **4**(3) pp 1614-1619

[6] Spectrum Monitoring Evolution Recommendation ITU-R SM.2039 2016 International Telecommunication Union (Geneva, Switzerland)

[7] Cotton M, Wepman J, Kub J, Engelking S, Lo Y, Ottke H, Kaiser R, Anderson D, Souryal M and Ranganathan M 2015 An Overview of the NTIA/NIST Spectrum Monitoring Pilot Program *IEEE Wireless Communications and Networking Conference Workshops (WCNCW)*(New Orleans, LA, USA) pp 217–222

[8] Rajendran S, Calvo-Palomino R, Fuchs M, Bergh B V, Cordobés H, Giustiniano D, Pollin S and Lenders V 2018 Electrosense: Open and Big Spectrum Data *IEEE Communications Magazine* **56**(1) pp 210-217

[9] Pfammatter D, Giustiniano D and Lenders V 2015 A Software-Defined Sensor Architecture for Large-Scale Wideband Spectrum Monitoring *Proceedings of the 14th International Conference on Information Processing in Sensor Networks* (Seattle, Washington) pp 71–82

[10] Ball D, Naik N and Jenkins P 2017 Spectrum Alerting System Based on Software Defined Radio and Raspberry Pi *Sensor Signal Processing for Defence Conference (SSPD)* (London, UK) pp 1-5

[11] Stewart R W, Crockett L, Atkinson D, Barlee K, Crawford D, Chalmers I, McLernon M and Sozer E 2015 A low-cost desktop software defined radio design environment using MATLAB, simulink, and the RTL-SDR*IEEE Communications Magazine* **53**(9) pp 64-71

[12] Nika A, Zhang Z, Zhou X, Zhao B Y and Zheng H 2014 Towards commoditized real-time spectrum monitoring *Proceedings of the 1st ACM workshop on Hot topics in wireless* (*Maui, Hawaii, USA*) pp 25-30

[13] Helbet R, Monda V, Bechet A C and Bechet P 2020 Low Cost System for Terrestrial Trunked Radio Signals Monitoring Based on Software Defined Radio Technology and Raspberry Pi 4 *International Conference and Exposition on Electrical And Power Engineering (EPE)* (Iași, Romania) pp 438-441

[14] Helbet R, Bechet P, Monda V, Miclăuș S and Bouleanu I 2021 Low-Cost Sensor Based on SDR Platforms for TETRA Signals Monitoring *Sensors* **21** 3160

[15] Degefa Assefa T, Kralevska K and Jiang Y 2016 Performance analysis of LTE networks with random linear network coding *39th International Convention on Information and Communication Technology, Electronics and Microelectronics (MIPRO)*(Opatija, Croatia) pp 601-606

[16] ANCOM 2019 ANCOM position on the granting of rights of use of radio frequencies available in 694-790 MHz, 790-862 MHz, 1427-1517 MHz, 2500-2690 MHz, 3400-3800 MHz and 24,25- 27,5 GHz frequency bands

[17] Sârbu A, Miclăuș S, Bechet P and Bouleanu I 2019 Using Persistence Spectrum for Realistic Exposure Assessment to Time Varying Radiofrequency Signals *International Conference KNOWLEDGE-BASED ORGANIZATION* **25**(3) pp 57-62

IOP Conf. Series: Materials Science and Engineering 1254 (2022) 012010 doi:10.1088/1757-899X/1254/1/012010

Microwaves propagation through a small fire curtain at 3.55 GHz and 6.26 GHz and the impact on vital signs detection by using a four-element patch antenna array

S Miclăuş[1], D Vatamanu[1], R Mitran[2] and C Neghină[3]

[1]"Nicolae Bălcescu'' Land Forces Academy, Sibiu, Romania
[2]Stratec Biomedical, Cluj-Napoca, Romania
[3]"Lucian Blaga" University, Sibiu, Romania

Email: simo.miclaus@gmail.com

Abstract. The impact of the presence of a fire curtain not wider than 7 cm on an air path 150 cm long of two continuous waves at frequencies of 3.55 GHz and 6.26 GHz was quantified. Two identical patch array antennas with four linear elements were used at emission and reception and they were designed and characterized specifically for this purpose. The magnitude and phase of the transmission coefficient were measured by a vector network analyzer in all the experimental sessions. Comparing the air path and the air-plus-fire path of the waves, it resulted that fire induced shifts in magnitude and phase that depended on frequency and on the experimental setup. The heating of the antenna could have contributed to the overall observed effects. The detection of the human heartbeat by Doppler radar technique was not much influenced by the presence of fire, but the amplitude of the detected beats was increased. The array antennas produced noisy signals for the vital sign detection and the origin of this behaviour was hypothesized.

1. Introduction

The effect of fire presence on radiofrequency and microwaves propagation path is a topics of interest for both wireless communications and radar applications, mainly in case of disasters. One of the first modern extensive research on this subject dates back in 2008 [1], [2] when the authors studied the attenuation and phase shift of signals on a wide band, starting from high frequency range up to microwaves X band, at propagation thru moderate intensity bushfire. It was demonstrated that the attenuation is generally larger in the ultra-high frequency band and in X-band than at high frequencies. Attenuations of 0.03 dB/m were reported at 30 MHz in grass fires, and of 0.05 dB/m at 151 MHz respectively. In the (10-12.5) GHz range, attenuations of 4.45 dB/m were reported. Phase shift was also observed at propagation thru fire, for example of 30° at 30 MHz and of 29.3^{o} at X band frequencies. In moderate-intensity vegetal fires it was observed that the attenuation depended on both the frequency and on the plume temperature. The hottest zones crossed by the signals induced the largest attenuations, as compared to the colder zones of the fire.

In 2009 the author of thesis [3] approached the study of the frequency range (100 MHz – 1 GHz). The combustion-induced plasma was identified to be the main cause of the observed effects. The signal strengths of frequencies lower than 600 MHz were reduced by the fire presence, while the larger frequencies showed either a reduction or an enhancement of the signal magnitude passed through fire. The enhancement was explained by diffraction phenomena which contributes to the recovery of the signal. The frequency range (100-180) MHz presented attenuations of (3-6) dB in outdoor small fires, while the (400-450) MHz range showed an attenuation of (10-15) dB for the same fire geometry and structure. It was also underlined the role the turbulences of fire have on the apparition of Doppler spread effects of the signals. The temperature and the fuel properties also had a great impact on the attenuation levels, as well as the flame depths. Generally, in [3] it was concluded that prediction of propagation through fire can hardly be solved by computations and need extensive experimental work.

In 2010 a2.4 GHz signal was used to test propagation thru fire [4]. The authors reported signal strength degradation and wireless link quality deterioration. They quantified the attenuations at different fire curtain heights and at different distances between emitting and receiving antennas.

Electric characteristics of the plasma that may affect propagation thru fire were investigated by simulations [5-7] when the refractive index variation was quantified along a large frequency band. More recently the authors of [8] analysed the attenuation of a signal thru fire at 700 MHz using the Full-Stack Model and Transmission Line Model (TLM).Other simulations [9] used microwaves at 6.78 GHz to simulate the possibility to detect fire presence behind a wall based on a 2x2 microstrip patch antenna. The authors of [10] aimed at measuring the attenuation of the signals in the X-band at (7 - 9.5) GHz due to vegetation fire. The main parameters followed were the ionisation changes and the electric conductivity modifications of the fire. Values of (0.35 - 0.9) dB attenuation were emphasized in this frequency range. Later in 2016 the same authors [12] carried out measurements at 151 MHz over a 590 m path in a moderate intensity fire and found a signal attenuation of 0.43 dB. In [13] the behaviour of the frequency range (350-400) MHz at thru-fire propagation when different fuel types were used was investigated. Concentrations and types of the smoke were also analyzed from the perspective of their radio wave attenuation. They concluded that the attenuation was not affected by the flame itself, but rather by the smoke – for example an attenuation of 1.16 dB was reported at 300 MHz and of 5.43 dB at 360 MHz. A first-order exponential function was deduced between the concentration of smoke and the signal attenuation.

Another research vista was the development of antennas for fire propagation studies. Papers [14-15] present designed antennas at 2.4 GHz for early fire detection destined to wireless sensor networks and analysed the influence of fire on antennas radiation pattern. Authors of [16] prepared an experimental setup based on designed wideband antennas in (0.6-8) GHz range to characterize the radio links functioning impairment by small-scale indoor fire.

Radar detection thru fire was also traced. For example, authors of [17] started in 2015 some experimental studies for search-and-rescue radar in firefighting by studying microwave propagation through walls. Recently, [18] checked the possibility of using radars to obtain images thru fire at 79 GHz and 300 GHz by experimental set-up while determining specific attenuations during 30 seconds of radiation travel.

On this background, present work proposed approaching the analysis of the influence that a small fire curtain (indoor fire) has on the propagation and on the detection of the heartbeat of a person. Staring from designing and building a patch antenna array specially destined, we aimed at: a) the quantification of the magnitude attenuation of the transmission coefficient and its phase shift at two frequencies of continuous waves, namely at 3.55 GHz and 6.26 GHz; the study compared the situation of air propagation with the one in which on the same path a thin and low-intensive fire curtain is present; b) checking the degradation of the heartbeat Doppler radar detection at 3.55 GHz when the wave travels through the same small fire, as compared to air propagation. The role of the antenna characteristics was considered important in describing the magnitude of the expected behaviour shifts of the travelling waves passing thru fire versus usual air.

2. Materials, measurement system and procedure

Two identical linear arrays composed of four patch antennas were designed, fabricated and characterised by simulation and measurements. The geometry and dimensions of one array is shown in Figure 1. CST Microwave Studio was the environment in which the model antennas were developed and characterized. To compare the voltage standing wave ratio (SWR) between simulations and measurements, we used a vector network analyzer (VNA) model ZNB from Rhode & Schwarz. We aimed at obtaining arrays with high directivity and good efficiency in the extremely high frequency range in several narrow bandwidths. To check the gain value, the direction of the main radiation lobe and the half-power beam-width (HPBW) values, we used a small electric (E) near-field probe model PBS1-E from Aaronia (visible in Figure 2 – down-right), an angle meter and a signal analyzer model FSV 3013 from Rhode & Schwarz. In this way, we had an experimental confirmation of the simulated radiation pattern.

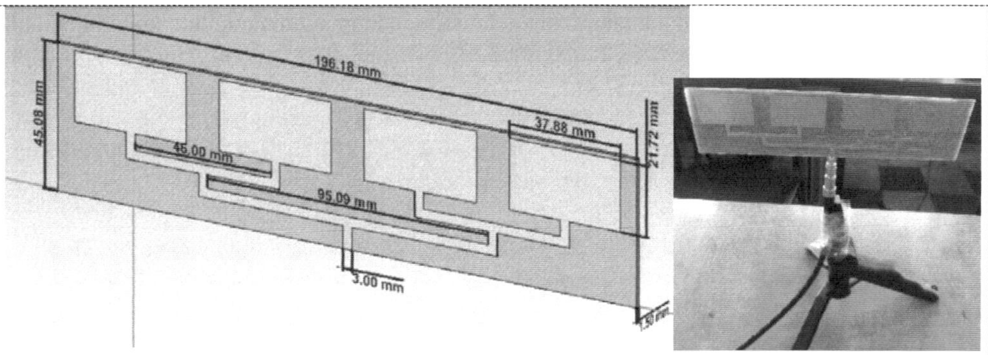

Figure 1. Geometry and dimensions of the modelled antenna array (left) and its photo (right).

The two antenna arrays were used in the experimental setup with two objectives:

- Determination of magnitude- and phase-shifts of the transmitted signals due to propagation thru a small fire curtain as compared to the propagation thru air, by using the VNA;
- Detecting the human heart beat with the two antenna arrays thru air and comparatively thru a small fire curtain, by using the VNA (Figure 2, up).

The fire curtain had a depth varying between 3-7 cm and a height varying between 15-40 cm, it was a small indoor fire based on Zippo lighter fluid as a flame retardant and cotton-wool as a burning fuel.

The two antennas connected at the two ports of the VNA as emitting and receiving antennas were placed at a distance of 150 cm between each other while the source of the fire was positioned at half-distance in between, for the characterization of propagation thru fire experiments (Figure 2 – up-right).

For the heart beat detection experiment, the two antennas were placed at a 10 cm interval between each other's edges, and at 100 cm from the surface of the human chest (Figure 2 – down-left). The fire wall was placed parallel to the antenna surfaces at 40 cm from them.

Two resonance frequencies of the antennas were used in the experiments, respectively 3.55 GHz and 6.26 GHz. They were extracted from the SWR graphs while the radiation patterns at the two frequencies were computed, experimental-checked and considered proper for achieving the objectives of the proposed research.

The settings of the VNA for the attenuation thru air and fire study were: number of points = 100001 (maximum available); sweep time = 90 s;

The VNA settings for the Doppler radar detection of the heartbeat were: number of points = 10000; sweep time = 20 s; input power in the emitting antenna (port 1) = 6 dBm.

The scattering parameters S11 (reflexion coefficient) and S21 (transmission coefficient) were measured by the VNA. Magnitude of S11 provided us the SWR values of the antennas on the frequency band of interest. Magnitude and phase of S21 provided the attenuation information needed for study of the propagation thru fire versus air. Phase shift was provided using Doppler radar technique and expressing the S21 phase variation in time due to minute movements of the human chest while breathing or by holding up respiration – when only the heartbeat slightly moved the body surface.

During some of the tests, we checked the antenna surface temperature because we hypothesized that antenna heating itself (due to proximity of the fire) may influence its intrinsic parameters and in consequence the measured values of the scattering parameters. The temperature change was quantified by infrared (IR) thermography, making use of a small IR camera model Thermal Expert™ TE-Q1 Profrom i3system (47x25x16mm) connected to a smart phone through the USB type C interface. The distance between the IR camera and the antenna surface was of 75 cm (Figure 2 – up-right). Video captures with a rate of 9.4 frames per second were then processed.

After the data collection of S21 phase for the Doppler radar detection of the heartbeat, the data processing consisted of two steps: filtering and parametric spectral analysis.

In the first step the measured signal was low-pass filtered to eliminate spectral components with a frequency higher than 3 Hz. The filter transition band was 3-5 Hz while in the stop band (above 5Hz) the attenuation was higher than 60 dB. The resulting filter had the order 1814 and introduced a delay of approx. 907 samples (approximately 1.8 s).The filtered signal was segmented into segments with a length of 1000 samples (2 seconds) and an overlap of 800 samples. This procedure resulted in 41 signal segments, which were then processed individually.

For a 2 s signal segment, the maximum spectral resolution that could be obtained by the Fast Fourier Transform (FFT) would be 0.5 Hz which is not enough for detecting a pulse signal in range (0.5-2) Hz corresponding to human heart beat or a respiration signal in the range of (0.15-0.4) Hz.

Figure 2. Experimental set-ups for the two experiments: heartbeat detection (left column photos); attenuation and phase shift at propagation thru fire wall (up-right photo); checking the antenna main lobe direction at 6.56 GHz (down-right photo).

In the second step we used a parametric spectral analysis approach in order to increase the frequency resolution. This analysis needed a signal model in order to estimate the frequency of the spectral components. We assumed that the S21 phase measured by the VNA may be modelled as a sum of two sinusoidal signals (pulse and respiration), a clutter signal and noise. 1000 samples of a segment were used to estimate the amplitude, frequency and phase for the 2 sinusoids by minimizing a cost function. According to equation 1 (below), there are 6 unknown parameters that should be estimated: frequencies f1 and f2, amplitudes A1 and A2 and phases θ1 and θ2. We used a nonlinear least squares method in order to estimate these parameters based on 1000 measures samples of one segment, as indicated in [19]. The 2 sinusoids that were present in each signal segment corresponded to pulse and respiration. If the breathing signal was absent, then the signal segment would be composed of a sinusoid and the first harmonic. The model function was considered to be:

$$s(t) = A\sin(2\pi f_1 t + \theta_1) + A_2 \sin(2\pi f_2 t + \theta_2) + clutter + e \qquad (1)$$

The clutter was estimated by a second order function.

The S21 magnitude and phase data addressing the attenuation of the wave thru fire were statistically processed to get boxplots of their values distributions as time series over 90 seconds.

3. Results and discussion

3.1. Antenna array characteristics and distribution of electric field strength along the propagation path in air

Figure 3 presents the comparative values of SWR for the patch antenna array obtained by simulation and by measurement. It is observed that the array is a very good narrow multi-band antenna. The resonance frequencies of 3.55 GHz and 6.28 GHz were chosen for the present measurements. The first frequency is also used in 5G mobile communications. The 3D radiation patterns at the two frequencies are presented in Figure 4 where we used the gain representation in linear scale and not in dBi to better observe the very good directivity at both frequencies. At 3.55 GHz the maximum gain is 5.86 dBiand HPBW = 21.4°and at 6.26 GHz the maximum gain is 4.55 dBi and HPBW =15.2°. At 6.26 GHz, the main lobe is directed upwards therefore during experiments we inclined antennas to compensate this elevation change.

In Figure 5 the E-field strengths distributions in arrival planes or along the propagation path in air are indicated. The emitting antenna had an input power of 6 dBm. The two upper images in Figure 5 emphasize the distribution of E-field level in the plane of the receiving (Rx) antenna, at 150 cm apart from the emitting (Tx) antenna, at the two frequencies. The middle and the lower images in Figure 5 present the attenuation of the E-field strength along the path between Tx and Rx antennas with vector-fields. The attenuation of the E-field level is larger at 3.55 GHz than at 6.26 GHz so that significantly higher field levels are received at 6.26 GHz at the Rx antenna. This behaviour which is different from the theoretical prevision for far-field propagation conditions may be due to the Tx element-antenna couplings and to the computational settings.

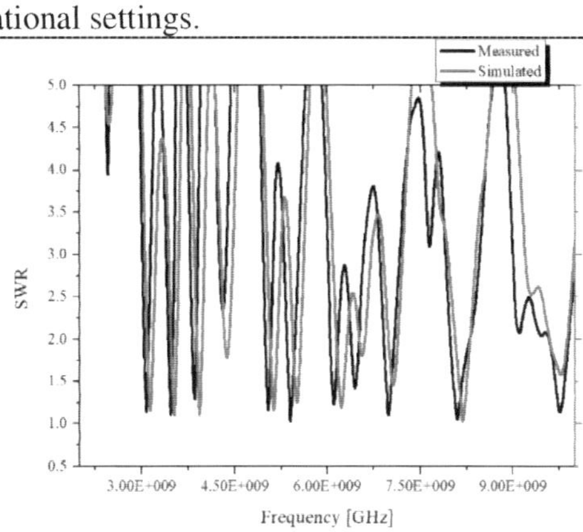

Figure 3. Standing wave ratio of the array antennas over a wide frequency band: comparison between the simulated and the real antennas.

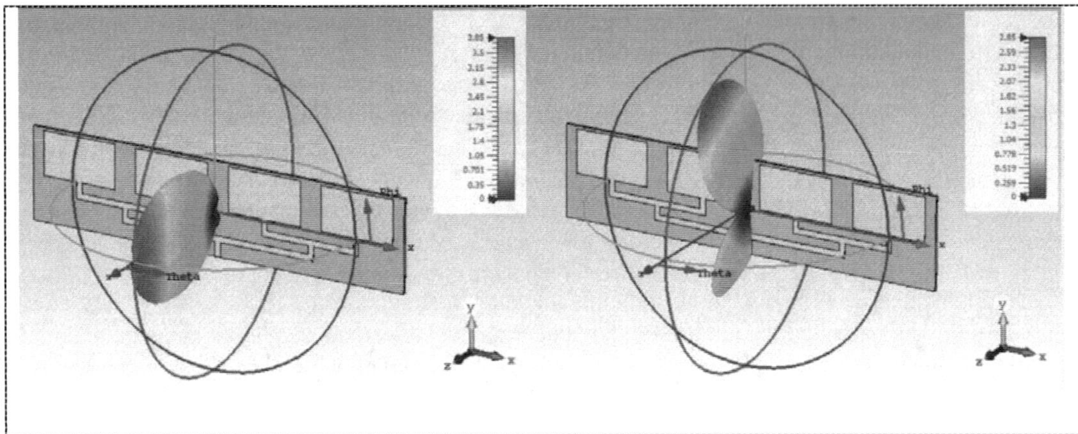

Figure 4. 3D radiation patterns of the antenna array at the frequencies of 3.55 GHz (left) and 6.26 GHz (right). Linear gains are represented to emphasize the high directivity of the array.

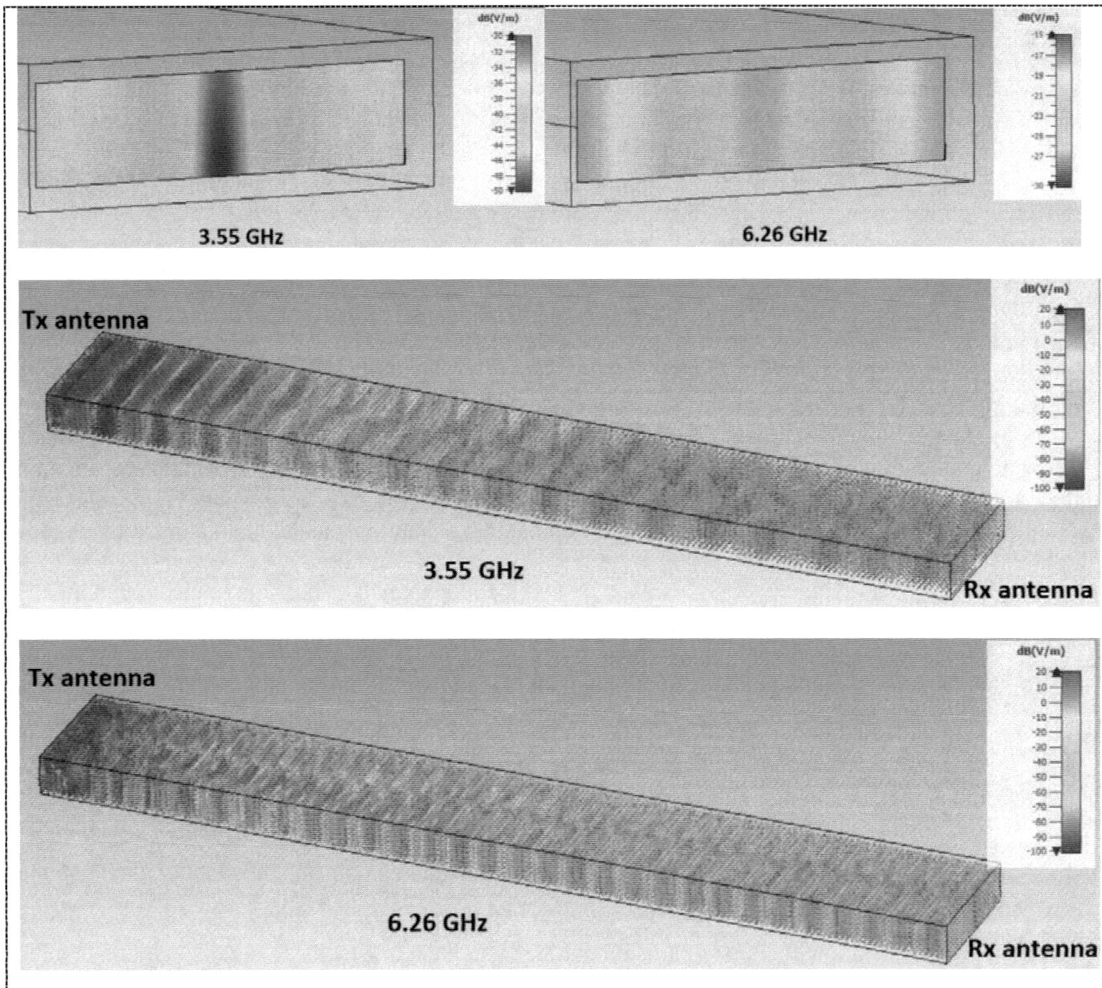

Figure 5. The distribution of the E-field levels in the plane of the receiving antenna (upper pictures). and in the space (air) between the two antennas (middle and lower pictures).

3.2. Propagation thru fire and the observed effects

Figure 6 (upper images) presents the variation of the S21 magnitude at propagation thru air at 3.55 GHz comparative with propagation thru the narrow fire curtain during the recording period of 90 s (equivalent to 100001 points on the analyzer). Three intensities of the fire were used: the most intense was fire 1 and the lowest and smallest was fire 3. If compared the propagation thru air (red trace) with the one thru the air+ fire curtain it is observed that at 3.55 GHz the magnitude of transmission is reduced with about 1 dB by the fire. However, the question raised if this decrease of transmission was not due to temperature increase of the antenna body due to its proximity to the fire. Boxplots of the distributions of transmission coefficient were also computed during 90 s of combustion, which was not constant in intensity, but diminishing in time at every experiment. Larger interquartile intervals (much spreading) were observed in case of fire transmission (right graphs of Figure 6).

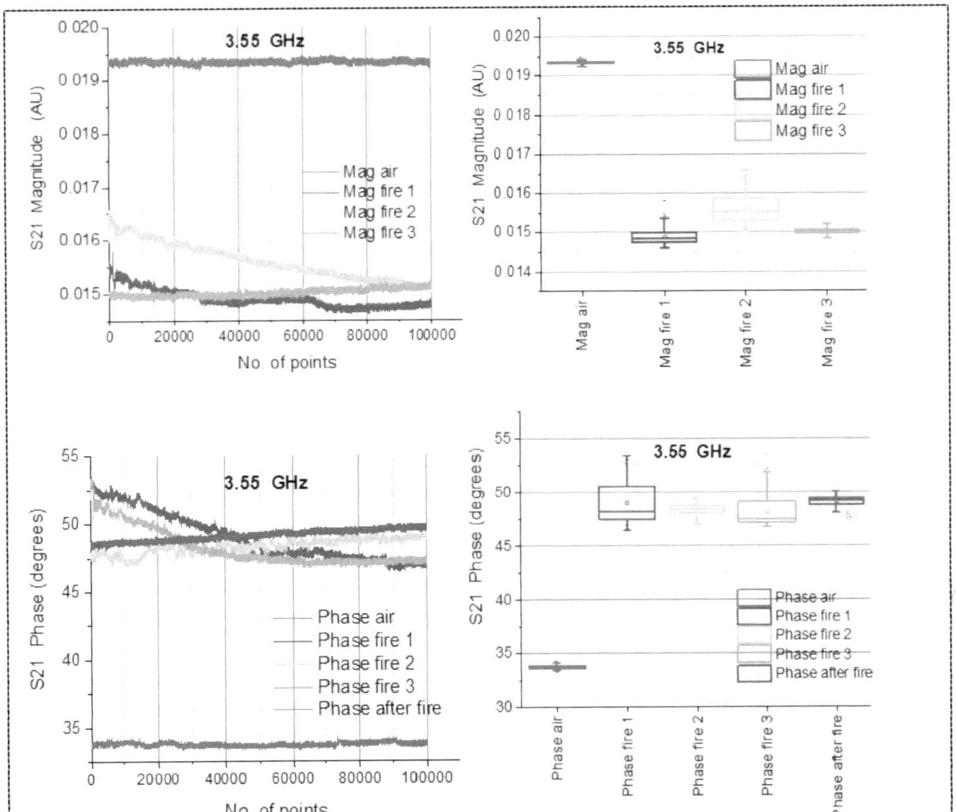

Figure 6. Magnitude (upper images) and phase (lower images) of S21 transmission coefficient at propagation thru air and comparatively thru fire wall of the wave of 3.55 GHz.

Figure 6 (lower images) refer to the effect of propagation thru air+fire reflected by the phase shifts of the transmission coefficient at 3. 55 GHz. The phase increased in all repeated experiments (fire 1 to fire 3). To observe the effect of just the receiving antenna temperature, a recording was made just after the fire was ended and in Figure 6 it is called "phase after fire". It can be seen that it was very similar to the ones recorded while combustion was on. It indicates that either the smoke that remained had an impact, or the antenna heated and remained warmer. The phase of the transmission increased on average with 41.3% at propagation thru fire than thru air. Boxplots of data collected during 90 s also indicate that the fire presence conducted to the spreading of phase values (Figure 6, lower-right graph).

Figure 7 shows the effects of the of passing thru fire at the second frequency, 6.26 GHz. Much smaller variations of both the magnitude and the phase of S21 coefficient are remarked. Moreover, either positive or negative shifts of both indicators were present in the case of fire presence reported to the case of non-fire path. This case may be correlated with the observations reported in [3] where the authors hypothesized that the diffraction may be responsible of such an effect.

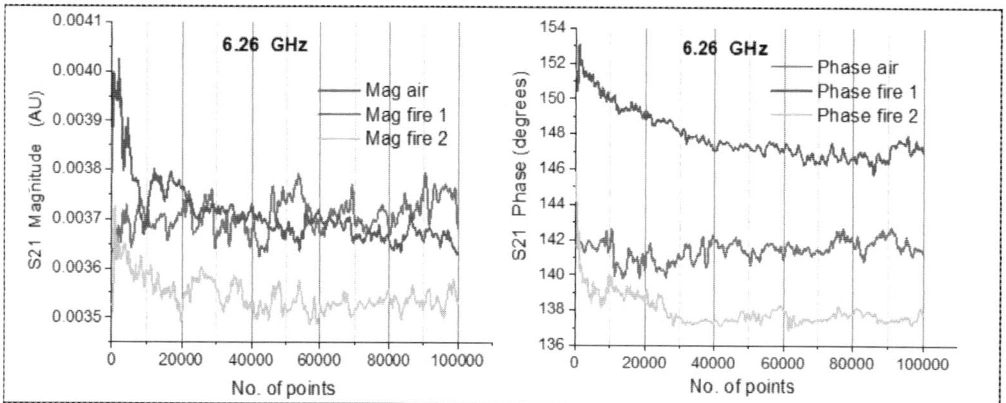

Figure 7. Magnitude and phase shifts of the transmission coefficient at passing thru the small fire curtain or thru the air of the wave at 6.26 GHz.

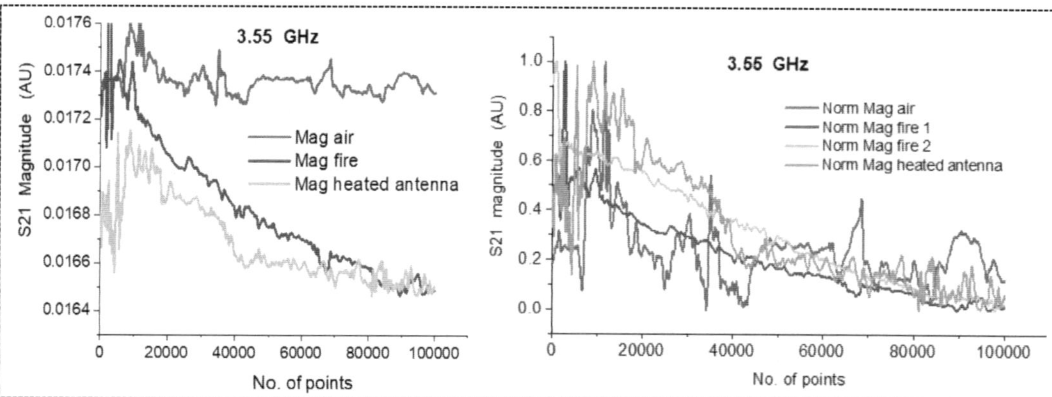

Figure 8. Temperature effect on the antenna parameters compared with the effect of propagation thru fire wall (left image); Normalized evolution of magnitude of transmission in time, comparatively for air, fire propagation and for heated antenna and air propagation (right image).

To check the influence of the antenna heating on the S21 magnitude, we made another experimental session in which the antenna was heated from a separate source and the propagation path was air. This case is observed in Figure 8 (left graph) as "Mag heated antenna". It is seen that very reduced difference exists in the magnitude of the transmission modification due to antenna heating and due to fire curtain passing, respectively. This observation, that heating the antenna conducts to change of its parameters, is in concordance to new results reported in [20].

Figure 8 (right graph) emphasizes another aspect observed in the propagation at 3.55GHz in a second experimental session, the fact that normalized magnitude data indicate a much smoother variation in time of the transmission coefficient when the fire was present than when only air was passed by the signal. The case "Norm Mag heated antenna" corresponds to the heated antenna case, but the microwaves propagation was thru air. It is concluded that fire do not necessarily conduct to much noisy propagation, and this depends on the fire turbulences presence [3].

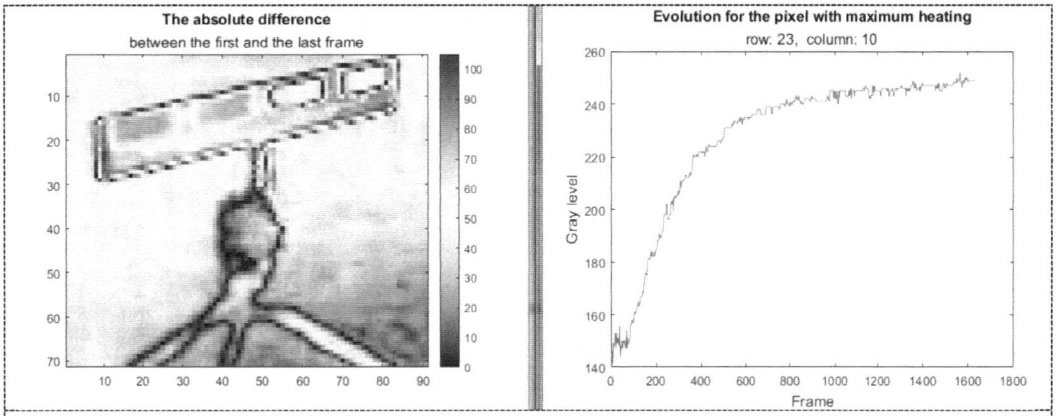

Figure 9. Heating of the RX antenna positioned at 75 cm from the fire curtain as captured by the IR camera. Absolute difference of temperatures between fire start and stop the thru-fire transmission is shown (left image) together with the temperature dynamics (grey scale) of the most heated pixel in the antenna picture (right image).

We have also checked the heating of the antenna due to fire proximity with the IR camera shown in Figure 2 (upper right). Due to the low resolution and technical problems related to temperature display, we could only retrieve the temperature increase indicated by the grey scale values modification. Indeed, it is clear that the temperature is changing during the thru-fire transmission, and future experiments will have to take this behaviour into consideration. So when antennas are positioned close to the fire curtain and some thermic drift of the antenna's temperature is expected due to proximity to the heating source, this effect should be extracted from the overall result. For the present experiments, an example of the grey level dynamics of one pixel in the IR image of the antenna is presented in Figure 9. It is observed that at some point in time the temperature is stabilizing and this corresponds to the decrease of the fire intensity.

3.3. Heartbeat detection

Figure 10 emphasizes the measured phase (noisy blue curve) and superimposed the filtered phase of S21 coefficient (orange curve) for one of the recorded samples of vital sign detection.

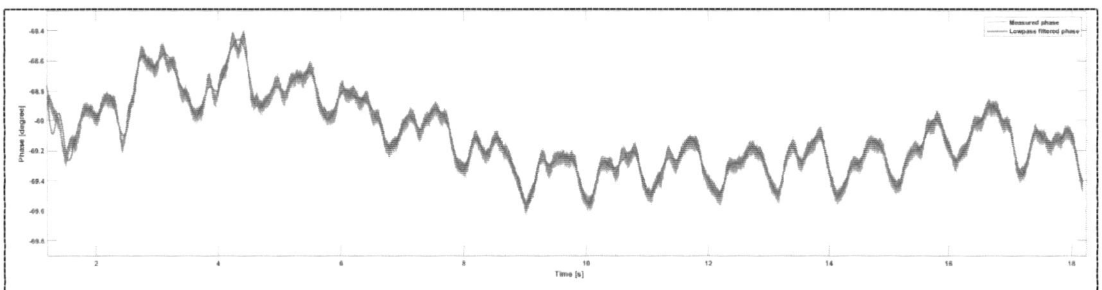

Figure 10. A sample of the raw S21 phase variation in time (blue) and the filtered one (red).

Figure 11 shows an example of frequency-amplitude distributions of the extracted spectral components of the heartbeat. The first and the second spectral components of the recorded human pulse during all radar detection tests are seen. The human pulse frequency values detected with the couple of array antennas at 3.55 GHz in the present session of tests were a bit striking at first site, mainly because they were spread due to the noisy recorded signal. If compared to some of our previous results reported in [21], when a single patch antenna was used, the present antenna solution seems worse. At a deepen analysis we identified the possible cause, which seems to be the mutual coupling between the patches of the array. Electromagnetic field coupling between elements of the array, either at Tx antenna or at Rxantenna may be expected, mainly by such a short path. The ratio between the array dimension and the path length is 0.19. Moreover, the 10 cm interval between Tx and Rx antennas edges conduct to a ratio between the antenna dimension and this space interval of 1.9. The electromagnetic field is shifted with different beams cases and the coupling cannot be neglected anymore. This phenomenon's effect is worth studying in a future work.

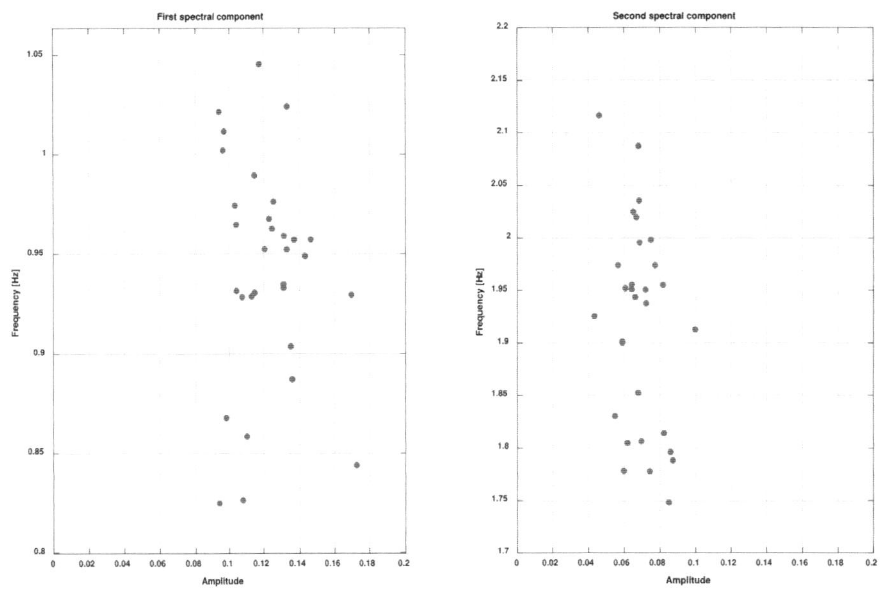

Figure 11. The distribution of the amplitudes and frequencies of the first two spectral components of the heartbeat detected by processed Doppler radar signal at 3.55 GHz in a series of tests.

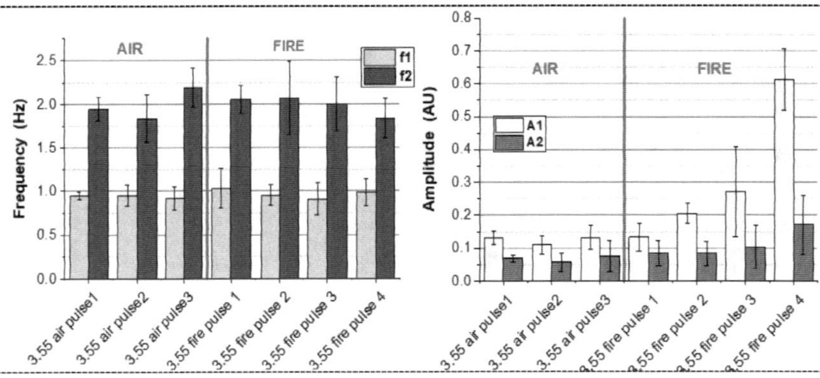

Figure 12. The distribution of the mean and standard deviation values of the frequency and amplitude of the heartbeat detected in a series of tests, compared for in-air propagation and thru-fire propagation.

With the collected data over many pulse recordings we made statistics of vital sign detection in air versus when fire curtain was passed and obtained the graphs of amplitude of the two spectral components (A1 and A2) and of their two frequencies (f1 and f2), as observed in Figure 12. The average values and their standard deviation is presented per each case during the 20 s of recording. It may be observed that the fire effect was minimal on the frequency of the heartbeat detection but the amplitude of the pulse was better detected at propagation through fire. This interesting result converges with the reported results of reference [3] which demonstrated that an enhancement of the signal magnitude passed through fire may be expected in some cases and it is due to diffraction phenomena.

4. Conclusions

Two identical linear arrays made of four patch antennas were designed, characterized and produced with the aim of using them in a wave propagation study and in a Doppler radar test of human heartbeat detection. They showed very good directivity and gain at the chosen frequencies.

Two continuous waves were transmitted indoor by the highly directive antennas on a distance of 150 cm, either in air or by interposing at middle distance of a small fire curtain - not wider than 7 cm, with the aim of comparing the magnitude and phase variation of the transmission coefficient at two different frequencies, and between the two propagation environments. Heartbeat detection was also analysed comparatively, at propagation thru air and thru air plus fire curtain. The overall objective was the assessment of the presence of a low-intensity and small-dimension fire on the wave propagation path and the quantification of the effects.

It resulted that both magnitude and phase of the transmission suffer a shift when in the path a fire zone is present, but the values are different at the two used frequencies.

Human pulse detection with this type of antenna array is affected by noise but can be extracted with parametric spectral analysis. Extracting the frequency of the pulse is not different at passing thru air or air plus fire, but the amplitude of the beats can be amplified by the fire presence.

The heating of the antenna due to fire proximity must not be neglected in the future work. The noisy signals collected by Doppler radar may be due to the electromagnetic coupling effect between the elements of the array due to the short distances between the antennas and of the reflection path.

References

[1] Mphale K 2008 Radiowave propagation measurements and prediction in bushfires *University of Botswana Phd Thesis*

[2] Mphale K M, Luhanga P V C and Heron M L 2008 Microwave attenuation in forest fuel flames *Comb and Fl.* 154 pp 728-739

[3] Boan J A 2009 Radio propagation in fire environments *PhD Thesis*

[4] Dissanayake C M, Halgamuge M N, Moran B and Farrel P 2010 The signal propagation effects on IEEE 802.15.4 radio link in fire environment *Fifth Inter.Conf. on Inf. and Aut. for Sust.*pp 411-414

[5] Faria S, Leonor N, Fernandes C, Felicio J, Salema C and Caldeirinha R 2020 Radiowave propagation modelling in the presence of wildfires: initial results *Eur. Conf. on Ant. & Prop. (EuCAP)* pp. 1-5

[6] Faria S, Vala M, Coimbra P, Leonor N, Felicio J, Fernandes C, Salema C and Caldeirinha R F 2020 Comparative study of computational electromagnetics applied to radiowave propagation in wildfires *URSI GASS Conf.* pp 1-4

[7] Faria S, Vala M, Coimbra P, Leonor N, Felicio J, Fernandes C, Salema C and Caldeirinha R F S 2021 Radiowave propagation modelling of dual wildfire front spreading over hilly terrain at 700 MHz *Telecoms Conference* pp 1-6

[8] Faria S, Vala M, Coimbra P, Felicio J, Leonor N, Fernandes C, Salema C and Caldeirinha 2021 Analytical studies of refractive index variation in pine needles media under lidfire conditions *Eur. Conf. on Ant. & Prop. (EuCAP)* pp 1-5

[9] Özkan Alkurt F, Bağmancı M, Karaaslan M, Bakır M, Altıntaş O, Karadağ F, Akgöl O, Ünal E 2018 Fire detection behind a wall by using microwave techniques *AIP Conference Proceedings* **1935**, 060002

[10] Letsholathebe D and Mphale K 2015 Absorption of mirowaves in low intensity eucalyptus litter fire *J. of Electr. Anal. and Appl.* **7** pp 217-224

[12] Letsholathebe D, Mphale K M, Chimidza S and Heron M L 2016 Radio wave propagation experiment in sugarcane fire enviroments *J. of Elec. Anal. and App.* **8** pp 124-131
[13] Li Y, Yuan H, Lu Y, Xu R, Fu M, Yuan M and Han L 2017 Experimental studies of electromagnetic wave attenuation by flame and smoke in structure fire *FireTechnol* 53 pp 5-27
[14] Oliviera T E S, Reis J R, Vala M and R F S Caldeirinha 2021 High performance antennas for early fire detection wireless sensor network at 2.4 GHz *IEE-APS Top. Conf. on Ant. And Prop. In Wir. Com. (APWC)* pp 158-163
[15] Oliviera T E S, Vala M, Faria S, Reis J R, Leonor N and Caldeirinha R F S 2022 A study on small sensor node antenna performance camouflaged under grassland fire *Int. of th. Tech. and Appl. Conf* 641 pp 86-95
[16] Coimbra P, Faria S, Vala M, Felicio J, Fernandes C, Leonor N, Caldeirinha R and Salema Carlos 2021 Experimental setup for radio characterization of fire at microwave frequencies *Telecoms Conference* pp 1-4
[17] Temme A K 2015 Experimental studies of microwave propagation through fires for through-wall, search-and-rescue radar in firefighting *PhD Thesis*
[18] Bystrov A, Daniel L, Hoare E, Norouzian F, Cherniakov M and Gashinova M 2021 Experimental evaluation of 79 and 300 GHz radar performance in fire environments *Sensors* 21
[19] Stoica P, Moses R 2005 Spectral Analysis of Signals *Prentice Hall* USA
[20] Bhattacharjee M, Nikbakhtnasrabadi F and Dahiya R 2021 Printed chipless antenna as flexible temperature sensor *IEE Int. of. Thi. J* **8** pp 5101-5110
[21] Vatamanu D, Iftode C and Miclăuş S 2021Microstrip patch and Sierpinski fractal antennas: simulations and experimental characterization for vital signs detection with Doppler radar technique *Indian J Sci and Technol* **14**(18) pp 1477-1487

Using infrared imaging for surface dosimetric measurements during microwaves exposures

C Neghină[1], A Cotoi[2], R Voina[3] and S Miclăuş[2]

[1]"Lucian Blaga" University, Faculty of Engineering, Sibiu, Romania
[2]"Nicolae Bălcescu" Land Forces Academy, Sibiu, Romania
[3]Technical University of Cluj-Napoca, Faculty of Electronics, Telecommunications and IT, Cluj-Napoca, Romania

E-mail: cotoi.alina@armyacademy.ro

Abstract. Infrared (IR) thermography enabled by a small camera attached to a mobile phone is used in the present research for the determination of the heating dynamics of biological liquid samples exposed to 2.3 GHz continuous wave. The methodology is presented and analysed and some relevant examples are Figured out. Differences in electromagnetic field dissipation are nicely emphasized proving that microwaves dosimetry applied to surfaces, if enough resolution and frame frequency is enabled, may represent a valuable tool in observing heating dynamics of the exposed surfaces even at the microscale.

1. Introduction

Dosimetry of microwaves (MW) frequencies is required in all research areas where quantifying of biological effects of electromagnetic field exposure is the objective. With increase of the frequency, the penetration depth of MW in biological objects decrease, so that from a few GHz upwards, the penetration depth in matter may decrease to a few millimeters. This is also the case of the new technologies belonging to the 5th generation (5G) of mobile communications. Therefore, the need of a volumetric assessment of the absorbed dose of radiation decreases and the superficial dose might suffice for extremely high frequencies. In this regard, one dosimetric technique that may be used in the case of thermal effects of MW irradiation is the infrared (IR) thermography. By extracting the slope of the heating curves of the monitored area of an irradiated object, the specific absorption rate (SAR) of energy deposition is computed and compared against regulated limits. In this way, the energetic regimen of electromagnetic field absorption is very well characterized.

It is worth mentioning that not too many references on this topic presently exist. For example, in [1] a wave of 10GHz was used for the exposure of rats and their skin was monitored with an IR camera. IR thermography revealed an increase in the average skin surface temperature and provided insights of the molecular responses after irradiation. In [2] IR thermography used as a noninvasive technique allowed correlating the metabolic changes with shifts of SAR values below the thermal breakdown levels after exposing hamsters to 900MHz field. Authors of [3] used IR thermography to record the temperature increase of the skin of two free rats while being irradiated at 1.8 GHz [3] to follow some biological parameters endpoints.

Papers [4-7] followed the effects of MW in different ranges at the human ears and skin level when using a mobile phone or wearable devices and used IR thermography as a dosimetric tool.

In [4] for instance, the focus was the quantification of the thermal effect of a mobile terminal emission over the human auricle when it was operating at 900MHz, 1800 MHz and 2250MHz. The skin temperature increase in the cheek and ear regions due to the absorption of MW energy from three handheld mobile phones in different situations was investigated in [5]. The thermographic information was used in another study that aimed at following the heating produced by a microstrip antenna of a wearable device radiating at 1.8 GHz, 2.4 GHz, 5 GHz and 8.9 GHz [6]. An experimental validation based on IR thermography was applied in [7], where the authors identified the highest temperature rise on the forearm after several minutes of exposure at 1.8GHz radiation. At cellular level, small changes in heat dissipation could be detected with the help of an IR camera and they were compared for two type of cells exposed to MW [8].

Another application type was provided in [9], where an antenna operating at 2.45GHz acted as an electromagnetic wave source and an IR camera offered images of the temperature distribution on an absorption screen situated at different distances from the antenna. The authors of [10] used the IR method to obtain a complete characterization of the near field structure (from 0.8 to 2.3GHz) of an antenna, and they proved that the method provided accurate and replicable radiation signatures.

The majority of the papers in the literature described the methodology of IR thermography instrumentation very synthetic, without analyzing the advantages and the limitations in using it as a dosimetric tool for MW exposures and without commenting its capabilities. It is this paper's aim to investigate a cheap method to describe the superficial dose of MW radiation absorbed in biological liquid samples and to provide an overview of the dynamics of the phenomena during the electromagnetic exposure. Such a method provides a reliable tool for proper dosimetry in the study of interactions of biosystems with thermal levels of electromagnetic fields.

2. Materials and method

The proposed method of MW dosimetry based on IR thermography is applied to three liquid samples exposed to thermal levels of a continuous wave. Each sample of 10 ml volume was composed of an extract of vegetal material in 70% ethanol, at a concentration of 1g/10 ml. The organic liquids contained minced leaves (sample 1), fresh mushrooms (sample 2), and dried mushrooms (sample 3), as shown in Figure 1.

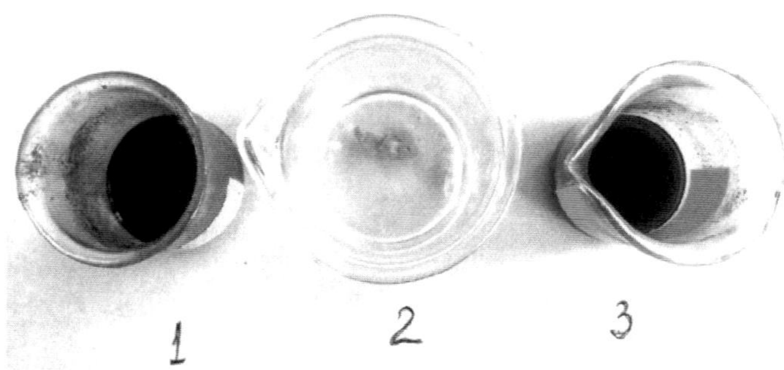

Figure 1. The organic liquid samples with:
1 – minced leaves, 2 – fresh mushrooms, 3 – dried mushrooms.

2.1. Microwaves exposure

The samples were exposed one by one in a shielded enclosure to a signal of 2.3GHz emitted by a horn antenna model BBHA 9120A – Schwarzbeck. The signal was provided by a generator model SM 300 (Rhode & Schwarz) connected to a radiofrequency power amplifier (model Ophir 5120). The electric field strength (E) incident to the liquid sample was measured in air by EMSense 10 isotropic field probe connected to the EM Center 7007-200 (ETS-Lindgren). The E-field levels were set in the range (750-1150) V/m. The duration of each exposure was of 10 minutes. The components of the MW exposure setup are observed in Figure 2.a. The initial temperature was the room temperature for each sample, and the inclination of the containers with the samples during experiments was at an angle of approximately 45°, as seen in Figure 2.b., in order for the IR camera to "see" the liquid's surface. The antenna opening and the IR camera objective faced the liquid surface.

2.2. Temperature Monitoring with IR Camera

A small infrared camera from i3system (47x25x16mm), model Thermal Expert™ TE-Q1 Pro (Figure 2.c.), was used for the thermographic imaging of the liquid surface during the electromagnetic exposures. The camera had the following specifications:

- Sensor resolution of 384x288;
- Measured wavelength 8-14μm;
- Thermal pixel of 110592;
- Pixel pitch of 17μm;
- Sensitivity of 0.08°C;
- Frame rate < 9Hz;
- Scene range temperature: -10°C to 250°C;
- Lens of 6.8mm f1.3.

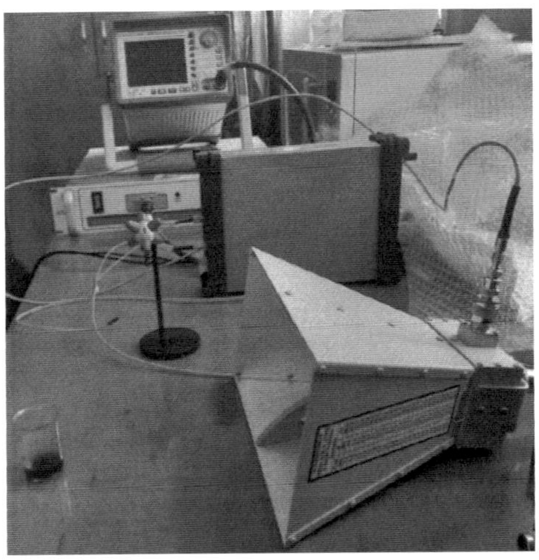

Figure 2.b. Position of samples during experiments.

Figure 2.a. The experimental setup.

Figure 2.c. The Thermal Expert™ TE-Q1 Pro [11].

The camera was connected directly to a smart phone through the USB type C interface, and positioned at the distance of 30cm from the samples. This large distance was dictated by the intense E-field the samples were in, which, at nearer distances, interfered with the correct electronic functionality of the camera.

The software used to take the thermographic images was Thermal Expert™ for Android (Figure 3.a.), with the following attributes:

- Colour palettes (12 colour maps);
- Storage images (IR, visible with snapshot and video);
- Digital zooming;
- Different image views;
- Report (generates .pdf report);
- Different temperature measurements (alarm, min/max, point/line/rectangle/circle/profile);
- Level Span function (High definition).

IOP Conf. Series: Materials Science and Engineering 1254 (2022) 012011 doi:10.1088/1757-899X/1254/1/012011

The color palette used for this paper was the WhiteHot palette, which can be described as a gray scale palette, in which hot values are represented in light pixels (pure white is the hottest), and cold values are represented in dark pixels (pure black is the coldest). For each liquid sample, a video of 600 seconds was taken, with the frequency of 9 frames per second, which led to a total of 5400 frames for each of the three samples.

An example of an IR image taken from one of the samples (sample 2) can be seen in Figure 3.b. as a dark disk in the upper side, surrounded by a light circle, which is the container walls (glass). The object of interest occupied a small percentage area of the image and was not centered because of two reasons: a) the field strength emitted by the antenna was high and we had to place the camera so as to reduce to minimum the induced currents and interferences upon smartphone's and the camera's electronics; b) the samples were kept in a shielded enclosure, which had only a small lid for the camera's objective positioning at that distance. The image resolution was 1065/540 pixels.

Figure 3.a. Screenshot of the Thermal Expert software.

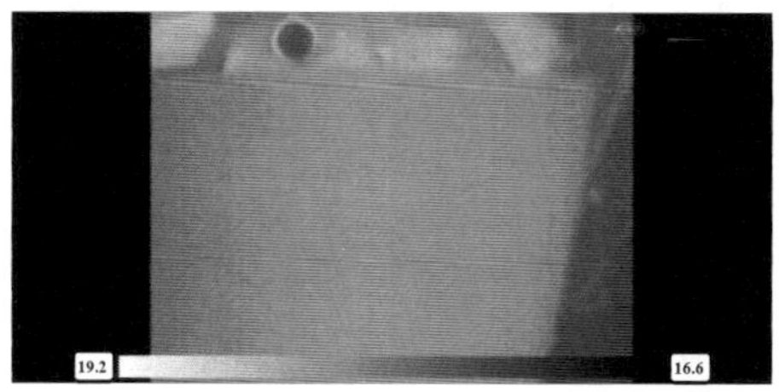

Figure 3.b. Infrared image of sample 2.

2.3. Matlab Image Processing

In digital thermal images, for each pixel there is a correspondence between its value and the surface temperature of the area represented by the pixel. Although the thermal camera can be set to different color palettes, for this experiment the simple 8-bit grayscale representation has been considered because it is more straightforward to use in subsequent computations. The thermographic camera also provides for each image a scale, mapping the grayscale levels to temperature in a linear way, with black (pixel value 0) representing the lowest temperature and white (pixel value 255) representing the highest temperature. Figure 4 shows an example of a thermal image together with its temperature scale in the base.

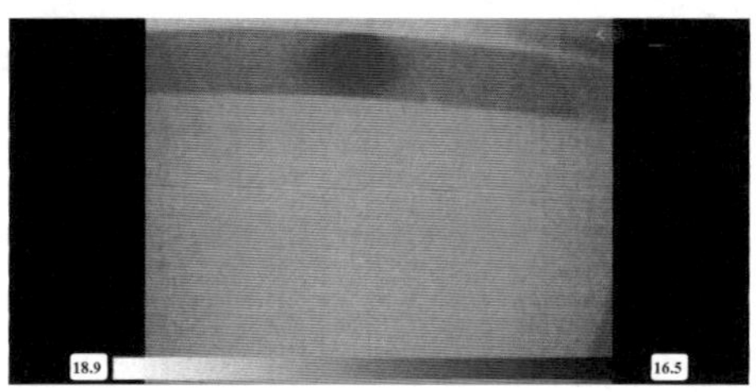

Figure 4. Frame from one of the videos used in our experiments (sample 1).

The processing algorithm has been built on several reasonable or justifiable assumptions regarding the environment and the object of interest. The capturing device (the camera) has a fixed position relative to both the object of interest and the background objects during the entirety of the experiment, thus all frames display the object of interest in the same spatial region of the image (ROI). The processing is therefore done on a cropped region containing the object of interest. The cropping coordinates are currently input manually for the first frame and maintained throughout each experiment (Figure 5).

IOP Conf. Series: Materials Science and Engineering 1254 (2022) 012011 doi:10.1088/1757-899X/1254/1/012011

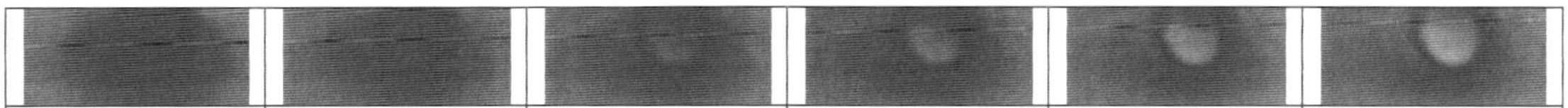

ROI from frame 1	ROI from frame 1000	ROI from frame 2000	ROI from frame 3000	ROI from frame 4000	ROI from frame 5000

Figure 5. ROIs from different frames

Furthermore, the background objects do not exhibit significant temperature changes. The temperature for the object of interest is always in between the lowest and largest background temperature, as can be checked in the histograms of any frame. Figure 6 shows the histograms for the first and last frame.

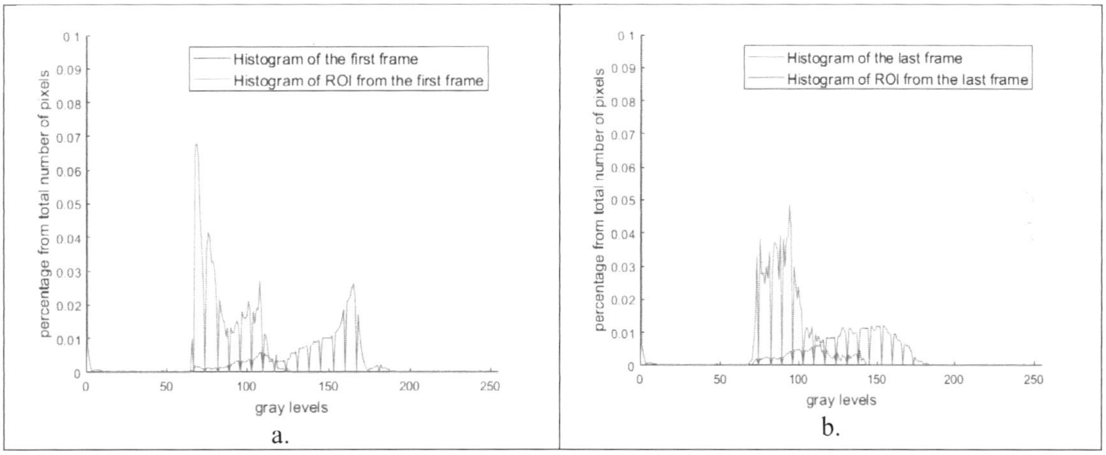

Figure 6. Histograms for first frame (a.) and last frame (b.).

The conversion from gray level to temperature value was done using a linear transformation as in Figure 7 and described mathematically by equation (1):

$$T_x = (T_{max} - T_{min}) \cdot G_x/255 + T_{min} \quad (1)$$

where:

- T_x: temperature of pixel x;
- T_{max}: maximum temperature from the frame;
- T_{min}: minimum temperature from the frame;
- G_x: gray level value of pixel x.

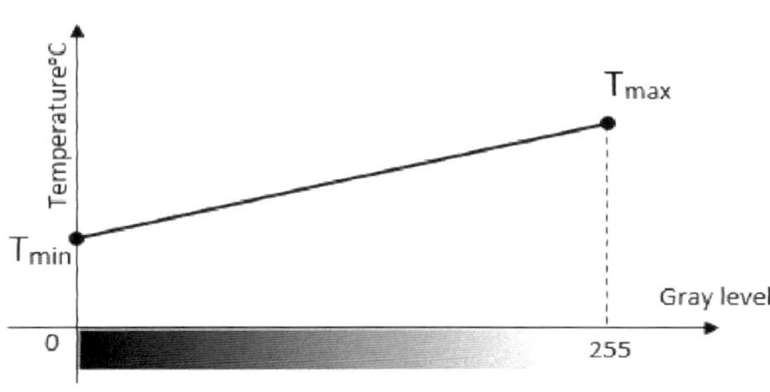

Figure 7. Linear conversion between gray level and temperature.

The minimum (T_{min}) and maximum (T_{max}) values of the temperature have been chosen as minimum and maximum in each video: for sample 1 15°C and 18.9°C; for sample 2 14.5°C and 19.2°C; for sample 3 12.2°C and 19.6°C. Further improvements would consider the actual limits given by the scale for each frame.

2.4. Statistical analysis of the results
Based on the temperature data dynamics for each pixel, a set of computations were made using Originlab software [12]. The temperature distribution per pixel during 10 minutes of irradiation could be provided in the form of boxplots representations. The heating curves per pixel, per area of interest or per entire irradiated surface may be calculated, either momentary or averaged on any time period. In this way, local and periodical determination of SAR may be obtained, based on the slope of the heating curves.

3. Results and discussions
To demonstrate the method's capabilities, the IR images in the frames were processed. A type of result is exemplified in Figures 8 - 10, for the three organic samples. The maximum heated pixel was extracted for analysis and its heating curve is represented in the left side of Figures 8.a. – 10.a. Specific temperature dynamics can be observed for each sample.

Differences between temperatures of corresponding pixels in the selected frames are represented together as heat maps of the ROI in Figures 8.b. – 10.b., showing the differences between the first and the last frame for each sample. The temperature evolution corresponding to a single pixel is a one-dimensional signal. Figures 8.b. – 10.b. represent the evolution of temperature for the pixel with the most extreme difference between the first and last frame of each sample.

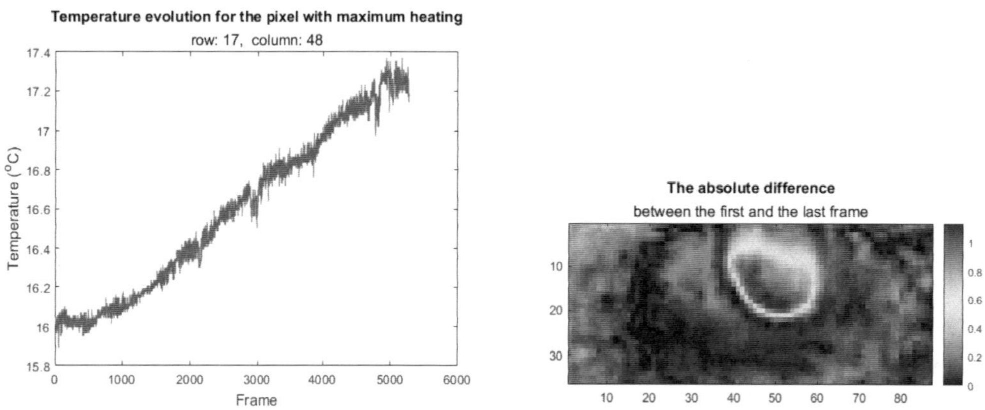

Figure 8.a. Temperature evolution for the pixel with maximum heating and **b.** the absolute difference for sample 1.

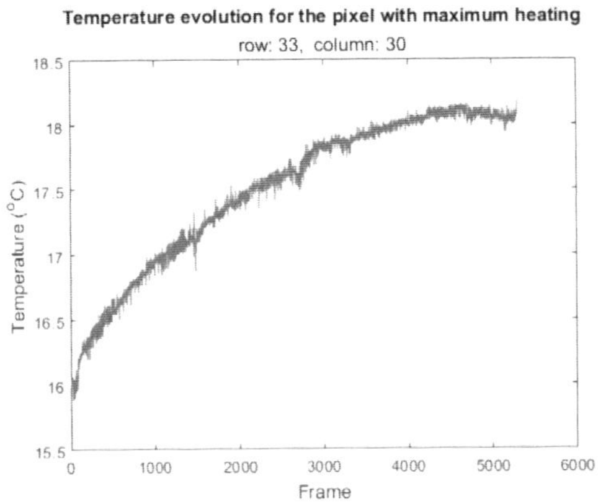

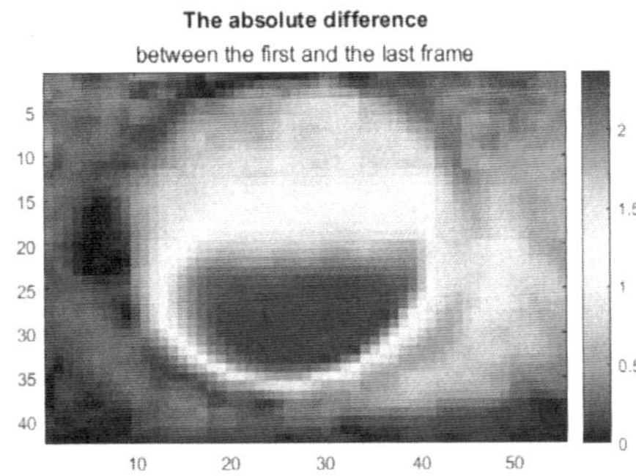

Figure 9.a. Temperature evolution for the pixel with maximum heating and **b.** the absolute difference for sample 2.

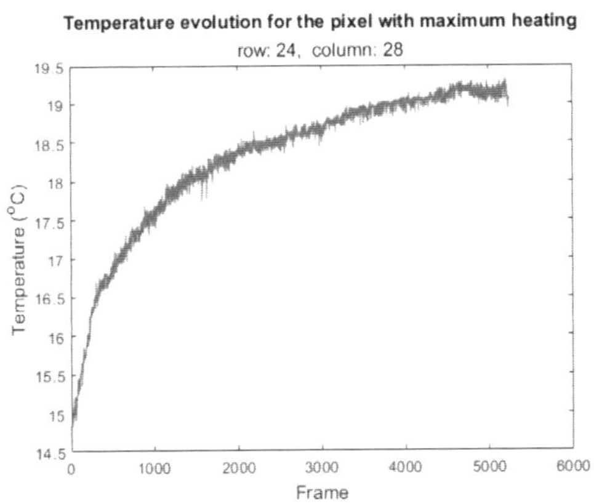

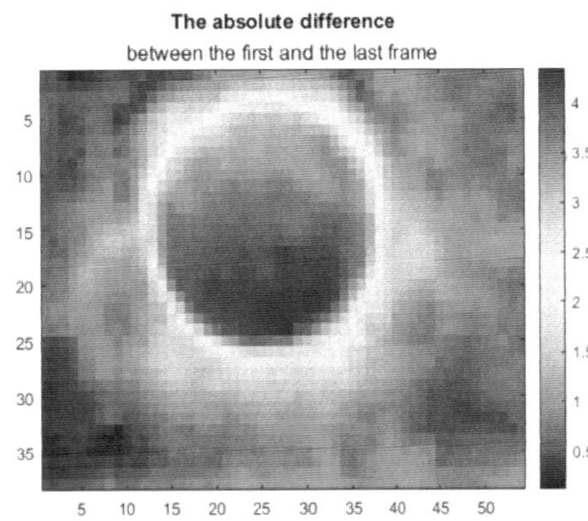

Figure 10.a. Temperature evolution for the pixel with maximum heating and **b.** the absolute difference for sample 3.

In Table 1 we present the maximum temperature variations for each sample after 10 minutes of irradiation and the corresponding incident E-field strengths on the samples. It may be seen that the temperature increase depends on the intrinsic properties of the sample. Indeed, as SAR is defined as:

$$SAR = \frac{\sigma * E\_int^2}{\rho} = c * \frac{\Delta T}{\Delta t} \tag{2}$$

where σ is the electrical conductivity of the sample, E_int is the internal E-field strength in the sample, ρ is the mass density, c is the heat capacity, ΔT is the temperature increase and Δt is the time duration of the exposure. So, not gaining access to the internal E-field strengths, the SAR is nicely indicated by the temperature slope of the sample. Temperature slope may not be constant during the exposures, and this can be seen for example if we compare Figures 8.a. and 8.b. At some point in time, due to thermal exchanges with the environment, the heating slope decreases. Such information is also valuable for identifying biological effects and mechanisms of interactions.

Table 1. Maximum variations over the exposed areas after 10 minutes of irradiation.

Sample number	ΔTmax(°C)	E (V/m)
Sample 1	1.2	1150
Sample 2	2.5	750
Sample 3	4.0	890

Figure 11 (upper line) shows temperature distributions over the surface of sample 1 at three different instances in time: at the beginning (left), after 286s of exposure (center) and after 600s of exposure (right). All the Figures are clockwise rotated with 90 degrees from the original positions. The images were obtained by interpolating the values between pixels based on the processed matrix data. In this way location of the warmest area can be observed and the scale of temperatures easy observable. Matrix data allowed obtaining the boxplot representations in the lower line of Figure 11. In this way median, mean and interquartile ranges of temperature distributions on column data could be available. All other mathematical processing can be easily performed. Integral or derivative computations can provide even more information. For example, average temperature rate per field strength can be obtained, $(\Delta T/\Delta t)/E$, (in °C/s per V/m) or respectively SAR/E (in W/kg per V/m) can be easily computed further on.

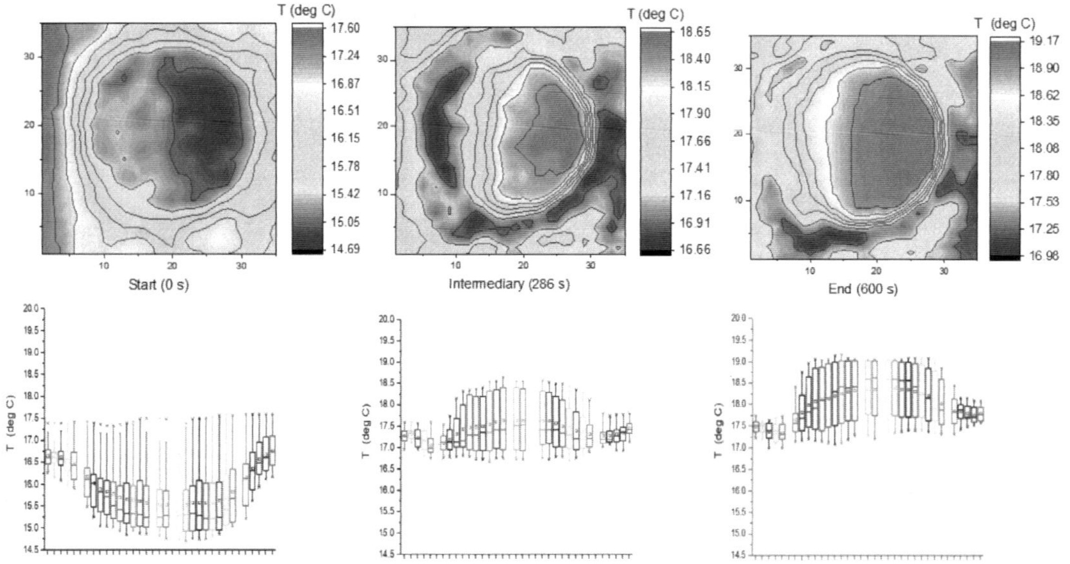

Figure 11. Reconstructing thermal images through interpolation based on matrix of temperature data and boxplot representations of temperature values spreading over the period of MW exposure along vertical lines of the images.

4. Conclusions

A series of biological and biochemical experiments are driven to follow the MW interaction mechanisms with matter. It is always very important to have the complete information about temperature rise at the smallest dimension possible, as a consequence of irradiation. The heating interval and rate could further make the difference between thermal, a thermal and non-thermal effects of exposure. Momentary and average heating rates are also valuable means in identifying correct mechanisms of interaction.

As supported by the experiments, the superficial dosimetric assessment of the MW field absorption in various dielectric samples could be properly accomplished by using a low-cost IR camera for capturing the radiation, followed by a relatively straightforward image processing algorithm for detecting the temperature variation in the region of interest. In the case of continuous MW exposure, the presented hardware and method form a valuable dosimetric tool, enabling the quantification of the thermal shift at pixel level which, depending on the camera distance to the object of interest and on the sensor resolution, may even represent a fine spatial resolution. The sensitivity of 0.08° C limits the detection of smaller thermal drifts, but this level seems to be sufficient for an incident field of tens of V/m. With regard to the object of interest, an IR camera can only see the surface of the object, and thermal dynamics within the object cannot be detected through IR imagery, leaving some uncertainty about which amount of the surface temperature comes from the direct MW exposure and which comes from internal heat conduction.

There are also limitations to the presented method. Related to the camera outputs, the temperature range of each frame could not be extracted independently of the visual output. For the experiments, the temperature of the object of interest is always around the middle of the observed temperature range and the temperature of the background objects is relatively unchanged, which allowed the assumption that the extreme temperatures remain constant throughout each experiment and only need to be read for the initial frame. However, if the temperature of the background objects is changing significantly or the object of interest is at one of the extremes, the algorithm would need to be adjusted to read the temperature range on a frame basis. Furthermore, in the presented results, the temperature comparison has only considered the first and last frames of each experiment, but the algorithm could be improved by searching for relevant frames to be used in the comparison, according to local temperature maxima and minima within the region of interest.

References

[1] Verma S, Keshri G K, Karmakar S, Mani K V, Chauhan S, Yadav A, Sharma M and Gupta A 2021 *Effects of Microwave 10 GHz Radiation Exposure in the Skin of Rats: An Insight on Molecular Responses* DOI: 10.1667/RADE-20-00155.1

[2] Taberski K, Klose M, Grote K, Ouardi A E, Streckert J, Hansen V W and Lerchl A 2014 *Noninvasive Assessment of Metabolic Effects of Exposure to 900 MHz Electromagnetic Fields on Djungarian Hamsters (Phodopus sungorus)* DOI: 10.1667/RR13646.1

[3] Wang X, Xia C, Lu L, Qi H and Zhang J 2021 *Electromagnetic Exposure Dosimetry Study on Two Free Rats at 1.8 GHz via Numerical Simulation* DOI: 10.3389/fpubh.2021.721166

[4] Bauer J, O'Mahony Ch, Chovan D, Mulcahy J, Silien Ch and Tofail S A M 2018 *Thermal effects of mobile phones on human auricle region* DOI: 10.1016/j.jtherbio.2018.11.008

[5] Lahiri B B, Subramaniam B, Soumya C, Jayakumar T and Philip J 2015 *Infrared thermography based studies on mobile phone induced heating* DOI: 10.1016/j.infrared.2015.04.010

[6] Karthik V and Rama Rao T 2018 *SAR Investigations on the Exposure Compliance of Wearable Wireless Devices Using Infrared Thermography* DOI: 10.1002/bem.22133

[7] Karthik V and Rama Rao T 2016 *Investigations on the Biological Effects of Wearable Antennas at UHF and UWB Frequencies: A Thermal Perspective* DOI: 10.1109/INDICON.2016.7838907

[8] Umberto L, Grazzini G, Montrucchio B, Grisolia G, Borchiellini R, Gervino G *et al* 2015 *Constructal thermodynamics combined with infrared experiments to evaluate temperature differences in cells* DOI: 10.1038/srep11587

[9] Muzaffar K, Tuli S and Koul S 2013 *Infrared Thermography for Electromagnetic Field Pattern Recognition* DOI: 10.1109/IMaRC.2013.6777742

[10] Prost D, Issac F and Romier M 2019 *Imaging electric and magnetic near field of radiating structures by infrared thermography EMC Europe* Barcelone Spain ffhal02364926

[11] *Thermal Expert – High Performance Thermal Imaging webpage* Available from: https://www.thermalexpert.eu/english-1/product-overview/

[12] *Originlab - Graphing for Science and Engineering* Available from: https://www.originlab.com/

IOP Conf. Series: Materials Science and Engineering 1254 (2022) 012012 doi:10.1088/1757-899X/1254/1/012012

Numerical evaluation of the induced voltages in an underground pipeline using the distributed source approach

L Czumbil, M Crețu, A Ceclan, L Dărăbant, D Șteț and DD Micu

Electrotehnics and Measurements Department, Electrical Engineering Faculty, Technical University of Cluj-Napoca, G. Baritiu street, no. 26-28, Cluj-Napoca, Romania

E-mail: Mihaela.Cretu@ethm.utcluj.ro

Abstract. This paper presents the distributed source analysis approach for treating the inductive coupling of power transmission lines with buried pipelines. The evaluation of the induced voltages in a pipeline which share the same path with a power line for a monophasic short circuit on the line is performed and also the analytical expression for current density induced in soil is calculated. The paper describes a practical case of subdivision of the zone of influence in sections (a circuit model) and a precise calculation of the induced voltages in different points based on the measured data, is performed in Mathcad, using special interpolation functions.

1. Introduction

Due to environmental and economic reasons, mainly related to construction costs and to their impact on local ecosystems high or medium voltage power lines are forced to coexist in a relatively narrow distribution corridor with gas, petrol or water pipelines with no access to new separate rights-of-way [1-3]. Figure 1, shows such a situation when an underground metallic pipeline is in the close proximity with an overhead power line. Similar situations are a common practice and could occur over several kilometers [4].

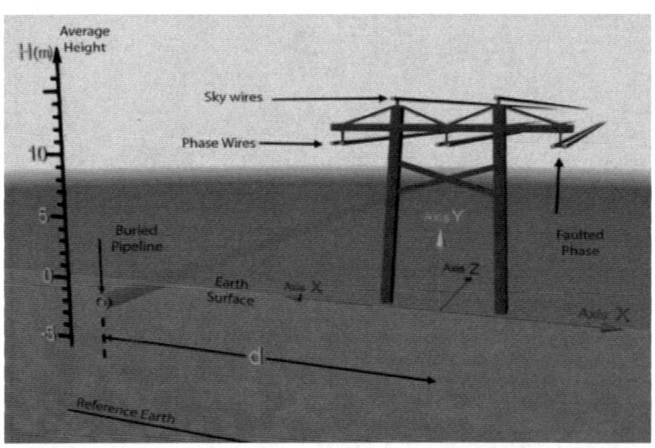

Figure 1. An underground gas pipeline next to a 400 kV / 50 Hz HVPL [4]

The metallic structures situated in the neighborhood of the power lines can catch some of the energy transported by the power line phase conductors, especially in unfavorable circumstances, such as power line fault conditions or long parallel exposures [5, 6].

Induced currents and voltages in the metallic pipeline placed in the vicinity of high voltage power lines could be determined through electromagnetic interference calculations consisting in case of underground pipelines in inductive and conductive coupling mechanisms analysis. The inductive coupling deals with phenomenon related to the magnetic fields produced by the power line, while the conductive coupling deals the galvanic connection that could appear through earth in case of phase to ground or lightning stroke related power line fault combined with very close proximity within the powerline and the underground pipeline [7, 8]. Therefore, these coupling mechanisms could be investigated separately and the obtained results to be combined together [4, 7].

For the inductive coupling analysis, a hybrid field theory/circuit theory approach could be applied as in [4] obtaining an equivalent circuit model as presented in Figure 2.

Figure 2. Equivalent electrical circuit approach [4]

In case of conductive interference calculation that deals exclusively with buried conductors, first the involved conductors are subdivided into small enough segments with respect to wavelength and the length of analyzed ground network. A finite elements field theory approach could then be applied to relate currents and voltages in the segments is such a way that the required parameters to be obtained for a similar equivalent electrical circuit model to the one presented in Figure 2, involving the internal and external impedances of the created conductor segments. Therefore, once again the circuit theory approach could be applied to compute induced voltages and leakage currents in the exposed underground pipeline.

As a result, the circuit theory is applied by the authors; the potentials at the segment end points and, also the longitudinal and leakage currents in every conductor segment are determined.

Calculation of the induced voltages on the pipelines is considered in two steps: first, the determination of the electromotive forces (E-EMF) induced along the pipeline and second, the potential difference between the pipeline and the earth, resulting from these EMF's is calculated.

2. The distributed source analysis approach

This analysis treats inductive coupling a theory called distributed source analysis. The pipeline and its surrounding earth is seen as a lossy transmission line, characterized by the propagation constant γ, and its characteristic impedance, Z_C. The inductive coupling from a nearby AC power line is modeled through a distributed voltage source function, $E_x(s)$, along the pipeline, where $E_x(s)$ is the longitudinal driving electric field parallel to the pipeline right of way [9].

Figure 3 presents an equivalent elementary circuit of the investigate pipeline section, which except for the voltage source is identical to the per unit length circuit modal of an electrical transmission line with the same definitions of series impedance $\underline{Z} = R + j\omega L$, and shunt admittance $\underline{Y} = G + j\omega C$.

For harmonically varying signals ($e^{j\omega t}$), the differential equations of the voltage and current distribution along the pipeline element are identical to those for the classical electrical transmission lines (except for the terms containing E_x). Therefore, assuming the terminating impedances of the pipeline to be $\underline{Z_A}$ at $x = x_A$ and $\underline{Z_B}$ at $x = x_B$ $(x_B > x_A)$, after some mathematical manipulations the solution for induced currents and voltages will have the following form:

$$\underline{I}(x) = \left(K_1 + P(x)\right) \cdot e^{-\gamma \cdot x} + \left(K_2 + Q(x)\right) \cdot e^{\gamma \cdot x}$$
$$\underline{V}(x) = Z_C \cdot \underline{I}(x) = Z_C \cdot \left(\left(K_1 + P(x)\right) \cdot e^{-\gamma \cdot x} + \left(K_2 + Q(x)\right) \cdot e^{\gamma \cdot x}\right) \quad (1)$$

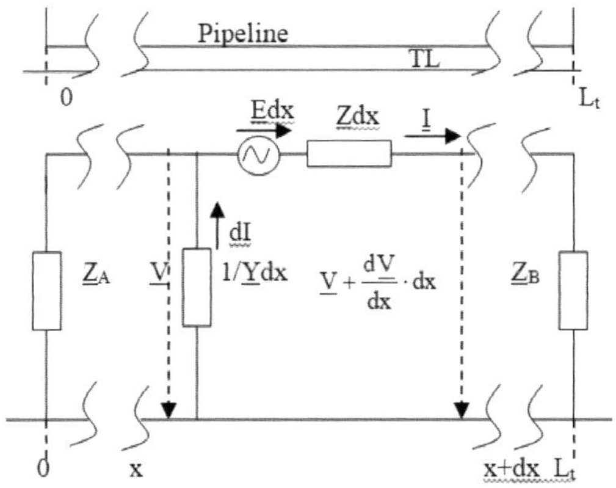

Figure 3. Equivalent elementary pipeline circuit

where:

$$P(x) = \frac{1}{2 \cdot \underline{Z_C}} \cdot \int_{x_A}^{x} \left[e^{\gamma \cdot s} \cdot \underline{E_x}(s)\right] ds$$
$$(x) = \frac{1}{2 \cdot \underline{Z_C}} \cdot \int_{x}^{x_B} \left[e^{-\gamma \cdot s} \cdot \underline{E_x}(s)\right] ds \quad (2)$$

and the coefficients resulting for the boundary conditions at ends being [10]:

$$K_1 = v_1 \cdot e^{\gamma \cdot x_A} \cdot \frac{v_2 \cdot P(x_B) \cdot e^{-\gamma \cdot x_B} - Q(x_A) \cdot e^{\gamma \cdot x_B}}{e^{\gamma \cdot (x_B - x_A)} - v_1 \cdot v_2 \cdot e^{-\gamma \cdot (x_B - x_A)}}$$
$$K_2 = v_2 \cdot e^{-\gamma \cdot x_B} \frac{v_1 \cdot Q(x_A) \cdot e^{\gamma \cdot x_A} - P(x_B) \cdot e^{-\gamma \cdot x_A}}{e^{\gamma \cdot (x_B - x_A)} - v_1 \cdot v_2 \cdot e^{-\gamma \cdot (x_B - x_A)}} \quad (3)$$

with the reflection coefficients:

$$v_1 = \frac{\underline{Z_A} - \underline{Z_C}}{\underline{Z_A} + \underline{Z_C}}$$
$$v_2 = \frac{\underline{Z_B} - \underline{Z_C}}{\underline{Z_B} + \underline{Z_C}} \quad (4)$$

The series impedance ($\underline{Z} = R + j\omega L$ - zero sequence series impedance of the pipeline) of a buried pipeline could be determined using the Carson form where Z_{int} is the pipeline internal impedance:

$$\underline{Z} = \underline{Z_{int}} + \frac{\omega\mu_0}{8} + j\frac{\omega\mu_0}{2\pi} \cdot \ln\left(\frac{1.85}{a' \cdot \sqrt{\gamma^2 + j\omega\mu_0 \cdot \left(\frac{1}{\rho} + j\omega\varepsilon\right)}}\right) \quad (5)$$

As a result, the general expression of the induced voltages at a point x ($x = 0$ and $x = L_t$) on the parallel zone line-pipeline will be:

$$\underline{V}(x) = \frac{E_0 \cdot \underline{Z_A} \cdot \underline{Z_B}\left[\mathrm{ch}(\gamma \cdot x) - \mathrm{ch}\left(\gamma \cdot (L_t - x)\right)\right] + \underline{Z_C} \cdot \left[\underline{Z_B} \cdot \mathrm{sh}(\gamma \cdot x) + \underline{Z_A} \cdot \mathrm{sh}\left(\gamma \cdot (L_t - x)\right)\right]}{\gamma \cdot \left(\underline{Z_A} \cdot \underline{Z_B} + \underline{Z_C^2}\right) \cdot \mathrm{sh}(\gamma \cdot L_t) + \underline{Z_C} \cdot \left(\underline{Z_A} + \underline{Z_B}\right) \cdot \mathrm{ch}(\gamma \cdot L_t)} \tag{6}$$

The analytical expression for current density in soil is determined in a closed form solving the Helmholz equation with the method of variable separation for a quasistationary magnetic regime:

$$\underline{J}(x) = \underline{I} \cdot \int_0^\infty \left[\frac{\lambda - \sqrt{\lambda^2 + \gamma \cdot h^2}}{\pi \cdot h^2} \cdot e^{\frac{y \cdot \sqrt{\lambda^2 + \gamma \cdot h^2}}{h}} \cdot \cos\left(\frac{\lambda}{h} \cdot x\right) \cdot \boldsymbol{e^{-\lambda}}\right] \boldsymbol{d\lambda} \tag{7}$$

3. More precise evaluation of the induced voltages

The evaluation of the induced voltages starts by dividing the area of interest in a certain number of sections, in order to find the voltages at many positions along the pipeline. Figure 3 illustrates an example of a circuit model.

Each section is reduced (represented) by a π cell submitted to the influence of the electromotive force calculated according the method developed in [4] for normal operation (steady state condition) and fault conditions. The ideal generators introduced in the equivalent circuit simulates in fact the presents of the electromagnetic interference between power line and pipeline [4]. For every section are introduced a pair of ideal current and voltage generators. If $Y_k(x_k)$ are the pipeline admittance in r.u., in section k, then the transverse current circulating in the pipeline, produced by the ideal current generator, are $\underline{j}(x_k) = \underline{Y}(x_k) \cdot \underline{U}(x_k)$.

The transverse admittance depends on the location of the point $\boldsymbol{P}$ having the x_k^i coordinate:

$$\underline{Y_k^i} = \begin{cases} \frac{y_k \cdot l_k^i}{2} + \underline{\widetilde{Y_k^\iota}} \;(i = 1); \; \frac{y_k \cdot \left(l_k^{i-1} + l_k^i\right)}{2} + \underline{\widetilde{Y_k^\iota}} \;\; (i = 2, \ldots, n_k) \end{cases} \tag{8}$$

where $\underline{\widetilde{Y_k^\iota}}$ is the admittance which take into account every load connected in point i.

The transverse current produced by the ideal current generator from cell i at section k depends on the location of the point $\boldsymbol{P}$ having x_k^i coordinate on O_x axes:

$$\underline{J_k^i} = \begin{cases} \int_0^{\frac{x_k^i}{2}} \underline{j_k}(x_k) dx_k \;\; (i = 1); \; \int_{\frac{x_k^i + x_k^{i-1}}{2}}^{\frac{x_k^i + x_k^{i+1}}{2}} \underline{j_k}(x_k) dx_k \;\; (i = 2, \ldots, n_k) \end{cases} \tag{9}$$

The values of the induced voltages at the end of every cell and the induced current (the longitudinal current which circulate in the pipeline) are obtained solving the above system of equations for every cell:

$$\begin{cases} \left(\underline{Z_{0,1}} + \underline{Z_1} + \underline{Z_{1,2}}\right) \cdot \underline{I_1} - \underline{Z_{1,2}} \cdot \underline{I_2} = \underline{E_1} \cdot L_1 \\ -\underline{Z_{1,2}} \cdot \underline{I_i} \cdot \left(\underline{Z_{1,2}} + \underline{Z_2} + \underline{Z_{2,3}}\right) \cdot \underline{I_2} - \underline{Z_{2,3}} \cdot \underline{I_3} = \underline{E_2} \cdot L_2 \\ \quad \ldots \\ \quad \ldots \\ \quad \ldots \\ -\underline{Z_{n-2,n-1}} \cdot \underline{I_{n-2}} + \left(\underline{Z_{n-2,n-1}} + \underline{Z_{n-1}} + \underline{Z_{n-1,n}}\right) \cdot \underline{I_{n-1}} - \underline{Z_{n-1,n}} \cdot \underline{I_n} = \underline{E_{n-1}} \cdot L_{n-1} \\ -\underline{Z_{n-1,n}} \cdot \underline{I_{n-1}} + \left(\underline{Z_{n-1,n}} + \underline{Z_n} + \underline{Z_{n,n+1}}\right) \cdot \underline{I_n} = \underline{E_n} \cdot L_n \end{cases} \tag{10}$$

4. Study of a practical case

The modelling and evaluation of the induced voltages in a pipeline which is located on the same path with a power line is done for a monophase short circuit on the line and presented in Figure 4.

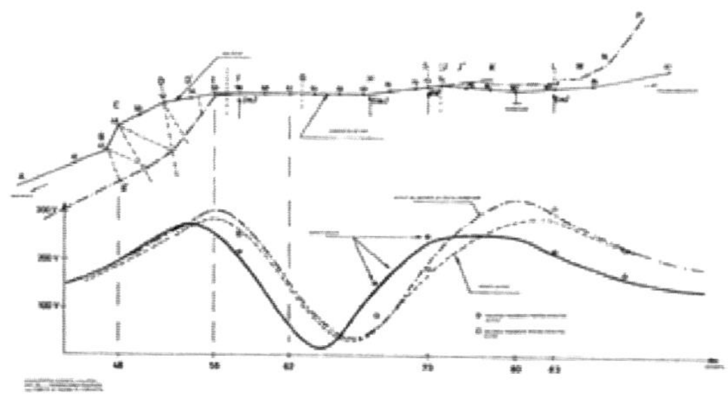

Figure 4. Line-pipeline that share the same path

The results could specify accurately the resistance and the adduction current in the pipe. At extremity L_1, with an electromotive force E_1, the influence of the line is equivalent with a voltage source with a value equal to the value of the voltage obtained with the formula obtained in section 2 and for $\underline{Z_B} = \propto$ and $\underline{Z_A} = \underline{Z_0}$:

$$\underline{V_B} = \underline{E_1} \cdot \frac{\underline{Z_A} \cdot (\mathrm{ch}(\gamma \cdot L_1) - 1) + \underline{Z_0} \cdot \mathrm{sh}(\gamma \cdot L_1)}{\gamma \cdot \left(\underline{Z_A} \cdot \mathrm{sh}(\gamma \cdot L_1) + \underline{Z_0} \cdot \mathrm{ch}(\gamma \cdot L_1)\right)} = -\frac{\underline{E_1}}{\gamma} \cdot (1 - e^{-\gamma \cdot L_1}) \tag{11}$$

The Z_A impedance is the circuit pipeline/soil impedance located at the left side of the A point. At the extremity L_2, with an electromotive force E_2, the equivalent voltage source is the sum of the sources given by the influence of AB and BC:

$$V_C = V_B \cdot e^{-\gamma L_2} + V_C^* \tag{12}$$

$$V_C^* = E_2 \cdot \frac{\underline{Z_B} \cdot (\mathrm{ch}(\gamma \cdot L_2) - 1) + \underline{Z_0} \cdot \mathrm{sh}(\gamma \cdot L_2)}{\gamma \cdot (\underline{Z_B} \cdot \mathrm{sh}(\gamma \cdot L_2) + \underline{Z_0} \cdot \mathrm{ch}(\gamma \cdot L_2))} = -\frac{E_2}{\gamma} \cdot (1 - e^{-\gamma \cdot L_2}) \tag{13}$$

with: $\underline{Z}_B = \underline{Z}_0$; $\underline{Z}_A = \underline{Z}_B = \underline{Z}_0$, so:

$$\underline{V_C} = \frac{\underline{E_1}}{\gamma} \cdot (1 - e^{\gamma \cdot L_1}) \cdot e^{-\gamma \cdot L_2} + \frac{\underline{E_2}}{\gamma} \cdot (1 - e^{-\gamma \cdot L_2}) \tag{14}$$

The induced potentials in the right and left side of the line are calculated and then, the superposition method is applied. The induced voltage is calculated using some predefined interpolation functions in Mathcad. Figure 5 shows the induced voltage calculation in different points on the pipeline for the measured data on S73 and S83.

Is observed that the interpolation spline function with linear end conditions is more precise than the linear interpolation polynom.

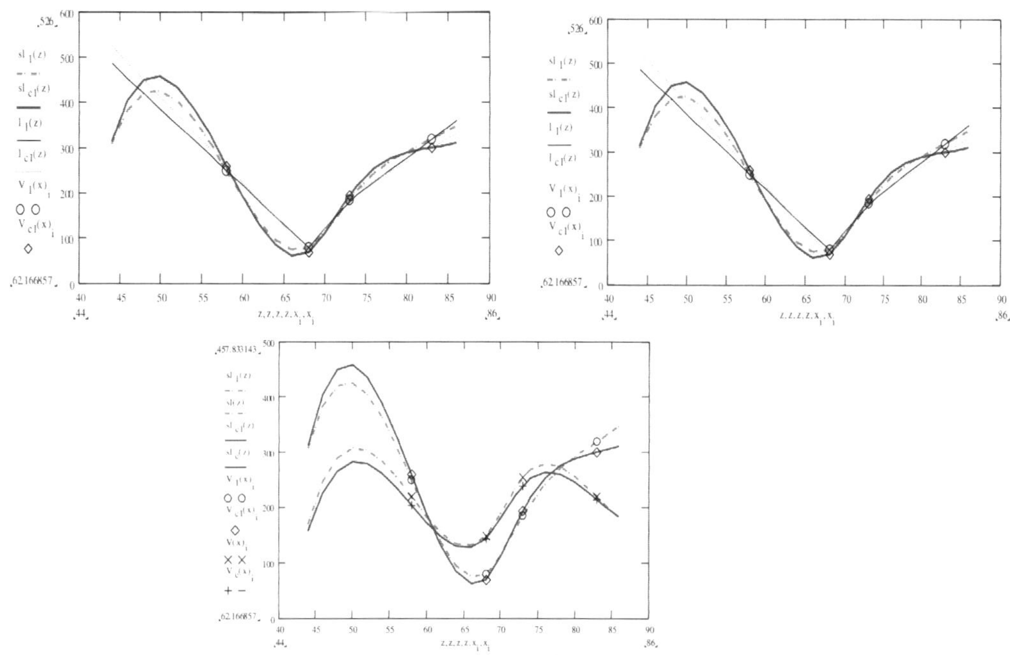

Figure 5. The induced voltages in pipeline. Interpolation functions

5. Conclusions

In the paper are evaluated the induced voltages in pipeline using the distribute source analysis approach. After the subdivision method is applied the superposition method used to compute the induced voltages. A practical case of subdivision of the zone of influence in sections is presented and a precise evaluation of the induced voltages is made using three kind of interpolation function. It is observed that the linear interpolation function gives the results close to the results measured for the induced voltage.

Acknowledgement

This work was supported by a grant of the Ministry of Education and Research, CNS/CCDI- UEFISCDI, project number PN-III-P3-3.6-H2020- 2020-0129/PPH no.44/2021.

References

[1] Lucca G and Zucchelli L2002 Electromagnetic interference produced on a complex metallic pipeline network by a fault in a power substation *CIGRE* Paris pp 36-01

[2] Popoli A Sandrolini L and Cristofolini A 2021 Reduction in the Electromagnetic Interference Generated by AC Overhead Power Lines on Buried Metallic Pipelines with Screening Conductors, *Electricity* **2** pp 316–329.

[3] Kim H S, Min H Y., Chase J G and Kim H2021 Analysis of Induced Voltage on Pipeline Located Close to Parallel Distribution System *Energies* **14** 8536

[4] Czumbil L, Christoforidis C G, Micu D D, Ceclan A, Stet D and Pop O 2011 A user – friendly Software Application for Induced A.C. Interference Evaluation *Proceedings of the 46th International Universities' Power Engineering Conference (UPEC)* 5-8 September Soest Germany

[5] Czumbil L, Mureşan A and Micu D D2020 Induced Currents and Voltages in Underground Metallic Pipelines due to nearby Power Lines *Virtual User Conference*

[6] Siecoban R C, Iobagel A M, Martis R, Nicu A I, 2015 Permanent Magnet Synchronous Machines for Small Energy-Efficient Applications *Proceedings of the E-HEALTH and Bioengineering Conference* Iasi, Romania.

[7] Popoli A, Cristofolini A and Sandrolini L 2021 A numerical model for the calculation of electromagnetic interference from power lines on nonparallel underground pipelines, *Math. Comput. Simul.*

[8] Racasan A, Munteanu C, Topa V, Pacurar C and Hebedean C 2017 Analysis and improvement techniques for the transfer function of a planar low - pass filter *Environmental engineering and management journal* **15** (12) pp 2579-2586

[9] Micu D D 2003 Numerical model for the calculation of the induced voltage in an underground pipeline *ANCME Gent*, pp 179-191.

[10] Laborelec Brussel 1990 *Resultats des measures et interference sur la ligne Neufchateau/Villers-sur-Semois e la canalization de gaz distrigaz Ben-Ahin/Athus* Brussel Research Report

IOP Conf. Series: Materials Science and Engineering 1254 (2022) 012013 doi:10.1088/1757-899X/1254/1/012013

Review on solar air heater hybrid systems thermal performances

V-C Ifrim, L-D Milici and C Bejenar

Faculty of Electrical Engineering and Computer Science,
Ștefan cel Mare University of Suceava, Suceava, Romania

E-mail: ifrimpv@gmail.com, dam@usm.ro, bejenar.ciprian@gmail.com

Abstract. As conventional fossil fuel systems release toxicants into the atmosphere, humanity is increasingly moving towards sustainability, towards using products that do not cause negative effects on the environment, have a long service life and offer multiple benefits. Solar air heating systems are increasingly used in household applications, even in industrial ones at times. The principle of operation of systems of this type is established on the existence of solar radiance, without this the panels have little or no efficiency at all, so they are limited to sunny periods. One of the methods of increasing the impact and performance of these systems is the realization of hybrid systems that offer several benefits with regard to heating the air in rooms, heating water and ventilating spaces. The meaning of this article is to present a summary of the existing literature, in the sphere of solar hybrid air heating systems, on the methods of improvement and the performance achieved.

1. Introduction

Solar air heating systems are technologies that rely on thermal energy. Within the panel the energy captured from the sun is absorbed by a surface and utilized to heat a fluid. Inside the panel there is an increase in temperature due to the influence of solar radiation and the isolated environment. This process is called greenhouse effect. The thermal energy captured by the absorbent material is transferred to the passing air flow through the system, thus achieving a heat transfer. Airflow convection can be natural or forced with the help of fans.

Although the energy used by this equipment is free of charge, some researchers have calculated their economic performance using the method of life savings analysis [1]. The method takes into account the value over time of money. Thus, the life-cycle cost (LCC) of a SAH model is outlined by the sum of the expenses linked with its period of operation in the current amount of money. The current value of the gain (PW) can be calculated with the formula:

$$PW = \sum_{N=1}^{N} \frac{Solar\ Economies}{(1+d)^N} \quad (1)$$

The Solar Economies of a system are computed each year by adding up the fuel, maintenance and tax savings associated with each year, by analogy with a conventional fossil fuel system. The letter N represents the number of each year that the solar savings are calculated for and d signifies the market discount rate in percent [1].

2. Solar Air Heater performance parameters

In order to optimize the solar air heating systems, their performance can be numerically calculated and experimentally measured considering the parameters exposed in Figure 1.

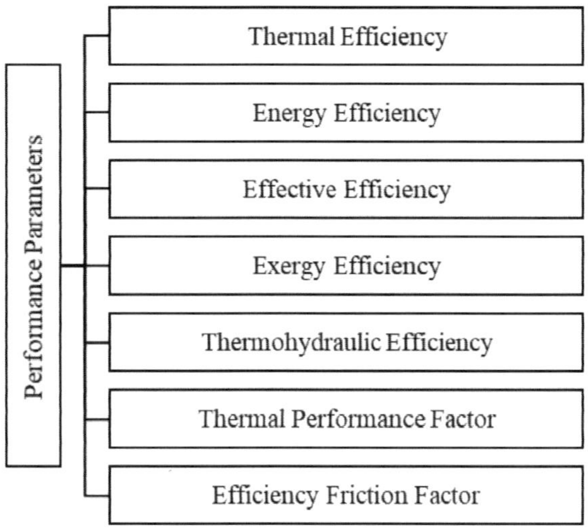

Figure 1. Performance parameters of Solar Air Heater Systems.

The thermal efficiency (η_{TE}) associated to SAH systems [2] can be assessed with the following equation:

$$\eta_{TE} = \frac{q_u}{I} = \frac{F_R[I(\tau\alpha)_e - U_L(T_i - T_a)]}{I} \qquad (2)$$

In the above expression (2) q_u represents the useful heat flow; I symbolize the intensity of the solar radiance on the absorption plate; F_R signify the temperature dissipation factor; $(\tau\alpha)_e$ is the effective transmission-absorption factor; U_L is the global coefficient of temperature loss; T_i is the temperature of the air and T_a the ambient temperature.

The energy efficiency (η_{EE}) of systems based solar air heater [3] shall be calculated with the following expression:

$$\eta_{EE} = \frac{Q_u}{A_{SAH} I_s} = \frac{m_a c_p (T_{out} - T_{in})}{A_{SAH} I_s} \qquad (3)$$

The elements in the formula (3) represents the following: Q_u is the temperature obtained usefully, m_a represents the mass flow rate of air; c_p signifies the specific thermal outcome of the air; T_{out} and T_{in} are the SAH temperatures of the air at the outlet and at the entrance respectively; A_{SAH} is the area of the equipment and I_s represents the solar radiation.

The effective efficiency (η_{Eff}) of solar air heating systems [4] is defined by the following relationship:

$$\eta_{Eff} = (Q_u - \frac{P_m}{C_j})/A_{SAH} I_s \qquad (4)$$

In equation (4) Q_u is the temperature obtained usefully; P_m represents the power required for airflow in W; C_j is the thermal conversion factor, with the suggested value of 0.2 in literature; A_{SAH} is the area of the equipment and I_s represents the solar radiation.

The exergy efficiency (η_{Exr}) of solar air heating systems [5] is defined by the following relationship:

$$\eta_{Exr} = 1 - (EX_{loss}/EX_{inlet}) \quad (5)$$

The elements of the above relation are representing: EX_{loss} the loss of exergy in the equipment and the EX_{inlet} the exergy admitted in the equipment.

The thermohydraulic efficiency (η_{Th}) of solar hot air heating systems [2] can be evaluated with the following equation:

$$\eta_{Th} = (Nu/Nu_s)/(f/f_s)^{\frac{1}{3}} \quad (6)$$

The Nu mentioned in the above equation (6) represents Nusselt's number for the proposed configuration; Nu_s is Nusselt's number for a conventional solar air heater; f represents the friction factor of the introduced configuration; f_s is the friction factor of a conventional solar air collector.

The parameter of thermal performances (TRP) assigned to SAH systems [2] shall be calculated using the following equation:

$$TRP = (T_o - T_i)/I_R \quad (7)$$

T_o mentioned in the formula (7) represents the temperature of the air at the outlet of the test section of the analyzed system; T_i signifies the temperature of the fluid (air) at the inlet to the test section of the analyzed collector and I_R represents the intensity of solar irradiance on the absorbing plate.

The friction improvement factor (EFF) for solar hot air heating systems [2]shall be calculated using the following equation:

$$FEF = f/f_s \quad (8)$$

For the above relation (8) f represents the friction factor for the considered configuration and f_s symbolizes the friction factor for a conventional solar air collector system.

3. Hybrid Solar Air Heater Systems

In order to improve the thermal performances of the solar air heating systems, researcher tried to integrate them with other renewable systems. Following the analysis of the research in which the following text“ solar air heater” has been encountered at least once during the interval of 2017-2021 from the “Web of Science” database, hybrid systems that contain solar air heating systems are distinguished through the types related in Figure 2 and Table 2. These categories are strictly based on the selected articles and do not comprehend the research period before 2017.

The absorption heat pump system based on semi-open membrane was developed for water heating applications [6]. This type of system has been developed and tested in the National Laboratory of Oak Ridge, Florida, USA in an environmental room that allows cold and warm atmospheres to carrying out various tests. Sensors of various types are also present. The component, most significant, of the system, is the absorption surface made of polymer, chemically treated to be hydrophobic, which determines the effectiveness of a SAHs. Heat and mass exchange is improved using this type of surface absorption and the surface-to-volume ratio increases twice by half the weight in distinction to previous studies.

Although the model presented is an experimental one and the manual valve control limits the introduction of a fixed air flow, the absorber surface allows more latent energy recuperation which brings on a higher thermal capacity of the solar collector. The use of ionic liquids prevents corrosion and crystallization of the absorption surface. This type of system has a water temperature at the outlet equivalent to 56°C and a coefficient of performance (COP) equal to 1.2.

A report of systems improvements acquired in this study compared with the results from a previous one can be seen in Table 1:

Table 1. Comparison between two SAH heat pump system.

Absorber characteristics		**Generation 1 (2017) [7]**	**Generation 2 (2019) [6]**
Number of panels used		4	7
Area of Active Surface (m^2)		0.42	0.92
Volume of Active Plate (m^3)		0.0084	0.008
Active Surface		50	115
	Water side	Metal (Cu/SS)	Polycarbonate
Materials	Sol-Water interface	SS sheet	CNC, thermal bonded
	Fins	1.5 mm	1 mm
Size of the Membrane Pore		1 µm	1 µm
		Not bonded on fins	Thermally bonded on fins

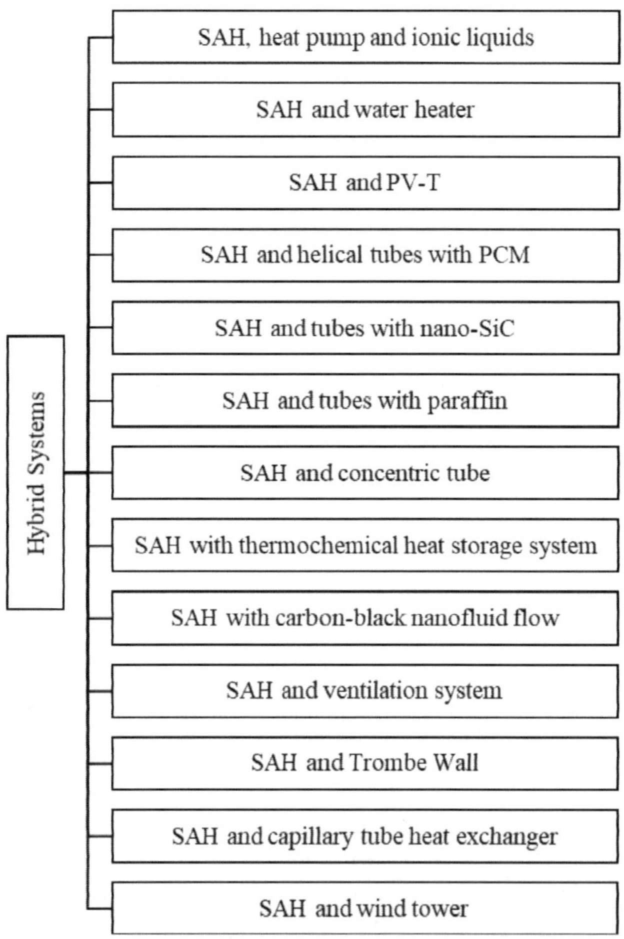

Figure 2. Classification of Hybrid Solar Air Heater Systems.

The purpose of a solar air heater and a water heater, both integrated in a hybrid system is to provide the pair of hot air and hot water at the same time [8]. The system can also only be used for hot air or for hot water. The numerical results of the conducted study indicated that 19.9% of the energy can be preserved only by covering the panel at night. The system performed, behaves efficiently for small air flows, having during the day a thermal efficiency equal to maximum of 82.8% for a jet air of 0.020 kg/s. The dual system has higher thermal efficiency in comparison with a simple water heating system. For the analyzed region, climate of Delhi, India, the maximum temperatures during the winter period were 56°C for water and 50°C for air and the maximum temperatures during the summer were 89.6°C for water and 81.1°C for air. However, this system is limited even ineffective for renewed, rainy or foggy days. These limitations can be improved by adding an electrical resistance.

The hybrid system realized by integrating a photovoltaic panel into a solar air heating system was experimentally investigated under natural meteorological conditions [9]. Manufactured system, PV/T, consists of a single-pass air channel with double flow. The integrated photovoltaic panel takes the place of the absorption surface for the solar air heating system. To augment the performance of heat transfer amongst the photovoltaic module and the air flow were attached to the module, ribs with uneven cross-section. The PV/T module can be seen in Figure 3 and the characteristics of the used PV module are exposed in Table 2.

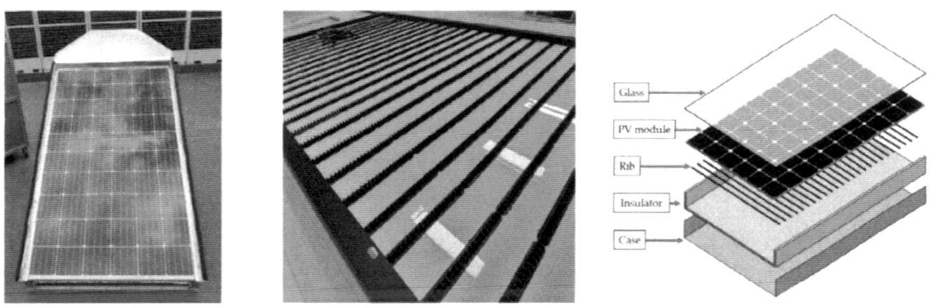

(a) Front view of the PV/T air collector (b) Transverse ribs installed at the back sheet of the PV module (c) Exploded view

Figure 3. Front view, absorber plate view and exploded view of PV/T[9].

Table 2. PV panel characteristics [9].

Parameter Label	Value
Output current at maximum power (A	9.46
Output voltage at maximum power (V)	32.75
Maximum output power (W)	310
Temperature coefficient (%/K)	-0.39
Electrical efficiency for maximum output power (%)	18.6
Number of cell (ea)	60
Module size (mm)	1670×1000×32

The experiment was performed for an air mass flow between 0.0198 kg/s and 0.07698 kg/s. As the air mass flow increased, it was found an improvement of the PV/T system in average thermal efficiency from 35.2% to 56.72% and in average electrical efficiency from 14.23% to 14.81%. The average total efficiency had values in the range of 49.44%-71.54% and represents the sum of electrical and thermal efficiencies. In conclusion, the maximum average total efficiency value for the hybrid system was 71.54% for the fluid mass flow rate of 0.07698 kg/s.

In order to improve a SAH system, a set of helical tubes loaded with material capable to store thermal energy, was integrated into the panel [10]. For the tests, two types of hybrid systems were made: one with tubes full of paraffin wax and one with tubes filled with a mix of Carbon powder added to paraffin wax. Both models of hybrid systems have been experimentally investigated underforced and natural convection airflow in the climatic environment of the city of Moradabad, India. The performance of the tube system filled with a mix of Carbon powder and paraffin wax were higher compared to the tube system filled with paraffin wax only. The thermal efficiency of the solar collector with a mix of material for energy storage was 79.10%, while the paraffin system achieved an efficiency of 57.41%. The maximum temperature at the exit of the solar system with tubes filled with Carbon powder added to paraffin wax, was 52.5°C.The values of the main parameters of the analyzed system indicate an optimal model for drying and heating operations of the premises.

For thermophysical performance, the hybrid system of SAH with pieces of aluminum and pipes filled with mix of nano-SiC and paraffin wax was realized [11]. The thermophysical features of the complex compound have been examined by adding diverse ratios of SiC nano-particles equal to 1, 2 and 3%. For the thermophysical tests the outcomes indicated a progress in the density and viscosity of the composite compound. Thermal conductivity improved in proportion to the mass of the added nano-particles to paraffin wax. When a 3% fraction of nano-SiC particles was mixed with paraffin wax, the increase in density and viscosity were 2% and 3.2%, respectively. Enhancement of thermal conductivity for the 3% fraction of nano-SiC particles within paraffin wax increased to 18.2%.

In order to mitigate the energy crisis and the problem of global warming, a solar air heater system is completed with exhaust tubes as an absorption surface and paraffin wax as a thermal storage component [12]. For a variation of the outside average temperature from -5.7°C to +36.2°C the efficiency of thermal storage during the analysis period varied from 56.1% to 67.5%. The values of 98.5% and 1268.8 W were acquired by the thermal efficiency and the average thermal extraction power, respectively. The energy conversion process and the numerical adjustments outcome that the average charging efficiency had a linear gradient with the normalized temperature variation. The average temperature at the output of the system and the average extraction power had a linear slope with the temperature at the inlet to the system and exponentially with the fluid mass flow. For high thermal performance, the system was completed with an integrated storage system specific to a building of 9.9 m^2. The benefits of the proposed system indicated annual energy savings and CO_2 reduction throughout its lifetime.

Concentric tube geometry was utilized to enhance the heat transfer process in a cooling system [13]. The paper presents mathematical modeling, simulation, design and manufacture of four solar systems containing concentric tubes for experimental analysis. For experimental performance, a removal factor was added, with the help of which the performance of the system increased to 68%.

Thermochemical materials include sorption and thermochemical reactions. This type of materials stores energy after the dissociation reaction and releases energy through a reverse chemical reaction. Thermochemical storage has the highest storage density, thus allowing the storage of a large amount of energy using small amounts of substances.

Storing thermal energy for a long time with the help of thermochemical is a promising and future method in the area of solar air heater systems. The method allows to store and release heat by reversible adsorption-desorption reactions, which in the direction of hydration are exothermic and in the reverse direction, of dehydration are endothermic. A 250-litre configuration based on the gas-solid reaction amongst water-zeolite 13X has been designed and investigated. Moist air is injected into a reactor filled with dehydrated component and due to the absorption of water vapor on the thermochemical material, heat is discharged. The reactor is made of four segments of 62.5 liters each that can be operated in several ways. The temperature is determined at distinct locations in the reactor to obtain a valid perspective on the effect of segmentation. Tests are carried out in cycles of hydration-dehydration, in various ways. The total efficiency was around 67% representing the energy liberated all along rehydration divided by the energy provided all along dehydration [14].

The solar air heating system with atmospheric aerosol-based nanofluid utilizing black carbon nanoparticles was analyzed for three types of absorption surfaces: triangularly corrugated, rectangular corrugated and sinuously corrugated in turbulent flow mode, Reynolds number varied in interval of 2500 and 4000. The outcome showed that for all periods of the year, the maximum performance evaluation factor for the sinus wavy absorption surface model was obtained. The greatest increase in temperature from input to output was obtained for SAH with a rectangular wavy absorption surface. For the sinusoidal wavy absorber surface system, the heighten in the volume fractions of the nanoparticles augments the coefficient of thermal performance to 1.9. The best thermal enhancement conditions were achieved at Reynolds number equal to 4000 and a 0.1% of volume fraction. For the solar system with rectangular wavy absorber surface, the use of nanofluid for the entire Reynolds range analyzed, leads to a reduction in the output temperature [15].

Ventilation of establishments during the winter period, particularly in areas with cold atmospheric conditions, demands a fairly amount of energy to preheat fresh ventilation air. To ensure the ventilation of low-cost houses, it is possible to preheat fresh air from the outside, before entering the house with the help of solar air heating systems. The system is made of solar air heaters, installed on the roof of the constructed dwelling, hot air diffusion ducts, indoor air traffic ducts and air circulation fans. In the warmth periods, the equipment can be used at night to circulate cold air from the outside through the interior space of the establishment. Such operation enables to cool the indoor air and the construction of the dwelling. As a consequence, the house accumulates the cold natural air from the outside at night to use it for daytime cooling [16].

The natural flow in a solar system was analyzed with a Thrombus wall and a Bell-Mouth channel. The thermal improvement factor was 1.37 and the thermohydraulic performance was 72%. The proposed and analyzed hybrid system, a Bell-Mouth channel heating system integrated with Trombe wall, has a low initial cost with insignificant operational and operating charges, therefore, the system has a short depreciation period equal to 4 months compared to a conventional fossil fuel-based active heating system [17].

In order to preheat the air and improve the comfort of an industrial building used as a factory, a system containing a transpired solar air heater with heat exchanger type capillary tube was experimentally realized [18]. The system allows the use of heated air based on solar radiation and a heat exchange system, when solar energy is missing or does not provide enough thermal energy. The main components of the realized system are: integrated plate for the heat exchange between SAH and the compressor (5), capillary tubes for heat exchange between cold air and hot water (2) and transpired solar air heater (1). All the components of the system can be observed in Figure 4.

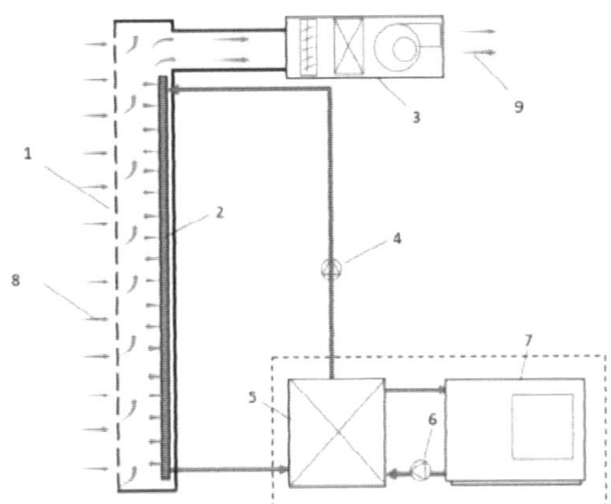

Figure 4. Transpired solar air heater with heat exchanger type capillary tube [18].

The other equally important components of the system are: the cold air inlet system (8), the hot air output system (9) with the module that performs the air flow (3), the air compressor (7), the integrated oil pump (6) that performs the heat exchange between the compressor and the integrated plate, and the integrated water pump that recirculates the water in the capillary tubes. The compressor used in the experiment is an industrial one that also offers the possibility of recovering thermal energy with the help of the recovery unit built into it. Thus, the built-in system achieves two heat exchanges: a heat exchange occurs amongst the thermal energy of the oil within the compressor and the water inside the capillary tubes through the compressor recovery module; the second heat exchange occurs between the energy in the capillary tubes and the air outside. The system works using two pumps that recirculate the water and the oil and a fan that performs the air flow. The system can also be viewed from the perspective of cooling the compressors in this way during operating, if there are several. The integration of a Building Management System (BMS) allows the monitoring and analysis of system performance. The capillary heat exchanger achieved an efficiency of up to 0.7, the average efficiency in using thermal losses was 52.3% and the maximum efficiency reached 85%. This study indicates that thermal energy, which is most often released into the environment, can be recovered and used for space heating [19-20].

Another research involving hybrid systems realized a combination between a wind tower and a SAH to assist a humidification-dehumidification system used to desalinate water which works in an open-air water cycle [21]. In this way, the use of fans is eliminated, the air flow for SAH being achieved through the wind tower. The system was numerically analyzed to test the cooling effect and the production of fresh water aiming the thermal comfort. The system was tested for different mass flows existing in the tower and for different heights of the tower. The height of the 5 m tower changed the humidity in the system from 40% to 80%, lowered the air temperature by 29%, achieved a 3.5 kW cooling effect and a 42% reduction in water necessity to desalinate. The air flow with values between 0.032 and 0.035 kg/s achieved for the open-air water cycle a maximum productivity of 4.2 kg/day. The airstream for water cooling in the dehumidifier with values from 0.035 to 0.038 kg/s achieved a desalination productivity of 218% for the closed-cycle and 233% for the open-air water cycle. For the open cycle system, an increase in productivity can be achieved by 110% if the absorber area of the solar air heating system is enlarged from 0.5 m^2 to 2 m^2.

For the hybrid systems exposed above, Table 3 contains the type of the system, references, a representation of the system, the operating and optimized parameters and the resulted efficiencies.

Table 3. Hybrid Solar Air Heater Systems.

No.	SAH Type/ Reference	Representation of the system	Operating Parameters	Optimized Parameters	Key Results
1	SAH, heat pump and ionic liquids [6]	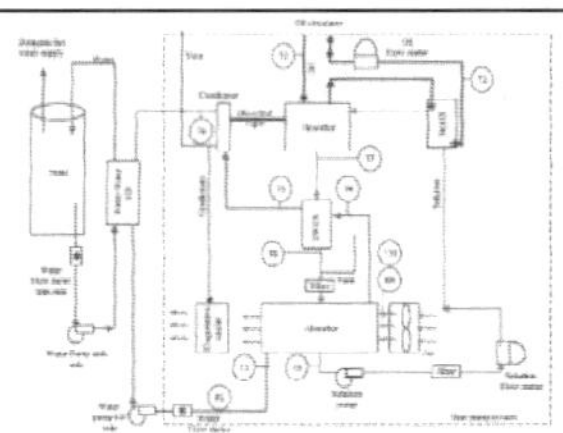	UEF test: 19°C, 46%RH; Space test: 21°C, 51%RH; Humid test: 25°C, 74%RH.	Uniform Energy Factor (UEF) test: 19°C, 46%RH;	Coefficient of performance (COP): 1.2
2	SAH and water heater [8]	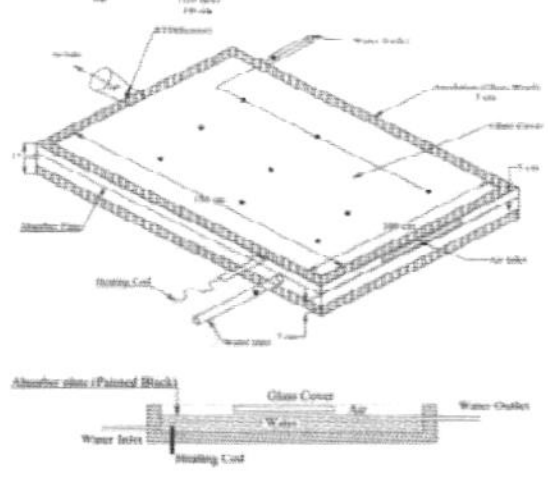	m = 0.005; 0.020 kg/s; I_R = 637; 1011 W/m^2.	m = 0.005 kg/s; I_R = 1011 W/m^2; T_{water} = 89.6°C; T_{air} = 80.1°C.	Thermal Efficiency: 82,8%.

3	SAH and PV-T [9]	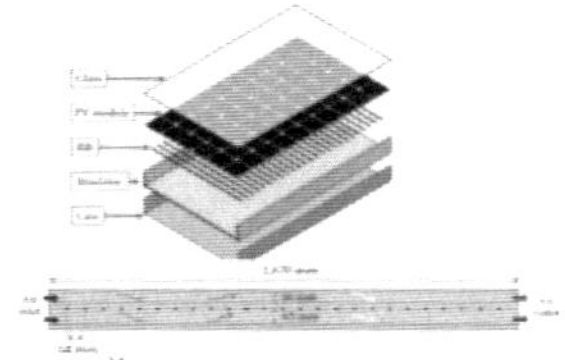	m = 0.01977-0.07698 kg/s; I_R = 186.40-638.52 W/m^2.	m = 0.07698 kg/s; I_R = 506 W/m^2; Total Average Efficiency: 71.54%.	Thermal Efficiency: 59.91%; Electrical Efficiency: 16.27%.
4	SAH and helical tubes with mixed PCM [10]	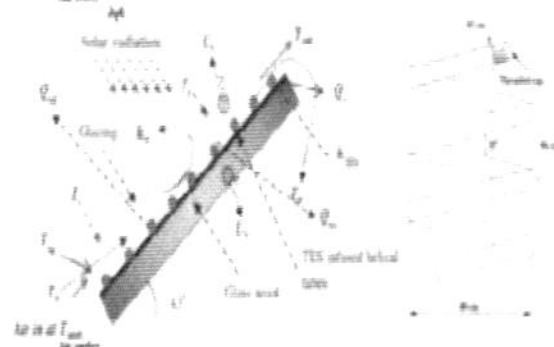	m = 0.061; 0.60 kg/s.	m = 0.60 kg/s; h = 411.05 W/m^{2}k; Q_u = 2412 W.	Thermal Efficiency: 79.10%.
5	SAH and tubes with nano-SiC [11]	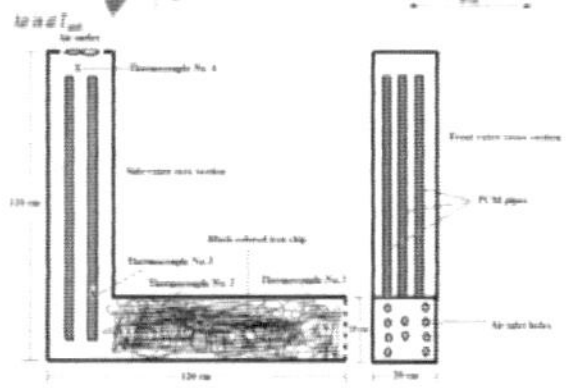	Nano-Sic = 1; 2; 3 wt%;	Nano-Sic = 1; 2; 3 wt%;	Thermal Conductivity 18.2%.
6	SAH and tubes with paraffin [12]	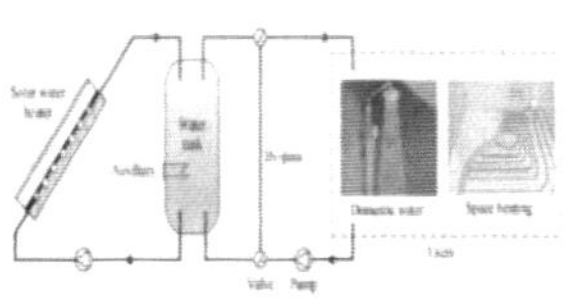	I_R = 684.0-889.0 W/m^2; T_{out} = -7.0-3.9°C.	m = 0.07801 kg/s; Solid $\eta_{st,m}$= 48.2%; Melting $\eta_{st,m}$= 76.3%; Liquid $\eta_{st,m}$= 41.9%.	Thermal storage Efficiency: 67.5%; Released Efficiency: 98.5%.
7	SAH and concentric tube [13]		-	-	Efficiency: 68%.
8	SAH with thermochemical heat storage system [14]	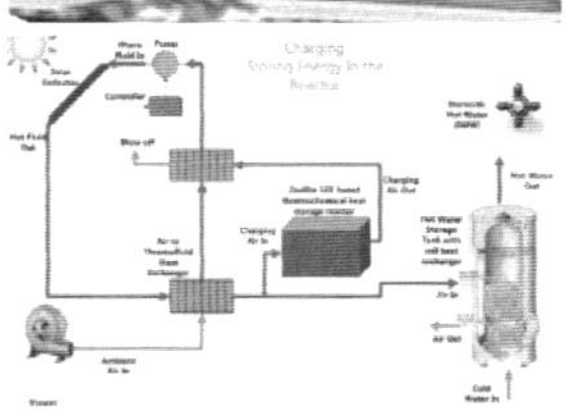	Zeolite 13XBF = 62.5; 125; 187.5; 250 L.	m = 0.05kg/s; $T_{reactor}$ = 20°C; Zeolite 13XBF = 62.5 L; Maximum Power = 0.9 kW.	Constant Effectiveness: 90%. Maximum Power (250 L): 3.6 kW.
9	SAH with carbon-black nanofluid flow [15]	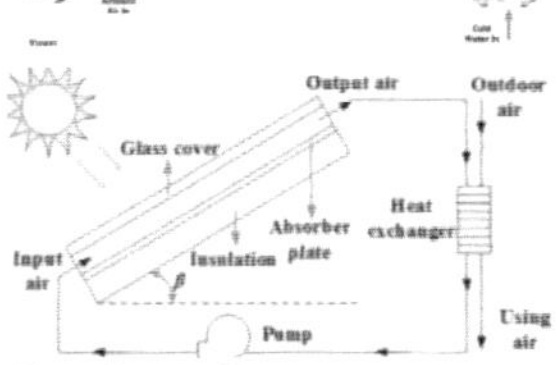	Re = 2500-4000; ϕ = 0.01-0.10%.	Re = 2500; ϕ = 0.10%.	Heat Performance Coefficient: 1.51.
10	SAH with ventilation system [16]	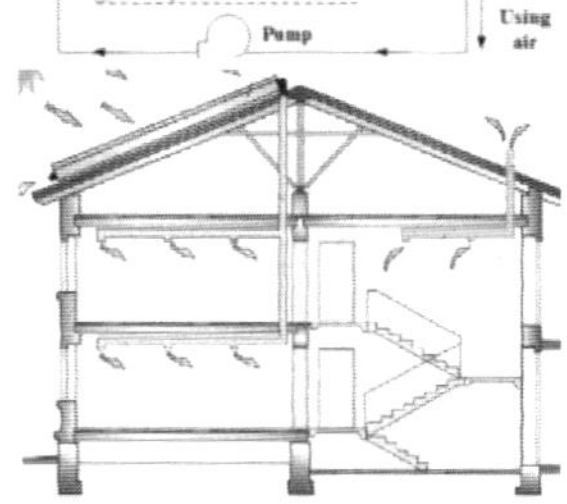	δ = 0.017; 0.034 m; ω = 0.5; 0.25 m/s.	l = 5 m; b = 4.7 m; δ = 0.034 m; ω = 0.25 m/s.	G_V: 0.0486 kg/s; Q: 1945 W.

11	SAH with Trombe Wall [17]	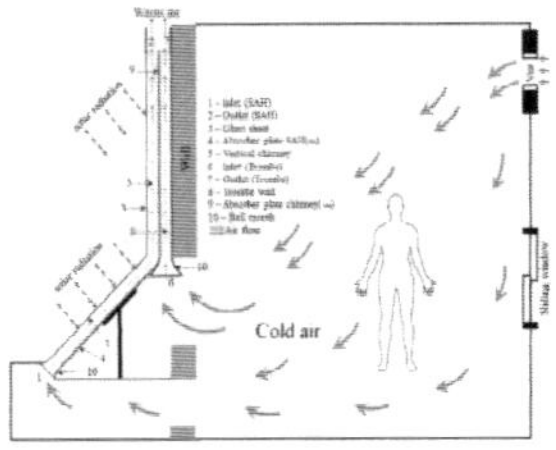	m = 0.016; 0.021; 0.024; 0.0696; I_R = 500-1100 W/m^2; h/R = 0.1-0.6;	I_R = 1000 W/m^2; h/R = 0.3; R = 100 mm;	Thermo-hydraulicEfficiency: 72%.
12	SAH with capillary tube heat exchanger [18]	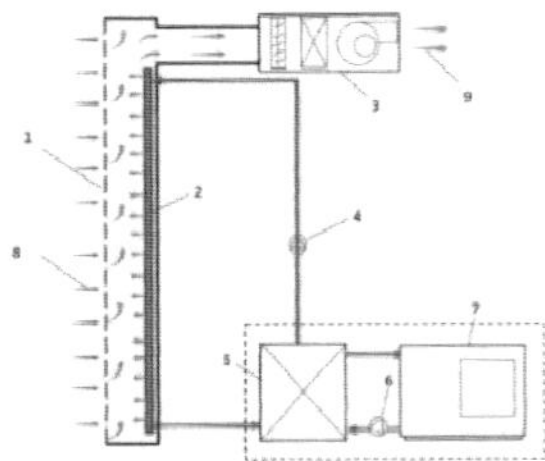	m = 0.00461; 0.005; 0.00689; 0.00919; 0.0115; 0.01381;0.01839 kg/s.	m = 0.005 kg/s;	Thermal Efficiency: 85%; Capillary tube Effectiveness: 0.7; Average waste Efficiency: 52.3%.
13	SAH and wind tower [21]	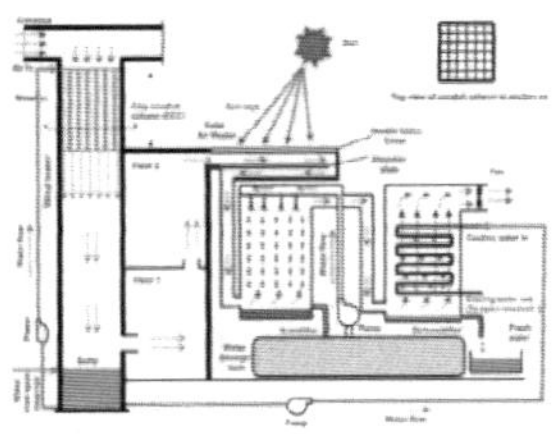	m = 0.032-0.038 kg/s.	h = 5 m; m = 0.032-0.035; 0.035-0.038 kg/s	Cooling effect: 3.5 kW; SAH Productivity: 4.2 kg/day; Dehumidifier Productivity: 200%.

4. Conclusions

Among the hybrid systems that consist of solar air heating systems mentioned in this review, the PV/T system [9] and the system with heat exchangers [18] are up-to-date and of high interest [22-23].

The large-scale PV/T system can solve three of the current problems in the field of renewable resources, namely: sustainable thermal energy with the help of SAH, electricity supplied by PV panels and ventilation of panels to reduce losses through the Joule effect. Imagine what we could achieve by combining this PV/T system with paraffin filled tubes [12]. The fourth problem would be accomplished, hot water at your discretion. The transpired solar air heater with heat exchanger type capillary tube offers and inspires the possibility of recovering and using a thermal energy, which would otherwise be lost. Most of the equipment that uses oil for cooling could be used to make an amount of thermal energy.

There we have two future perspectives: PV/T with paraffin filled tubes and power transformers providing thermal energy through heat exchangers.

Nomenclature

Meanings of Table parameters are the following:

IR – Solar radiation (W/m^2);
UEF – Uniform Energy Factor;
COP – Coefficient of performance;
m – Mass flow rate (kg/s);
h – Heat transfer coefficient (W/m^{2}k);
Qu – Energy released at constant volume (W).
Nano-Sic – nanoparticle ratio (wt%);
Solid $\eta_{st,m}$– mean efficiency when the PCM absorbs thermal energy;
Melting $\eta_{st,m}$–mean efficiency when the PCM changes its phase;
Liquid $\eta_{st,m}$– mean efficiency when the PCM releases thermal energy;
Re – Reynolds number;
ϕ – Nanoparticles fraction (CBNP) (%);

G_V – Quantity of ventilated airflow from outside (kg/s);
Q – Heater thermal productivity (W);
l – length, b – width, δ – height of double-glazed heater (m);
ω – air velocity (m/s);
h/R – bell-mouth factor range;
R – bell-mouth opening (mm);
h – column (tower) height (m).

References

[1] Kalogirou AS 2004 Solar thermal collectors and applications *Progress in Energy and Combustion Science* **30** 231295 https://doi.org/10.1016/j.pecs.2004.02.001

[2] Mund C,Rathore SK and Sahoo RK 2021 A review of solar air collectors about various modifications for performance enhancement *Solar Energy* **228** 140167 https://doi.org/10.1016/j.solener.2021.08.040

[3] Reddy J,Roy S,Das B and Jagadish2021 Performance evaluation of sand coated absorber based solar air collector *Journal of Building Engineering* **44** 102973 https://doi.org/10.1016/j.jobe.2021.102973

[4] Phu MN, Thao BP and Hap VN 2021 Effective efficiency assessment of a solar air heater having baffles spaced with different successive ratios *Case Studies in Thermal Engineering* **28** 101486 https://doi.org/10.1016/j.csite.2021.101486

[5] Vengadesan E and Senthil R 2020 A review on recent developments in thermal performance enhancement methods of flat plate solar air collector *Renewable and Sustainable Energy Reviews* **134** 110315 https://doi.org/10.1016/j.rser.2020.110315

[6] Chugh D,Gluesenkamp KR,Abu-Heiba A, Alipanah M,Fazeli A,Rode R,Schmid M,Patel V, Kand Moghaddam S 2019 Experimental evaluation of a semi-open membrane-based absorption heat pump system utilizing ionic liquids *Applied Energy* **239** 919927 https://doi.org/10.1016/j.apenergy.2019.01.251

[7] Chugh D,Gluesenkamp KR, Abdelaziz O and Moghaddam S 2017 Ionic liquid-based hybrid absorption cycle for water heating, dehumidification and cooling *Applied Energy* **202** 746754 https://doi.org/10.1016/j.apenergy.2017.05.161

[8] Somwanshi A and Sarkar N 2020 Thermal performance of a dual-purpose collector-cum-storage type air-water heater *Applied Thermal Engineering* **171** 115094 https://doi.org/10.1016/j.applthermaleng.2020.115094

[9] Choi HU and Choi KH 2020 Performance Evaluation of PV/T Air Collector Having a Single-Pass Double-Flow Air Channel and Non-Uniform Cross-Section Transverse Rib *Energies* **13** 2203 https://doi.org/10.3390/en13092203

[10] Saxena A,Agarwal N and Cuce E 2020 Thermal performance evaluation of a solar air heater integrated with helical tubes carrying phase change material *Journal of Energy Storage* **30** 101406 https://doi.org/10.1016/j.est.2020.101406

[11] Jawad QA,MahdyAMJ,Khuder AH and Chaichan MT 2020 Improve the performance of a solar air heater by adding aluminum chip, paraffin wax, and nano-SiC *Case Studies in Thermal Engineering* **19** 100622 https://doi.org/10.1016/j.csite.2020.100622

[12] Wang Z Diao Y Zhao Y Chen C Liang Land WangT 2020 Thermal performance of integrated collector storage solar air heater with evacuated tube and lap joint-type flat micro-heat pipe arrays *Applied Energy* **261** 114466 https://doi.org/10.1016/j.apenergy.2019.114466

[13] Acuña A,Velázquez N,Sauceda D and Aguilar A 2017 Modeling, Construction, and Experimentation of a Compound Parabolic Concentrator with a Concentric Tube as the Absorber *Journal of Energy Engineering* **143** 04016059 https://ascelibrary.org/doi/abs/10.1061/%28ASCE%29EY.1943-7897.0000416

[14] Gaeini M, Alebeek van R, Scapino L, Zondag H A and Rindt C C M 2018 Hot tap water production by a 4-kW sorption segmented reactor in household scale for seasonal heat storage *Journal of Energy Storage* **17** 118128 https://doi.org/10.1016/j.est.2018.02.014

[15] Sadripour S 2019 3D numerical analysis of atmospheric-aerosol/carbon-black nanofluid flow within a solar air heater located in Shiraz, Iran *International Journal of Numerical Methods for Heat & Fluid Flow* **29** 13781402 https://doi.org/10.1108/HFF-04-2018-0169

[16] Melikyan Z, Egnatosyan S and Egnatosyan N 2019 Developing of a ventilation system for residential houses with solar air heater *Web Conference* **97** 06017 https://doi.org/10.1051/e3sconf/20199706017

[17] Singh A P, Kumar A, Akshayveer and Singh O P 2021 Effect of integrating high flow naturally driven dual solar air heaters with Trombe wall *Energy Conversion and Management* **249** 114861 https://doi.org/10.1016/j.enconman.2021.114861

[18] Akhan H and Eryener D 2018 Building integrated solar air heating with waste heat utilization *Energy Conversion and Management* **157** 136145 https://doi.org/10.1016/j.enconman.2017.12.007

[19] Milici L D and Milici M R 2013 Consideration about the evolution of performance in nature and technology *European Journal of Science and Theology* **9** 157166 http://www.ejst.tuiasi.ro/Files/35/ygvrty16qmpk.pdf

[20] Rață M, Rață G, Milici D and Graur I 2009 An efficient solution of the step-down converter for students teaching *Elektronika ir Elektrotechnika* **91** 7780 https://www.eejournal.ktu.lt/index.php/elt/article/view/10606

[21] Sachdev T, Gaba V K and Tiwari A K 2020 Performance analysis of desalination system working on humidification-dehumidification coupled with solar assisted air heater and wind tower: Closed and open water cycle *Solar Energy* **205** 254262 https://doi.org/10.1016/j.solener.2020.04.083

[22] Cernușcă D, Milici D L and Poienar M 2016 Development of special actuators using smart materials *International Conference and Exposition on Electrical and Power Engineering* 267570 https://doi.org/10.1109/ICEPE.2016.7781404

[23] Atănăsoae P, Pentiuc R D, Milici D L, Olariu E D and Poienar M 2019 The cost-benefit analysis of the electricity production from small scale renewable energy sources in the conditions of Romania *Procedia Manufacturing* **32** 385389 https://doi.org/10.1016/j.promfg.2019.02.230

Measurements of Human Exposure to EMF from 4G systems: some experimental issues in urban environments

G Betta[1], D Capriglione[1], G Cerro[2], G Miele[1], D S˘uka[1, 3] and M Ruttner[4]

[1]Dept. of Electrical and Information Engineering, University of Cassino and Southern Lazio, Cassino, Italy
[2]Dept. of Medicine and Health Sciences "V. Tiberio", University of Molise, Campobasso, Italy
[3]Faculty of Electrical Engineering, University of East Sarajevo, East Sarajevo, Bosnia & Herzegovina
[4]A1 Telekom Austria AG, Vienna, Austria

E-mail: betta@unicas.it; capriglione@unicas.it; gianni.cerro@unimol.it; gianfranco.miele@unicas.it; darko.suka@etf.ues.rs.ba; manfred.ruttner@a1.at

Abstract. Assessments of human exposure to EMF in urban environments represent one of the most complex tasks because of the factors causing large temporal variations and spatial non-uniformities of the measurement results among which are increased number of RF sources, numerous reflections, and scattering objects. In this sense, human exposure to 4G systems became a real challenge due to its current wide spreading. Following the guidelines provided by the relevant technical standards in force, this paper describes the experimental results achieved by performing a long-term analysis for several days and weeks on three network operators and three frequency bands adopted in Italy for 4G communications. In particular, the stability of the 4G pilot signal levels, i.e. PBCH, is analyzed. Indeed, such a level is involved in the maximum extrapolation techniques which should always warrant worst-case and time-independent estimations of the maximum E field. The effects of the averaging time on raw data is evaluated for quantifying its effect on the possibility of reducing the variability of human exposure assessment based on the use of the extrapolation techniques. Another important issue that is related to exposure safety requirements is the determination of compliance boundaries from RBS transmitting antennas (when put into service), mainly estimated by measurements and calculations (or simulations) for the areas accessible by the general public. In this regard, here are illustrated examples for measurement and simulation approaches for a roof terrace exposure conditions at two different locations in Austria.

1. Introduction

The continuous spread of 4th generation (4G) cellular networks is in close connection with humans' growing need for using communication devices exploiting the 4G assigned frequencies. From one side, the high variations of 4G data and voice traffic (also due to the multipath propagation [1]) make reliable human exposure forecast very difficult. On the other side, an objective forecast of EMF (Electromagnetic Field) levels is important. To mitigate these issues and to control the impact of EMFs on the population, the standards (international and national ones) that limit the exposure of different RF-EMF (Radiofrequency-Electromagnetic Field) sources (both in the near- and far-field zones) are recently redefined [2-3]. Since most users are primarily exposed to the EMF radiation of their mobile phones (near-field exposure), EMF levels coming from them are of huge interest to the scientific community. In that sense, some factors (not only the SAR parameter) such as the device's instantaneous radiated power, a duty cycle of the signals, or its crest factor (CF) have to be taken into consideration during various operation situations. Some investigations aimed to identify the gaps in the time-amplitude variation of the emitted signals from devices connected to 4G-LTE (Long Term Evolution) and 5G-NR (New Radio) cellular network in very close proximity to the human body are made in the previous two years for completion of new metrics considering external physical quantities, such as Instantaneous Power

Density (IPD) [4-6]. In addition, a comprehensive understanding of the user exposure's dynamics could be achieved by applying the method based on CCDF (Complementary Cumulative Distribution Function) information in conjunction with a signal analyzer and a specific-developed software [7].

Similar to questioning the near-field exposure assessment methods, in more recent literature [8-11], it has been illustrated how factors such as instrument settings and time intervals in which measurements are recorded, can largely affect the measurement results and their repeatability. In some cases, such effects were comparable to or even higher than typical uncertainty components of the measurement chain. Additionally, in [8-11] has been proved how the variability of EMF levels during the day and week, can be considered as either bias and/or measurement uncertainty components, thus making the actual human exposure evaluation unpredictable or unreliable. In particular, starting from the technical standards in force [12-13], an improved measurement and post-processing procedure was proposed, under the assumption that a basic Spectrum Analyzer (SA) is adopted as a measuring instrument. The focus has been devoted to examining the pilot signals' power consistency (i.e., stability analysis of the PBCH, Physical Broadcast Channel) for three operators and three frequency bands.

Another important issue that is related to exposure safety requirements is the determination of compliance boundaries from RBS (Radio Base Station) transmitting antennas (when put into service), mainly estimated by measurements and calculations (or simulations) for the areas accessible by the general public. In principle, determination of the compliance boundaries can be determined by different methods described in [12]. Especially, in cases where measurements involve considerable efforts, usually in terms of time (i.e. long-term field analyses are not possible) different approaches could be taken. This considers the fact that the calculation of the safety distances from RBS antennas could be determined in advance. Such approach also involves the visual inspection of the site by checking whether there is a publicly accessible area within the compliance boundary or not. For this evaluation, a multi-stage procedure can be used, always considering relevant sources, as required by [12]. For example, the corresponding compliance distances are calculated according to the box-shaped compliance boundary as defined in Section 6 of [12]. This procedure has to be done for every new RBS and relevant upgrades. If accessible areas are identified in such cases, either an alternative position for the new RBS is planned, or compliance can be demonstrated using a simulation tool calculating the compliance boundary as an iso-surface considering precise antenna parameters and environmental conditions. However, the limitations of simulation tools primarily lie in the exact replication of the environment as a complete 3D scenario such as buildings, rooftops, trees, etc. Therefore, in individual cases, it is unavoidable to perform measurements. Besides the necessary access to the measurement point, the situation could become even more difficult in case of conducting measurements in private facilities such as houses, flats, private roof terraces, etc. In situations of detailed analysis of compliance it must be ensured by the suitable methods, given in [12], that the reference levels are complied with the safety limits at any time of a day (taking into account the averaging time of the relevant guideline or standard) and in any possible operating state of the RBS.

Since a long-term measurement is hardly possible in these situations, a constant part of the signal is measured (i.e. PBCH if basic SA is used or CRS (Cell Specific Reference Signal) in case of VNA (Vector Network Analyzer)), and the actual or theoretical maximum possible exposure is calculated based on it. In case the pilot signal PBCH is observed (as done in this paper) for EMF exposure assessment purposes, it is of great importance to derive its better stability using different trace processing and averaging techniques which will be described in detail in the following subsections.

2. Materials and Experimental procedure

2.1. *Testing scenarios for experimental long term-analysis*

For analyzing the RF-EMF emitted by a 4G LTE RBS it is necessary to focus the attention to the LTE downlink (DL) physical signal. This signal adopts an OFDM (Orthogonal Frequency Division Multiplexing) scheme which divides the transmission bandwidth into several subcarriers with equal spacing (15 kHz), and each subcarrier can be modulated with a different modulation scheme [14]. The smallest entity that can be scheduled in the frequency domain is the Resource Block (RB) and it is composed of 12 consecutive subcarriers, with a bandwidth equal to 180 kHz.

As DL transmissions regards, LTE standard defines many physical channels. Among them, the PBCH is considered for RF-EMF intensity evaluation, because the power associated with the PBCH is independent of the channel bandwidth of the LTE signal. This is due to the fact that it occupies a bandwidth of 6 RBs, approximately equal to 1 MHz, and it is at the center frequency of the LTE signal.

As the LTE human exposure evaluation concerns, [12] introduces an extrapolation techniques, for the assessment of the maximum exposure level from an LTE RBS, which makes use of a basicSA equipped with zero–span measurement mode for measuring the PBCH power (P_{pbch}).

The SA settings suggested by [12] for measuring the PBCH power are described below:

- set the center frequency equal to the LTE signal center frequency;
- set the frequency span equal to 0 Hz (zero–span mode);
- set the RBW (Resolution Bandwidth) equal to 1 MHz, for assuring to integrate the entire PBCH;
- set the sweep time equal to 70 µs times the number of trace points for assuring an integration time approximately close to the symbol duration for each trace point;
- set the RMS detector;
- set the max–hold functionality selected with a minimum time equal to 20 s.

Applied the above–mentioned settings, P_{pbch} is equal to the measured peak value of the displayed trace. It is worth noting that the PBCH measurement is closely connected to the maximum field extrapolation analytical formulation, reported in eq. 1.

$$E_{max} = \sqrt{nPBCH}E_{PBCH} \quad (1)$$

where n_{PBCH} is a constant multiplication coefficient taking into account the signal bandwidth and E_{PBCH} is the electric field associated to the acquired PBCH signal power.

The measurement set-up reported in Figure 1 has been adopted to perform the experimental campaign able to gather data coming from three MNOs (Mobile Network Operators, later simply named OP (Operators)), operating in Italy and simultaneously trasmitting from the same RBS. Each of the MNO has three frequency bands where to deliver its signal, 800 MHz (B1), 1800 MHz (B2), and 2600 MHz (B3), except for OP2 in the considered test-site at the second floor of a University building in Cassino (Italy). The set-up includes an active directional antenna, namely the Rohde & Schwarz HE300A log periodic antenna, a basic SA (Rohde & Schwarz FSH8 Handheld SA having a 9 kHz – 8 GHz frequency range, characterized by a measurement uncertainty lower than 1 dB with a 161 dBm sensitivity, and a PC (Personal Computer) running a Matlab™ script to get, process and remotely store (in a Cloud Space) acquired data. The acquisition process is designed to get one full week of data for each OP and each frequency band. The whole measurement campaign lasted 8 weeks. Measurements have been carried out according to the international standards regulating the applicability of ET (Extrapolation Techniques) [12-13].

2.2. *Testing scenarios for determination of the compliance boundary*

In terms of determination of the compliance boundary, measurement and simulation analyses were carried out at two different locations in Austria. The measurement was conducted at site on a rooftop in the north of Vienna (Figure 2) (observed RBS at the time was also equipped with GSM and UMTS in the 900 MHz and 2100 MHz frequency bands). The attic of the building was reconstructed into apartments and terraces were built. The terrace is located in the immediate vicinity (less than one meter) near one of the two poles used for the antennas. The initial inspection showed very quickly that the terrace was too close to the antennas and a reconstruction of the RBS was initiated. The pole of the antennas near the terrace was removed and the antennas were moved to the second pole, representing more distant antenna support (behind the satellite antenna, an old pole which was dismantled, is still visible in Figure 2). This reconstruction work could be evaluated directly on site, as unhindered access was still possible. This verification measurement took place on 27.09.2013 between 10:30 and 12:30. Due to the undisturbed conditions in the direct vicinity of the antennas and to be able to work in a time- saving manner, the individual measurement runs were reduced, so the 6-min measurements were not necessary. The exact position of the measurement point was determined on one hand basedon the positional relationships of the main beam directions and the other hand by a preliminary measurement to find the points of maximum exposure. At two points where the highest exposure was determined, measurements were made on a tripod at three heights (170 cm, 150 cm, and 110 cm) according to [15] (applicable at that time). Considering the measurement setup in this case, a hand-held SA Narda SRM-3006 was used together with a 3D isotropic antennas Narda 3502. This antenna covers a frequency range from 420 MHz to 6 GHz. The expanded measurement uncertainty of the instruments together with the 1.5 m long cable is in range +3.1 to -4.9 dB (confidence interval 95%) for the frequencies between 420 and 6 GHz.

During the measurements, overview measurements in the frequency range from 420 MHz to 6 GHz were performed to identify all the sources that may be present at the measurement site. In Figure 3 the result at a height of 150 cm is presented. Dominant signals were GSM at 900 MHz, GSM 1800 and UMTS at 2100 MHz. As can be seen, there was not traffic present at the time of measurement, so extrapolation to maximum possible values for checking compliance is necessary.

To make this measurement reproducible, the main measurement setting of SRM-3006 were as follows: RBW 500kHz, VBW was switched off, sweep time 1,081 sec for the isotropic measurement (there are three antennas inside the used 3502 antenna from NARDA, whereas the measuring instrument use one antenna after the other, therefore, having the antenna as quasi-isotropic), and measuring time less than one minute to have an overview and to save the time spent on the field. No other relevant source were identified (5% of the limit as defined in [15]) in this frequency range. Green line from Figure 3 represents an Average values averaged over 4 runs, whereas magenta line represents the Standard in force as defined in [16-18]. In parallel, a broadband measurement from 100 kHz to 6 GHz (using Narda NBM-550 Broadband Field Meter with the E-Field-Probe EF 0691) was performed. If these results match as far as possible, it can be assumed that no other relevant sources are overlooked. For an accurate assessment of mobile communications, frequency-selective measurements are carried out with the NARDA SRM-3006 in all downlink ranges of mobile communications that were present atthe measurement site and, at locations where necessary code-selective measurements are also carried out for UMTS, LTE and, more recently, for 5G.

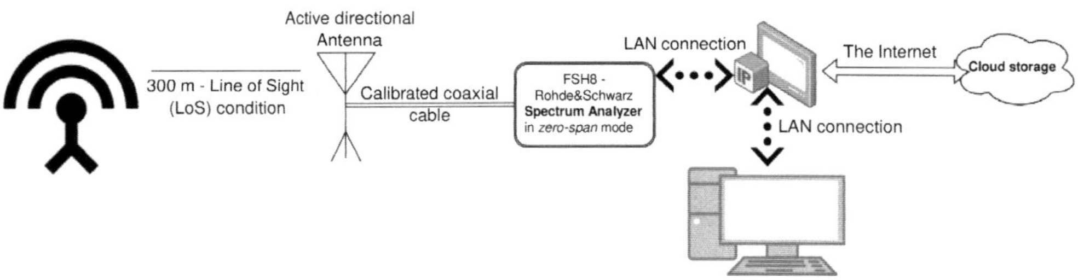

Figure 1. The adopted experimental set-up.

In the second example, a measurement was not possible because the site was in the planning phase. Therefore, the simulation analyses involved mobile communications systems from all three operators in Austria on the tower of a historic castle in the north of Lower Austria. This tower has a viewing platform that was originally closed for public access. However, since this platform is again being used for public access and compliance with the safety distances could thus no longer be guaranteed, the site was reconstructed. This had to be solved in close coordination between the radio-technology requirements and the monument protection authorities. The result, in terms of compliance with the reference levels, was a compromise to place the antennas on the highest possible position on a central mast. One operator left the antennas on the outside of the parapet to the viewing platform and two operators including A1 positioned the antennas on a central mast. To ensure compliance with the reference levels for all the operators in the planning phase, a simulation of the safety distances was carried out in advance, considering all the relevant sources. The software EMF Visual [19] was used for this purpose. It is an EM exposure simulation software that can accurately simulate exposure in both near- and far-fields of the antennas. It can also consider its environment and other sources of EMFs. The computation is based on the ray-tracing techniques. The user can select the mesh size of the computation volume and use a non-uniform grid in order to improve the resolution around the antennas and save computation time. Concerning a simulation, the environment of the antennas must be simulated as accurately as possible. In a first step it is necessary to define the complete 3D scenario. This means that at least the nearest environment surrounding the RBS must be reproduced three-dimensionally. It can be done directly in the software based on existing components provided with the software (buildings, ground profiles, etc.) or by rebuildingit in CAD (Computer-aided Design) software. Furthermore, the antennas of all operators (in this case 3 3 antennas) have to be positioned exactly like in reality. In addition, radio parameters (antenna type, maximum antenna input power per system, frequency range, down tilt) of all operators must be available.

2.3. Pilot signals stability analysis for evaluating the repeatability of the PBCH power measurement

To evaluate pilot signal power constancy, a suitable figure of merit, namely $R_{MAX,min}$, has been defined as in eq. 2.

$$R_{MAX,min} = 10log_{10}(\frac{\max(P_{pbch})}{\min(P_{pbch})}) \quad (2)$$

In eq. 2, P_{pbch} is the power obtained by acquiring the PBCH pilot channel. The meaning of the proposed figure of merit is to quantify the distance (in decibel, dB) between the maximum and minimum values experienced during a day-time acquisition. The lowest $R_{MAX,min}$, the better the signal stability and, consequently, the closer the theoretical hypothesis of PBCH signal constancy. In an ideal case, $R_{MAX,min}$ should be equal to 0 dB.

Figure 2. View of the rooftop terrace in the north of Vienna with RBS and measurement point chosen for compliance boundary assessment.

2.4. Experimental analysis for evaluating compliance boundary assessment

Before RBSs are put into service following the Product standard [20], it is necessary to conduct EMF compliance assessments to make sure that such RF sources satisfy relevant regulatory requirements on EMF exposure. The purpose of these assessments is to determine compliance boundaries always taking into account all relevant sources outside which the RF-EMF exposure is below the applicable exposure limits. The distance at which the exposure level is always below the limit is called the 'compliance distance'. The compliance distance may be based on the evaluation of several parameters such as field strength, power density, or SAR (primary for small cells, portable devices, and tablets). In either case, the compliance distance provides a conservative safety margin. Concerning the field distribution around RBSs, two main approaches, such as measurements and calculations (simulations), can be applied to get an insight into the EMF exposure levels in its immediate vicinity.

As for the measurement approach, EMF exposure can be determined by measuring the actual value of exposure and calculate the maximum possible theoretical or actual maximum exposure as defined in [12], whereas calculating the maximum usually includes post-processing.

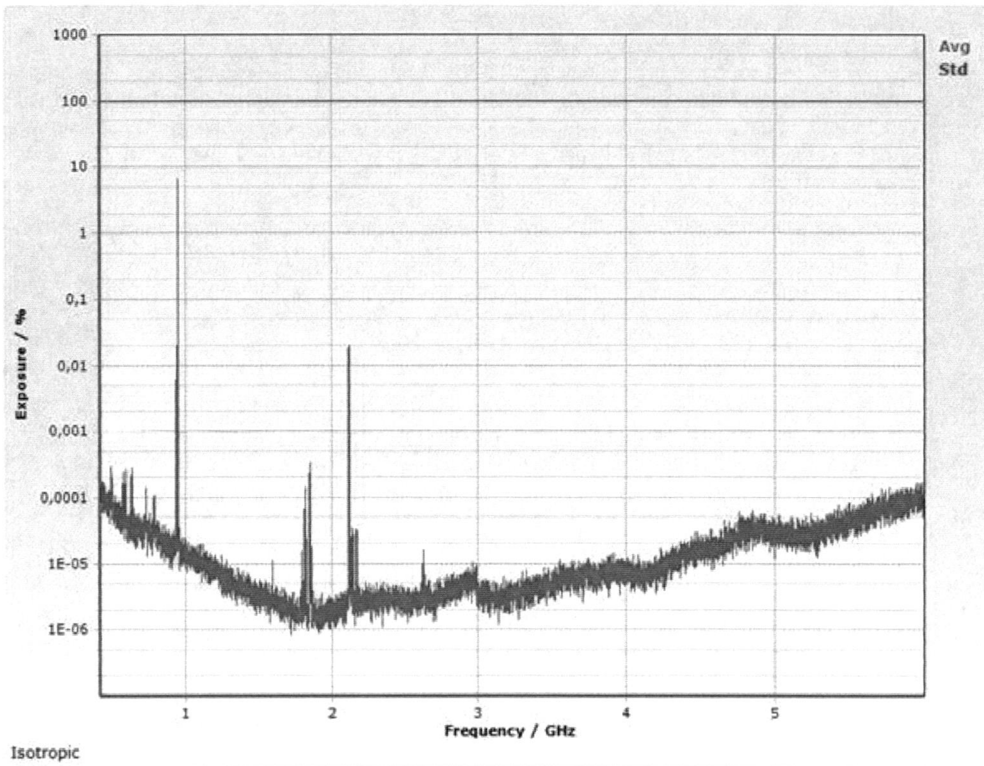

Figure 3. Spectrum overview at the measurement site located on the rooftop terrace in the north of Vienna.

Therefore, it is of great importance to derive better stability of traffic independent pilot signals using different trace processing and averaging techniques which are used in the calculations of compliance boundary exposure as well.

As for the calculation (simulation) approach, a multi-stage procedure is often used. This includes some of the following steps such as (i) Calculation of the safety distance (only as asingle distance) with the far-field formula (if no area with public access is near then no further step might be necessary), and (ii) Calculation of the compliance distances according to the box- shaped compliance boundary as defined in Section 6 of [12] (here also applies: if there is no public access within the box-shaped compliance boundary no further step might be necessary). However, if there is public access, then step (iii) Reconstruction and/or simulation of the RBSsite is necessary.

In situations of detailed analysis of compliance, it must be ensured that the reference levels are complied with the safety limits at any time of the day (taking into account the averaging time of the relevant guideline or standard) and in any possible operating state of the RBS.

3. Results and Discussion

For the sake of clarity, here are presented and discussed the results of long-term and compliance boundary measurements as well as the results of simulation analysis. In addition, conclusions that follow are based on the behavior observed during the whole measurement period.

Figure 4. Simulation analysis for two RBSs situated on the central mast of a historical castlein the north of Lower Austria.

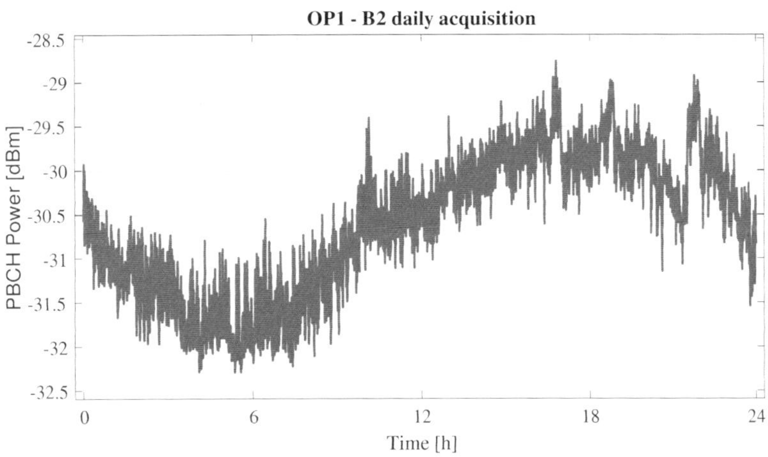

Figure 5. OP1–B2 daily acquisition - PBCH signal power

In particular, as for the experimental tests related to the PBCH signal power measurements, two kinds of analyses will be described: the former shows the behaviour of $R_{MAX,min}$ when applied on row data (collected each 20 s), whereas, the latter shows the effects of different averaging times on such a figure of merit.

Considering the determination of compliance boundary for the measurement site illustrated in (Figure 2), the measurement results were evaluated for maximum possible exposure and the

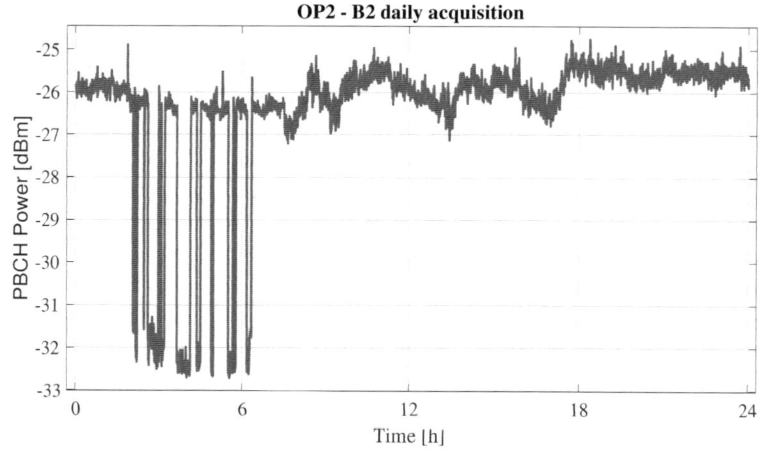

Figure 6. OP2–B2 daily acquisition - PBCH signal power.

measurement uncertainty specified by the manufacturer of the measuring device was added to the measurement results. After that, the average of the three measured values was taken. The results, in this case, were significantly below the reference values with a maximum of 19.38% of the reference level. Since access to the terraces was no longer possible after this detailed evaluation, the extension with LTE was carried out by calculations according to the far-field formula, considering the corresponding angular attenuations of the antenna according to [12]. Since the calculation results were also significantly below the reference values, the extension with LTE (in the 800 MHz frequency range) was carried out in the same way. The results of the simulation from Figure 4 has shown that a reference value exceedance is only given at a height of 3.5 m above the viewing terrace.

3.1. Raw Data Analysis

As previously described, for each combination operator-band, one whole week has been recorded for analyzing the behaviour of the PBCH signal power. As an example, Table I reports $R_{MAX,min}$ as its mean value over the week along with its range of variation. This parameter was calculated according to eq. 2 for every day of the week. The value out of parentheses represents the mean of the daily values for each day (considering 7 values in total), whereas the value in parentheses represents the range (difference in dB) between the highest and the lowest mean values during the week.

As it can be seen, in all cases, the observed values are far from the ideal one (0 dB), which highlights how such signal is not constant during the day and week. In particular, the most significant mean values of $R_{MAX,min}$ have been observed in the case of OP1-B1, OP2-B2, OP3-B2 and OP1-B3.

Furthermore, since the variation ranges are quite small in the most of cases, it also means that the observed values of $R_{MAX,min}$ seems quite systematic.

Since the PBCH signal power is directly linked to the maximum power expected by applying the eq. 1, the achieved results highlight that some problems could arise in the comparison with the emission limits in all cases in which $R_{MAX,min}$ is significantly far from the ideal value (i.e.0 dB).

In addition, the variations in PBCH transmitted power are subject to the operation modes of RBSs and the OPs sometimes adjust them to account for different traffic conditions, area coverage, etc. The variations in the received signal strength also happen due to the variation in PBCH transmitted power as well as to multipath propagation effects such as reflection, scattering, diffraction, especially in dense urban environments.

Table 1. $R_{MAX,min}$ [dB] on raw data: mean value (range value) in a week–time.

OP/B	B1	B2	B3
OP1	5.19 (3.27)	3.25 (1.76)	18.40 (0.49)
OP2	2.44 (1.62)	8.57 (1.91)	/
OP3	2.95 (0.54)	6.00 (0.51)	5.47 (1.92)

3.2. Analysis of the variability reduction due to the averaging process

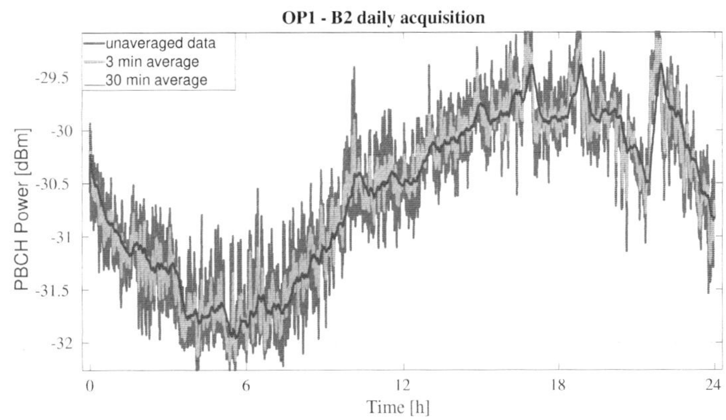

Figure 7. Effect of the averaging time on OP1-B2.

This section analyses the effects of the averaging time on $R_{MAX,min}$. ICNIRP guideline defines the averaging times to account for temporal variations of EMFs in tissues considering the local or whole-body exposure. Averaging effects smooth such variations, whereas higher exposure than the limit is permitted for short-term exposures [21]. Indeed, starting from behaviours like ones observed in Figures 5–6, it is obvious that the averaging could help in smoothing the traces,thus also potentially reducing the variability of resulting values. As an example, Figures 7–8 report such effects for several values of averaging time, in the case of OP1-B2 and OP2-B2. Moreover, Figures 9–11 report the evolution of the mean value (evaluated over the week) of $R_{MAX,min}$ and the related ranges versus the averaging time. As for averaging time values they have been selected on the basis of both practical considerations (about the time required for making a measurement) and also the typical values of time intervals considered in the definition of the ICNIRP emission limits. Looking at Figures 9–11, as expected, the smoothing effect due to the averaging time is evident because an increasing of the averaging time leads to a decreasing of $R_{MAX,min}$ (the traces of the acquisitions are better smoothed, and the results of measurements are more stable having the range values lower). Such trend is kept in all Figures, and such analysis may help in calculating the uncertainty regarding the result obtained

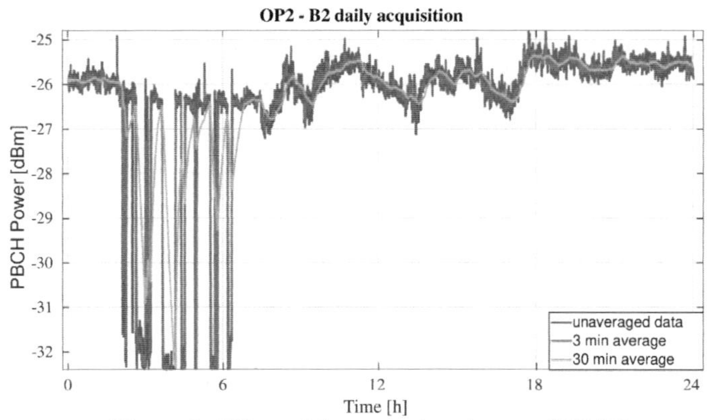

Figure 8. Effect of the averaging time on OP2-B2

in shorter observation time compared to ones at longer times. However, in many combination of Operator-Band, $R_{MAX,min}$ still remain significantly high concerning the expected experimental variability (typically less than 3 dB). This is particularly true for the case of OP1-B1, OP2-B2, OP3-B2, OP1-B3 and OP3-B3. As for the variability ranges of $R_{MAX,min}$ observed in Figures 9–11, they are not affected by the considered averaging time, which means that the variability over different days does not depend on the averaging time.

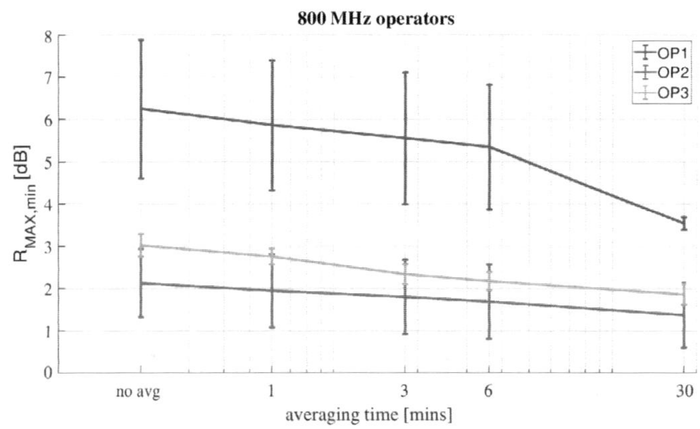

Figure 9. $R_{MAX,min}$ values at 800 MHz band: the averaging effect.

3.3. Issues related to determination of compliance boundary exposure

The examples given for the determination of the compliance boundaries show that both on-site measurements and simulations require considerable effort. In addition to the considerable effort required for the evaluation as well, the accessibility of the measurements is also of decisive importance. In recent decades, data traffic in mobile networks has been increased significantly, which is one of the reasons why more and more frequency ranges have been allocated to mobile communications. Nowadays in Italy and Austria, frequencies in the 700 MHz, 800 MHz, 900 MHz, 1500 MHz, 1800 MHz, 2100 MHz, 2600 MHz, and 3400 MHz bands are allocated to mobile communication.

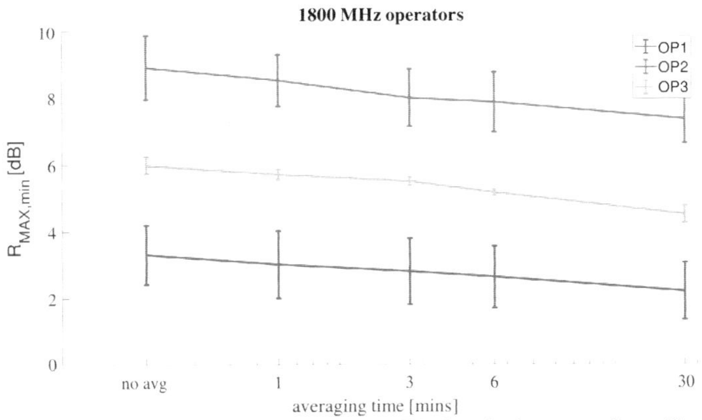

Figure 10. $R_{MAX,min}$ values at 1800 MHz band: the averaging effect

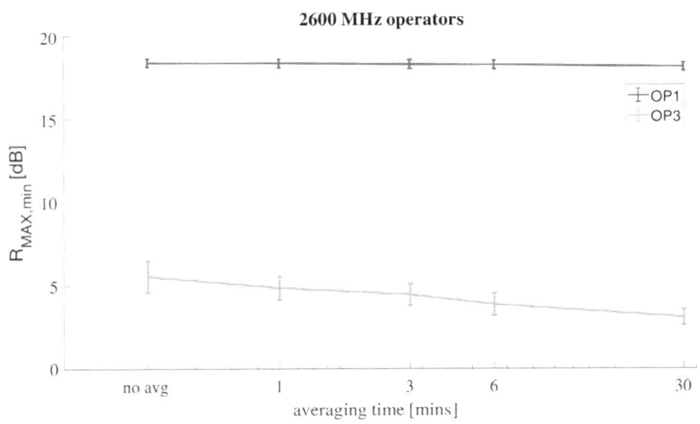

Figure 11. $R_{MAX,min}$ values at 2600 MHz band: the averaging effect

Most of these frequency bands are allocated on a technology-neutral basis, considering the fact that different technologies can be used within one frequency band. However, this was not the case in the past. For example, the 900 MHz and 1800 MHz frequency bands were used exclusively for 2G and the 2100 MHz frequency band for 3G only. When 4G was first introduced, the frequency ranges of 800 MHz and 2600 MHz were only used for LTE. Nowadays it is quite common that 4G and 5G technologies started using frequency bands previously occupied by 2G and 3G, whereas 3G is using frequency bands previously occupiedby 2G. Under realistic conditions, it is, therefore, possible for a RBS to have LTE at 800 MHz,GSM with one or more channels at 900 MHz, UMTS with two channels at 900 MHz, LTE with two channels at 1800 MHz, UMTS with one or more channels at 2100 MHz, LTE with two channels in 2600 MHz and also 5G at 2100 MHz and 3400 MHz. This corresponds to a total of 11 or more individual channels.

According to the normative specifications [16], each channel would have to be measured for 6 minutes. The measurement of more than one frequency band at the same time is only possible when performing a frequency-selective measurement. To decode the stable signal of UMTS, LTE, or 5G it is necessary to perform single measurements at every channel with special measurement equipment that can decode these parts of the signals, such as VNA or NARDA SRM-3006 SA using the appropriate operating modes and software upgrades necessary for decoding the signal. For the exact evaluation, it is necessary to measure, i.e. 11 6 min = 66 min. This represents the pure measurement time per measuring point without any post-processing. Since the spatial averaging over at least three points is required according to [12], this time is tripled again. In addition, there may be several operators at the same location, which must be taken into account as relevant sources. As a result, the pure measurement time at one location can extend up to one day. This represents a huge effort that is performed very rarely in practice. Since the accessibility of the measurement points is often not possible, a simulation in such cases must be used. Simulations must also be used if the site is still in the planning phase and measurements cannot yet be carried out as a result. However, in both cases the effort is considerable, so both can be evaluated in several steps, as mentioned above. Having that in mind, the number of measurements in some situations should be kept as low as possible to reduce the effort for the EMF assessment per site. As a consequence, the measurement time could be reduced to optimize the time necessary for the EMF exposure evaluation.

3.4. Aspects toward improving the reliability of the measurement results

New guidelines for limiting exposure to EMFs (100 kHz to 300 GHz) [21] define the reference levels for exposure, averaged over 30 min. Until now, it is unclear how this recommendation will be implemented in the standards for measurements such as [12]. If it will be necessary to apply such an approach, then 30-min measurement time could be very long in the real-life situations, and especially in the compliance boundary determination cases. The same issue according to measurement time and effort also arises for the measurements that are located further away from the RBSs. In the case of measurements for local residents or concerned citizens, the question often arises about the maximum possible exposure, or at what range of values the exposure can fluctuate within a day. For these reasons, if the question of safe compliance with the limit values is not the top priority case, then the measurements (frequency- or band-selective) could be made over shorter periods, and these results could be presented with the corresponding uncertainties. From all these aspects, the necessity to reduce the duration of measurements to be able to carry out them quickly and cost-efficiently are obvious.

Of course, on the other hand, the accuracy of such measurements must not suffer considering its ability to prove compliance with the reference levels. In a few cases, however, the measurements will be unavoidable. Since the accessibility to private areas generally represents a problem, in such cases simulations have to be used as well as in the planning phase of the RBSs. Although spatial representation of the real exposure situation could make calculation time running for several hours, they could, in many cases, represent very efficient evaluation method.

4. Conclusion

Measurements of human exposure to RF-EMFs generated by the 4th generation of cellular networks in urban environments is a very important topic in today's world since human life is heavily influenced by many technologies exploiting the 4G assigned frequencies. In that sense, the higher the number of RBSs, especially in complex urban environments, the more difficult the realistic exposure assessment. Despite specific guidelines used for the measurements of EMFs, there are still several issues to be discussed regarding the reliability of the measurement procedures and the evaluation of the measurement uncertainty. In general, to assess human exposure, some easier and faster approaches have been proposed in the literature and technical standards, among which Extrapolation Techniques (ETs) play the central role. They are thought to overestimate the maximum channel power on the basis of the supposed constancy of the level of a suitable pilot signal in the measurement point of interest, thus warranting a conservative approach concerning human exposure at that point. However, several experimental campaigns have proved how the hypotheses of the pilot signal power constancy cannot always be kept. Having that in mind, in some cases, supposed overestimation can turn into underestimation, thus losing the "conservative" feature and making the power measurement unreliable. For that reason, it is of great importance to achieve better stability of the pilot signals power as much as possible to have reliable and comparable results that are not heavily influenced by the time of measurement on a day or a week time scale.

Following the guidelines provided by the relevant technical standards in force, this paper describes the experimental results achieved by performing a long-term analysis for several days and weeks on three network operators and three frequency bands adopted in Italy for 4G communications.

The effects of the averaging time on raw data were evaluated to observe such an effect on reducing the variability of the PBCH power levels during the 24 h period per day. To these aims, a suitable figure of merit has been defined, namely $R_{MAX,min}$. In particular, as for the experimental tests related to the PBCH signal power measurements, two kinds of analyses were described: the former showed the behavior of $R_{MAX,min}$ when applied on row data (collected each 20 s), whereas, the latter showed the effects of different averaging times on such a figure of merit. The variation ranges of $R_{MAX,min}$ parameters evaluated over raw data sets of up to 18.40 dB found in the case of OP1-B3 (Table 1) justified the conducted analysis aimed at refining the measurement procedures and post-processing techniques given by the technical standards. Considering these effects, the results generally, significantly improved but in some situations, also high values of averaging time cannot assure an adequate smoothing of the collected traces. Therefore, these results suggest to further investigate on the possibility of improving the measurement procedures provided by the technical standards in force and their correct application on the field.

Another important issue that has been discussed in this paper was related to exposure safety requirements regarding the determination of compliance boundaries from RBS transmitting antennas, mainly estimated by measurements and calculations (simulations). Illustrated examples described the measurement and simulation approaches at two different locations in Austria for a roof terrace exposure condition. These procedures must be done for every new RBS and relevant upgrade.

As for the measurement approach, EMF exposure can be determined by measuring the actual value of exposure and calculating the possible theoretical or actual maximum exposure as defined in IEC 62232, whereas calculating the maximum usually includes post-processing.

As for the calculation (simulation) approach, a multi-stage procedure is often used (calculation of the compliance distance with the far-field formula, calculation of the compliance distances according to the box-shaped compliance boundary, simulation of the compliance boundary as an iso-surface for the RBS site). In any case, it must be ensured that the RBS is complied with the safety limits at any time of the day (taking into account the averaging time of the relevant guideline or standard) and in any possible operating state of the RBS. If this goal is not achieved, reconstruction will be necessary or access restrictions will have to be implemented. Considering the determination of compliance boundary in this paper, the measurement results were significantly below the reference values with a maximum of 19.38% of the Austrian reference level. In addition, the results of the simulation at the second location have shown that a reference value exceedance is only given at a height of 3.5 m above the viewing terrace.

Because the standards (international and national ones) that limit the RF-EMF exposure are recently redefined, it is still unclear how some steps such as 30-min measurement and averaging time will be implemented in real-life situations, especially in compliance boundary determination cases. The same issue according to measurement time and effort also arises for the measurements that are located further away from the RBSs.

4.1. Acknowledgments

This work is partially supported by the "Dipartimenti di Eccellenza", Italian Ministry of Education, University and Research funding programme.

References

[1] Miclăuş S, Bechet P and Iftode C 2013 *Measurement* **46** pp 1355–1362
[2] International Commission on Non-Ionizing Radiation Protection and others 2020 *Health physics* **118** 483–524
[3] 2019 *IEEE Std C95.1-2019 (Revision of IEEE Std C95.1-2005/ Incorporates IEEE Std C95.1-2019/Cor 1-2019)* 1–312
[4] Deaconescu D B, Buda A M, Vatamanu D and Miclăuş S 2022 *Engineering, Technology & Applied Science Research* **12** pp 8101–8106
[5] Miclăuş S, Bechet P, Helbet R, Miclăuş A and Sârbu A 2021 *2021 12th International Symposium on Advanced Topics in Electrical Engineering (ATEE)* (IEEE) pp 1–4
[6] Miclăuş S and Bechet P 2020 *Progress In Electromagnetics Research M* **89** pp 151–159
[7] Sârbu A, Bechet A, Bălan T, Robu D, Bechet P and Miclăuş S 2019 *Measurement* **134** pp 874–887
[8] Betta G, Capriglione D, Cerro G, Miele G, Salone D'Amata M, Suka D, Pejovic P and Simic-Pejovic M 2020 *IEEE Instrumentation & Measurement Magazine* **23** pp 5–13
[9] Bernieri A, Betta G, Capriglione D, Cerro G, Miele G and Salone D'Amata M 2021 *IEEE Transactions on Instrumentation and Measurement* **70** pp 1–13
[10] Betta G, Cerro G, Miele G, Salone D'Amata M and Capriglione D 2020 *2020 IEEE International Instrumentation and Measurement Technology Conference (I2MTC)* (IEEE) pp 1–6
[11] Betta G, Capriglione D, Cerro G, Miele G and Salone D'Amata M 2021 *IEEE Transactions on Instrumentation and Measurement* **70** pp 1–13
[12] IEC 2017 IEC 62232:2017 Standard | Determination of RF field strength, power density and SAR in the vicinity of radiocommunication base stations for the purpose of evaluating human exposure Standard IEC 62232:2017 International Electrotechnical Commission Geneva, CH
[13] IEC 2019 IEC TR 62669:2019 Case studies supporting IEC 62232 - Determination of RF field strength, power density and SAR in the vicinity of radiocommunication base stations for the purpose of evaluating human exposure Standard IEC TR 62669:2019 International Electrotechnical Commission Geneva, CH
[14] Gessner C 2011 *Long Term Evolution. A concise introduction to LTE and its measurement requirements* (Munich, Germany: Rohde & Schwarz GmbH)

[15] CENELEC 2008 EN50492. Basic standard for the in-situ measurement of electromagnetic field strength related to human exposure in the vicinity of base stations. Standard EN 50492:2008 CENELEC (European Committee for Electrotechnical Standardisation) Brussels, Belgium
[16] International Commission on Non-Ionizing Radiation Protection and others 1998 *Health physics* **74** pp 494–522
[17] 1999 1999/519/ec: Council recommendation of 12 july 1999 on the limitation of exposure of the general public to electromagnetic fields (0 hz to 300 GHz)
[18] Standards A 2017 OVE-Richtlinie R 23-1: 2017 04 01 - Elektrische, magnetische und elektromagnetische Felder im Frequenzbereich von 0 Hz bis 300 GHz – Teil 1: Begrenzung der Exposition von Personen der Allgemeinbev¨olkerung (deutsche Fassung)
[19] MVG 2022 https://www.mvg-world.com/en/products/rf-safety/public-rf-safety/emf-visual-software Simula- tion software EMF Visual Microwave Vision Group Paris, France
[20] CENELEC 2019 EN 50401:2019 Product standard to demonstrate the compliance of base station equipment with radiofrequency electromagnetic field exposure limits (110 MHz - 100 GHz), when put into service. Standard EN 50401:2019 CENELEC (European Committee for Electrotechnical Standardisation) Brussels, Belgium
[21] 2020 Guidelines for limiting exposure to electromagnetic fields (100 kHz to 300 GHz) Tech. Rep. ICNIRP 2020 Health Physics

Web business computing tool in support to photovoltaic prosumers

M (Nechifor) Roman[1], A Sălceanu[1]
[1]"Gheorghe Asachi" Technical University of Iasi, Romania, Dimitrie Mangeron Blvd. 67

E-mail: madalina.nechifor@student.tuiasi.ro, alexandru.salceanu@academic.tuiasi.ro

Abstract. Electricity is one of the most essential keys in the evolution of nowadays civilization. Nevertheless, huge levels of electric power production and consumption could cause a negative impact on the environment, such as poor air quality, greenhouse emissions, climate change, and ecological disasters. Consequently, today's society needs sustainable solutions that can meet the geopolitical, socio-economic and climatic difficulties that energy production requires. Understanding this, international development programs have been modified approaching the use of sustainable sources of energy, which are gradually becoming economically viable. Among the valid options, photovoltaic (PV) systems have reached both the general and industrial public, becoming a sustainable energy source. Given these things, to gain a boost and to maintain a vital position in the market, it is necessary for prosumers to be aware and to recognize the mobility of characteristics in the fields of finance sphere, but also the structure of indicators able to measure a specific reality. We are calling it "specific reality" because we do not have enough business intelligence instruments in this sector and according to Eurostat, in 2020 only 3.09% of the electricity production in Romania came from photovoltaic panels and the increase was below one percent compared with 2019 when it was only 2.98%. The output power of a solar installation can be measured in several ways, including the competitiveness of the photovoltaic installations. To manipulate and guide the performances in this segment, we need Success Assessment Tools, that check a collection of characteristics used to calculate the outcomes and profitability of a prosumer's movements. According to National Energy Regulatory Authority, in May 2020 there were only 514 prosumers in Romania (455 individuals, 59 legal entities). A huge untapped opportunity could be represented in Romania by individual households. According to a study custom-built by the European Commission, published in 2017, the theoretical Romanian potential would be about 3.36 million residential prosumers.

1. Introduction

For a sustainable and balanced house and society, from the electricity supply perspective, every citizen could capture the eternally renewable energy of the sun for producing electrical power. Technology helps us from this point of view, the price of photovoltaic panels went down, while conversion efficiency has significantly improved.

Prosumer is an abbreviated form that brings together the terms provider + consumer. Prosumer could be a legal entity or a natural person who is connected to the electricity grid not only as a consumer but also as a producer of energy (from renewable sources). Anyone might become a prosumer, being very important to be up to date with the latest regulations in the field, to have a profitable and sustainable system in the market. Thus, business intelligence systems help prosumers to store, analyze and forecast in a highly dynamic but unstable and rather unpredictable energy market [1].

Our country has set a goal of phasing out coal energy by 2032. This step is expected to increase the expansion of the sustainable energy domain, especially for wind turbines and solar photovoltaic installations. Romania's photovoltaic capacity is expected to triple in the period 2021-2030, increasing from 1.39 GW in 2020 to 4.25 GW in 2030, these forecasts being done by “Global Data”, a data and analysis company [2].

The scheme is expected to be launched before the end of 2022 and the Romanian government is counting on it to stimulate the development of renewable energy in the country. This will also provide opportunities for transport and distribution infrastructure developers, as the country would like to integrate renewable energy sources into its national grid [3, 4]. The rapid addition of solar photovoltaic capacity will help Romania increase its electricity production, thus reducing energy imports from neighbouring countries, which were high in 2019 and 2020. Under current legislation, prosumers are paid for the energy they deliver to the grid at the average (PZU) price for the previous year. In other words, in 2022, prosumers will receive about 0.55 RON/kWh for energy produced in addition to consumption and delivered to the grid [3]. Considering the above data, Romania encourages new prosumers installation. New or existing prosumers need help to maximize their energy and this study will focus on this topic.

KPI is the abbreviation for key production indicator, and it is a measurable value used to evaluate success at reaching the performance target. KPIs are a collection of metrics that concentrate attention on performance and are the most complex guidelines for the present and future development of the installation. When the collection is defined, the measurement structure becomes the second point of interest because each KPI will weigh according to the target performance and the relationships between them. Following the above data, the resulting System for Evaluating Performance is built based on a ranking structure of the Analytic Hierarchy Process (AHP) [5, 6]. The AHP weighting algorithm is used to determine the importance of each KPI.

Essential Standpoints (ES) and Important Success Features (ISF) represent two layers of metrics collection (KPIs) established by structuring the indicators in a decision tree, where the ESs are part of the main structure of the tree, spread out in ISFs.

Four ESs have been used and deployed on 26 ISFs and 39 KPIs, Table 1. The first point is represented by the Key Partnerships, which means the evaluation of the level of interaction done by the owner of a photovoltaic installation with the other interested parties. The second point is characterized by strategic management and processes, consisting of measuring the performance of the project [6, 7].

The third point is the innovation in the production of photovoltaic energy, the evaluation of the photovoltaic panels’ performance as innovative installation, the knowledge of the technology used and the prospecting of future scenarios. The fourth point is the monitoring of photovoltaic production by measuring the state of conservation and financial resources that are relevant to the photovoltaic sector.

The purpose of here presented instrument is to verify the relationship between the ambitious side view and the progress of photovoltaic energy in the energy matrix. Therefore, each question is composed of decision-maker and importance level, Table 1 [8, 9].

Table 1. Representation of Essential Standpoint (ES), Important Success Features(ISF) and Key Production Indicators (KPIs) using hierarchical structure of the Analytic Hierarchy Process(AHP)

Essential Standpoints (ES)	**Important success features(ISF)**	**Key Production Indicators (KPIs)**
Key Partnerships	Vendors	Partnerships with solar system vendors (the company that helped you install the panels or other similar companies)
	Similar Systems	Relationship with similar solar systems (neighbors/friends/acquaintances/business partners)

	Educational/Research Institutions	Relationship with organizations in photovoltaic sector
	Environmental Organizations	Relationship with environmental agencies (ONGs promoting renewable energy)
	Public Energy	Relationship with public authorities (Administration of the Environment Fund)
	Policy/Procedures	Political relations (political parties oriented towards environmental protection) to obtain support for a possible installation or extension of an installed system
	Entities available in PV sector	Relations with sectorial entities for support
	Marketplace	Relationship with the energy market, as a prosumer
		Relationship with the international energy market compared to the national market (support measures used in the EU vs. Romania)
Solar Energy Monitoring	Power Production	Monitoring of electricity generated
		Monitoring of consumed electricity
	Profitability and Incentives	Monitoring of national incentives and profitability
		Monitoring of international incentives and profitability
	Installation Features	Tracking the position of the photovoltaic system
		Tracking the orientation of the photovoltaic system
		Tracking the local climatic conditions
	Equipment Characteristics	Photovoltaic modules measurement
		Support structure monitoring and fixing measurement
		Tracking the condition of electrical devices of the photovoltaic system
	Equipment Maintenance	Monitoring the maintenance of the photovoltaic system
	Modelling of Data	Storage and control of all monitoring records
Management and Strategic Processes	Technological Documentation	Technological training in renewable energies (photovoltaic energy)
	Managerial Education	Courses/mentors in solar energy production
	Environmental Marketing	Spread of photovoltaic efficiency on the Internet
		Participation in fairs/events
	Technical Management	Advanced photovoltaic energy management techniques
	Capital and Access to Funds	Solid economic feasibility to upgrade the existing project
	Technical Support	Accessibility to technical assistance
	Personal	Access to qualified information and support

	Involvement	The level of confidence in the photovoltaic system
		Level of flexibility in the face of needs
Innovations in Energy Production	Sustainable Technologies	Innovation in sustainable technology
		Monitoring new trends in renewable energy
	Results	Recognition of innovative results coming from the installed system
		Prosumer satisfaction with results
		Last year's investment results
	Innovation	Innovation evaluation indicators
		The potential for expansion of the photovoltaic system
		Investing in energy efficiency
	Promptitude	Proactive position

The correspondence between various qualitative appreciations and the percentage weight is presented in Table 2.

Table 2. Importance levels of representation from the first Fundamental Viewpoint

Fundamental Viewpoint 1	**Key Partnerships**	
Point 1	**1.1. How is your key partnership with your solar system vendors?**	**Weight (%)**
P1	Very Effective	100
P2	Effective	75
P3	Intermediate	50
P4	Little Effective	25
P5	Non-existent	0

2. Web tool development and case study

Competitive diagnosis consists in identifying the performance of the photovoltaic installation in a relatively short period, by referring to the strategic position, consumption management, knowledge of the system, research and development, operations and logistics, opportunities, threats assessment and control. Based on this need a web tool has been built to assist the owners of solar power installation.

An online survey provides participants with the opportunity to communicate answers via Internet, in real-time, from the comfort of their home or workspace, usually by completing a questionnaire on an online webpage. These online tools can allow anyone to participate, small or medium enterprises (SME) or residential owners. The online tool has been developed by the authors of this paper and could be accessed by the web page of our "Sun project", at the link https://www.acoperisultausolar.com/prosumator-masurarea-competitivitatii [10].

In the first graphic part of the web page, four buttons allow the user to be redirected to the web forms built for data collection: “Prosumers Key Partnerships” (ES1), “Prosumers Energy Production Monitoring” (ES2), “Prosumers Management of Processes” (ES3), “Prosumers Innovation in Efficiency” (ES4), a section for each essential standpoint. All the four standpoints are in a single-page application, which does not require page reloading when the user clicks on one of the above buttons. He is routed to the correspondent poll where the interrogations are, each with five points of an answer, according to the research tool [5].

The data processing is done through the JavaScript language, being developed all the logical parts required to calculate the substitute ranks of each KPI, using mathematical formulas. Business intelligence involves performing complex calculations for engineering applications, with possibilities for interpretation and generalization of results. The last box, called "Results" holds four buttons, each one developed to generate graphical reports on the overall performance of all respondents' part of the study. Using this reporting interface, some examples being presented in Figures 1, 2, 3, 4, the owners can employ the provided graphs to solve any potential problems from the very beginning [7, 11].

This view allows an intuitive check-up of the overall situation, of the quantitative variables measured by category classes provided by all survey participants, using the graph of proportions. The visual management of the PV Solar System is aided by the presentation of the entire community outcomes via dashboards, simplifying the communication, distribution and evaluation of indicators at the national level, after centralizing data from more than half of prosumers in Romania.

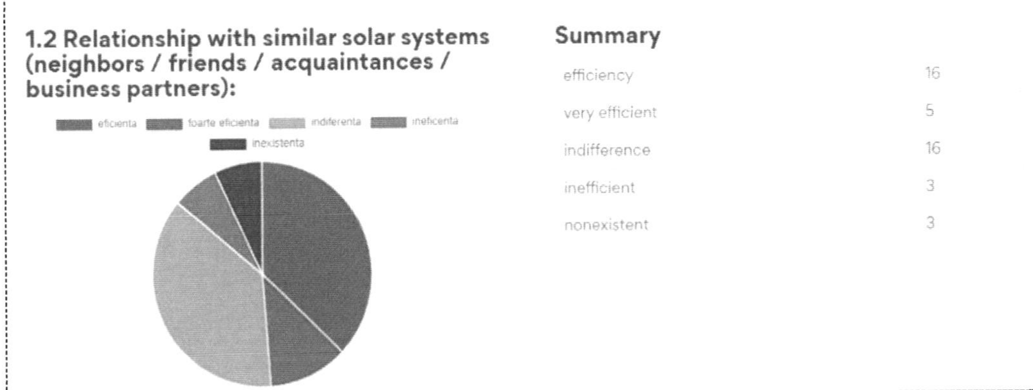

Figure 1. Caption of the web custom report generated for the first Essential Standpoint "Key Partnerships for Prosumers".

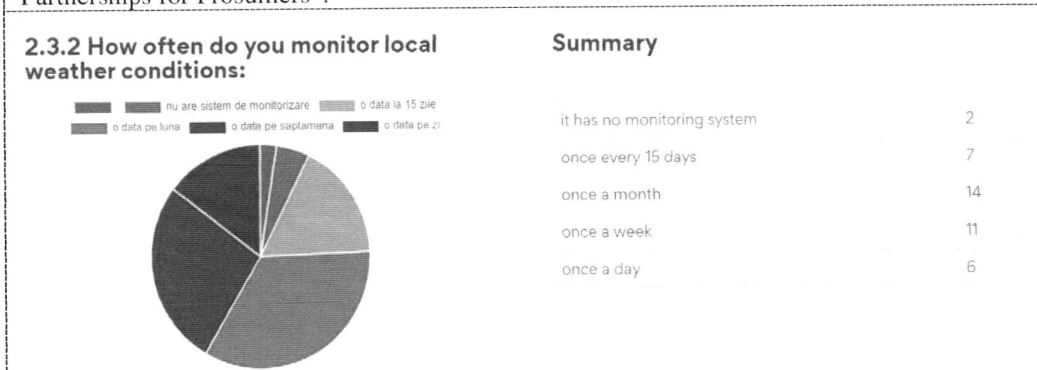

Figure 2. Caption of the web custom report generated for the second Essential Standpoint "Energy Monitoring Section for Prosumers".

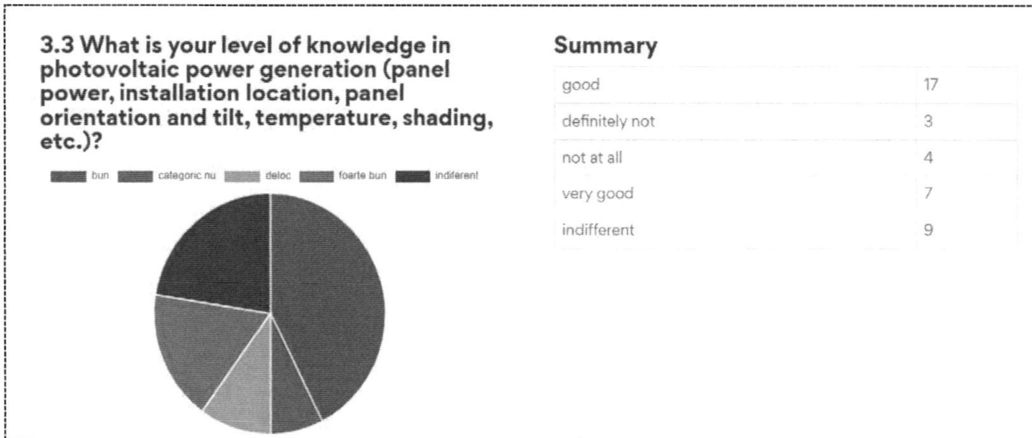

Figure 3. Caption of the web custom report generated for the third Essential Standpoint "Managerial Processes for Prosumers".

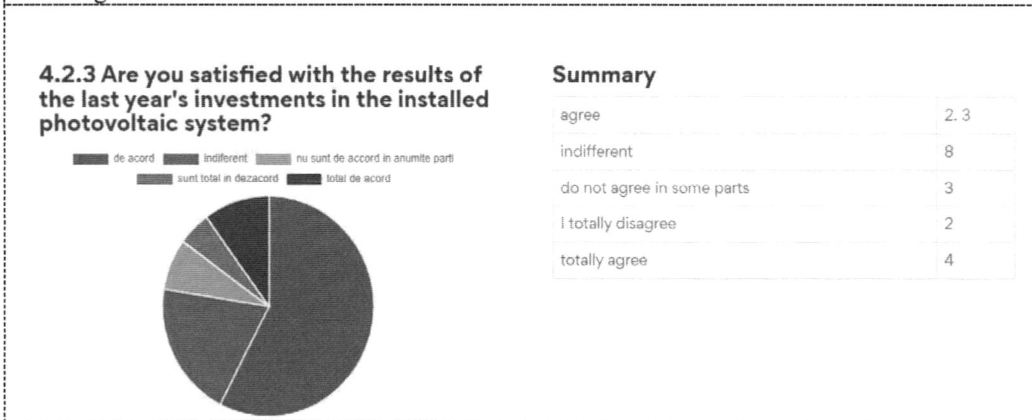

Figure 4. Caption of web custom report generated for the third Essential Standpoint "Innovation in Efficiency for Prosumers".

The "individual analysis" integrates, in comparison with the reports presented above, the substitution rates and the step-by-step decision diagram. The statement generated is personalized, individual this time and targets the current state of the prosumer, to help identify and implement the necessary controls to correct errors or problems. This report is made by quantifying the indicators, each performance indicator has a local substitution value, equivalent to its significance. With local replacement rates, global rates, the individual contribution of important success features and essential data views will be calculated. "Individual analysis in Excel format" was designed according to the same principles of mathematical calculation, this time translated into Excel formulas. Respecting the privacy criteria of users, there is the possibility to download and complete offline the questionnaire, without internet access [12].

Above the fundamental points of view is the result of global competitiveness. Because it is the primary level of the decision scheme, this result has a value of 100%, and its value is calculated as a function of additive aggregation of the local distribution rate for each of the four essential standpoints: ES1 (22.57%), ES2 (25.58%), ES3 (25.97%), and ES4 (25.97%) (25.87%). At the end of the web tool, the performance of each fundamental point of view will be illustrated together with the overall performance of the evaluated organization [13].

3. Study case: results and discussions

Sequentially, the tool was applied to 39 subjects. Out of the desire to stimulate sincere and objective answers, all 39 respondents have been promised and respected anonymity. With the proposed modelling and calculation, it can be seen from the results obtained that there is no clear relationship between the size of the solar system and the installation's diligent production. The web application gives important support to prosumers by evaluating the rank of installation effectiveness and in the same time, forecasting future scenarios and ruling.

During the design and development of this web-tool we have also based on the experience gained as an outcome of the realization and use of the photovoltaic power plant, within our Faculty of Electrical and Power Engineering, Technical University of Iasi, Romania [14 - 16].

First Fundamental Viewpoint, Key Partnerships, has on average a weight of 55% on all the questions except the one of UE relationship. The interviewed subjects have more trust in UE legislation and support than in Romania. This will confirm the YouGov poll where 90% would like the government to provide them with more information on how they can generate their own solar energy and would like more financial support.

The second Fundamental Viewpoint, Solar Energy Monitoring, has on average a weight of 56.81% and it is showing that for actual prosumers the monitoring it is important just once in a month, when they can see how much money will receive on the bill. To discover the full potential of the system or how to improve it, more frequent monitoring is required, using systems to track and store this amount of data. The prosumers should be informed, through publications that monitoring is important and they should plan to adopt one system in this direction. Third Fundamental Viewpoint, Management and Strategic Processes have on average a weight of 75%. This means that the interviewed prosumers follow the news and available communities in this sector.

Fourth Fundamental Viewpoint, Innovations in Energy Production, has on average a weight of 75% and focuses on innovation, an innovation that is not yet sufficiently supported by the Romanian authorities, even if the participants on the energy production would like to implement it. Obviously, only money received for the injected energy is not sufficient for prosumers to invest alone in new technologies.

The result of PV national competitiveness is 65.81% based on these 39 prosumers and it is proofing one more time that we have a lot of potential, but the big disadvantage is the initial financial support. The use of the web application in the existing community of photovoltaic installations brings more data in the system, data that allow the expansion of the research, through the perspective of the main goal, to create a business intelligence application for solar energy optimization with the aid of indicators that provide a current examination of photovoltaic energy efficiency. Mathematical modelling allows prosumers to pay attention to the productiveness of solar energy installation and the results generated could boost competitiveness by increasing the values of indexes, thus helping to prospect future scenarios and make decisions.

4. Conclusions

We have developed a web tool that can be used for free by accessing the web page of our "Sun Project", https://www.acoperisultausolar.com. The scientific research behind this web tool is focused on issues related to increasing the efficiency of photovoltaic power plants. The aim was to design and implement a system that will allow the owners of photovoltaic systems to identify (and consequently to fix) their weaknesses. The generation of photovoltaic energy is incorporated in a highly diligent market, a feature that makes photovoltaic power installations to try to reach the maximum efficiency.

The primordial purpose of this application is to determine the level of solar energy production's effectiveness in the prosumers' photovoltaic power industry, by using a series of information on the main indicators. Mathematical modelling allows the owners of the plant to track the accomplishment of solar installation and increase potential and production by investing to improve the values of indicators, helping to prospect future scenarios and making inspired decisions. The tool was built with the recognition of four essential standpoints, 26 important success features and 40 production indicators, figured in a closed interrogation format with multiple answers.

Our country's capability is for 3.36 million residential prosumers and despite this, no management calculation tools have been found in the specialized literature that could help to identify, measure and analyse these performance indicators on the Romanian energy market. With this motivation and intending to complete the study in a prospecting sense, the paper's key contribution is the development of a web calculation engine, based on performance measures, that would visually display the level of diligence of photovoltaic installations owned by prosumers.

References

[1] Eurostat, Data Browser, *Energy statistics - quantities, Complete energy balances, 2022*

[2] REN21, Global Status Report *2021 Renewable energy statistics.* Available from: https://www.ren21.net/reports/global-status-report/

[3] Hoarcă I C 2021 Romania's energy strategy 2020-2030 with the perspective of 2050, *Proc. International Conference on Electronics, Computers and Artificial Intelligence (ECAI)*, pp. 1-6 Available from: http://doi.org/ 10.1109/ECAI52376.2021.9515143

[4] Oprea S V and Bâra A 2017 Analyses of Wind and Photovoltaic Energy Integration from the Promoting Scheme Point of View: Study Case of Romania, *Energies*, **10** 2101 Available from: http://dx.doi.org/10.3390/en10122101

[5] Rosa C B, Rediske G, Rigo P D, Wendt J F M, Michels L and Siluk J C M 2018 Development of a Computational Tool for Measuring Organizational Competitiveness in the Photovoltaic Power Plants, *Energies*, **11** 867 Available from: http://dx.doi.org/10.3390/en11040867

[6] Wang C H and Chuang JJ 2016 Integrating decision tree with back propagation network to conduct business diagnosis and performance simulation for solar companies, *Decision Support Systems*, **81**C p 12–19 Available from: https://doi.org/10.1016/j.dss.2015.10.004

[7] Sampaio P G V and González M O A 2017 Photovoltaic solar energy: Conceptual framework *Renew. Sustain. Energy Rev.* **74** pp 590–601 Available from: https://doi.org/10.1016/j.rser.2017.02.081

[8] Solar Power Europe, *Global Market Outlook 2021-2025.* Available from: https://www.solarpowereurope.org/wp-content/uploads/2021/07/SolarPower-Europe_Global Market-Outlook-for-Solar-2021-2025_V1.pdf

[9] Eurostat (online) *Renewable energy statistics.* Available from: https://ec.europa.eu/eurostat/statistics-explained/index.php?title=Renewable_energy_statistics

[10] https://www.acoperisultausolar.com/prosumator-masurarea-competitivitatii

[11] D'Adamo I and Rosa P 2016 Current state of renewable energies performances in the European Union: A new reference framework *Energy Convers.Manag.* p 84–92

[12] https://www.acoperisultausolar.com/prosumatorf-analiz%C4%83-individual%C4%8

[13] Eltawil M A and Zhao Z M 2010 Grid-connected photovoltaic power systems: Technical and potential problems—A review. *Renew. Sustain. Energy Rev.* **14** 1, p 112-129 Available from: https://doi.org/10.1016/j.rser.2009.07.015

[14] Machidon D and Istrate M 2016 Photovoltaic power plant on the ENERED research platform of the electrical engineering faculty of Iasi *Proc. International Conference and Exposition on Electrical and Power Engineering (EPE),* pp 884-887 Available from: https://doi.org/10.1109/ICEPE.2016.7781463

[15] Oprea R, Istrate M and Machidon D 2017 Assessments on the ENERED photovoltaic power plant electricity production *Proc. International Conference on Modern Power Systems (MPS)*, p 1-4 Available from: https://doi.org/10.1109/MPS.2017.7974375

[16] (Nechifor) Roman M, Sălceanu A, Păuleț M and Machidon D 2021 Evaluation upon the Energy Resources of Photovoltaic Systems depending on their Location *Proc. International Conference on Electromechanical and Energy Systems (SIELMEN)*, p 253-256 Available from: https://doi.org/10.1109/SIELMEN53755.2021.9600326

IOP Conf. Series: Materials Science and Engineering 1254 (2022) 012016 doi:10.1088/1757-899X/1254/1/012016

Expected benefits and foreseen steps in creating energy communities in Romania

L Darabant

Technical University of Cluj-Napoca, 26-28 Baritiu Str., Cluj-Napoca, Romania

E-mail: Laura. Darabant@ethm.utcluj.ro

Abstract. In 2019 the EU updated its energy policy (by adopting the *Clean energy for all Europeans package*) to enhance the transition from fossil fuels towards green energy - implementing the commitments of the EU's Paris Agreement (adopted at the COP21 in December 2015) for reducing greenhouse gas emissions. The agreement on this new energy rulebook marked a significant step towards implementing the energy union strategy, published in 2015. Through this package, the EU has introduced the concept of *energy communities* in its legislation, notably as *citizen energy communities* and *renewable energy communities.* To incorporate the new regulations introduced by this package in the Romanian law, the Romanian government issued an Emergency Ordinance (143/December 28, 2021), that updates the Electrical energy and natural gas law (123/2012). The concept of *energy communities* is, from now on, part of our own legislation, and the creation of such communities in Romania can start, although there are still many questions regarding the effective steps to be taken in this process. This paper tries to emphasize the foreseen advantages of this new concept and how it could be implemented in our country.

1. Introduction

The new concept of *energy communities* enables the joint efforts of citizens to participate in the energy system and take ownership of their energy consumption. The expected benefits of such communities can be addressed in several points:

- The communities can contribute to the increased acceptance of the citizens regarding renewable energy sources (RES). As we all know, the new challenges of changing the traditional energy production with RES require a strong public support. We start by evaluating the status of public acceptance of RES in Romania. We acknowledge that steps were already taken towards increasing the engagement of citizens in the energy transition, but we estimate that being part of an energy community can enhance this commitment even more.
- The communities can make it easier to attract private investments in the clean energy transition. The installation of RES is always associated with significant investment efforts. All the countries in the EU implemented some incentives in this context, meant to stimulate the energy production from RES by facilities installed in individual households or small or medium enterprises (SMEs). Further on, we will try to assess the incentives in place in Romania and estimate their efficiency.
- The communities also have the potential to provide direct benefits to citizens by implementing energy efficiency measures and lowering the energy bills. Many efforts are in

place in Romania for advancing energy efficiency, in different sectors: public and residential buildings, public transportation, water and waste management, but also in industry (with a special focus on SME). But the energy communities could enhance even more the effectiveness of these actions and produce direct - and rapid - benefits for their members.

- By supporting citizen participation, energy communities can moreover help in providing flexibility to the electricity system through demand-response and storage. The differentiated cost per energy unit, according to specific time slots, is not often used in Romania by households. The time slots are usually only day and night and getting less expensive energy by night means paying more for the energy consumed during daytime (compared to the non-differentiated price of energy). But since the usage of RES is expected to grow rapidly, the required flexibility of the changed electricity system will push towards the differentiated costs, that stimulate demand response from the consumers. Therefore, preparing citizens for this change is very important, and energy communities can play a crucial role in this transition.
- Individually or through citizen energy communities, active prosumers will be able to participate in all markets, either by generating, consuming, sharing or selling electricity. Empowering renewable energy communities to produce, consume, store and sell renewable energy will also help advance energy efficiency in households, support the use of renewable energy and at the same time contribute to fighting poverty through reduced energy consumption and lower supply tariffs.

Energy communities can take any form of legal entity, for instance: an association, a cooperative, a partnership, a non- profit organization or a small/medium-sized enterprise. Citizens and other market players can team up and invest together in energy assets. This contributes to a more decarbonized and flexible energy system, as the energy communities can act as one entity and access all suitable energy markets, on a level-playing field with other market actors. All these regulations are now enforced in Romania, and the Technical University of Cluj-Napoca (TUCN), as an advanced research and education body, wants to contribute, through its specialists, to the creation of energy communities in the North-Western region of Romania.

2. Milestones in the creation of energy communities in Romania

To assess the regional acceptance of RES in Romania, we cite a paper by Jijie (2021), that investigates if the Romanian public opinion favours such an evolution [1]. The study is established on the north-eastern region of Romania. The researchers applied a survey to 649 households. The results suggest the existence of strong support for adopting RES, growing concerns about climate change, and efforts to save energy. The main the factors supporting electricity production based on RES are: concerns regarding climate change or various economic factors and behaviours oriented towards reducing energy waste. According to the same paper [1], several expected impacts of turning to RES to produce electricity were assessed through focused questionnaires, and responses were measured on a 5-point Likert scale, as seen in Figure 1.

We estimate, therefore, that the public acceptance of RES is, nowadays, even higher, as the current energy crisis, deeply accentuated by the Ukrainian war, showed the world how important it is to beneficiate from energetic security and energetic independence, both as a country and an independent household. Becoming a prosumer, being responsible for the production of most (if not all) the energy you need, is now more urgent than ever before, as the cost of energy exploded this year.

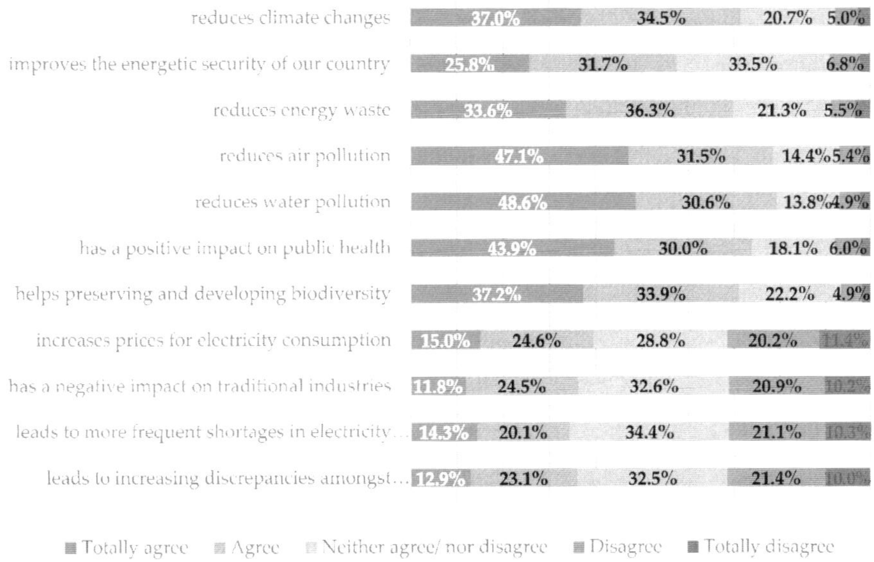

Figure 1. Perceived effects of turning towards RES–EP [1].

Regarding the required costs for RES installation, each country has in place a set of incentives to stimulate citizens or communities to invest in this sector. In Romania, the oldest program started in 2010 and is called Green House. It provided funds for the installation of heating systems that use renewable energy, and it is managed by the Administration of the Environment Fund (AFM). In 2019, a special section of the program called Green House Photovoltaics was launched, providing funds for PV installation. However, according to the information that Green Report obtained from ANRE (the Romanian regulatory agency for the energy sector) [3], at the end of November 2021, 13.182 prosumers were registered in Romania. As a percentage, however, the increase is significant, considering that, at the end of February 2021, there were only 1.732 prosumers. The total power of the electricity generation installations from renewable sources with Pi <100 kW (Pi means installed power) belonging to the prosumers was 79.434 kW. The amount of electricity produced and delivered by prosumers at the price provided by law was 26.048.302 (kWh). Although in place for 12 years now, this incentive did not prove to be highly effective, as the available funds were extremely limited. Because the financing program has not achieved its goal of supporting citizens to produce their own energy, those who want to install PV systems must do so from their own funds. Moreover, the price of energy regulated by law, offered to prosumers for the energy delivered in the network, was 3 or even 4 times lower than the purchase price from the supplier, making the investment disadvantageous and discouraging, precisely because of this large price discrepancy. On average, the depreciation of a photovoltaic system, without financial support, takes about 15 years.

However, the government promised and delivered the emergency ordinance 143/2021 amending the Energy Law, easing the bureaucratic path from a simple energy consumer to a prosumer, while providing an incentive for those who will deliver the surplus green energy produced in the network. It stipulates that electricity suppliers are required, at the request of prosumers, to make a quantitative compensation in the invoice of the prosumers, respectively to invoice only the difference between the amount of energy consumed and the amount of energy produced and delivered in the network. However, some of the prosumers were dissatisfied [4], considering that, although they deliver energy in the system, the energy they consume will have a much higher price, because it will also contain several taxes. ANRE is to elaborate the Methodology regarding the quantitative compensation provided in art. 73 ^ 1 para. (9) of Law no. 123/2012, with subsequent amendments and completions, within the deadline for issuing the regulations provided by law (they have not, so far). Through this measure, the government values more the energy produced by the citizens. Although the law does not set a target for the production capacity of energy through RES owned by prosumers, considering the new incentives in place, ANRE estimates the number of prosumers by the end of 2022 to 30.000 [3].

The previously mentioned drawback of the law, the fact that the energy consumed will have a much higher price, because it will also contain several taxes, including the costs of energy transportation in the grid, can be addressed through the creation of energy communities. In several EU countries, the households located within a certain distance from a prosumer (500m in Spain) can consume the locally produced energy without paying the transportation costs [8]. This is the main reason people are motivated to legally create energy communities, and a similar regulation in our country could stimulate the association of citizens for energy and environmental benefits.

SMEs can also be part of the energy communities. They could be stimulated by the City Councils to invest in PVs (or other RES) installed on the rooftops of publicly owned buildings. This energy transition partnership, between SMEs and local authorities, could be organized in a certain legal form, like the contribution that SMEs have to the financing of cultural events or parks maintenance (reference city: Cluj-Napoca). They could share the revenue from the energy produced by those RES, and jointly contribute to the wellbeing of the community.

Regarding the flexibility of the energy system, Demand Response (DR) programs encourager customers to reduce the peak power and energy consumption in response to a price signal or financial incentive [6]. Typically, the request to reduce peak demands is made for a specific period on a specific day, which is referred to as a Demand Response event (DR event). A demo pilot site was implemented during 2016–2019 in 4 public buildings of the TUCN- Figure 2 – and it tested four different automated and/or manual scenarios for DR. This pilot project was part of a Horizon 2020 European Project, no. 696114/2016, called DR BOB - Demand Response in Blocks of Buildings (www.dr-bob.eu).

The buildings have different occupational characteristics and occupants' behavior, different operating regimes, different administration mode, different operating programs, and different energy use profile. The testing scenarios included voluntary involvement of students, academic and administrative staff [2].

Peak power reduction in the educational buildings was, during one event, of 15,6% of the baseline consumption of the time slot. For the swimming pool complex (where less people were involved), the reduction was much larger (77%). One part of the electricity savings can be due to the awareness campaign organized by the pilot site. As stated before, preparing citizens for the required flexibility of the energy system that includes RES [10] is particularly important, and TUCN now has the necessary experience in training people and raising awareness regarding this problem [5, 7].

Another important topic for communities is the efficient use of energy. This is widely considered as the backbone of sustainable development in the future and there is still room for improvements concerning energy efficiency in various domains. TUCN is part of consortium that developed an educational & training program through the Horizon 2020 European Project no. 847132/2019, called A holistic framework for Empowering SME's capacity to increase their energy efficiency. The program is focusing on increasing the skills and competencies of the energy professionals employed by SMEs and the energy efficiency measures implemented by the SMEs, using self-created financial tools. Energy professionals and decision-making people must work together to obtain the best results for their organization, and TUCN contributes to making this happen [9].

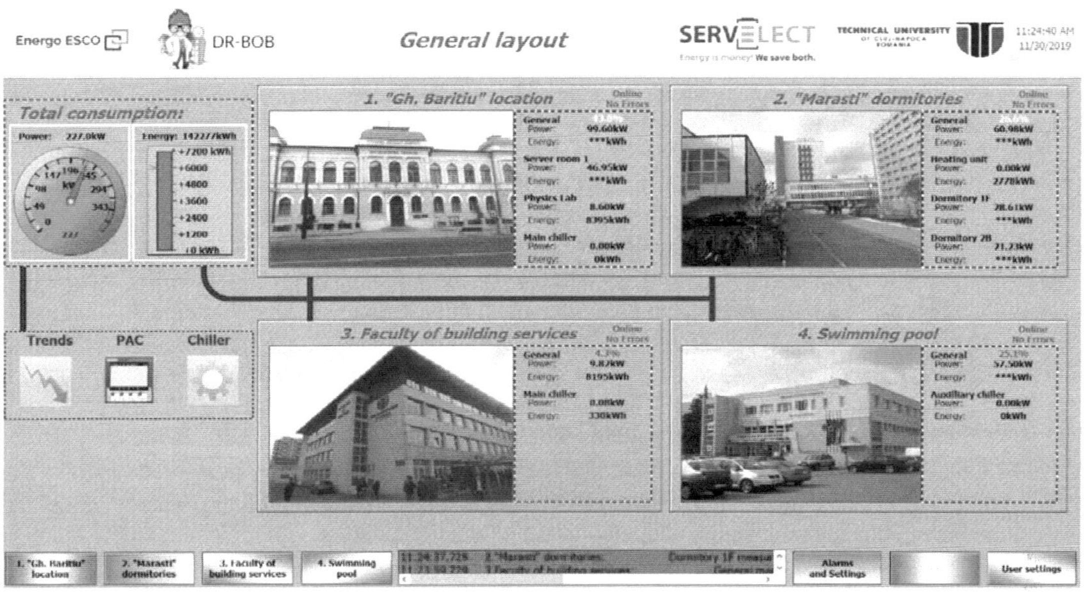

Figure 2. The interface of the monitoring systems implemented on four buildings belonging to the TUCN. Daily energy consumptions in (kWh) and absorbed power in (kW) were measured and stored for further data processing [2].

3. Conclusions

The foreseen steps into creating energy communities in Romania include:

- Choosing the appropriate form of legal entity: an association, a cooperative, a partnership, a non- profit organization or a small/medium-sized enterprise.
- Investing, together or individually, into the most efficient RES for the specific region of the country (i.e., PVs are suited for all regions, but wind energy can be exploited in Dobrogea or other regions), taking advantage of all the incentives in place provided by the policy makers.
- Organizing, inside the communities, training and awareness rising events, that could enhance energy efficiency or the capacity of the community to self-adjust and balance energy consumption and production. The excess production can be stored or sold into the grid.

Other incentive mechanisms are expected to be created by policy makers, but the communities themselves can participate in the co-creation of these, based on their acquired experience and their needs.

Another important support in the creation of energy communities can be accessed by participating with concrete projects in the call HORIZON-CL5-2022-D3-01-08, *Supporting the action of consumers in the energy market and guide them to act as prosumers, communities, and other active forms of active participation in the energy activities.* A similar goal is pursued by JPI Urban Europe, through the call *Positive Energy Districts and Neighbourhoods for Sustainable Urban Development.* The EU is quite committed to facilitate the implementation of the *Clean energy for all Europeans package,* and financing mechanisms are in place for communities, in different forms, if properly accessed by the end-users.

Acknowledgment This work was supported by a grant of the Ministry of Education and Research, CNS/CCDI-UEFISCDI, project number PN-III-P3-3.6-H2020- 2020-0129/PPH no.44/2021.

References

[1] Jijie D-T, Maxim A, Roman T and Roşcovan M 2021 Public Acceptance and Support of Renewable Energy in the North-East Development Region of Romania *Energies* **14**(18) 5834

[2] Jurj D I, Czumbil L, Bârgăuan B, Ceclan A, Polycarpou A and Micu D D 2021 Custom Outlier Detection for Electrical Energy Consumption Data Applied in Case of Demand Response in Block of Buildings *Sensors* **21**(9) p 2946

[3] https://green-report.ro/romania-30-000-prosumatori-finalul-anului/

[4] https://www.greenpeace.org/romania/articol/5955/prosumatorii-abandonati-si-de-actualul-guvern-pretul-pentru-energia-produsa-de-acestia-ramane-derizoriu/

[5] Crețu M, Czumbil L, Bargauan B, Ceclan A, Berciu A, Polycarpou A, Rizzo R and Micu D D 2020 Modelling and Evaluation of the Baseline Energy Consumption and the KPIs in TUCN Buildings within a Demand Response Programme – A Case Study *IET Renewable Power Generation Journal* **14**(15) pp 2864-2875 DOI: 10.1049/iet-rpg.2020.0096

[6] Crețu M, Ceclan A, Czumbil L, Şteţ D, Bârgăuan B and Micu D D 2019 Key Performance Indicators (KPIs) for the Evaluation of the Demand Response in the Technical University of Cluj-Napoca Buildings *28th International Conference on Modern Power Systems (MPS2019*, Cluj-Napoca Romania 21-23 May WOS:000612401900138

[7] Crețu M, Czumbil L, Bârgăuan B, Şteţ D, Ceclan A, Polycarpou A, Rizzo R and Micu D D 2019 Modeling and Forecasting Energy Demand in TUCN Buildings *International Conference on Clean Electrical Power (ICCEP2019)* Otranto Italy 2-4 July WOS:000620331100038

[8] Pacurar C, Topa V, Giurgiuman A, Munteanu C, Constantinescu C, Gliga M *et al* 2021 High Frequency Analysis and Optimization of Planar Spiral Inductors Used in Microelectronic Circuits *Electronics* **10**(23) 2897

[9] Stet D, Czumbil L, Ceclan A, Cirstea S, Muresan A Jurj D *et al* 2021 Educational and Training Program to Increase SME's Energy Efficiency Skills *56th International Universities Power Engineering Conference (UPEC)* DOI: 10.1109/UPEC50034.2021.9548263

[10] Farkas T, Czumbil L, Cretu M, Darabant L, Stet D, Ceclan A *et al* 2021 Assessment of the Romanian Pilot Site Energy Consumption Indicators and Technical Prerequisites in the Implementation of the RE-COGNITION Horizon Project *MPS Proceedings*

IOP Conf. Series: Materials Science and Engineering 1254 (2022) 012017 doi:10.1088/1757-899X/1254/1/012017

Influence of the Geometrical Parameters of a Planar Yagi-Uda Antenna on its Performances

C Constantinescu, C Păcurar, C Munteanu, A Giurgiuman, S Andreica and M Gliga

Faculty of Electrical Engineering, Department of Electrotechnics and Measurements, Technical University of Cluj-Napoca, George Baritiu Street no. 26, Cluj-Napoca, Romania

E-mail: Claudia.Constantinescu@ethm.utcluj.ro

Abstract. The paper follows the construction of a planar Yagi Uda antenna and the influence of the geometrical parameters on its characteristic parameters. According to the studies conducted on classical Yagi Uda antennas, modifying their geometrical parameters can lead to the improvement of the antenna function. Thus, the authors will consider modifying the number of directors, the distance between the fed dipole and the first director, the distance between the directors and the length of the directors in an attempt to determine their influence on the antenna's characteristic parameters. All the results are compared and the similitude between planar and classical antenna is highlighted. Some of the most important antenna parameters were analyzed namely S parameters, gain, directivity, frequency bandwidth and radiation pattern.

1. Introduction

Antennas can be defined as a system of conductor materials, that makes it possible for the signals to propagate from the emitter to the receptor on the shortest way. Usually, antennas are constructed from conductor materials such as copper or aluminum, their efficiency depending on the antenna parameters. The classical Yagi Uda antenna was invented in Japan and the first published results can be found in 1928. This type of antenna has high conductivity, are good as receivers their manufacturing costs are low and their gain and directivity have good values. Their constructive elements are reflector, dipole and directors as can be seen in Figure 1 a). Their advantages are the fact that they are robust and resistant due to their construction, they are easy to mount on poles and can be used for receiving low intensity signals [1-4].

a)

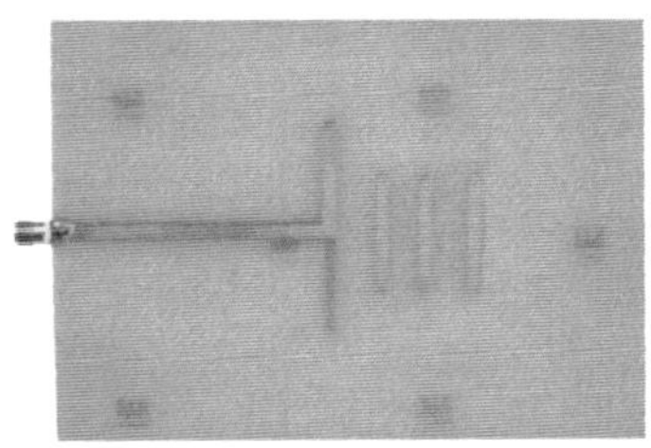

b)

Figure 1. Classical Yagi Uda antenna (a) and planar Yagi Uda antenna(b)[5].

Their disadvantages on the other hand made the researchers to consider another construction method for this type of antenna, namely the fact that their gain is small and in order to increase it the dimensions of the antenna must be increased and for lower values of the frequency this becomes practically impossible.

The first Yagi Uda planar antenna was proposed by Qian in 1998.In the following years, the researchers proposed numerous configurations using different feeding types and different shapes of the antenna's constitutive elements and concluded that each design has its own advantages and disadvantages due to the fact that even though an antenna has one of its specific parameters improved, another one can be negatively affected, thus an optimum can be reached only considering a single application at a time [5-9].

As all the planar antennas, the planar Yagi Uda antenna has as advantage its small dimension and mass, low manufacturing costs, and its ease of fabrication. They are used in wireless communication, radar systems, portable systems for direction detection, RFID (Radio Frequency Identification) readers [6-8].

The planar Yagi Uda antenna is easily placed in small devices, but one must consider its effect through radiation on the other elements of the systems, thus an electromagnetic study of the devices must take place [4,10-16].

2. Construction of the antenna implemented in the simulation program

The analyzed planar structure was modeled with the help of the numerical modeling program HFSS from ANSYS as it can be observed in Figure2.The antenna consists of a flexible dielectric, polyimide, with a relative permittivity of 3.5[17].On the upper side of this dielectric the dipole and directors are constructed, while on the other side there is the reflector. The initial dimensions in mm of all the elements can be seen in Figure 3.

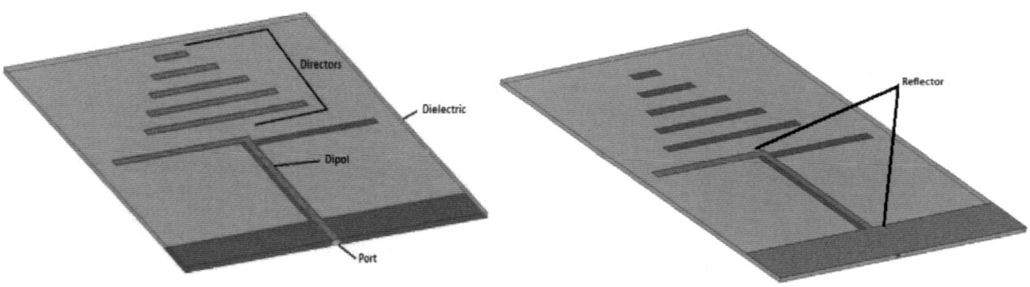

Figure 2. Planar Yagi Uda antenna with 5 director microstrip lines structure.

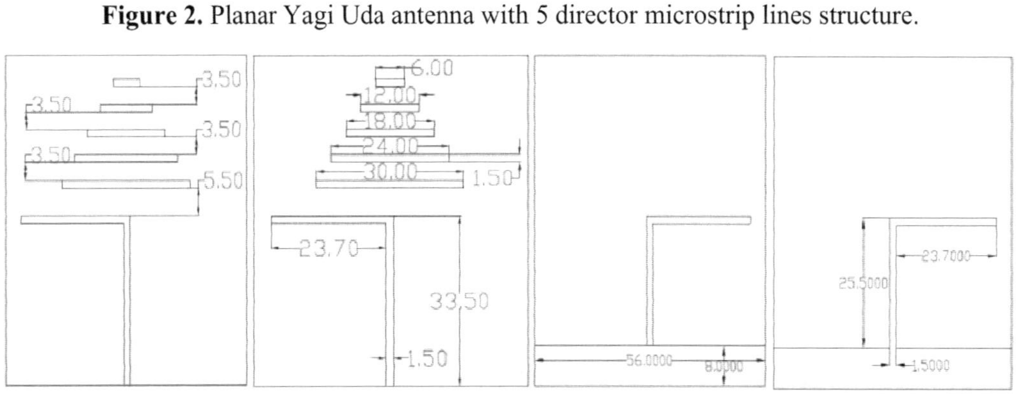

Figure 3. Dimensions of the conductor elements of the planar Yagi Uda antenna with 5 director microstrip lines.

The antenna was analyzed in the frequency range between 1 and 5 GHz. The influence of different constructive parameters like the distance between the feeding element and the first director, the distance between the directors, the length of the directors and the number of directors is determined, and some conclusions are drawn, all compared in the same time with the influence of these parameters on a classical Yagi Uda antenna.

The specific parameters of the antenna followed in this study are the radiation pattern, gain and directivity, and S parameters [16,18-20].

It has to be specified that the term S parameters is referring to the parameter S11 which is determined with the help of the numerical modelling program Ansys HFSS for the studied structures.

3. Influence of the antenna construction on the specific parameters

Starting from the original structure, the geometrical characteristics of the directors were modified in order to determine their influence on the antenna functionality using Ansys HFSS.

3.1. Influence of the distance between the feeding element and the first director

The distance between the feeding dipole and the first director was increased from 3.5 mm to 9.5 mm, being varied below and above the value of the distance considered for the initial structure. Considering the maximum gain and directivity, it can be stated that the maximum value is obtained for a distance of 9.5 mm between the feeding element and the first dipole. It can be stated that the maximum directivity and gain increase with the increase of the distance. In Figure 4 are represented the four analyzed structure with the distances of 3.5 mm, 5.5 mm, 7.5 mm and 9.5 mm between directors and feed.

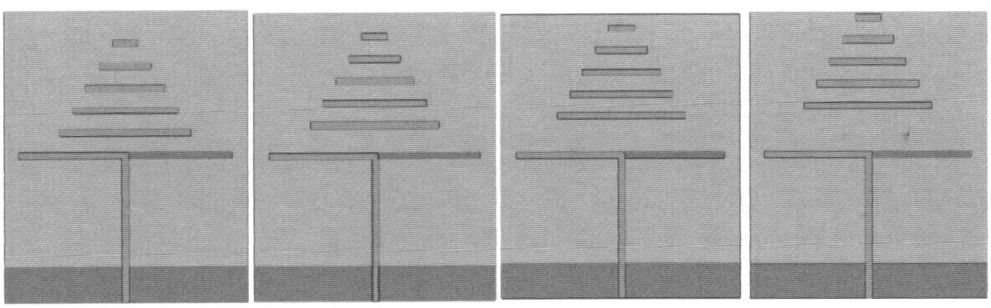

(a) (b) (c) (d)

Figure 4. The four considered structures for the study: (a) 3.5 mm, (b) 5.5 mm, (c) 7.5 mm,(d) 9.5 mm.

In Table 1 the maximum directivity and gain value for the considered structures are presented.

Table 1. Influence of the distance between the feeding element and directors on gain and directivity.

Distance between the feeding element and first director [mm]	Maximum directivity value	Maximum gain value
3.5	4.0853	4.0659
5.5	4.2013	4.1817
7.5	4.2763	4.2675
9.5	4.3519	4.3459

Considering the S parameters, each of the structures has its own operational bandwidth. The S parameters representation can be observed in Figure 5, while in Table 2 the bandwidth for each structure is determined as a percentage. It can be stated that the structure with the smallest distance between the feeding element and the first director has the widest bandwidth. The bandwidth decreases with the increase of the distance up to one point and after that there is a possibility of a slight increase again. The resonant central frequency is maintained at the same value of approximately 2.6 GHz.

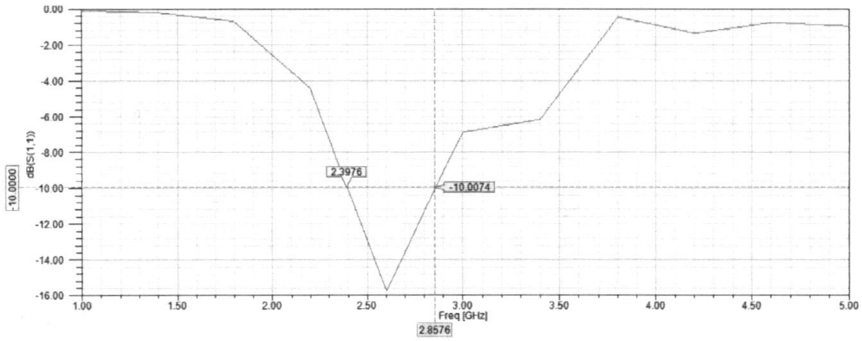

(a)

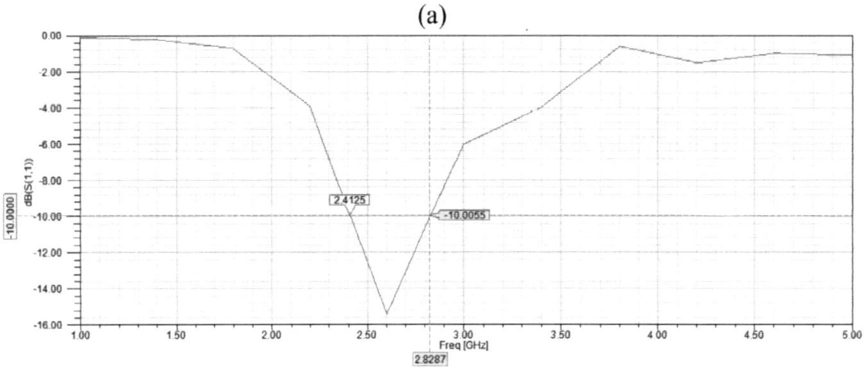

(b)

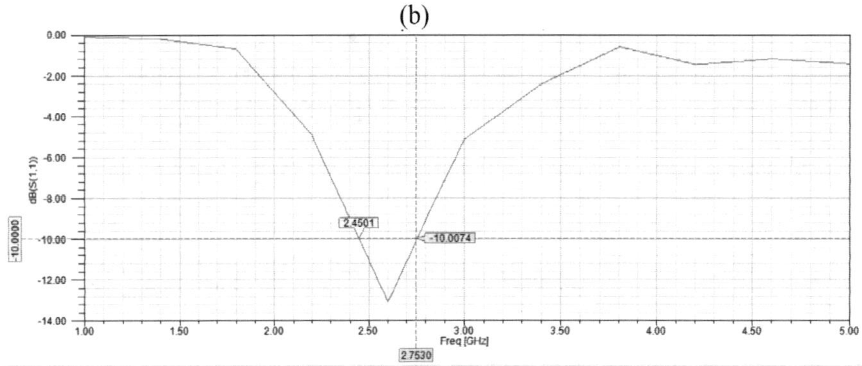

(c)

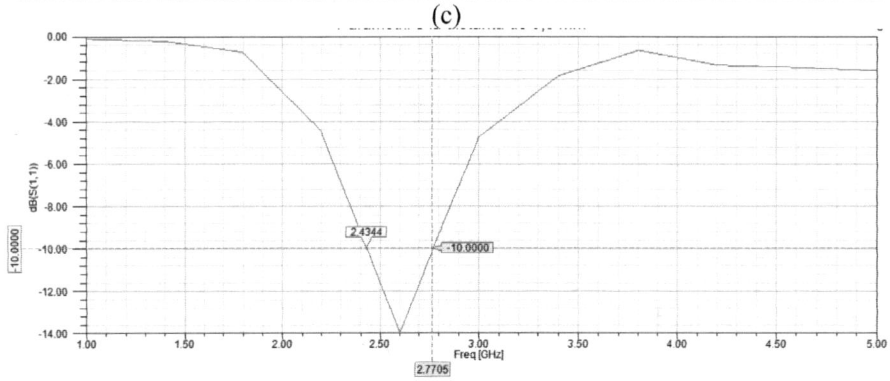

(d)

Figure 5. S parameters representation for the 4 structures: (a) 3.5 mm, (b) 5.5 mm, (c) 7.5 mm, (d) 9.5 mm.

The bandwidth for each of the four studied structures is calculated as a percentage and can be seen in Table 2.

Table 2. Influence of the distance between the feeding element and the first director on the bandwidth.

Distance between the feeding element and first director [mm]	Bandwidth [%]
3.5	17.51
5.5	15.88
7.5	11.645
9.5	12.92

The radiation pattern of such an antenna remains the same throughout the study due to the elements role, namely to conduct the field along the directors and prevent it to go in the direction of the reflector, the same as for a classical Yagi Uda antenna. Small differences appear considering the maximum value of the radiation, but they are practically unnoticeable. The highest values are obtained for the distance between directors and feeding of 9.5 mm (Figure 6).

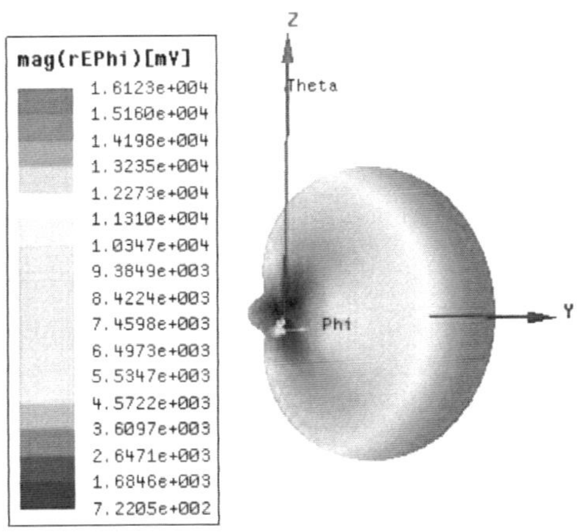

Figure 6. Radiation pattern for the antenna with distance between feeding and directors of 9.5 mm.

3.2. Influence of the number of directors

In this study, the number of directors was varied, thus resulting the structures from Figure 7, with 3, 4 and 5 directors. The maximum gain and directivity of these structures were obtained, and the results are shown in Table 4. As it was expected and following the same pattern as the one of a classical Yagi Uda antenna, the maximum values were obtained for the highest number of directors.

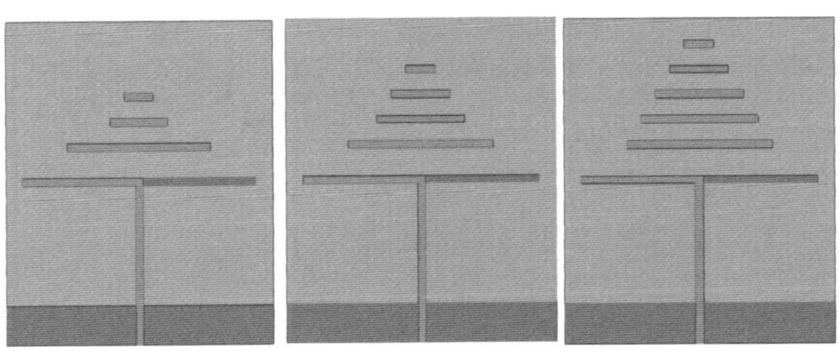

(a) (b) (c)

Figure 7. The three considered structures for the study: (a) 3 directors, (b) 4 directors, (c) 5 directors.

In oder to compare the directivity and gain maximum values for the three structures, their obtained values are presented in Table 3.

Table 3. Influence of the number of directors on the gain and directivity.

Number of directors	Maximum directivity value	Maximum gain value
3	3.8981	3.8944
4	4.0751	4.0658
5	4.2013	4.1817

As in the previous cases, the scattering parameters were obtained and the bandwidth of each of the considered structures was determined. Thus, from Figure 8 it can be concluded that the largest bandwidth is obtained for the antenna with 5 directors, and, as it can be seen from Table 4, the bandwidth has almost tripled as a percentage in comparison to the one obtained for the 3-director structure. The central frequency is maintained at the same value of 2.6 GHz.

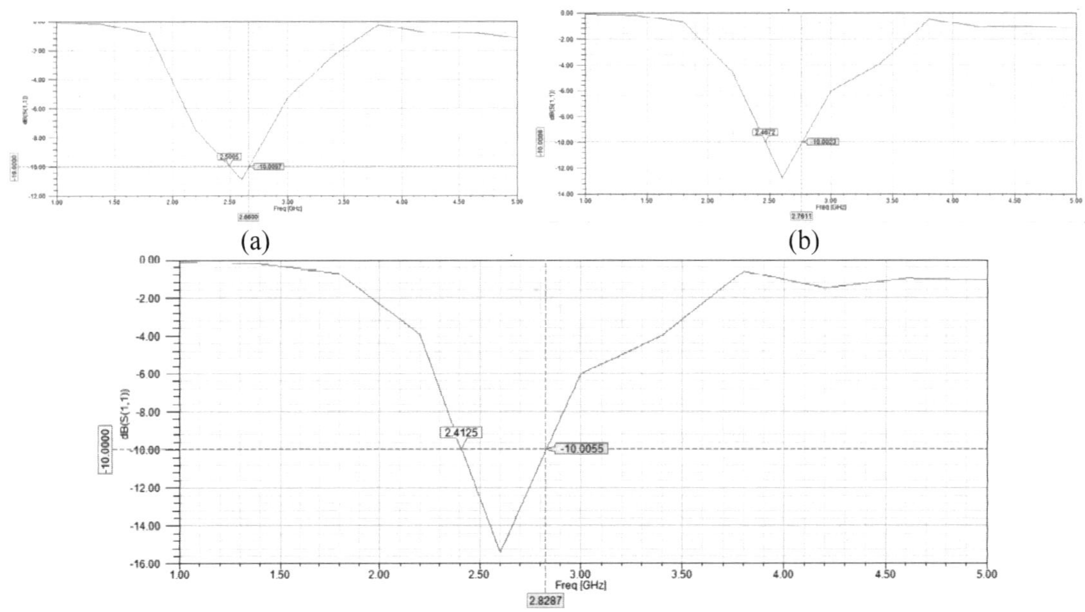

(a) (b) (c)

Figure 8. Scattering parameters for the structure with: (a) 3 directors, (b) 4 directors, (c) 5 directors.

The bandwidths of operation for the three different antenna structures are presented in Table 4.

Table 4. Bandwidth obtained for the 3 analyzed structures with different director number.

Number of directors	Bandwidth [%]
3	6.18
4	11.24
5	15.88

As it was expected, the maximum value of the radiation increases with the increase of the number of directors, as it can be seen in Figure 9.

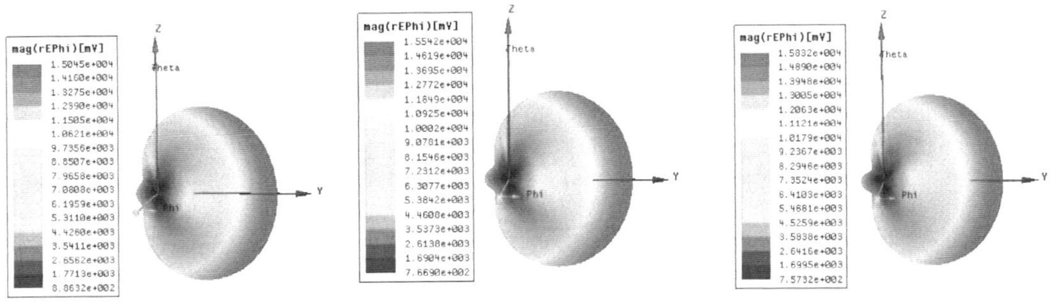

(a) (b) (c)

Figure 9. Radiation pattern obtained for the structures with:(a) 3 directors, (b) 4 directors, (c) 5 directors.

3.3. Influence of the distance between the directors

The distance between the directors has a great impact on the characteristic parameters of the antennas. Thus, we wanted to determine for this specific application the impact of modifying this distance between the directors and considered 3 distances, namely 2.5 mm, 3.5 mm and 4.5 mm.

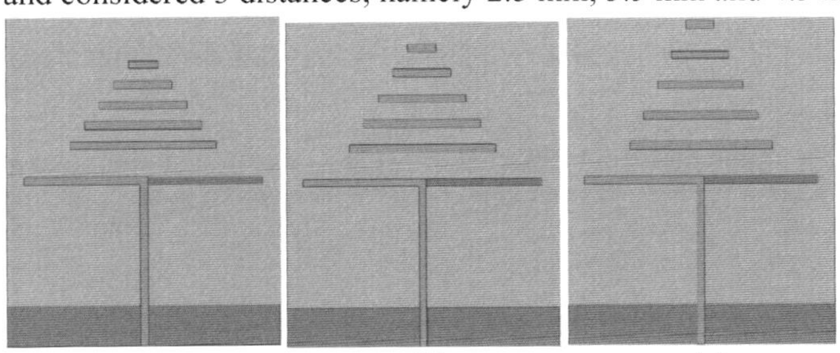

(a) (b) (c)

Figure 10. Structures with different distances between directors: (a) 2.5 mm, (b) 3.5 mm, (c) 4.5 mm.

After determining the maximum gain and directivity, it was observed that the directivity and fain increase with the increase of the distance between directors, thus from this point of view a larger value of the distance is what we want.

Table 5. Influence of the distance between the directors on the gain and directivity.

Number of directors	Maximum directivity value	Maximum gain value
2.5	4.0287	4.0241
3.5	4.2013	4.1817
4.5	4.3005	4.3167

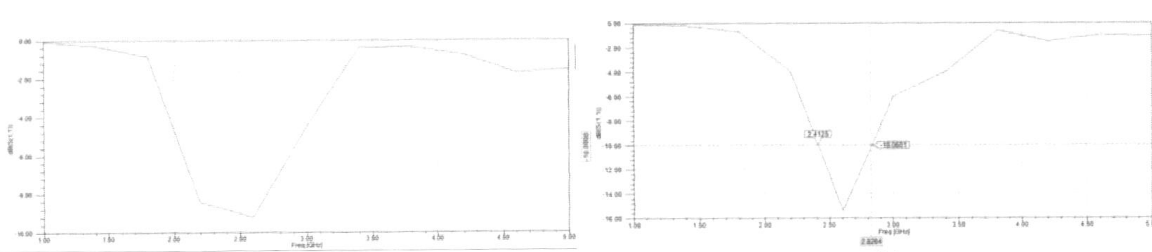

(a) (b)

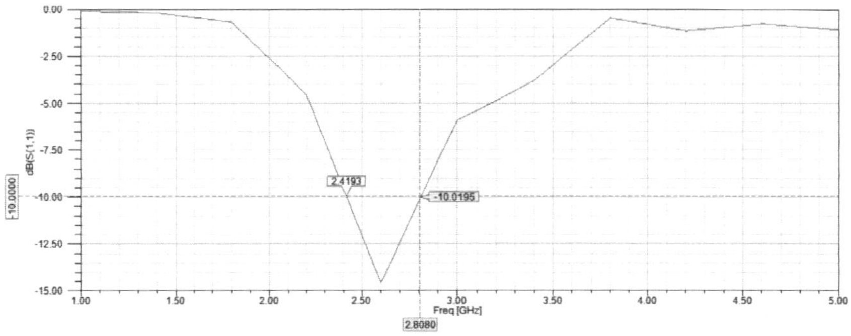

(c)

Figure 11. Scattering parameters for the structure with distance between directors: (a) 2.5 mm, (b) 3.5 mm, (c) 4.5 mm.

After determining and analyzing the S parameters for the different director distance, it was observed that the first structure, the one with the distance of 2.5 mm does not function in the analyzed frequency range. The second structure has the wider operational bandwidth, but the third structure's bandwidth is large also. It also can be stated that the bandwidth does not vary linearly with the distance between directors. The central frequency is found at the same value of approximately 2.6 GHz.

Table 6. Bandwidth obtained for the 3 analyzed structures with different distance between directors.

Distance between directors	Bandwidth [%]
2.5	-
3.5	15.8
4.5	14.87

Analyzing the radiation pattern for the three structures, it can be stated that the allure is not affected by the distance of the conductors and the maximum value is increased a bit with the increase of the distance between directors (Figure 12).

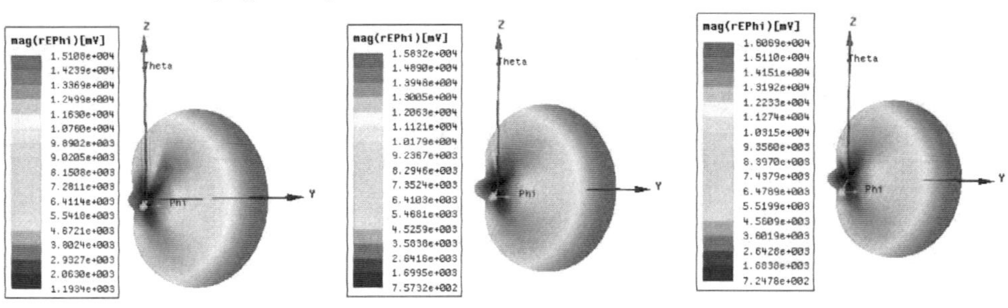

(a) (b) (c)

Figure 12. Radiation pattern obtained for the distance between the directors: (a) 2.5 mm, (b) 3.5 mm, (c) 4.5 mm.

3.4. *Influence of the length of the directors*

It was considered the initial structure as the beginning point of this study. The director lengths of this structure were decreased with 5 mm (Figure 12(a)) and also increased with 10 mm (Figure 12(c)). After using HFSS for modelling and analyzing these structures, the results show an increase of the directivity and gain with the increase of the director lengths. Even though the increase of the lengths is beneficial, one must keep in mind when designing such an antenna that the dimension of a director is always smaller than the one of the feeding elements and also the dimension of the dielectric is an impediment.

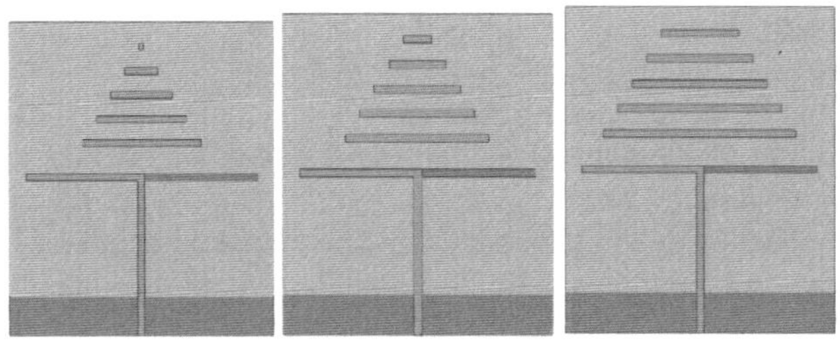

(a) (b) (c)

Figure 13. Structures with different distances director lengths: (a) initial - 5 mm, (b) initial, (c) initial + 10 mm.

Table 7. Influence of the director lengths on the gain and directivity.

Length of directors	Maximum directivity value	Maximum gain value
-5 mm from the initial structure length	4.0543	4.0318
Length from the initial structure	4.2013	4.1817
+10 mm to the initial structure length	4.8706	4.8313

Even though the increase of the director length was benefic to the directivity, when considering the bandwidth, the lowest value of 7.18 % according to Table 8 is obtained for the structure with the highest length value. The highest bandwidth value is obtained for the initial structure. The radiation maximum value is also obtained for the structure with the longest directors, and in this case, there is a visible shrinking of the back lobe, thus the antenna has lower losses [21].

Table 8. Bandwidth obtained for the 3 analyzed structures with different director lengths.

Length of directors	Bandwidth [%]
-5 mm from the initial structure length	11.06
Length from the initial structure	15.88
+10 mm to the initial structure length	7.18

Analyzing the radiation pattern, it can be stated that the increase of the director length leads to a reduction of the secondary posterior lobe, fact which improves its functionality, also slightly increasing the radiation value.

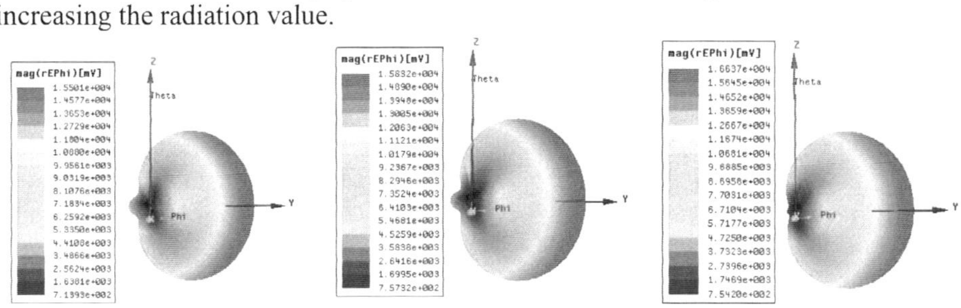

(a) (b) (c)

Figure 14. Radiation pattern obtained for the structures with different director lengths: (a) initial - 5 mm, (b) initial, (c) initial + 10 mm.

4. Conclusions

After the four studies conducted on the planar Yagi Uda structures some conclusions were drawn. It was observed that the maximum directivity and gain increase once new director elements are added to the structure. The operating bandwidth and the radiation pattern are also the highest for the structure with the highest number of directors.

Modifying the distance between the 5 directors it was determined that the gain and directivity maximum value is obtained for the highest distance between directors considered. The directivity linearly increases with the distance between directors as it was expected and in accordance with the classical Yagi Uda operation. The bandwidth unfortunately decreases with the increase of the distance between directors, thus an optimum for every application should be determined.

When considering the length of the directors, the directivity is increased with the increase of the director length, while the bandwidth does not linearly depend on this parameter modification. The radiation is better when the length of the directors is higher, resulting in a reduced secondary lobe, thus reduced losses.

The main conclusion is that the theory regarding the conventional Yagi Uda antenna can be applied to the planar Yagi Uda antenna. The specific parameters of these antennas can be easily modified through geometrical modifications.

References

[1] Ochala I and Okeme I C 2011 Design and Implementation of a High-Gain Compound Yagi Antenna *Advances in Applied Science Research, Pelagia Research Library* **2** (6) pp 41-50

[2] Balanis C A 2005 *Antenna theory Analysis and design*, Third Edition, Hoboken, (New Jersey: John Wiley&Sons, Inc.)

[3] Păcurar C *et al* 2020 High Frequency Analysis of the Influence of YagiUda Antenna on the Human head *2020 International Conference and Exposition on Electrical and Power Engineering (EPE)* pp 370-375 doi: 10.1109/EPE50722.2020.9305622

[4] Constantinescu C *et al* 2020 High Frequency Analysis of the Vivaldi Antenna Parameters *2020 International Conference and Exposition on Electrical And Power Engineering (EPE)* pp 376-381 doi: 10.1109/EPE50722.2020.9305674

[5] Pasternak M and Pietrasiński J 2017 Planar Uda-Yagi antenna for imaging SAR system Proc. SPIE 10418 *XI Conference on Reconnaissance and Electronic Warfare Systems* 104180I (20 April 2017)

[6] Abbosh A I, Babiceanu R F, Al-Rizzo H, Abushamleh S and Khaleel H R 2013 Flexible Yagi-Uda Antenna for Wearable Electronic Devices *IEEE Antennas and Propagation Society International Symposium (APSURSI)* 7-13 July pp 1200-1201

[7] Chahat N, Zhadobov M, Le Coq L and Sauleau R 2012 Wearable endfire textile antenna for on body communications at 60 GHz *IEEE Antennas and Wireless Propagation Letters* July 2012 vol II pp 799-802

[8] Jianying L, Yichen Z, Xin Y, Lulu C, Panpan Z, Mengjun W 2016 Benging effects on a flexible Yagi-Uda antenna for wireless body are network*7th Asia-Pacific International Symposium on Electromagnetic Compatibility (APEMC)*17-21 May pp 1001-1003

[9] Răcășan A, Munteanu C, Topa V, Păcurar C and Hebedean C 2014 Filter geometry optimization for the conduction electromagnetic interferences suppression *2014 International Conference on Optimization of Electrical and Electronic Equipment (OPTIM)* pp 46-51 doi: 10.1109/OPTIM.2014.6850930

[10] Răcășan A, Munteanu C, Topa V, Păcurar C and Hebedean C 2014 Electromagnetic interferences suppresion in planar integrated devices *2014 International Symposium on Electromagnetic Compatibility* pp 940-945 doi: 10.1109/EMCEurope.2014.6931038

[11] Kraus J D 1992 *Electromagnetics* (New York, McGraw-Hill Book Co.)

[12] Kumar H and Kumar G 2017 Compact planar Yagi-Uda antenna with improved characteristics *11th European Conference on Antennas and Propagation (EUCAP)* pp 2008-2012

[13] Mesquita M, Pires N, and Moreira A A 2013 Influence of Deformations on the Matching of a Flexible Dual-Band Antenna *Proceedings of 7th European Conference on Antenna and Propagation* (Gothenbure: Sweden) April 8-12 pp 3283-3285

[14] Constantinescu C, Munteanu C, Păcurar C, Giurgiuman A, Gliga M and Andreica S 2019 High Frequency Analysis of Bowtie Antennas *2019 11th International Symposium on Advanced Topics in Electrical Engineering (ATEE)* pp 1-6 doi: 10.1109/ATEE.2019.8724972
[15] Zheng G, Kishk A, Glisson A and Yakovlev A 2004 Simplified feed for modified printed Yagi antenna *Electronics Letters* **40** No. 8 pp 464-466
[16] Constantinescu C, Munteanu C, Păcurar C and Răcășan A 2018 Influence of the Patch Antenna Feeding on their Parameters *2018 International Conference and Exposition on Electrical and Power Engineering (EPE)* pp 0235-0240 doi: 10.1109/ICEPE.2018.8559782
[17] Constantinescu C, Păcurar C, Giurgiuman A, Munteanu C, Andreica S and Gliga M 2021 Numerical Modelling and Analysis of Circular Patch Antenna Array for Further Use Determination *2021 9th International Conference on Modern Power Systems (MPS)* pp 1-6 doi: 10.1109/MPS52805.2021.9492557
[18] Giurgiuman *et al* 2020 The Analysis, Modelling and Comparison between Circular and Rectangular Patch Antennas *2020 International Conference and Exposition on Electrical and Power Engineering (EPE)* pp 382-387 doi: 10.1109/EPE50722.2020.9305549
[19] Sharma G, Sharma A N, D Ashish and P K Singhal 2012 Yagi-Uda Antenna For L-Band Frequency Range *International Journal of Engineering and Technology, Science Publishing Corporation,* **1**(4) pp 315-320
[20] Zhang H S, Xiao K, Qiu L and Chai S L 2014 Wide band E-shape wearable antenna for wireless body network *IEEE International Wireless Symposium (IWS 2014)* 24-26 March
[21] Hebedean C, Munteanu C, Răcășan A and Păcurar C 2012 Optimum geometry for planar structures regarding their loss factor *2012 International Conference and Exposition on Electrical and Power Engineering* pp 693-698 doi: 10.1109/ICEPE.2012.6463851

The influence of the patch antennas emissions on the human head

C Păcurar, V Ţopa, A Giurgiuman, C Munteanu, C Constantinescu, M Gliga and S Andreica

Faculty of Electrical Engineering, Department of Electrotechnics and Measurements, 26-28 George Baritiu Street, Cluj-Napoca, Romania

E-mail: claudia.pacurar@ethm.utcluj.ro

Abstract. The influences of the antenna's emissions on the human body are to be investigated because in our days, as a simple example, the mobile phone is permanently in use, on day or night. The long-time human exposure at the electromagnetic field radiations is obvious and there exist the possibility that these can affect the human health. The influences of the circular patch antenna's emissions on the human head are analyzed at high frequency in this paper. First the patch antenna is modelled and simulated to extract its specific parameters and to represent the emitted electric and magnetic field around it. Next the human head is considered near to the patch antenna in position that imitates the mobile phone at ear and the electric field, the magnetic field, and the specific absorption rate (SAR) are represented to determine the antenna's emissions at its operating frequency. So, the main aim of the paper is the high frequency analyzes of the circular patch antennas to investigate the influence of their emissions on the human head. The results obtained are compared and reported to the standards and legislation in force on human occupational exposure, to verify that the limits are met.

1. Introduction

In everyday life we are in constant contact with electromagnetic waves. This is due to the continuous development of mobile networks, communications, data transfer via satellite or radar, these being just some examples.

In this context, the influences of the electromagnetic fields emissions on the human body are to be investigated. In our days we use frequently devices that incorporate antennas for different personal or professional activities. For example, the mobile phones are permanently in use, functioning on day and not only, very often even on night, so that the long-time human exposure at their electromagnetic field's emissions can affect the human health.

In this paper, the influences of the circular patch antenna's electromagnetic fields emissions on the human head are analyzed, considering that this type of antennas can be found in the devices that are very often used by people. In our research activities different types of antennas were high frequency analyzed to find their parameters [1-7].

The antenna can be a device with which a high frequency energy is radiated, case in which it is an emission antenna or can be a device which allows the capture of energy from a high frequency electromagnetic field and introduces it into a receiving installation, in this case it is an antenna receiver. Because the antennas are made of conductive material, under the action of radio waves the movement of electrons in the conductor is determined and an electric current is created. Due to its variation, a variation of the electric field created by the current is determined, which causes a magnetic field [2, 8-9].

The analysis by high frequency numerical modelling of the circular patch antenna, to determine SAR, electric and magnetic fields on the human head, by three-dimensional numerical modelling, will be performed using ANSYS-HFSS (High Frequency Structure Simulator).

A circular patch antenna is implemented in the program. First, only the antenna is modelled to determine its specific parameters and its specific field quantities. Next, the antenna is modelled in the presence of the human head to analyze the influence of the antenna's emissions (SAR, electric field, and magnetic field) on it.

The obtained results are compared and reported to the standards and legislation in force on human occupational exposure, to verify that the limits are met.

An order has been issued at national level, regulating the levels of exposure of the population to electromagnetic fields. This document transposes the recommendation made by the European Council and must be applied to all member states [10-12].

2. High frequency analyses of circular patch antenna

The circular patch antenna is high frequency analyzed to find its specific parameters and to represent its emitted electromagnetic field. The circular patch antenna with the configuration detailed in Figure 1 is implemented in the ANSYS-HFSS for high frequency 3D numerical modeling.

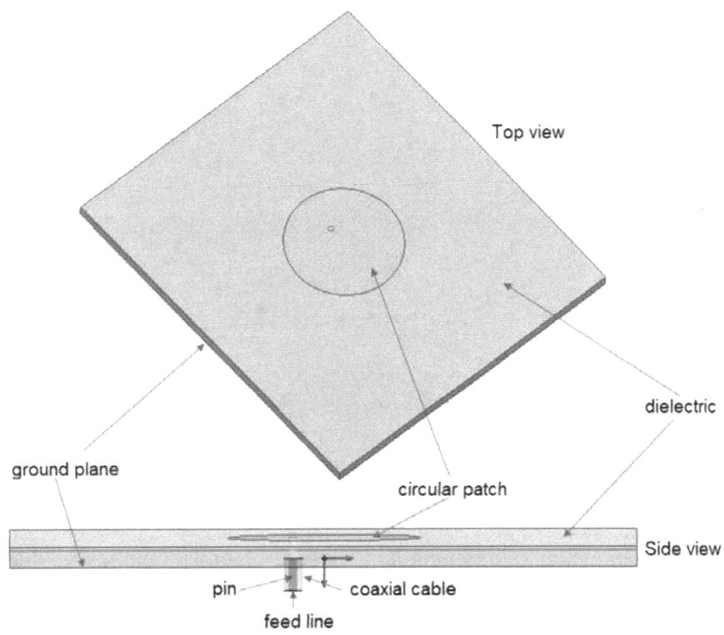

Figure 1. The circular patch antenna.

The patch antennas consist out of the following layers: very thin conductor layer for the patch of the antenna, dielectric layer on which the patch is placed, and a ground plane placed on the bottom of the dielectric layer. The feeding type for the considered circular patch antennas is the coaxial cable feed. The antenna was designed for the frequency 3.5 GHz. The antenna's principal components dimensions are patch radius 1.55 cm and dielectric 10x9x0.32 cm.

2.1. Antenna's parameters

The circular patch antenna's parameters are determined first: peak directivity, peak gain, peak realized gain, radiated power, accepted power, radiation efficiency and so on. The parameters of the analyzed

IOP Conf. Series: Materials Science and Engineering 1254 (2022) 012018 doi:10.1088/1757-899X/1254/1/012018

circular patch antenna are presented in Figure 2. The peak directivity is 5.33, the radiated power is 10.27 mW and the antenna has a good radiation efficiency.

Antenna Parameters:

Quantity	Freq	Value
Max U	3.5GHz	4.3634 mW/sr
Peak Directivity		5.3362
Peak Gain		5.5269
Peak Realized Gain		5.4134
Peak System Gain		5.4134
Radiated Power		10.276 mW
Accepted Power		9.9211 mW
Incident Power		10.129 mW
System Power		10.129 mW
Radiation Efficiency		1.0357

Maximum Field Data:

rE Field	Freq	Value	At(Theta,Phi)
Total	3.5GHz	1.8138 V	-90deg,90deg
X		1.8138 V	-90deg,90deg
Y		1.1043 V	-90deg,0deg
Z		98.324 mV	-40deg,90deg
Phi		1.8138 V	-90deg,90deg
Theta		201.21 mV	-2deg,90deg
LHCP		1.2832 V	-90deg,90deg
RHCP		1.2819 V	-90deg,90deg
Ludwig3/X dominant		1.8138 V	-90deg,90deg
Ludwig3/Y dominant		1.1043 V	-90deg,0deg

Figure 2. The parameters of circular patch antenna.

2.2. The directivity of the antenna

The circular patch antenna's directivity as 3D diagram and as representation with the antenna are represented in Figures 3. From the tridimensional diagram is deduced that the principal lobe of the directivity has only a direction proving that the antenna is a unidirectional one.

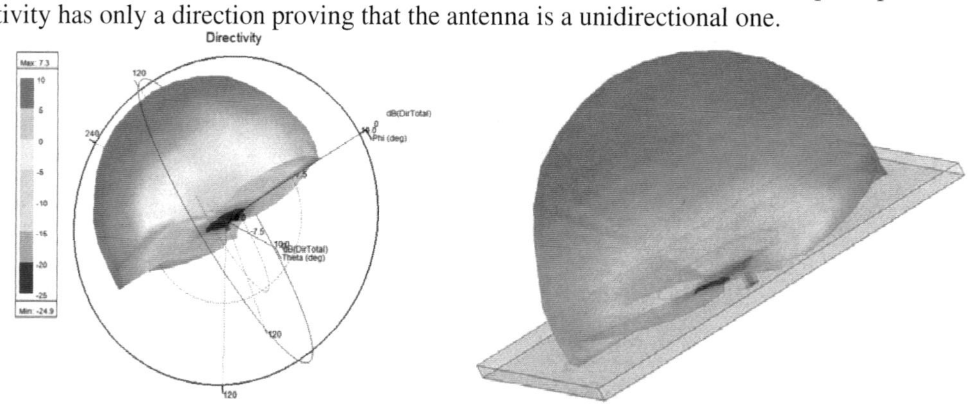

(a) diagram (b) representation with antenna

Figure 3. The 3D directivity diagram of circular patch antenna.

2.3. The S parameters of the antenna

The simulated S parameters of the circular patch antenna supplied with coaxial cable are represented in Figure 4. Analyzing the S parameters in the frequency domain 1-6 GHz, obtained for the circular patch antenna it can be note that the resonant frequency of the antenna is 3.5 GHz.

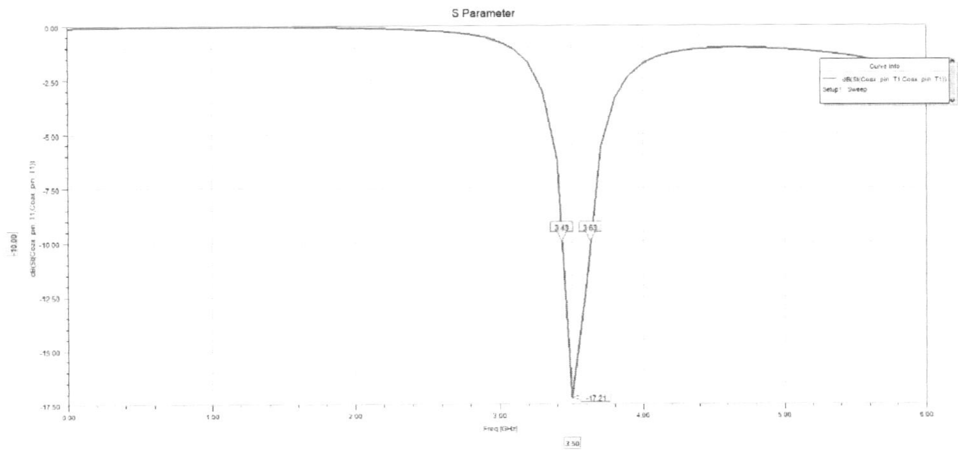

Figure 4. The S parameters of circular patch antenna.

The plot shows that the antenna operates in the frequency range between 3.43 GHz and 3.63 GHz and resonating at 3.5 GHz. Using the data from this plot, we can calculate the bandwidth of the circular patch antenna. First, with the formula:

$$f_{avg} = \frac{f_1 + f_2}{2} \tag{1}$$

is calculated the average frequency. The band width will be:

$$b_w = \frac{f_2 - f_1}{f_{avg}} \cdot 100, \ \% \tag{2}$$

So, using the data from S parameters plot, $f_1 = 3.43$ GHz and $f_2 = 3.63$ GHz (Figure 4). For the circular patch antenna simulated, the average frequency is $f_{avg} = 3.53$ GHz and the band width is $b_w = 5.66$ %.

2.4. The electric field emitted by the antenna

To analyze the patch antenna's emissions, the electric field must be represented and compared with the limits of the standards in force.

In Figure 5 is represented the propagation of the electric field around the planar patch antenna, in the entire 3D model top and bottom view. The field is radiated in the emission direction of the antenna and the maximum filed value is 3241 V/m.

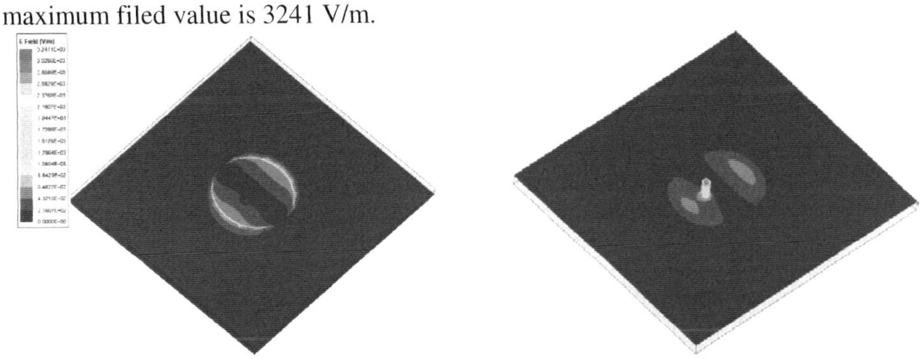

(a) top view (b) bottom view

Figure 5. The electric field of circular patch antenna.

In Figure 6 is represented the propagation of the electric field around the planar patch antenna, in the entire 3D model, from top and bottom views using limit 0-200 V/m for a better visualization of the field distribution around antenna and comparison with the standards enforced.

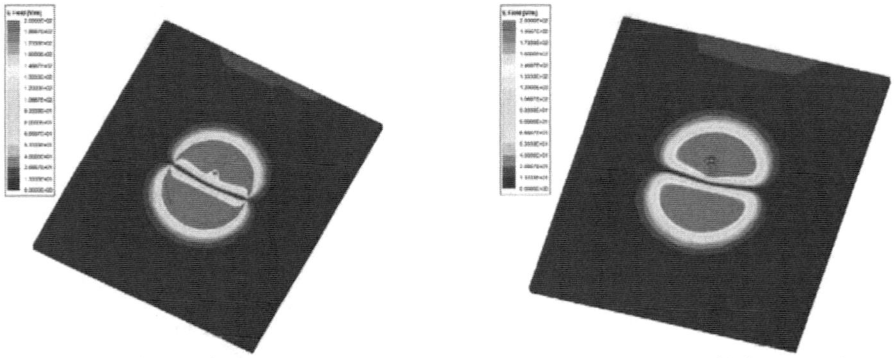

(a) top view (b) bottom view

Figure 6. The electric field of circular patch antenna.

In Figure 7 is represented the same electric field in XY plane, a cross-section plane of the antenna to see how it penetrates in the area around it for the same limits.

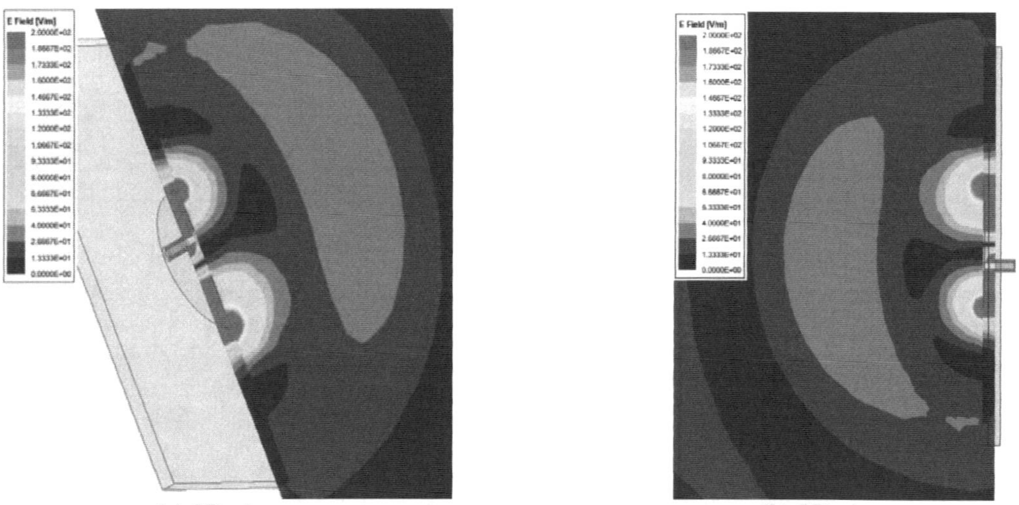

(a) 3D view (b) 2D view

Figure 7. The electric field of circular patch antenna in XY plane.

We use the limits 1-200 V/m in the electric field representations from Figures 6 and 7, for a better visualization of the field distributions in and around the circular patch antenna. The same limits will be used for the electric field in all the analyzed cases, and obvious in the final comparison.

2.5. The magnetic field emitted by the antenna

In Figure 8 is represented the propagation of the magnetic field around the planar patch antenna, in the entire model of the antenna. The maximum values are found in the antenna area, in its immediate vicinity and at a certain distance from it in its operating direction.

As in the case of electric field, the magnetic field decreases in intensity gradually by moving away from the antenna as we expected to be normal.

In this figure can be noted that the magnetic field in the entire model is more intense in the coaxial cable area, with values around 6.73 A/m.

IOP Conf. Series: Materials Science and Engineering 1254 (2022) 012018 doi:10.1088/1757-899X/1254/1/012018

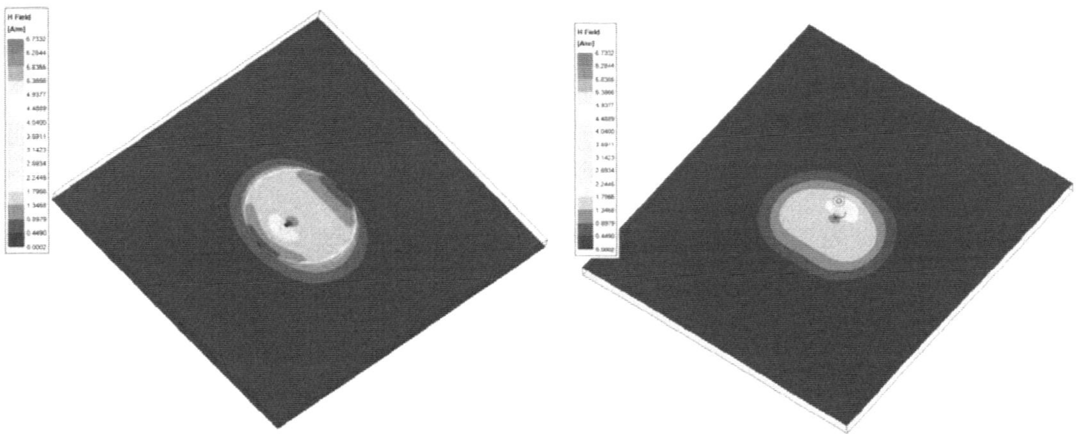

(a) top view (b) bottom view

Figure 8. The magnetic field of circular patch antenna in the entire model.

We use the limits for magnetic field representation to be 1-5 A/m for a better visualization of the field variation and the magnetic field is represented with these limits in Figures 9 and 10.

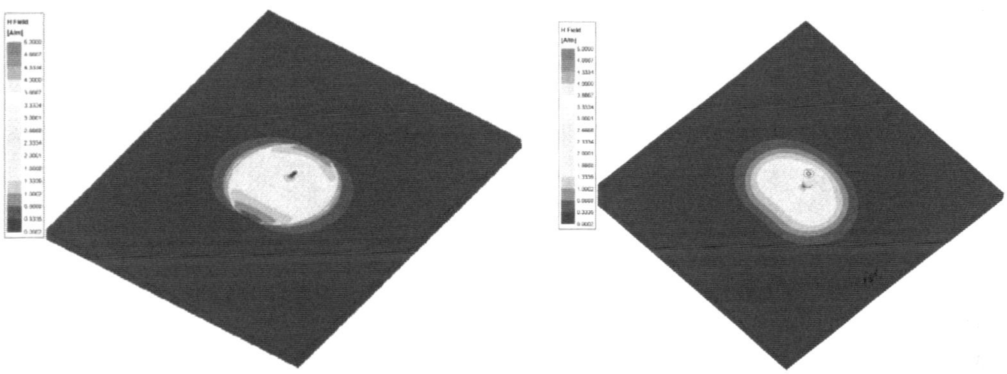

(a) top view (b) bottom view

Figure 9. The magnetic field of circular patch antenna in the entire model.

The magnetic field intensity distribution is even more obvious in Figure 10, in the cross-section plane, XY plane.

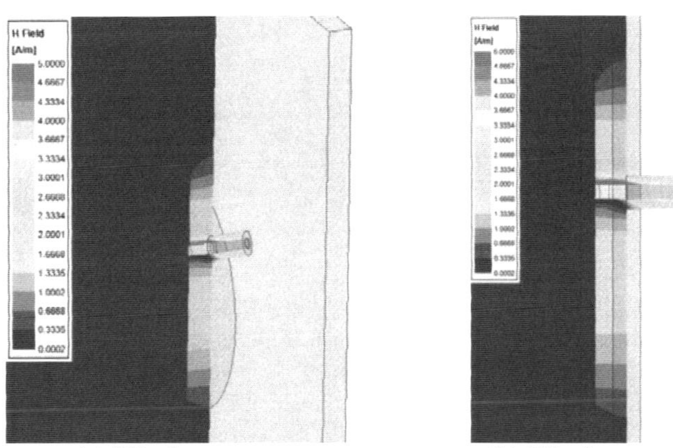

(a) 3D view (b) 2D view

Figure 10. The magnetic field of circular patch antenna in the XY plane.

3. The influence of circular patch antenna on the human head

The antenna is modelled in the presence of the human head to analyze the influence of the antenna's emissions (SAR, electric field, and magnetic field) on it. For this simulation, the female head phantom from the ANSYS-HFSS's Components Libraries is considered (Figure 11).

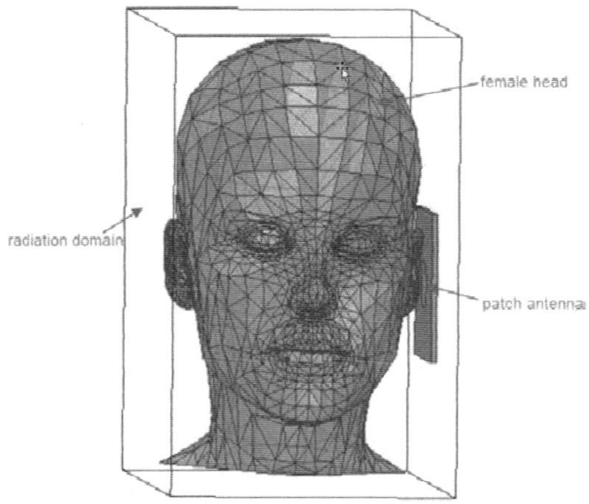

Figure 11. The circular patch antenna in presence of female head.

The position of the antenna was chosen such as to imitate the mobile phone in use, in the left side of the head, close to the ear, as can be seen in Figure 11. The patch antennas are very often used in mobile phone, for this reason we analyze the antenna's emissions influence on the human head.

3.1. Antenna's parameters

It is interesting to also investigate how the human head can affect the antenna's parameters if it is placed in the direction of the antenna's emission. For this, we extract the antenna's parameters in presence of the head, and we detailed them in Figure 12.

Antenna Parameters:

Quantity	Freq	Value
Max U	3.5GHz	996 uW/sr
Peak Directivity		10.446
Peak Gain		2.1606
Peak Realized Gain		1.2317
Peak System Gain		1.2317
Radiated Power		1.1982 mW
Accepted Power		5.7929 mW
Incident Power		10.162 mW
System Power		10.162 mW
Radiation Efficiency		0.20684

Maximum Field Data:

rE Field	Freq	Value	At(Theta,Phi)
Total	3.5GHz	866.59 mV	90deg,0deg
X		174.18 mV	-58deg,90deg
Y		866.18 mV	90deg,0deg
Z		114.61 mV	-92deg,90deg
Phi		866.18 mV	90deg,0deg
Theta		301.78 mV	180deg,90deg
LHCP		612.77 mV	90deg,0deg
RHCP		612.77 mV	90deg,0deg
Ludwig3/X dominant		174.18 mV	-58deg,90deg
Ludwig3/Y dominant		866.18 mV	90deg,0deg

Figure 12. The circular patch antenna' parameters in presence of female head.

3.2. The directivity of the antenna in presence of the human head

The directivity of the antenna in the presence of the human head is represented in Figure 13. Also for this case, the 3D directivity characteristic as diagram (Figure 13 (a)) and as representation with the model (Figure 13 (b)) are represented.

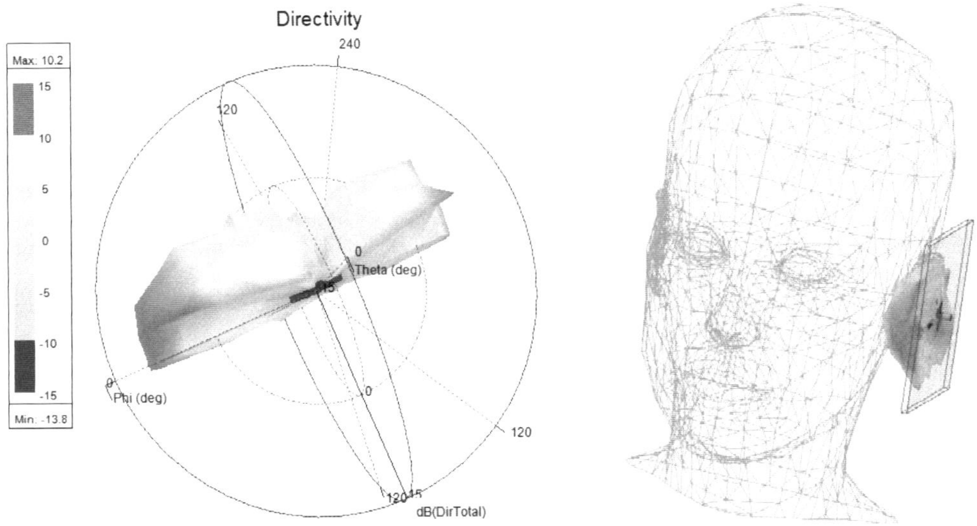

(a) diagram (b) representation with the model

Figure 13. The 3D directivity of circular patch antenna in presence of the female head.

3.3. The S parameters of the antenna in presence of the human head

The S parameters of the circular patch antenna in the presence of the female head are represented in Figure 14, in the same frequency range 1-6 GHz. As can be noted, the plot shows that the antenna operates in the frequency range between 3.67 GHz and 3.89 GHz and is resonating at 3.8 GHz.

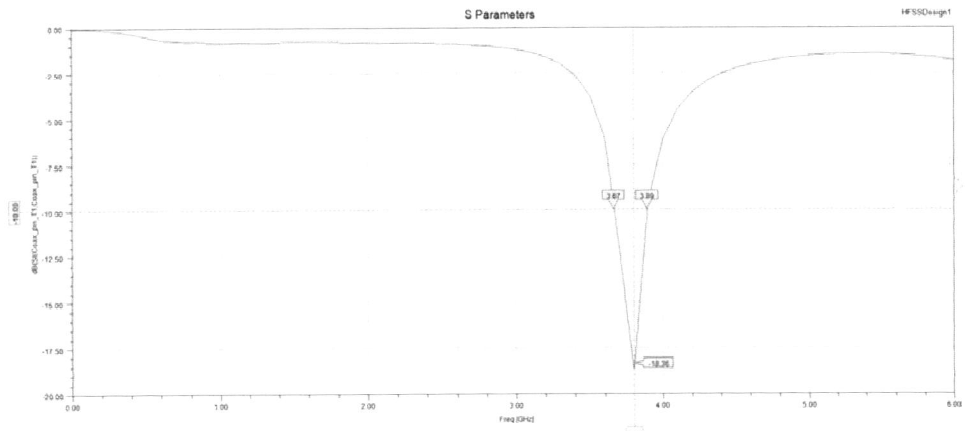

Figure 14. The S parameters for the circular patch antenna in presence of female head.

From the S parameters representation, are extracted the data $f_1 = 3.67$ GHz and $f_2 = 3.89$ GHz (Figure 14). By replacing these values in expressions (1) and (2), the average frequency is calculated to be $f_{avg} = 3.78$ GHz, and the bandwidth is $b_w = 5.82$ %.

3.4. The SAR Field

The specific absorption rate (SAR) is represented in the full model as 3D view and a zoom with the ear for a better visualization of the distribution in Figure 15. It has the maximum value 2.54 W/Kg that exceed the limits imposed by the Order no 1193/2006.

IOP Conf. Series: Materials Science and Engineering 1254 (2022) 012018 doi:10.1088/1757-899X/1254/1/012018

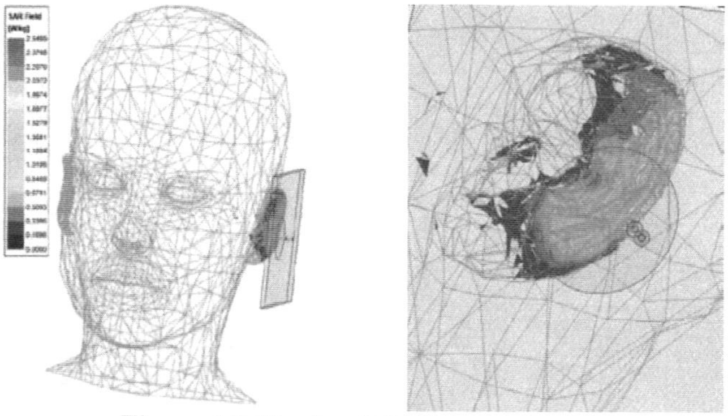

Figure 15. The local SAR in the model.

To see how the SAR field penetrate the human head a cross section plane representation is presented in Figure 16 and can be observed also here how the field penetrate the human ear.

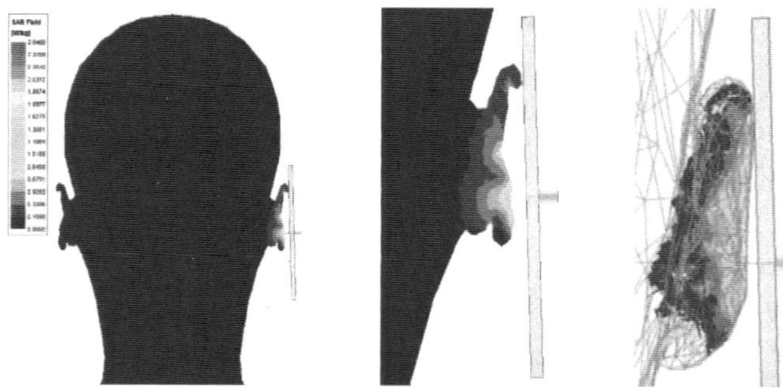

Figure 16. The local SAR in the YZ plane.

For a better interpretation the SAR variation on a line that goes from the antenna through the human head from left ear to the right one, a line of 15 cm length is represented in Figure 17.

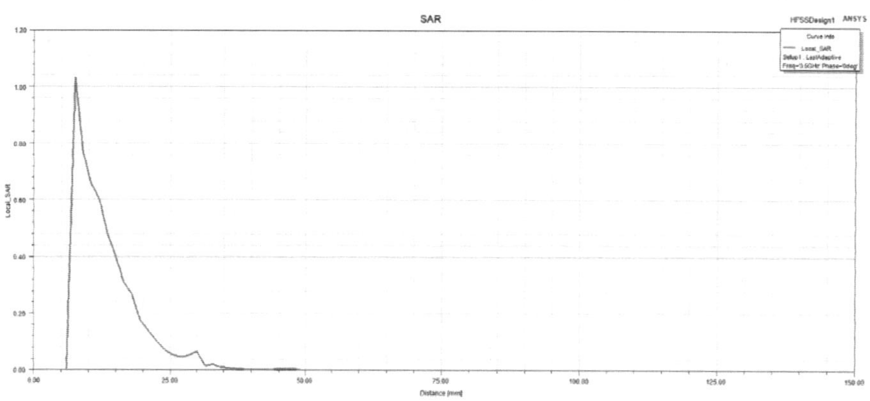

Figure 17. The local SAR on a line transversal from antenna to the head.

On the considered line the maximum SAR value is 1.033 W/kg at a distance 7.5 mm in the ear flag and it decrees after 50 mm inside the head.

3.5. The electric field emitted by the antenna in presence of the human head

Now the electric field in the human head is investigated. The antenna is analyzed in the presence of the human head at its operating frequency of 3.5 GHz.

So, a tridimensional representation of the electric field in the head is presented in the entire model, and as a zoom view in Figure 18.

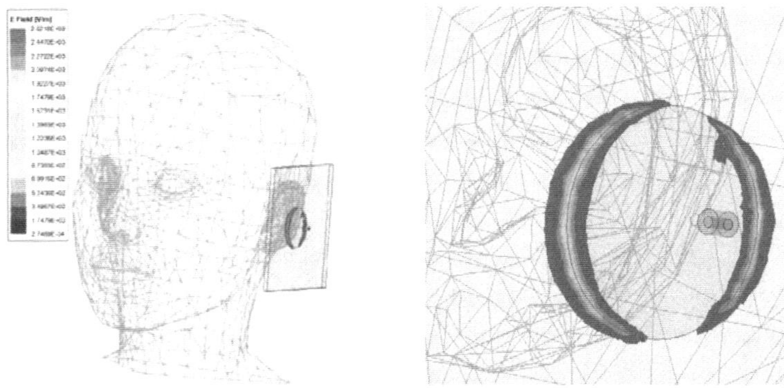

Figure 18. The electric field propagated by the patch antenna on the head.

We use the same limits in representation from 0-200 V/m so that at conclusion to compare the results of our analyses and the electric field with these imposed limits is represented in Figure 19.

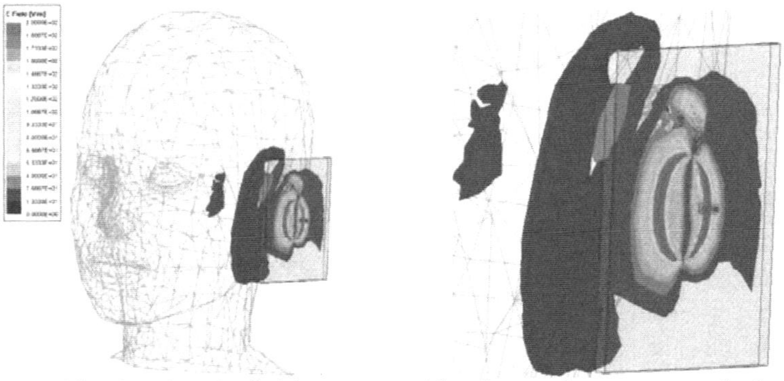

Figure 19. The electric field propagated by the patch antenna on the head.

The maximum values of the field are in the antenna as well as in the human ear as can be observed in the distribution.

To see the electric field inside the head, a cross-section plane is considered and the field distribution in this plane is presented in Figure 20, plotting the distributions in the same limits of 0-200 V/m for a more detailed visualization.

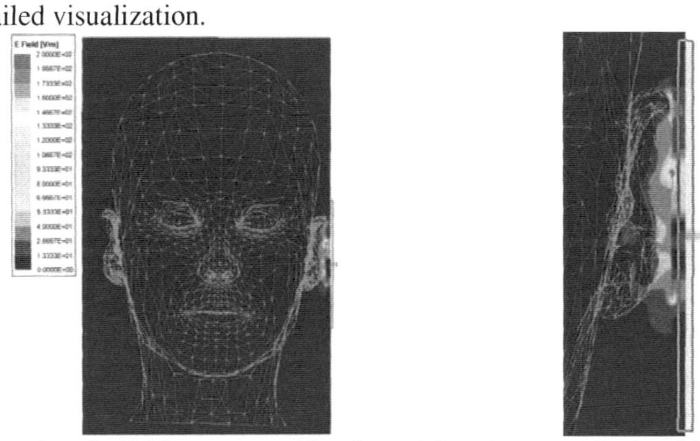

Figure 20. The electric field propagated by the patch antenna on the head in the YZ plane.

The electric field is also represented on the line from the antenna through the head from left ear to right one in Figure 21, and it can be observed that the maximum 42.65 V/m is detected in the ear area.

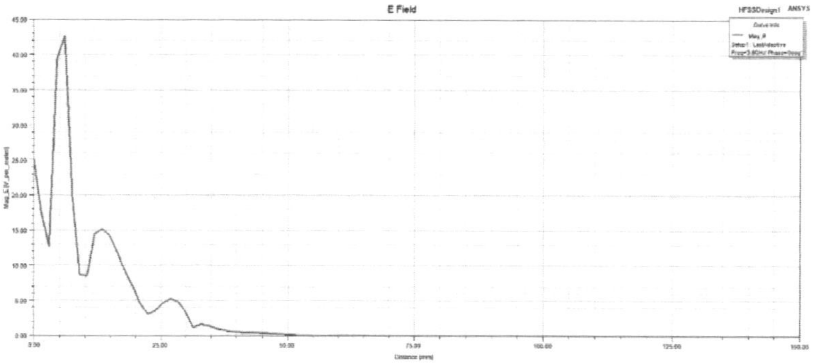

Figure 21. The graphic variation of the electric field on a line transversal from antenna to the head.

3.6. The magnetic field emitted by the antenna in presence of the human head

The magnetic field intensity on the head surfaces is represented in Figure 22. The magnetic field intensity maximum value is 0,367 A/m.

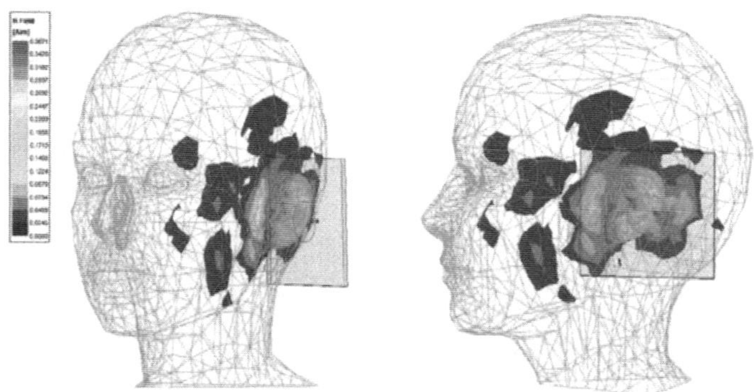

Figure 22. The magnetic field emitted by the patch antenna on the head.

To see how deep the magnetic field penetrates the head, a cross-section plane representation of the magnetic field is detailed in Figure 23.

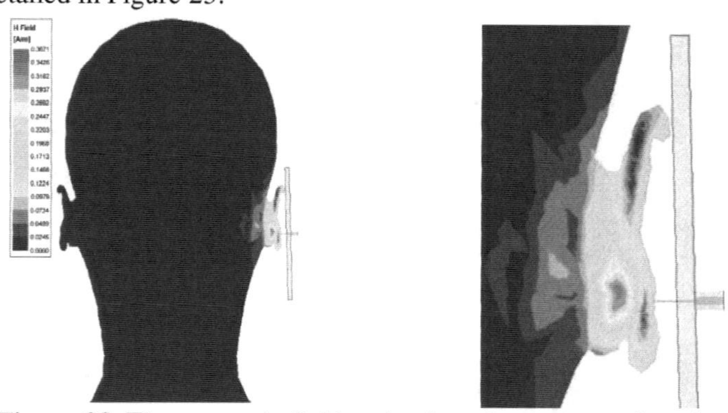

Figure 23. The magnetic field emitted on the cross-section plane.

The magnetic field is also represented on the line from the antenna through the head from left ear to right one in Figure 24, and it can be observed that the maximum of 0.316 A/m are found on the ear flag, at 13.5 mm inside the ear, and the values decrease after 30-40 mm. Compared with the limits [10] concerning the limitation of exposure of the general population to electromagnetic fields from 0 Hz to 300 GHz, it is found that in our case the limit is exceeded.

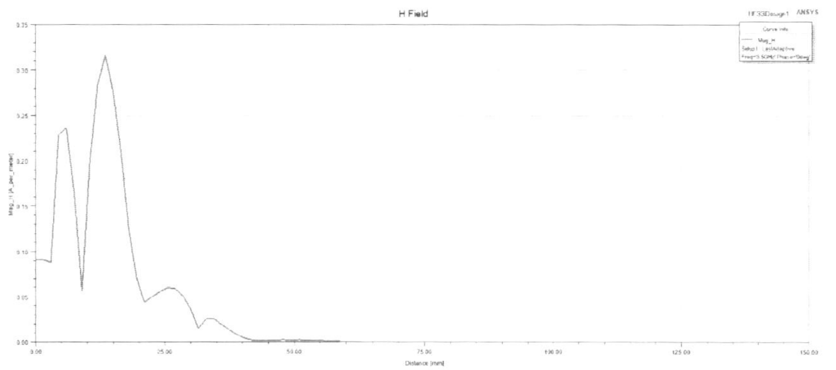

Figure 24. The graphic variation of the magnetic field on a line transversal from antenna to the head.

4. Comparation of the results

The comparation of the results obtained for the case of antenna without head with those obtained for the antenna with head is presented to underline the influence of the antenna's emissions on the human head and also the influence of the human head on the antenna's parameters and fields.

The antenna's parameters were extracted for both cases using the ANSYS-HFSS software program dedicated to high frequencies analysis. The antenna's parameters are significantly influenced by the presence of the human head. The antenna's peak directivity increases significantly in the presence of the head and the peak gain, radiated power and the radiation efficiency decrease significantly.

Comparing the directivity diagrams of the circular patch antenna without head with that of the antenna in presence of the human head it is found that the peak directivity increases, and the human head alters the antenna's directivity.

In the Table 1 are detailed for comparison the results obtained for the two analyzed cases: the bandwidth, directivity, and maximum values for electric and magnetic field.

Table 1. Comparison of results for antenna with and without head.

	Patch antenna without head	Patch antenna with head
Band width	5.66%	5.53%
Directivity	5.33	10.44
Electric Field	3241 V/m	2621 V/m
Magnetic Field	6.73 A/m	0.367 A/m

5. Conclusions

The influences of the circular patch antenna's emissions on the human head and the influence of the human head on the patch antenna's parameters and fields were high frequency analyzed in this paper.

The antenna's parameters are significantly influenced by the presence of the human head as was highlighted in the paper.

The maximum SAR value is 2.54 W/kg, and the normal limits are 2 W/kg [2-3]. The legal limits are exceeded.

The maximum values of the electric field are found in the tip of the ear and in a point placed in that area it is around 146 V/m. Normal exposure limits are 140 V/m, so there is a little bit exceeded.

The maximum values of the magnetic field are found in the base area of the ear, and it is 0.36 A/m. Normal exposure limits are 0.16 A/m, so here the legal limits are exceeded.

We intend to construct this antenna and to test it experimentally, to measure the S parameters, SAR, the electric and magnetic field in our future research activities.

References

[1] Păcurar *et al* 2014 Analysis of the Patch Antennas at High Frequency *Acta Electrotehnica Journal* **55** (3-4) pp 169-173

[2] Păcurar *et al* 2020 High Frequency Analysis of the Influence of Yagi-Uda Antenna on the Human head *EPE 2020* pp 370-375 doi: 10.1109/EPE50722.2020.9305622
[3] Constantinescu *et al* 2021 Numerical Modelling and Analysis of Circular Patch Antenna Array for Further Use Determination *MPS 2021* pp 1-6 doi: 10.1109/MPS52805.2021.9492557
[4] Constantinescu *et al* 2020 High Frequency Analysis of the Vivaldi Antenna Parameters *EPE 2020* pp 376-381 doi: 10.1109/EPE50722.2020.9305674.
[5] Giurgiuman *et al* 2020 The Analysis, Modelling and Comparison between Circular and Rectangular Patch Antennas *EPE 2020* pp 382-387 doi: 10.1109/EPE50722.2020.9305549
[6] Constantinescu C, Munteanu C, Păcurar C, Giurgiuman A, Gliga M and Andreica S 2019 High Frequency Analysis of Bowtie Antennas *ATEE 2019* pp 1-6 doi:10.1109/ATEE.2019.8724972
[7] Constantinescu C, Munteanu C, Păcurar C and Răcăşan A 2018 Influence of the Patch Antenna Feeding on their Parameters *EPE 2018* pp 0235-0240 doi 0.1109/ICEPE.2018.8559782
[8] Spindler E 1983 *Antene* (Bucharest: Bucharest Technical Publishing House)
[9] Payal P, Goud R M and Kaur K 2014 Design of Yagi-Uda Antenna using Microstrip Circuit, International Journal of Computer Applications *Computer Science*
[10] Minister of Public Health 2006 *Order no. 1193 of September 29, 2006, for the approval of the Norms regarding the limitation of the exposure of the general population to electromagnetic fields from 0 Hz to 300 GHz.*
[11] The Government Of Romania 2016 *Decision No. 520/2016 of 20 July 2016 on the minimum safety and health requirements regarding the exposure of workers to risks generated by electromagnetic fields*
[12] Commission of the European Communities 2008 *Council Recommendation of 12 July 1999 on the limitation of exposure of the general public to electromagnetic fields (0 Hz to 300 GHz) (1999/519/EC)*
[13] *User's guide – High Frequency Structure Simulator* 2015
[14] Răcăşan A, Munteanu C, Ţopa V, Păcurar C and Hebedean C 2014 Electromagnetic interferences suppresion in planar integrated devices *International Symposium on Electromagnetic Compatibility* pp 940-945 doi: 10.1109/EMCEurope.2014.6931038
[15] Răcăşan A, Munteanu C, Ţopa V, Păcurar C and Hebedean C 2014 Filter geometry optimization for the conduction electromagnetic interferences suppression *OPTIM 2014* pp 46-51 doi: 10.1109/OPTIM.2014.6850930.
[16] Kumar A 2013 Rectangular Microstrip Patch Antenna using 'L' Slot Structure *Journal of Research in Electrical and Electronics Engineering* **2**(2) pp 15-18
[17] Sukla A and Biswas M 2012 CAD Model to Compute the Resonant Frequency of Cavity Backing Triangular Patch Antenna *Antenna Test and Measurement Society* pp 1-4

Minkowski fractal antenna based on 3D printing

M Richterova[1], J Olivova[1], M Popela[2] and V Blazek[1]

[1] Department of Communication Technologies, Electronic Warfare and Radiolocation. University of Defence, Kounicova 65, 662 10, Brno, Czech Republic
[2] Department of Radio Electronics, Brno University of Technology, Technicka 10, 616 00, Brno, Czech Republic

E-mail: marie.richterova@unob.cz

Abstract. This paper describes the manufacturing of a Minkowski fractal antenna using a 3D commercial printer. The Minkowski fractal antenna was chosen for its simplicity and practical versatility. The paper also describes a manufacturing procedure for Minkowski fractal antenna by using metalic sprays. The design of the Minkowski fractal antenna in the MATLAB application using the Antenna Toolbox extension is also described, including 3D printing procedures, post processing procedures (plating) and practical testing of its functionality. The measured results are compared to simulations and then analysed.

1. Introduction

Fractal antennas belong to the group of small broadband multi-frequency antennas. Fractal antennas can be used in modern communication technologies. Fractal antennas are expected to be used in the construction of small satellites, where not only size but also low weight is crucial.

The paper presents the possibilities of using 3D printing in the production of Minkowski fractal antennas. Currently, the development of 3D printing technologies, both in the field of printers and in the field of 3D printing materials, presents new and interesting possibilities for the use of this technology for various applications in the field of communication technologies. 3D printing is the process of creating a three-dimensional solid object from a digital file, where by the resulting printed object is gradually created by laying down (printing) continuous layers of material until the entire object is completed. There are many 3D printer manufacturers around the world who are interested in developing and producing different types and sizes of 3D printers.

The Prusa i3 MK3S+ 3D printer manufactured by Prusa Research was used to produce the Minkowski fractal antenna. This 3D printer has a print area of 25 x 21 x 21 cm with the ability to adjust the print quality (print height) between 0.05 - 0.35 mm.

In order to print the Minkowski fractal antenna on the 3D printer, it was necessary to model (design) parts of the antenna in a program designed for 3D modeling. Nowadays, many software programs are available for drawing 3D objects. The software Onshape (freeware) was chosen for modelling the Minkowski fractal antenna. The Onshape is a computer design software system supplied over the Internet through a model "Software as a Service" [1]. The Figure 1 shows modelled Minkowski fractal antenna in Onshape program.

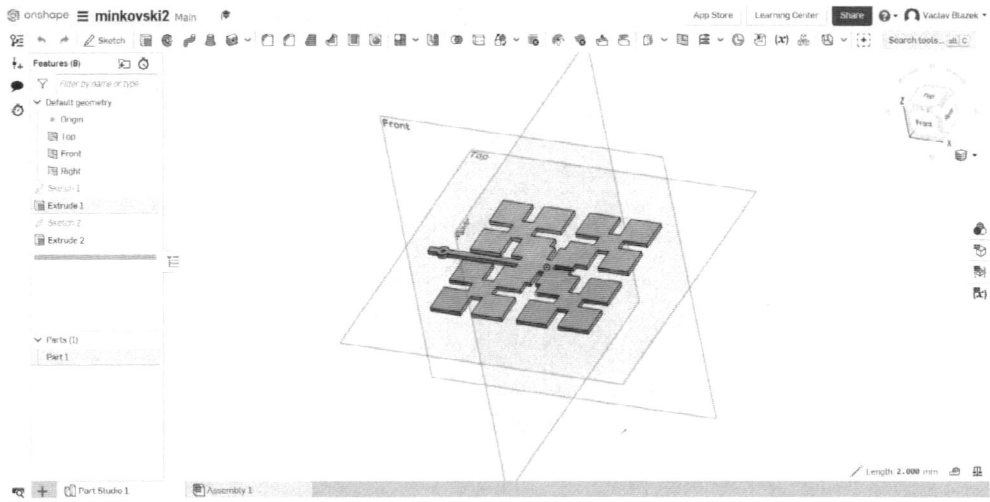

Figure 1. Modelled Minkowski fractal antenna in Onshape.

2. Analysis of Minkowski fractal antennas

A fractal is a type of mathematical shape that is infinitely complex. Essentially, a fractal is a pattern that repeats forever, and each part of a fractal, no matter how close or far away you are, looks very similar to the whole picture [2].

A conventional wired antenna has limited bandwidth. In practical applications, a wired antenna cannot always be used because its physical dimensions are too large. This is the reason why a suitable replacement for wired antennas has been sought for applications where large bandwidth is required simultaneously with small antenna size [3].

The fractal concept can be used to physically reduce the size of the antenna using fractals such as Koch curve, Koch snowflake, Minkowski fractal and Sierpinski triangle. The fractal antenna concept can be used to achieve multiple antenna resonances and increase the bandwidth of each individual band due to the self-similarity of the fractal in terms of its geometry [3].

The procedure for constructing the Minkowski fractal is illustrated in Figure 2.

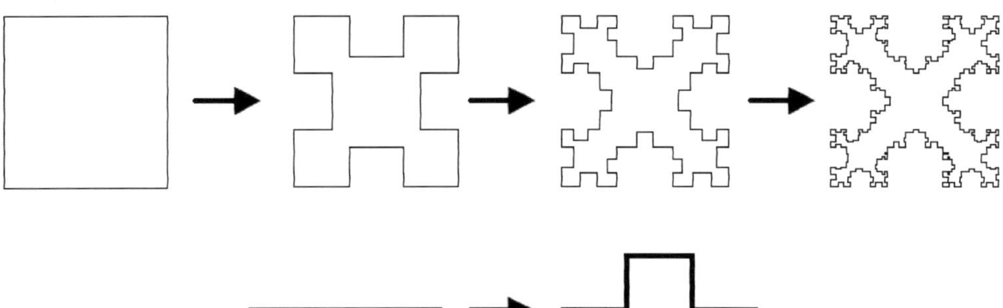

Figure 2. Minkowski fractal and its iterations from 0 to 3.

The zero iteration of the Minkowski fractal is a square that is formed by four zero iterations of the Minkowski curve. The zero iteration of the Minkowski curve is a straight line. At each higher iteration an octal generator is applied to each segment of the curve, this procedure produces the next iterations of the fractal [4].

The fractal geometry defines a structure with long lengths that fits into a compact region. This property of fractals can be used to design miniaturized resonant antennas due to their reduced physical length. The shape of the fractal is formed by an iterative mathematical process [5].

Simulations of the properties of the Minkowski fractal antennas of the 1st and 2nd iterations were performed in the MATLAB programming environment using the Antenna Toolbox.

The first and second iterations of the Minkowski fractal were chosen because of the fabrication of Minkowski fractal antennas using 3D printing. Higher iterations of the Minkowski fractal cannot be used to produce a planar antenna based on the Minkowski fractal because it would not be possible to maintain the desired fractal shape during printing.

The power radiation pattern for vertical and horizontal polarization and the impedance of the Minkowski fractal antenna were modeled in the Antenna Toolbox in order to subsequently verify the characteristics of the fabricated antenna by measurements in an anechoic chamber and using a Rohde & Schwarz analyzer.

The center frequency for the zero iteration of the Minkowski fractal antenna was chosen to be 1.8 GHz. The base length of the antenna for the zero iteration of the Minkowski fractal corresponds to 0.5*wavelength, so the basic size of the patch antenna is 8.33 x 8.33 cm. The antenna thickness was chosen to be 2 mm to allow the designed antenna to be fabricated using 3D printing followed by metal spray coating. In order to connect the SMA feed connector to the antenna, it was necessary to design the ground plane to maintain mechanical strength. The dimensions of the ground plane are basic antenna length + 40mm and basic antenna length + 16mm, the size of the ground plane is 12.33 x 9.33 cm. For simulating the designed antenna in Antenna Toolbox, FR4 dielectric with the properties of ε_r = 4.8 and thickness of 6 mm was used. The antenna impedance was calculated in the frequency band from 1 GHz to 5 GHz with increments of 5 MHz.

Subsequently, a planar fractal antenna was designed for the 1st and 2nd iteration of the Minkowski fractal, see Figure 3.

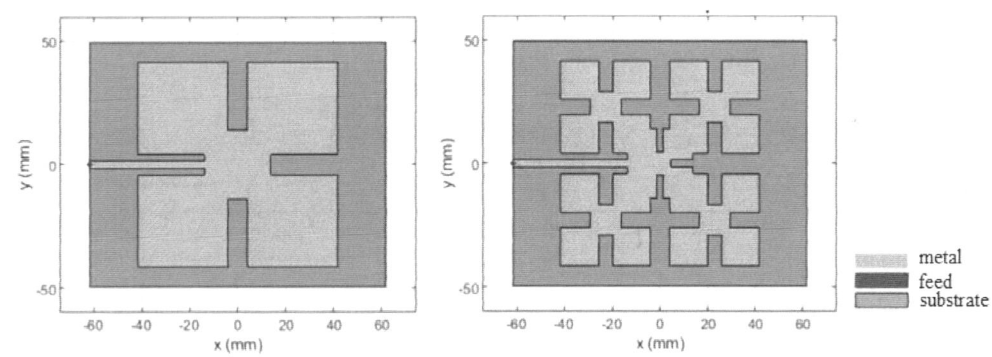

Figure 3. The 1st and 2nd iteration of the Minkowski fractal antenna.

For the selected resonant frequencies, the power radiation patterns were calculated and displayed, see Figure 4 and Figure 5. The impedance of the designed antennas was calculated in order to compare with the results obtained by measurements.

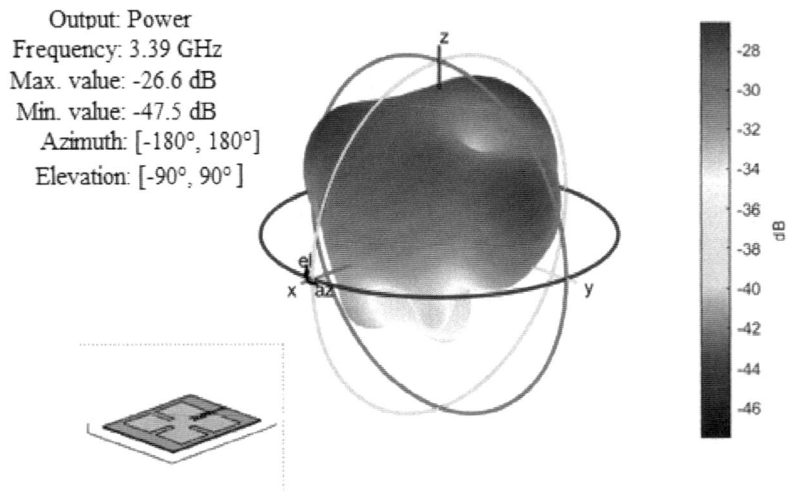

Figure 4. Power 3D radiation pattern of 1st iteration of the Minkowski fractal antenna.

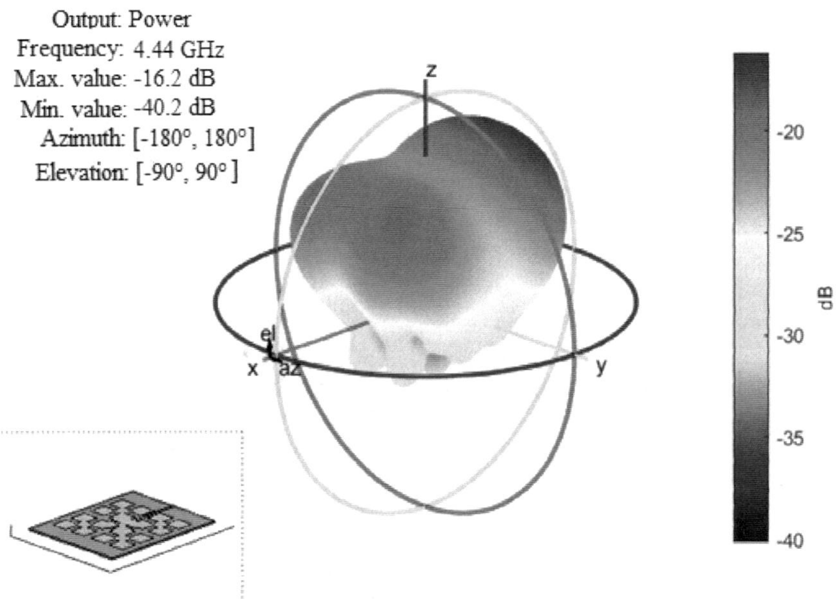

Figure 5. Power 3D radiation pattern of 2nd iteration of the Minkowski fractal antenna.

3. Manufacturing of Minkowski fractal antennas

The polylactide filament, commonly known as PLA, is one of the most popular materials used in desktop 3D printing. It is the default filament choice for most extrusion-based 3D printers because it can be printed at low temperatures and does not require a heated substrate. PLA is easy to print, a very inexpensive material that can be used to create parts for a wide range of applications. The PLA filament one of the most environmentally friendly filaments on the market today. The advantages of PLA are low cost, it is a tough, strong and very durable material. Disadvantages include low heat resistance, PLA is not suitable for exposure to direct sunlight [6].

After performing the simulation in Antenna Toolbox, it was necessary to create 3D objects in the online programming environment Onshape. Subsequently, the data file for 3D printing can be implemented in the PrusaSlicer software environment.

PLA filament was used to 3D print the planar Minkowski fractal antennas, because the fabricated antennas should not be mechanically stressed. The nozzle temperature was chosen to be 210 °C and the heated substrate temperature 60 °C. Both temperatures are within the recommended functional range for a 3D printer.

After printing the antennas on the 3D printer, the antennas were sprayed with black paint and then the surface was smoothed so that the copper spray adhered well to the surface. Each copper part of the antenna was covered with a total of four layers of spray paint with a copper content of 99.9%. After the copper layers were applied, the individual antenna components were glued together and a 50 Ω SMA connector was attached to measure the power radiation pattern. The SMA connectors were glued with conductive glue and painted with conductive varnish to achieve an optimal conductive connection. The fabricated planar fractal antennas are presented in Figure 6.

Figure 6. Planar Minkowski fractal antennas 3D printing.

4. Antenna performance measurements and results

In order to compare the radiation patterns of the simulated Minkowski fractal antennas and the printed Minkowski fractal antennas, it is necessary to perform the measurements in an anechoic chamber to eliminate the negative influence of the surrounding environment on the experiment. Tests of the printed antennas were performed in an anechoic chamber. The Figure 7 shows the test setup the anechoic chamber.

IOP Conf. Series: Materials Science and Engineering 1254 (2022) 012019 doi:10.1088/1757-899X/1254/1/012019

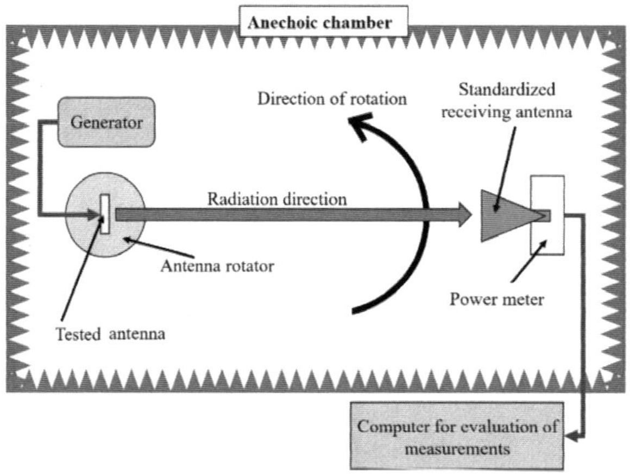

Figure 7. Test setup in the anechoic chamber.

The proposed planar Minkowski fractal antennas based on 3D printing fabricated (see Figure 6) and their power radiation patterns were measured in an anechoic chamber (see Figure 8).

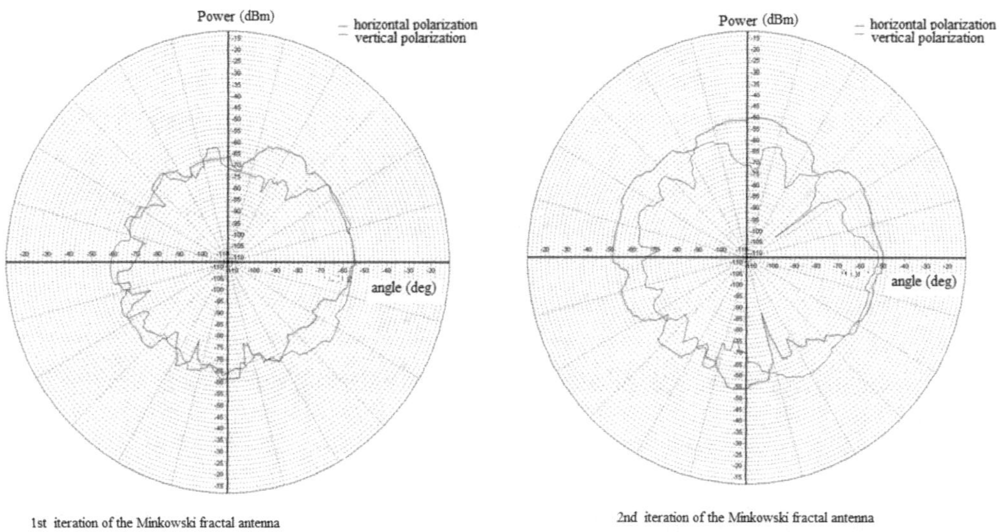

Figure 8. Power radiation pattern of 3D printing Minkowski fractal antennas for 3.3875 GHz.

The impedances of the designed Minkowski fractal antennas and the standing wave ratio were measured using a vector analyzer ZNLE6-Vector Network Analyzer manufactured by Rohde & Schwarz with frequency range from 1MHz to 6 GHz. The results show the resonant frequencies of the 1st iteration of the Minkowski fractal antenna (see Figure 9) and the resonant frequencies of the 2nd iteration of the Minkowski fractal antenna (see Figure 10).

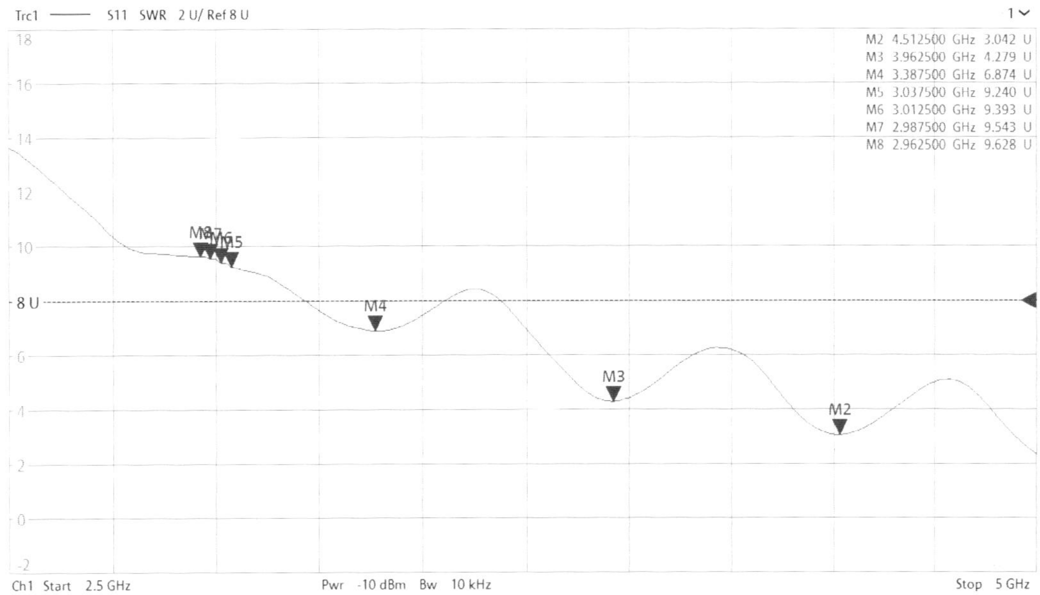

Figure 9. Standing wave ratio (SWR) of the1st iteration of 3D printing Minkowski fractal antenna.

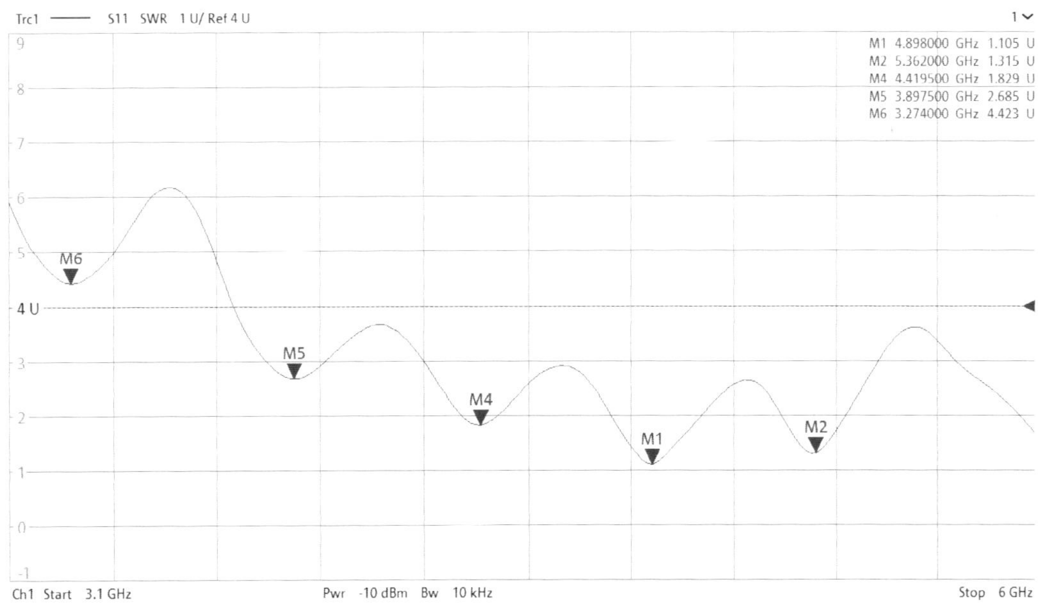

Figure 10. Standing wave ratio (SWR) of the 2nd iteration of 3D printing Minkowski fractal antenna.

By measuring the impedance and standing wave ratio of the printed fractal antennas using a vector analyzer, the resonant frequencies in the frequency band from 2 to 6 GHz were determined, see Figure 11.

The resonant frequencies obtained by the simulations were compared with the measured values of the resonant frequencies, see Table 1.

IOP Conf. Series: Materials Science and Engineering 1254 (2022) 012019 doi:10.1088/1757-899X/1254/1/012019

Table 1. Simulated and measured resonant frequencies.

1st iteration of the Minkowski fractal antenna		2nd iteration of the Minkowski fractal antenna	
Simulated resonant frequency (GHz)	Measured resonant frequency (GHz)	Simulated resonant frequency (GHz)	Measured resonant frequency (GHz)
3.2172	2.9625	3.8889	3.2740
3.3929	2.9875	4.2778	3.8975
3.5687	3.0125	4.4444	4.4195
3.9788	3.0375	5.0000	4.8980
4.2717	3.3875	5.6111	5.3620
4.3889	3.9625		
4.8283	4.5125		

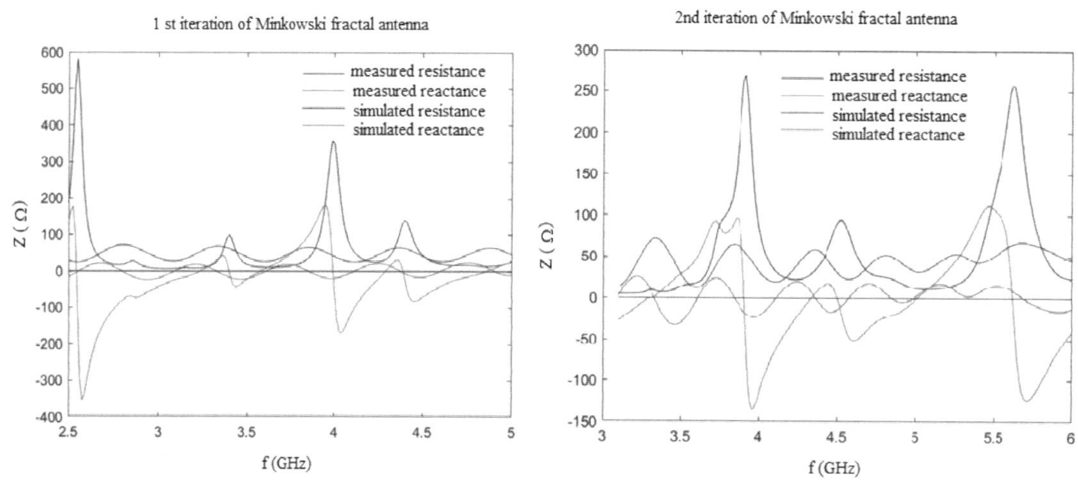

Figure 11. Impedance of the 1st and 2nd iteration of 3D printing Minkowski fractal antenna.

It can be seen in the Figure 11 that the simulated resonant frequencies of 3.8889 GHz and 4.4444 GHz and the measured resonant frequencies of 3.9195 GHz and 4.4195 GHz are very similar. The other measured and simulated resonant frequencies are also close to each other. There is a difference between the simulated values of the resonant frequencies and the measured values of the resonant frequencies for the 1st and 2nd Minkowski fractal antenna interactions (see table 1). The difference is due to the fact that in the simulations FR4 was used as the dielectric with the properties listed in the dielectric catalogue in the Antenna Toolbox. The printed Minkowski fractal antennas were made of PLA filament. The dielectric properties of that PLA filament are not tabulated, so it was not possible to include these properties in the simulations. Measuring the dielectric properties of 3D printing filaments is the subject of our further research.

5. Conclusion

The measured results were compared with the simulation results. There is a difference between the results obtained by simulations in MATLAB and the results obtained by measurements on the vector analyzer and in the anechoic chamber. If the specific procedure of manufacturing planar fractal antennas based on 3D printing is followed, the measurement results can be expected to be close to the simulated values.

Simulated Minkowski fractal antennas with dielectric properties of filaments for 3D printing would contribute significantly to the accuracy of the results obtained by simulations and measurements.

The obtained results show that planar Minkowski fractal antennas are suitable for applications where miniature dimensions and low device weight are required. The designed and realized antennas can be used for military applications as well as for civilian applications, e.g. for small satellites and for 5G communication technologies.

Acknowledgments

The work presented in this article was supported by specific research of the University of Defence entitled "Modern Antenna Systems and Methods of their Management", and by the Czech Republic Ministry of Defence – the University of Defence development program "AIROPS".

References

[1] Prusa Slicer introduction & download Available from: https://www.prusa3d.com/prusaslicer/#_ga=2.201229934.1128987098.1626340328-1785738773.1626340328
[2] Balanis C A 2016 *Antenna Theory Analysis and Design* (Hoboken, New Jersey: John Wiley & Sons)
[3] Yang X, Chiochetti J, Papadopoulos D and Susman L 1999 Fractal Antenna Elements and Arrays *Applied Microwave & Wireless* pp 34-46
[4] Riddle L 2021 Sierpinski Carpet (Larryriddle.agnesscott [online]) Available from: https://larryriddle.agnesscott.org/ifs/carpet/carpet.htm
[5] Singh A and Kumar Singh M 2014 Design and Simulation of Miniaturized Minkowski Fractal Antennas for microwave applications *International Journal of Advanced Research in Computer and Communication Engineering* Ghaziabad India **3**(1) pp 5309-5311
[6] PLA Simplify 3D (online) Available from: https://www.simplify3d.com/support/materials-guide/pla/

IOP Conf. Series: Materials Science and Engineering 1254 (2022) 012020 doi:10.1088/1757-899X/1254/1/012020

Pre-compliance testing of the shielding effectiveness of 3D printed multiwall carbon nanotube/ethylene-vinyl acetate copolymer composites in the 1 – 7 GHz range

D B Deaconescu[1], D Vatamanu[2], S Miclăuş[2], F Stan[3] and I Sandu[3]

[1]Doctoral School of Electrical Engineering, Technical University of Cluj-Napoca, Cluj-Napoca, Romania
[2]Department of Communications, IT & Cyb. Defense, "Nicolae Bălcescu'' Land Forces Academy Sibiu, Romania
[3]Center of Excellence Polymer Processing, Dunărea de Jos University of Galati, Galati, Romania

E-mail: deliadeaconescu@yahoo.com, davidvatamanu@yahoo.com, simo.miclaus@gmail.com, felicia.stan@ugal.ro, laurentiu.sandu@ugal.ro

Abstract. Copolymer composites with 1-, 3- and 5 % weight concentration of carbon nanotubes (CNT) with average diameters of 9.5mm were fabricated using three different printing orientations: 0°, alternative ± 45°, and alternative +0°/- 90°. Shielding effectiveness (SE) of samples of 100x100x1 mm was measured with near-field probes connected to a vector network analyser based on a pre-compliance procedure. Both electric- and magnetic-shielding properties could be investigated. Using a reference material of a recognised producer, we emphasized comparative shielding properties in the (1-7) GHz range. Low SE values were generally observed for all samples, not exceeding 12.5 dB, but they were dependent on both the printing direction and CNT wt%. The measurements indicate that frequencies larger than 7 GHz may show better shielding properties. Also, indications of the best combination between CNT concentration and printing direction were deducted.

1. Introduction

In the last years, due to the development of wireless telecommunication, military equipment and electronic devices, electromagnetic pollution, such as noise, radiation pollution and electromagnetic interference, has raised the concerns over not only the human health but also the safety and proper functioning of electronic devices and systems. To assure smooth functioning and life span of the electronic devices, efficient electromagnetic interference (EMI) shielding is required [1]. Metals and their alloys are ideal candidates for EMI shielding, but the EMI shielding applications are limited especially due to their low corrosion resistance and relatively high weight [1]. Therefore, in the last years, researches have been devoted to developing polymer-based composites that can be used for EMI shielding applications [1-3]. Therefore, to obtain conductive polymer-based composites with EMI shielding properties, various conductive fillers were added into polymers including carbon black, carbon nanotubes and graphene [3].

Carbon nanotubes (CNTs) exhibit an exceptional combination of mechanical, electrical and thermal properties [4-9], and, thus, are extensively used as fillers for the preparation of polymer/CNT nanocomposites. Depending on the CNT content, these composites can be used in various emerging applications, including electrostatic dissipation, device shielding, energy storage, sensors and electronics [1-3, 10-19].

The shielding effectiveness (SE) of polymer/CNT nanocomposites depend on many factors including electrical conductivity, plate thickness, and frequency [1-3, 8,16-17]. On the other hand, the electrical conductivity of polymer/CNT nanocomposites depends on the amount of CNTs and, thus, follow the principle of percolation theory– above the percolation threshold the nanocomposites behave like conductors [4, 6].

Researches on EMI shielding behaviour of polymer/CNT nanocomposites have shown that EMI effectiveness increases with CNT loading due to formation of effective conductive networks [1-3, 8-9, 16-17]. Moreover, it was shown that the SE of conventional conductive polymer-based composites can be significantly improved if a small amount of carbon nanotubes is added [1, 3, 14, 16-17].

In order to obtain superior conductive properties and therefore high SE, a perfect conductive network must be obtained at low percolation threshold [1]. In addition to the CNT loading, the electrical conductivity network depends on the manufacturing process, such as injection moulding [20, 21] or 3D printing [22-27], which induces orientation in the conductive network. In general, the CNT network is oriented in the polymer flow direction under high shear rates [20-21].

3D printing technology can be used to fabricate structures with different internal configurations that can effectively absorb and reflect the electromagnetic waves [22-27]. However, 3D printing of polymer/CNT composites is not straightforward, especially at high CNT concentration because CNT tends to agglomerate obstructing the nozzle and therefore the processability [22-27].

The development of flexible electronic devices requires the use of EMI shielding materials with high flexibility, and attempts have been made to develop flexible polymer-based nanocomposites for such applications [10, 13-15, 18-19]. Because of their high flexibility and low processing temperature, ethylene-vinyl acetate copolymer/CNT composites [28-30] are a good candidate for constructing flexible interface shielding structures. However, the SE of the EVA/CNT composites has not been investigated.

Thus, in this study, the EMI shielding performances of 3D printed ethylene-vinyl acetate copolymer (EVA) filled with carbon nanotubes (CNT) have been investigated. EVA/CNT samples with 1, 3 and 5 weight fraction (wt%) of CNT were fabricated by 3D printing and the EMI shielding behaviour was analysed. The results of pre-compliance testing of SE values of a set of planar samples characterized by using a pair of small near field probes and a vector network analyser (VNA) are presented and the measured values are compared against the ones corresponding to a producer's characterized shield. The correlation between EMI properties, carbon nanotube loading and 3D printing direction is also discussed.

2. Materials, measurement system and procedure

The composites of ethylene-vinyl acetate copolymer /carbon nanotubes (EVA/CNTs) used in this study were purchased from Nanocyl S.A. (Belgium) in the form of pellets with 1, 3, and 5 wt% of CNTs. The carbon nanotubes were multiwalled carbon nanotubes (Nanocyl 7000, Nanocyl S.A. (Belgium) with average diameter of 9.5 nm and average length of 1.5 μm [31]. According to the manufacturer, EVA/CNT composites were obtained by dilution of the EVA2001 masterbatch (Nanocyl S.A., Belgium) containing 20 wt.% of CNTs [32]. The EVA copolymer was EVATANE® 20-20 (Arkema) with 20 wt.% of VAC and a melt flow index 17 g/10 min (190°C, 2.16 kg) [33].

The filaments of EVA/CNT composites with 1.75 ± 0.5 mm diameter were fabricated by extrusion using a single screw extruder (NEXT 1.0 Advanced Silver, 3devo). Samples of 100 mm×100 mm×1 mm were fabricated by 3D printing using a Prusa MK3S printer (Prusa Research s.r.o., Czech Republic).

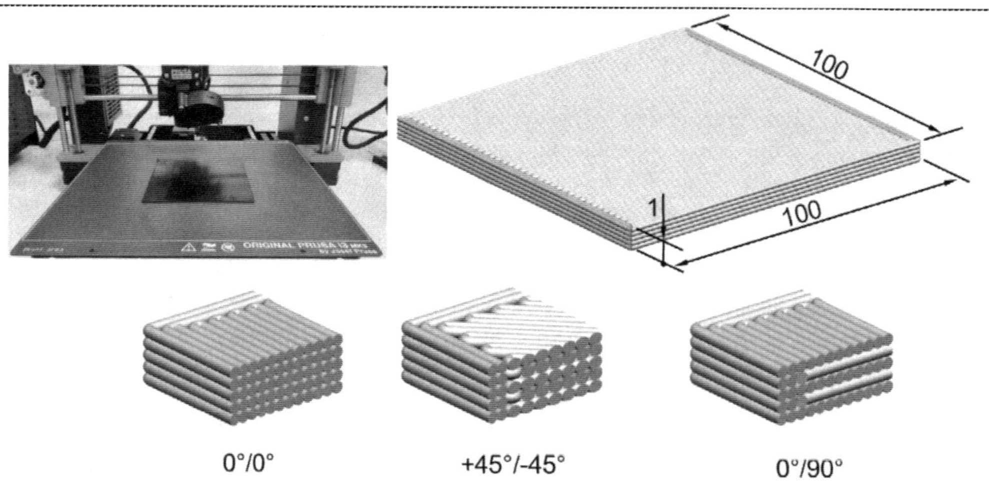

Figure 1. Geometry (in mm) of the 3D printed sample and 3D printing directions.

In order to investigate the effect of CNT orientation on the electromagnetic properties, three different groups of samples were fabricated using three different printing orientations: 0°, alternative ± 45°, and alternative +0°/- 90°, as shown in Figure 1. The printing parameters were the following: layer thickness 0.2 mm, nozzle diameter 0.6 mm, printing speed 20 mm/s, and in fill 100%. The extrusion nozzle temperature was optimized depending on the CNT loading, 215-225°C. It should be noted that the 3D printing process was optimized depending on the CNT wt%.

From the SE assessment perspective, two are the parameters of interest: the printing orientation and CNT wt.%. Therefore, notation of a sample as "0_1%" means that printing direction was 0° and the CNT wt% was 1%.

The SE characterizes the material by expressing the ratio between the transmitted power and the incident power propagating through a material and is simply expressed as [34]:

$$SE = 10 \log (P1/P2) \qquad (1)$$

where: P1 is the power generated by the electromagnetic (interference) source and incident to the material and P2 is the power passed through the shielding material.

The SE is due to three mechanisms of power attenuation in a material: attenuation by reflection (R), absorbent attenuation (A) and attenuation caused by multiple reflections on the frontiers (M). As SE is expressed in dB, it can be represented by the summation of these three contributions:

$$SE = R + A + M \qquad (2)$$

A number of standardized techniques and procedures are presently used to quantify the SE of materials, either in near field or in far field conditions of shield positioning nearby the emitting antenna [35-40]. Non-standard methods of measurement are also used mainly when comparisons with well-characterized materials is available or for some specific needs of the shield [41–45].

In this paper, a non-standardized procedure was applied for the SE measurement using near field probes [46]. Both electric (E) and magnetic (H) field strengths can be assessed with this method. It is especially useful to compare different materials or different thickness of the same material.

Particularly, in this study, a producer-characterized material, namely Aaronia Shield Ultra (Aaronia AG, Germany [47]) was used as a reference material which will be noted as Ref 2. It is a half transparent conductive radiofrequency shielding-fabric with high damping factor of up to 70dB in the (0.01 – 10) GHz range.

In order to assess the SE of E-field, a pair of identical E-field probes (sniffers) (Model PBS-E1, Aaronia AG, Germany) was used (Figure 2a). Magnetic field measurement was also used to evaluate shielding by reflection mechanism. Reflection is related with the currents induced in the shield and the magnetic field created by those currents.

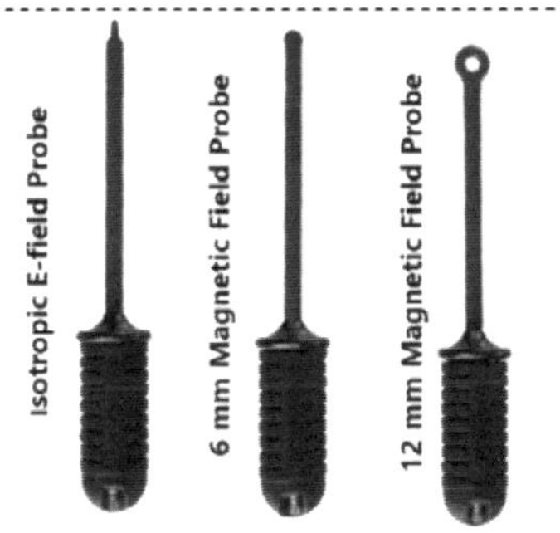

Figure 2a. Aaronia near-field probes used in the measurements.

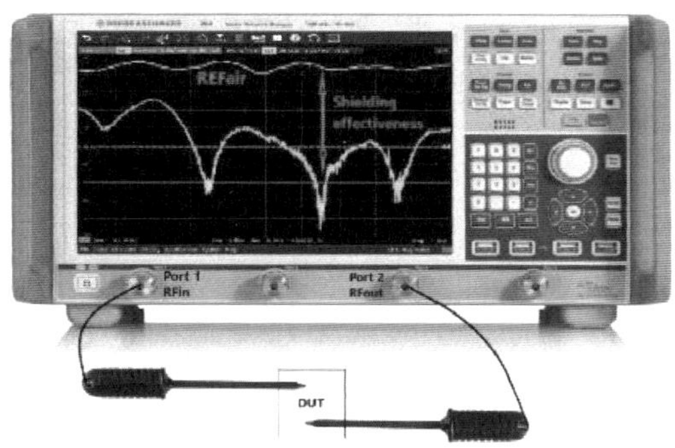

Figure 2b. Schematic measurement procedure with a pair of probes connected to the VNA ports and the device under test (DUT) / sample interposed between the field-probes.

In this regard, a pair of identical H-field probes (Models PBS-H1 and PBS-H2, Aaronia AG, Germany) was used as shown in Figure 2a. All the probes are passive near-field probes. Each pair was connected to the two ports of a vector network analyser (VNA) (Model ZNB 40, Rhode & Schwarz, Germany). A schematic set-up is presented in Figure 2b. Shielded coaxial cables were used to connect the probes and the VNA calibration (through-open-short-match) was always made prior to each measurement in a certain frequency band. The resonances obtained during the measurements were removed at the post-processing step with a smoothing algorithm based on the adjacent averaging method.

The EVA/CNT samples were inserted one by one between the probes as shown in Figure 3.The frequency band of testing was (1 – 7) GHz and the frequency step of the sweep was set to 0.5 MHz. A comparison between measured and producer's data (theoretical) for the reference material shield, Ref 2, was made in the first step. We have chosen an accepted difference of maximum 5 dB to consider the present method as reliable. Figure 4 illustrates the comparative SE curves for the reference material. The distances (d) between the sample and the field probe were chosen so that in all cases $d/\lambda < 0.2$, where λ is the wavelength, so all the measurements were made in the near field and the distance between probes was fixed. The transmission coefficient S21 between port 1 and port 2 of VNA was measured with and without the sample, and the difference resulted in exactly SE (dB).

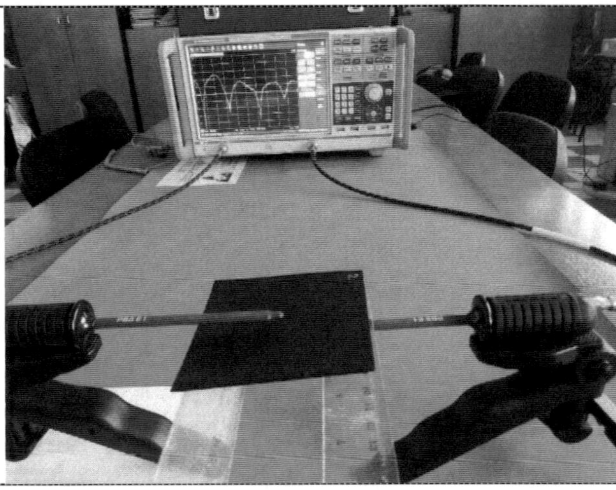

Figure 3. Experimental measurement set-up with the EVA/CNT sample inserted between the PBS-E1 field probes.

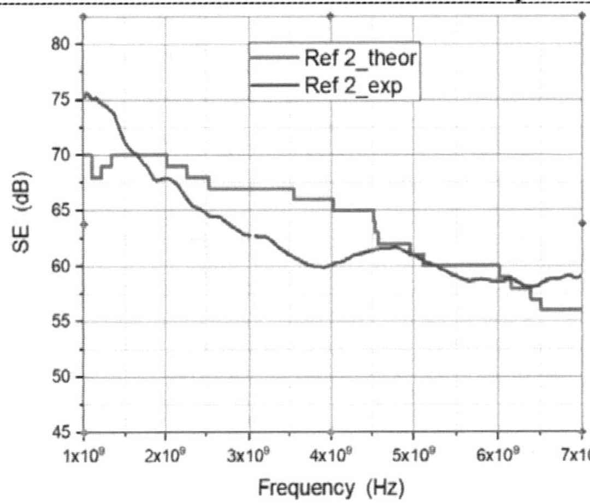

Figure 4. Characterizing the reference material, Aaronia Shield Ultra: SE values measured with the present method (Ref 2_exp) versus SE values provided by the producer (Ref 2_theor).

3. Results and discussion

Figure 5a and Figure 5b present a comparison of the SE values in the band (1-4) GHz and (4-7) GHz, respectively, for the EVA/CNT composites and the reference material Aaronia Shield Ultra (red curve). The first main observation is that all the EVA/CNT samples present a SE lower than 12 dB along the whole frequency range and that a (45-70) dB difference in the SE is observed between the reference material (very good shield) and all the samples. Some of the samples have practical no shielding properties in the (1-4) GHz frequency range, for example the sample 45_3% and 0_1%. However, from 5.5 GHz upward, the shielding behaviour is changing or even inter-changing between materials, as observed in Figure 5b. The sample 90_5% for example, suddenly increases its SE from 6.8 GHz upward becoming the best for shielding, while in all the lower frequencies its SE is almost the lowest. The SE values observed in the (6.5-6.7) GHz range indicate that the behaviour of these EVA/CNTs composites may drastically change at frequencies higher than 7 GHz. Generally, in the (1-6) GHz range, SE is low, some differences can be observed and they maintain in-between the samples. For frequencies higher than 6.5 GHz it seems that significant changes of shielding mechanisms are happening and the SE values at the limit of 7 GHz investigated here indicate a possible large increase of the EMI shielding for higher frequencies.

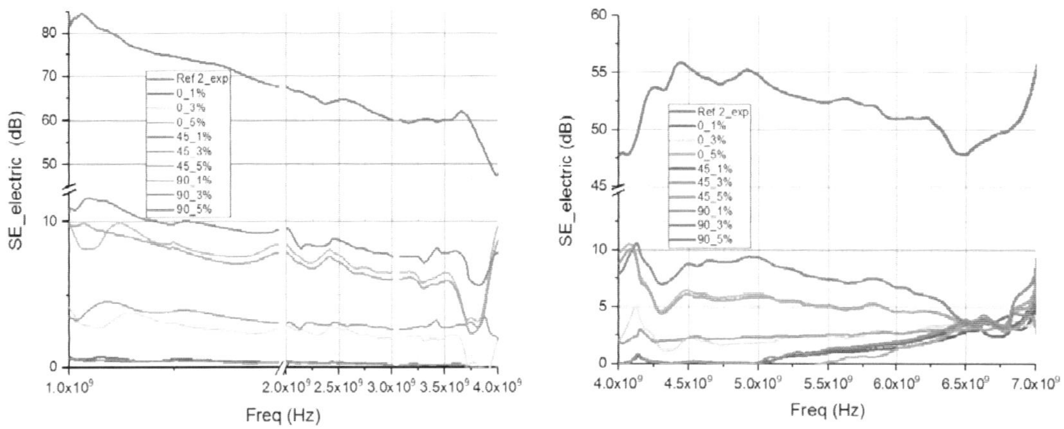

(a) (b)

Figure 5. The shielding effectiveness of EVA/CNT composites in the range of (a) (1 – 4) GHz and (b) (4 – 7) GHz. Reference Aaronia shield is shown for comparison.

Figure 6 depicts the magnetic component of the SE that quantifies the shielding by reflection mechanism. The currents induced in the shield produce a magnetic field themselves and this is captured by the H-field probe. The investigated band was just (0.5-2.5) GHz, because higher frequencies produced very noisy results. In this frequency range, the SE is nearly independent on frequency. Regarding the effect of CNT wt.%, it can be observed that some 3D printed samples present practically null SE. As compared to the Ref 2 material from Aaronia, lower SE values are observed, and the differences increase at higher frequencies.

The shielding performance of polymer/CNT composites depend on the CNT wt% and the interconnected CNT network, which is directly influenced by the printing direction. During the 3D printing process, due to high shear rates, carbon nanotubes are oriented in the printing direction. Figure 7 schematically indicates the orientation of carbon nanotube network during the 3D printing process. For the 0°/90° and +45°/-45° printing directions, an interconnected CNT network is expected to be formed during the 3D printing, providing the premise for enhancing the EMI shielding performances.

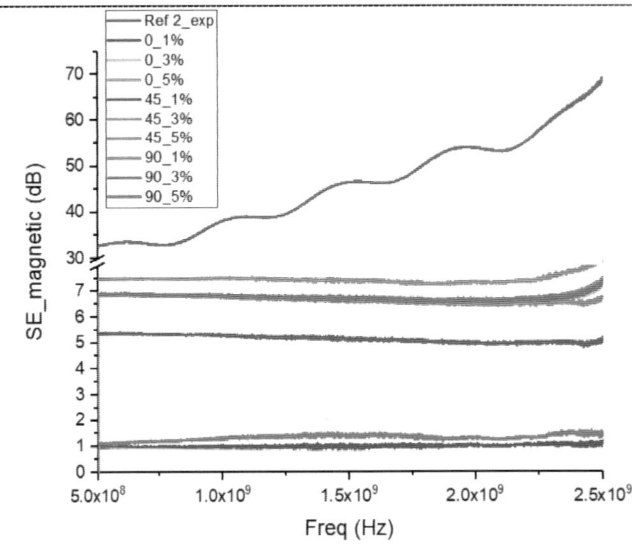

Figure 6. Reflexion shielding properties of the EVA/CNTs composites in the (0.5 – 2.5) GHz range. Reference Aaronia shield is shown for comparison for the reflexion part of the shielding.

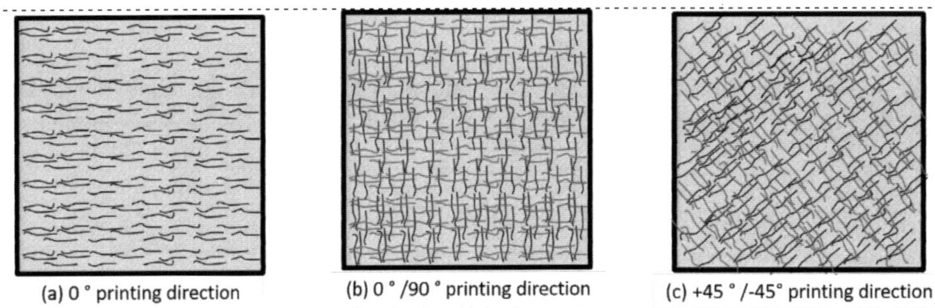

Figure 7. Schematic representation of the CNT network orientation during the 3D printing process.

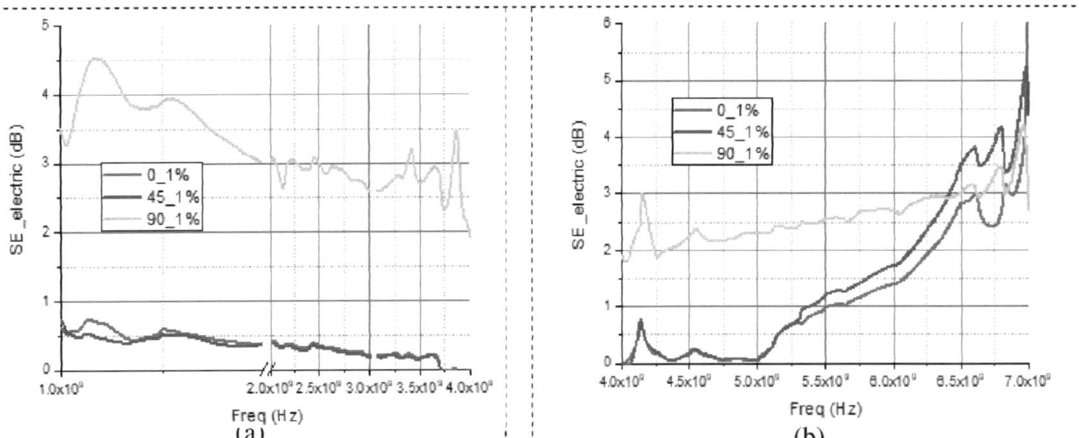

Figure 8. Effect of 3D printing direction on the SE of EVA/CNT composite with 1wt% in the range of (1-4) GHz (a) and (4-7) GHz (b).

Figure 8 shows the effect of printing direction on the SE values of EVA/CNT composite loaded with 1wt%, based on E-field measurements. It can be seen that the SE value is much larger for the 0°/90° printing direction than for the other two directions, which show very similar but lower shielding properties (Figure 8a and Figure 8b). For the composite with 3 wt% of CNTs, the printing direction has a significant effect on the SE as shown in Figure 9. This could be explained by the fact that addition of more carbon nanotube into the EVA matrix improves the density of the CNT network, especially for the 0°/90°printing direction. However, the composite loaded with 5 wt% of CNTs, the samples printed with 0°/90°direction exhibits the lowest SE value as compared with the other two directions, which exhibit nearly the same shielding properties (Figure 10).

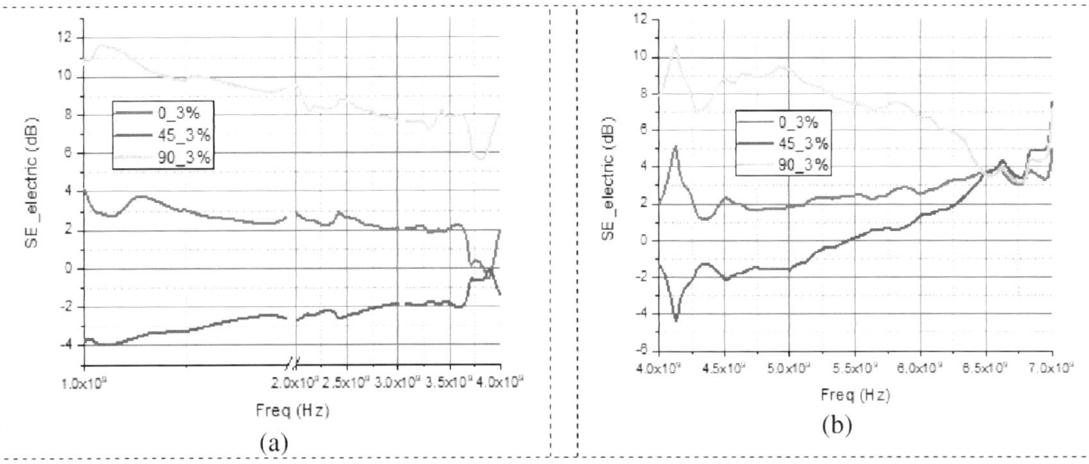

Figure 9. Effect of 3D printing direction on the SE of EVA/CNT composite with 3 wt% in the range of (1-4) GHz (a) and (4-7) GHz (b).

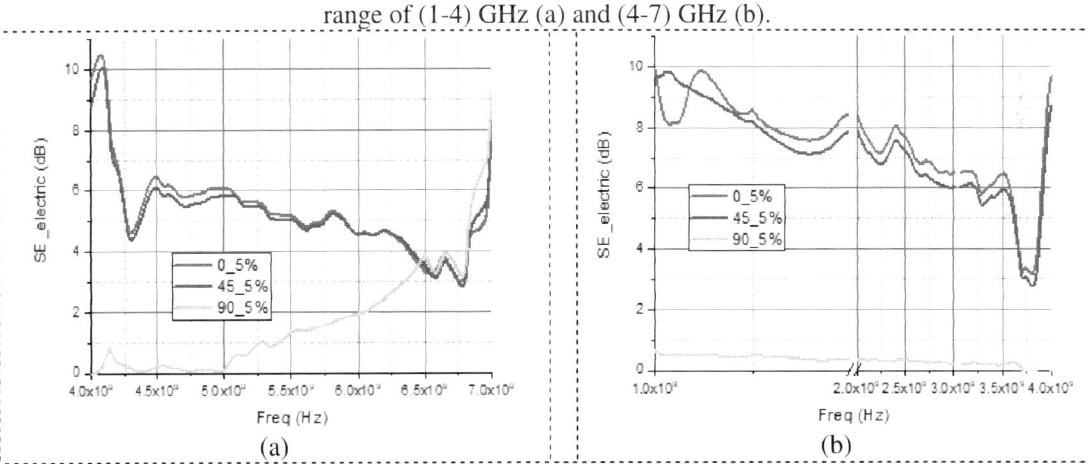

Figure 10. Effect of 3D printing direction on the SE of EVA/CNT composite with 5 wt% in the range of (1-4) GHz (a) and (4-7) GHz (b).

Figures 11, 12 and 13 illustrate the effect CNT wt% on the EMI shielding behaviour of EVA/CNT composites based on E-field measurements. It can be seen that for the 0° printing orientation, the EVA/CNT composite loaded with 5 wt% exhibits the highest SE value followed by the composite with 3 wt% and 1 wt%, as shown in Figure 11. The same SE behavior was observed for the EVA/CNT composite loaded with 5 wt% printed with +45°/-45° direction. This behaviour may be associated with higher electrical conductivity at higher nanotube loading due to the formation of a percolating CNT network, leading to more absorption of the electromagnetic waves. However, for 0°/90° printing direction, the composite with 3 wt% exhibits the best SE properties.

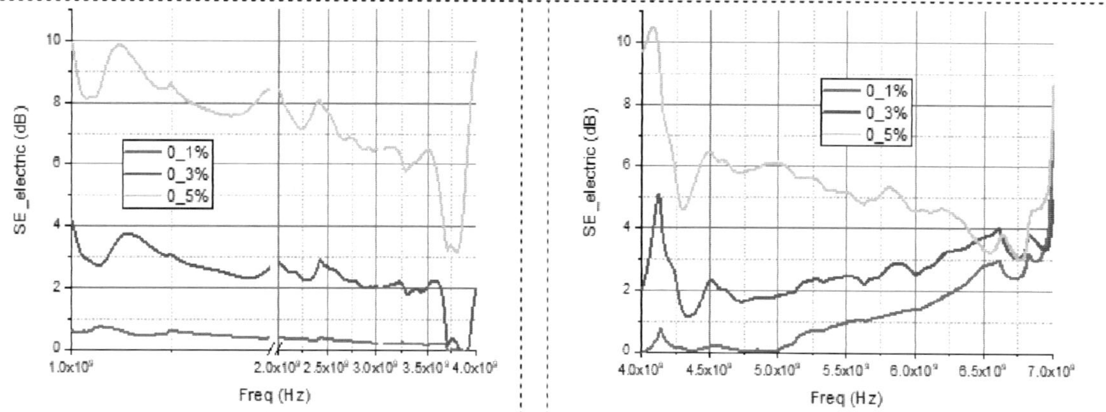

(a) (b)

Figure 11. SE for EVA/CNT composites for 0° printing direction in the range of (1-4) GHz (a) and (4-7) GHz (b).

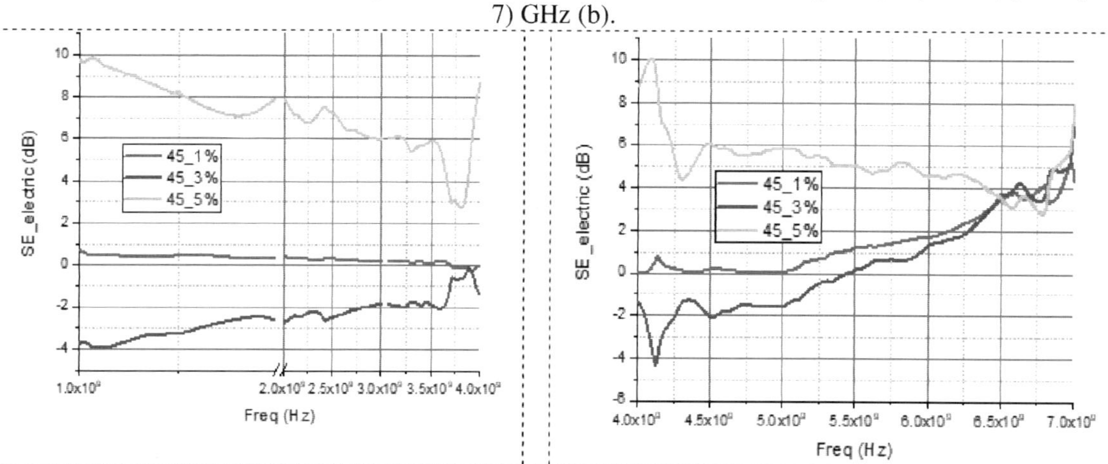

Figure 12. SE for EVA/CNT composites for +45°/-45° printing direction in the range of (1-4) GHz (a) and (4-7) GHz (b).

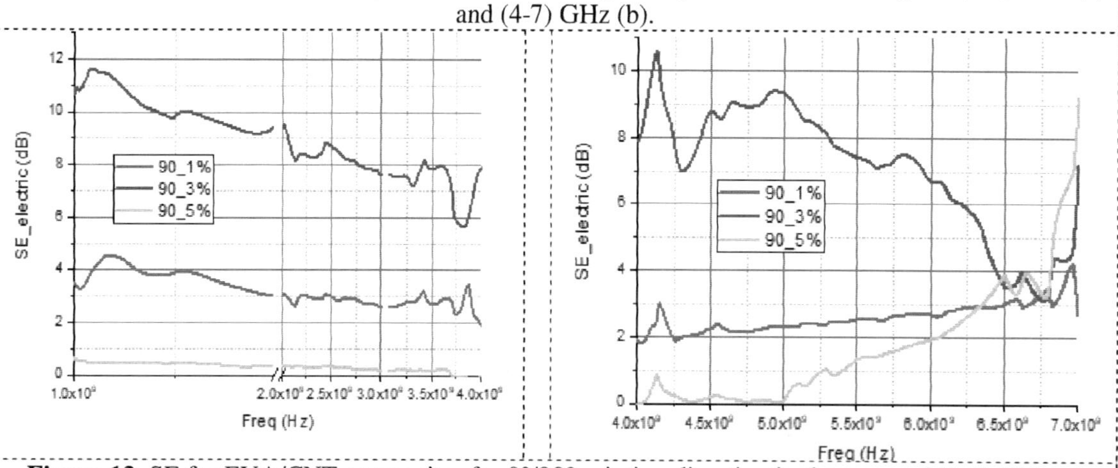

Figure 13. SE for EVA/CNT composites for 0°/90° printing direction in the range of (1-4) GHz (a) and (4-7) GHz (b).

4. Conclusions

This paper reported preliminary results on the EMI shielding effectiveness of 3D printed EVA/CNT composites in the 1 – 7 GHz range. Samples of EVA/CNT composite with 1, 3 and 5 wt% were 3D printed with three different printing directions (0°, alternative ± 45°, and alternative +0°/- 90°), and the SE behaviour characterized by using a pair of small near field probes and a vector network analyser. Based on the experimental results, the following main conclusions could be drawn: The EVA/CNT composites exhibited low SE properties in the 1 – 7 GHz range, dependent on both the printing direction and CNT wt%. Overall, the SE value increased with increasing CNT loading due to the formation of a continuous conductive network that spans the entire sample. However, a strong interaction between the CNT loading and printing direction was observed. A limitation of this study is that it was conducted in a reduced frequency range. Thus, future work will address the SE behaviour at frequencies higher than 7 GHz. Also, future work will address the effect of sample thickness and internal structure as the absorption of the incident electromagnetic waves may increase with increasing the sample thickness.

References

[1] Sankaran S, Deshmukh K, Ahamed M B, Khadheer Pasha S K 2018 Recent advances in electromagnetic interference shielding properties of metal and carbon filler reinforced flexible polymer composites: A review *Composites Part A: App. Sc. and Manufc.* **114** pp 49-71
[2] Bhattacharjee Y, Biswas S and Bose S 2020 Thermoplastic polymer composites for EMI shielding applications *Materials for Potential EMI Shielding Applications* pp 73-99
[3] Liu H, Wu S, You C, Tian N, Li Y and Chopra N 2021 Recent progress in morphological engineering of carbon materials for electromagnetic interference shielding *Carbon* **172** pp 569-596
[4] Ebbesen T W, Lezec H, Hiura H, Bennett J W, Ghaemi H F and Thio T 1996 Electrical conductivity of individual carbon nanotubes *Nature* **382** pp 54-56
[5] Coleman J N, Khan U, Blau W J and Gun'ko Y K 2006 Small but strong: A review of the mechanical properties of carbon nanotube–polymer composites *Carbon* **44** pp 1624-1652
[6] Bauhofer W and Kovacs J Z 2009 A review and analysis of electrical percolation in carbon nanotube polymer composites *Comp. Sci. and Tech.* **69** pp 1486-1498
[7] Balandin A A, Ghosh S, Bao W, Calizo I, Teweldebrhan D, Miao F and Lau C N 2008 Superior thermal conductivity of single-layer Graphene *Nano Letters* **8** pp 902-907
[8] Mittal G, Dhand V, Rhee K Y, Park S J and Lee W R 2015 A review on carbon nanotubes and graphene as fillers in reinforced polymer nanocomposites *J. of Ind. and Eng. Chem.* **21** pp 11- 25
[9] Jouni M, Djurado D, Massardier V, Boiteux G 2017 A representative and comprehensive review of the electrical and thermal properties of polymer composites with carbon nanotube and other nanoparticle fillers *Polymer International* **66** pp 1237-1251
[10] Das N C, Khastgir D, Chaki T K and Chakraborty A 2000 Electromagnetic interference shielding effectiveness of carbon black and carbon fibre filled EVA and NR based composites *Composites Part A: App. Sc. and Manuf.* **31**(10) pp 1069-108
[11] Al-Saleh M H and Sundararaj U 2009 Electromagnetic interference shielding mechanisms of CNT/polymer composites *Carbon* **47** pp 1738-1746
[12] Kim Y Y, Yun J, Kim H I and Lee Y 2012 Effect of oxyfluorination on electromagnetic interference shielding of polypyrrole-coated multi-walled carbon nanotubes *J. of Ind. and Engin.Chem.* **18** pp 392-398
[13] Joseph N, Janardhanan C and Sebastian M T 2014 Electromagnetic interference shielding properties of butyl rubber-single walled carbon nanotube composites *Comp. Sc. and Tech.* **101** p 139
[14] Kwon S, Ma R, Kim U, Choi H R, Baik S 2014 Flexible electromagnetic interference shields made of silver flakes carbon nanotubes and nitrile butadiene rubber *Carbon* **68** pp 118-124

[15] Polley D, Barman A and Mitra R K 2014 EMI shielding and conductivity of carbon nanotubepolymer composites at terahertz frequency *Optics Letters* **39** pp 1541-1544
[16] Raulo A, Suin S, Paria S, Khatua B B 2015 Expanded graphite (EG) as a potential filler in the reduction of percolation threshold of multiwall carbon nanotubes (MWCNT) in the PMMA/HDPE/EG/MWCNT nanocomposites *Polymer Composites* **37** pp 2070-2082
[17] Al-Saleh M H 2016 Electrical EMI shielding and tensile properties of PP/PE blends filled with GNP:CNT hybrid nanofiller *Synthetic Metals* **217** pp 322-330
[18] Ravindren R, Mondal S, Nath K and Das N C 2019 Investigation of electrical conductivity and electromagnetic interference shielding effectiveness of preferentially distributed conductive filler in highly flexible polymer blends nanocomposites *Composites Part A: App. Sc. and Manuf.* **118** pp 75-89
[19] Retailleau C, Alaa Eddine J, Ndagijimana F, Haddad F, Bayard B, Sauviac B, Alcouffe P, Fumagalli M, Bounor-Legaré V and Serghei A 2022 Universal behavior for electromagnetic interference shielding effectiveness of polymer based composite materials *Comp. Sc. and Tech.* **221** pp 109351
[20] Stan F, Roşculeţ R T and Fetecău C 2019 Direct Current Method with Reversal Polarity for Electrical Conductivity Measurement of TPU/MWCNT Composites *Measurement* **136** pp 345-355
[21] Stanciu N V, Stan F, Sandu I L, Fetecău C and Ţurcanu A M 2021 Thermal, Rheological, Mechanical, and Electrical Properties of Polypropylene/Multi-Walled Carbon Nanotube Nanocomposites *Polymers* **13** pp 187
[22] Chizari K, Arjmand M, Liu Z, Sundararaj U and Therriault D 2017 Three-dimensional printing of highly conductive polymer nanocomposites for EMI shielding applications *Mat. Tod. Comm.* **11** p 112-118
[23] Viskadourakis Z, Vasilopoulos K C, Economou E N, Soukoulis C M and Kenanakis G 2017 Electromagnetic shielding effectiveness of 3D printed polymer composites *App. Phy.* **123** p 736
[24] Gnanasekaran K, Heijmans T, van Bennekom S, Woldhuis H, Wijnia S de With G, Friedrich H 2017 3D printing of CNT- and graphene-based conductive polymer nanocomposites by fused deposition modeling *App. Mat. Tod.* **9** pp 21-28
[25] Schmitz D P, Ecco L G, Dul S, Pereira E C L, Soares B G, Barra G M O and Pegoretti A 2018 Electromagnetic interference shielding effectiveness of ABS carbon-based composites manufactured via fused deposition modeling *Mat. Tod. Comm.* **15** pp 70-80
[26] Stan F, Stanciu N V, Constantinescu A M and Fetecău C 2021 3D Printing of Flexible and Stretchable Parts Using Multiwall Carbon Nanotube/Polyester-Based Thermoplastic Polyurethane *J. of Manuf. Sc. and Engin. Trans. of ASME* **143** 051002
[27] Wang Y, Fan Z-W, Zhang H, Guo J, Yan D-X, Wang S, Dai K and Li Z M 2021 3D-printing of segregated carbon nanotube/polylactic acid composite with enhanced electromagnetic interference shielding and mechanical performance *Mat. & Des.* **197** 109222
[28] Stan F, Stanciu N-V and Fetecău C 2017 Melt Rheological Properties of Ethylene-Vinyl Acetate/ Multi-Walled Carbon Nanotube Composites *Comp. Part B Eng.* **110** pp 20-31
[29] Roşculeţ R T, Stan F and Fetecău C 2018 On the strain sensing of EVA/MWCNT composite Mat.Pl. **55**(3) pp 274-278
[30] Stanciu N V, Stan F, Sandu I L, Susac F, Fetecău C and Rosculeţ R 2019 Mechanical, Electrical and Rheological Behavior of Ethylene-Vinyl Acetate/Multi-Walled Carbon Nanotube Composites *Polymers* **11** p 1300
[31] Nanocyl Technical Data Sheet: NC7000™ Edited 12th July 2016
[32] Nanocyl Technical Data Sheet: Plasticyl ™ EVA2001 Edited 26th January 2016.
[33] EVATANE® 20-20 Product Data Sheet
[34] Celozzi S, Araneo R and Lovat G 2008 *Electromagnetic shielding* (John Wiley & Sons)
[35] IEEE 299-2006 IEEE Standard Method for Measuring the Effectiveness of Electromagnetic Shielding Enclosures
[36] IEC-50147-1 - Measuring the Effectiveness of Electromagnetic Shielding Enclosures

[37] ASTM F3057-14 - Measuring the Effectiveness of Electromagnetic Shielding of Glazing/Glasses
[38] ANSI/SCTE 48-3 - Measuring the Electromagnetic Shielding Effectiveness of Cables and Connectors InGtem Cell
[39] ASTM D4935-18 Standard Test Method for Measuring the Electromagnetic Shielding Effectiveness of Planar Materials
[40] Marvin A C, Dawson L, Flintoft I L and Dawson J F 2009 A Method for the Measurement of Shielding Effectiveness of Planar Samples Requiring No Sample Edge Preparation or Contact *IEEE Transactions on Electromagnetic Compatibility* **51**(2) pp 255-262
[41] Miclăuș S, Barbu-Tudoran L, Bechet P and Baltag O 2018 Characterizing and Testing A Fabric Containing Amorphous Glass-Coated Ferromagnetic Microwires for its Magnetic and ElectricShielding Properties in the Frequency Range (1-3) GHz *Scientific Bulletin AFT* 2(**46**) pp 94-102
[42] Miclăuș S, Bechet P, Paljanos A, Mihai George, Pătru I, Aron A M and Baltag O 2016 Shielding Effectiveness Of Some Conductive Textiles And Their Capability To Reduce The Mobile Phones Radiation *Proceedings of the International Conference Knowledge-Based Organization XXII* No 3 pp 524-530
[43] Apreutesei A-L, Curteza A, Paljanos A, Miclăuș S, Mihai G, Aron A M, Bechet P and Baltag O 2016 Radiation reduction capabilities of some woven fabrics with metallic yarns attached to mobile phones emitting in 2G- and 3G- communication standards EPE Iasi
[44] Miclăuș S, Mihai G, Aron A M, Mitrescu C, Bechet P and Baltag O 2017 Shielding efficiency of a fabric based on amorphous glass-covered magnetic microwires to radiation emitted by a mobile phone in 2G and 3G communication technologies *Review of "Nicolae Bălcescu" Land Forces Academy* XXII nr. 4(**88**) pp 289-297 (Sibiu, Romania)
[45] Roșu G, Velicu V, Boitan A, Mihai G, Tuta L and Baltag O 2022 On the electromagnetic shielding properties of carbon fiber materials *El. Eng. & Elect.* 1 pp 38-43
[46] Mediano 2017 A Evaluate shielding effectiveness with your VNA https://incompliancemag.com/article/evaluate-shielding-effectiveness-with-your-vna/
[47] Aaronia Shield Ultra datasheet https://aaronia.com/shielding-materials/aaronia-shield-ultra/

THE EFFECT OF 1.74 AND 2.3 GHZ MICROWAVE EXPOSURE OF *AILANTHUS ALTISSIMA* LEAF EXTRACTS ON THEIR ANTIOXIDANT CONTENT AND ACTIVITY

M D Cocîrlea[1], S Miclăuş[2] and S Oancea[1]

[1]Department of Agricultural Sciences and Food Engineering, "Lucian Blaga" University of Sibiu, 7-9 Dr. I. Rațiu Street, 550012 Sibiu, Romania
[2]Department of Communication and IT, "Nicolae Bălcescu" Land Forces Academy, 3-5 Revoluției Street, Sibiu, Romania

E-mail: simona.oancea@ulbsibiu.ro

Abstract. Microwave (MW) irradiation in reactors with input power levels of 200-1000 W has been frequently applied for the extraction of natural compounds from vegetal material. The aim of the present research was to investigate the effect of much lower MW input powers (< 12 W) delivered through antennas into shielded enclosures on mixtures of ethanol solution-*Ailanthus altissima* leaves (frozen and oven-dried), at different frequencies (1.74 and 2.3 GHz) and exposure times (0.5, 1 and 3 h). The phenolic content and antioxidant activity were quantified after the exposures. Significant differences between the antioxidant content/ activity and the type of raw material were found, in favour of frozen samples. After MW treatment, the bioactive content and activity increased in frozen *Ailanthus* leaf extracts compared to control. No significant effect was observed in case of extracts prepared from oven-dried samples at 60°C. The ATR-FTIR analysis indicates no significant structural changes after the irradiation. These results revealed a better extractability of phenolic compounds following MW exposure in particular at 2.3 GHz.

1. Introduction

In the last few years, the efforts of reducing the negative impact of various invasive plant species by using them in applications that support the quality of the environment and human health has become more and more widespread [1]. Nevertheless, their potential to be used in pharmaceutical and industrial processes, as well as their economic value, has not been sufficiently studied [2].

The tree of heaven (*Ailanthus altissima*) is a woody species native from East Asia [3],which causes serious environmental impact in Europe, North America [4] and South Africa [5], exhibiting characteristics of a very aggressive invasive plant [6], such as withstanding large variations in the amount of light available in the environment [4].

The chemical composition of *Ailanthus altissima* leaves consists in several compounds with bioactive potential, such as alkaloids, flavonoids, tannins, quercetin, isoquercetin, phenolic acids (gallic, caffeic, chlorogenic, *p*-coumaric, ferulic) [7,8] that makes them attractive for research. The

Ailanthus altissima extracts showed antioxidant [7, 9], antimicrobial [7] antitumor properties [10], and cytotoxicity, as well [9].

Microwave-assisted extraction (MAE) represents a modern and efficient method of extracting phenolic compounds and flavonoids from various plant sources [11], operating at lower energy and shorter process time compared to those of conventional methods [12]. The use of microwave (MW) energy for extraction of natural compounds is strongly related to heating as a result of ionic conduction and dipole rotation which occur in the solvent and in the sample [13]. Thus, the efficiency of MAE and heating of materials is depending on the solvent and the type of compounds in the sample. Higher temperatures may negatively affect the thermosensitive compounds, such as phenolics. However, MAE is considered an environmentally friendly technology [14] leading to extracts with lower level of cytotoxicity [11] but a higher protein and sugar content [15].The quality of the process proved to be significantly influenced by various factors, mainly solvent composition (aqueous methanol and ethanol solutions), temperature, extraction time, irradiation power (200-600 W for closed systems), sample moisture [13, 16].

This study aimed to investigate the effect on the phenolic content and antioxidant activity following the exposure of mixtures of ethanol solution-*Ailanthus altissima* leaves(frozen and oven-dried) to MW,at various process parameters, frequencies(1.74 and 2.3 GHz), incident power densities (1155-2400 W/m^2)and exposure durations (0.5, 1 and 3 h). Additionally, Fourier-Transform Infrared (ATR-FTIR) analysis was performed to evaluate the potential structural changes of the investigated samples.

2. Materials and methods

2.1. Collection, drying and storage of raw material

The leaves of the tree of heaven collected in October 2021, from Vâlcea county, Romania, were divided in two parts: one part was kept in the freezer and the other one was subjected to dryingat 60°C using a convection oven (UFE 400, with forced air circulation, Memmert, Germany), until a final moisture content of 6% was achieved as determined with a moisture analyser (MAC 210 Radwag, Polonia). After drying, the leaves were grinded and stored in the refrigerator in labeled zippered bags, until analysis.

All reagents were of analytical grade.

2.2. Preparation of ethanol extracts of frozen and dried leaves

The frozen leaves were previously blended to obtain a paste texture. Mixtures of 70% ethanol and *Ailanthus* leaves(20 ml/2 g) were exposed to different microwave irradiation conditions: frequencies (1.74 and 2.3 GHz), incident power densities (1155-2400 W/m^2), incident electric field strengths (660 and 950 V/m) and exposure durations (0.5, 1 and 3 h). Control samples without MW treatment of frozen and oven-dried leaves were used for comparative studies.

2.3. MW irradiation of mixtures of ethanol solution-Ailanthus altissima leaves

Each mixture was placed at a distance of 20 cm from the antenna edge on its maximum radiation direction, so that at both investigated frequencies, 1.74 GHz and 2.3 GHz, respectively, the ratio distance (d) perwavelength (λ), was kept at $d/\lambda < 0.2$. This setting ensured that all exposures took place under far-field conditions. The experimental exposure system presented in Figure 1 was composed of a signal generator model SM 300 (Rhode & Schwarz), a radiofrequency power amplifier (model Ophir 5120), a directive horn antenna (model BBHA 9120A, Schwarzbeck), and a shielded enclosure where the samples were inserted for irradiation. The emission antenna was placed at the entrance of the enclosure, covering it almost totally. The electric field strength incident to the liquid sample was measured in air using the EM Sense 10 isotropic field probe connected to the EM Center 7007-200 (ETS-Lindgren). Data were collected and recorded using the Probe-View V software.

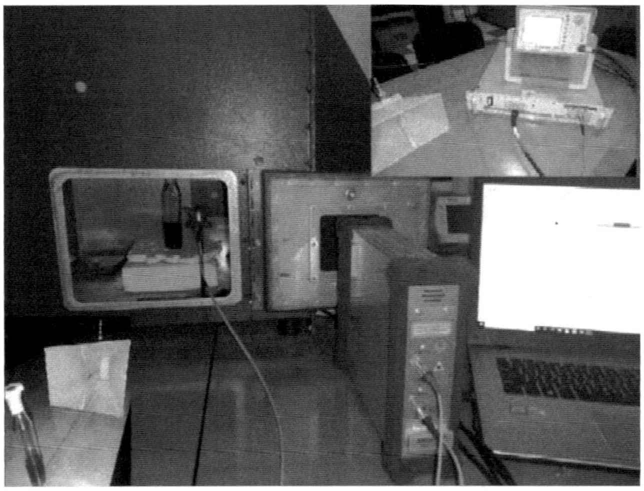

Figure 1. Components of the exposure and measurement experimental system: signal generator and amplifier, horn antenna, shielded enclosure, tri-axial field-probe with fibre optic cable and the liquid sample in the glass container.

An average value was computed along the exposure periods and provided as a variable of the exposure. The temperature measurements inside the irradiated samples indicated the power absorption/deposition in the sample, being done with a fiber optic temperature probe (model TS2, Weidmann-Optocon) connected to the Optocon center model OEM (Figure 2). The tip diameter of the temperature probe was 1.3 mm. Sample exposures were performed without agitation, so that the local temperature was indicated by the small probe. This type of temperature probe is based on a GaAs sensor so is non-conductive, inert to MW, provides high accuracy and stability, generating repeatable and accurate results. The temperature recordings were made by using the Probe View V software with a sampling period of 2 s. The dynamics of the temperature evolution due to MW absorption in the samples was recorded continuously during 30 minutes (Figure 3) and then extrapolated during the longest exposure duration of 3 h.

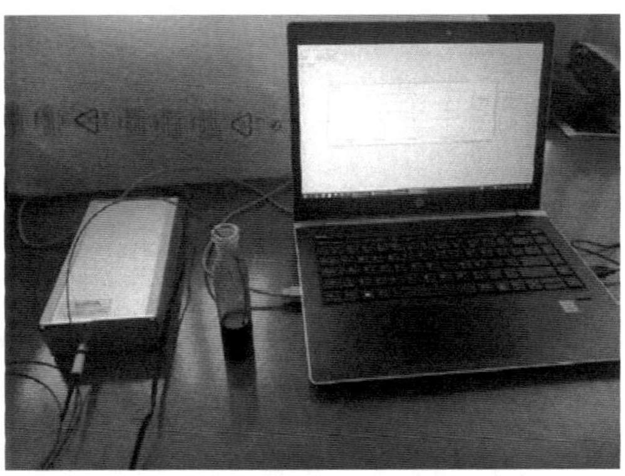

Figure 2. The set-up of the out-of-the enclosure temperature measurement using the fluoroptic probe inserted in the mixture.

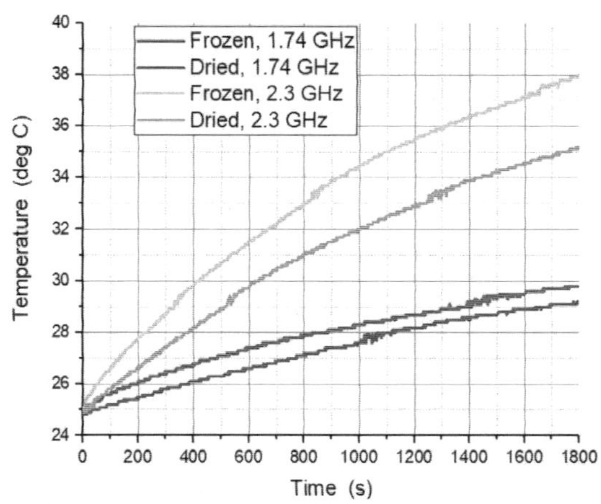

Figure 3. Temperature dynamics of MW exposure of the samples at 1.74/ 2.3 GHz and 660/ 950 V/m field strength.

Temperature control enabled the knowledge of the temperature differences between the initial and the final point of the irradiation, and the momentary and average rate of electromagnetic power deposition in the liquid.
Two sets of 6 experimental runs were performed for each type of material, frozen and oven-dried *Ailanthus altissima* leaves. For each set of experiments, non-MW irradiated sample (control) was used.

2.4. Determination of total phenolic content, total tannin content and antioxidant activity of leaf extracts
After exposure, mixtures were centrifuged at 8000 rpm, 4°C for 10 min using the refrigerated centrifuge (Universal 320, Hettich, Germany). The total phenolic content was determined by the Folin-Ciocâlteu method [17] using the spectrophotometer Specord 200Plus UV-Vis (Analytik Jena, Germany). The results were expressed as mg of gallic acid equivalents per 100 g dry weight (mg GAE/100 g DW).The total tannin content was determined spectrophotometrically by the vanillin assay [18] and expressed as mgcatechin equivalents per 100 g dry weight.The total antioxidant activity was measured using the Ferric Reducing Antioxidant Power (FRAP) assay [19] and expressed as mg ascorbic acid equivalents per 100 g dry weight (mg AAE/100 g DW).

2.5. ATR-FTIR analysis of leaf extracts
Fourier transform infrared (FTIR) analysis was carried out using the ALPHA FTIR spectrometer (Bruker, Germany) with the combined QuickSnapTM sampling modules and Zn Se ATR (Attenuated Total Reflectance) with a resolution of 4 cm^{-1}. An average of 32 scans was recorded in the ATR mode.

2.6. Statistical analysis of the results
Duplicate experiments were done and results were expressed as mean values±standard deviation (SD).The statistical analysis was performed using the Kruskal-Wallis H test, Dunn's Test and Generalized Linear Model (GLM) in R x64 4.0.3. [20].

3. Results and discussions
3.1. The influence of MW irradiation on the content of phenolics, tannins and antioxidant activity of Ailanthus altissima leaf extracts
Ethanol extracts of frozen and oven-dried *Ailanthus altissima* leaves were evaluated for the antioxidant compounds and activity, in relation to MW exposure under different conditions, frequency (1.74 and 2.3 GHz), exposure time (0.5, 1 and 3 h), incident electric field strength (660 and 950 V/m). The temperature dynamics of MW irradiation (Figure 3) indicated that neither of the exposures produced a final temperature of the mixtures of ethanol solution-*Ailanthus altissima* leaves higher than 43°C. The results are shown in table 1.

Table 1. The content of phenolic and tannin compounds, and antioxidant activity of leaf 70% ethanol extracts of *Ailanthus altissima* exposed to MW at 1.74 and 2.3 GHz for different irradiation time.

Experimental parameters	Control sample (untreated)	Exposure duration (h)	Frequency (GHz)	Incident electric field strength (V/m)	MW treated sample
Extracts of *Ailanthus altissima* frozen leaf					
Total phenolics (mg GAE/100 g DW)	1652.47±26.51	0.5	1.74	660	1741.39±21.16
		1	1.74	660	1684.86±42.85
		3	1.74	660	1744.86±22.73
		0.5	2.3	950	1690.89±35.73
		1	2.3	950	1768.01±34.26
		3	2.3	950	1880.18±15.41

Total tannins (mg catechin/ 100 g DW)	401.43±5.71	0.5	1.74	660	464.50±2.74
		1	1.74	660	439.89±7.44
		3	1.74	660	448.12±9.06
		0.5	2.3	950	403.93±1.96
		1	2.3	950	469.11±9.55
		3	2.3	950	424.89±4.94
Antioxidant activity (mg AAE/100 g DW)	1566.21±28.95	0.5	1.74	660	1647.05±28.90
		1	1.74	660	1391.25±55.81
		3	1.74	660	1518.17±43.28
		0.5	2.3	950	1506.40±29.41
		1	2.3	950	1583.65±20.84
		3	2.3	950	1694.75±31.28
Extracts of *Ailanthus altissima* dried leaf					
Total phenolics (mg GAE/100 g DW)	1158.35±19.55	0.5	1.74	660	1143.37±16.47
		1	1.74	660	1178.64±13.07
		3	1.74	660	1152.19±21.45
		0.5	2.3	950	1138.83±18.35
		1	2.3	950	1196.86±9.49
		3	2.3	950	1161.55±18.09
Total tannins (mg catechin/ 100 g DW)	447.41±8.71	0.5	1.74	660	437.66±6.33
		1	1.74	660	442.81±6.41
		3	1.74	660	440.26±7.63
		0.5	2.3	950	463.23±6.62
		1	2.3	950	428.27±6.64
		3	2.3	950	432.94±8.97
Antioxidant activity (mg AAE/100 g DW)	857.96±21.31	0.5	1.74	660	847.65±26.81
		1	1.74	660	842.97±28.71
		3	1.74	660	871.46±19.27
		0.5	2.3	950	841.66±29.22
		1	2.3	950	877.86±21.23
		3	2.3	950	864.85±23.19

Regarding the bioactive content of frozen *Ailanthus altissima* control (non-irradiated) samples, the obtained results showed significant levels of phenolic compounds, good antioxidant activity and low content of condensed tannins. Other authors found much lower phenolic content of *Ailanthus altissima*70% ethanol leaf extracts obtained either by reflux (716 mg GAE/ 100 g DW) or ultrasound-assisted extraction (553 mg GAE/ 100 g DW) [21]. As reported by other authors, the use of 99.5% methanol determined a better extractability of phenolic compounds by maceration of fresh *Ailanthus altissima* leaves (247±8.4 mg GAE/g DW) [7]. Regarding the oven-dried leaves at 60°C, the obtained total phenolic content of control 70% ethanol extracts (1158.35±19.55mg GAE/100g DW) was found lower than other reported values, *e.g.* 158.72±0.38 mg GAE/g DW for 50% ethanol extracts obtained by reflux from *Ailanthus altissima* leaves dried at 35°C [22]. The results show that extraction method (maceration, reflux, ultrasound-assisted), solvent type and drying temperature of raw materials highly influence the extractability of phenolic compounds.

As noticed in table 1, the type of sample highly influenced the extraction efficiency. The statistical analysis using the Kruskal-Wallis H test showed significant differences between the means of the total phenolic content (p-value = 0.001725)/antioxidant activity (p-value = 0.001745) andthe type of the sample – frozenor oven-dried, both performed Dunn's tests showing a better response in case of frozen

Ailanthus altissima leaves. In contrast, the drying of the leaves at 60°C did not significantly affect the means of the total content of tannins between samples. The boxplots presented in Figure 4 indicate how the data were distributed.

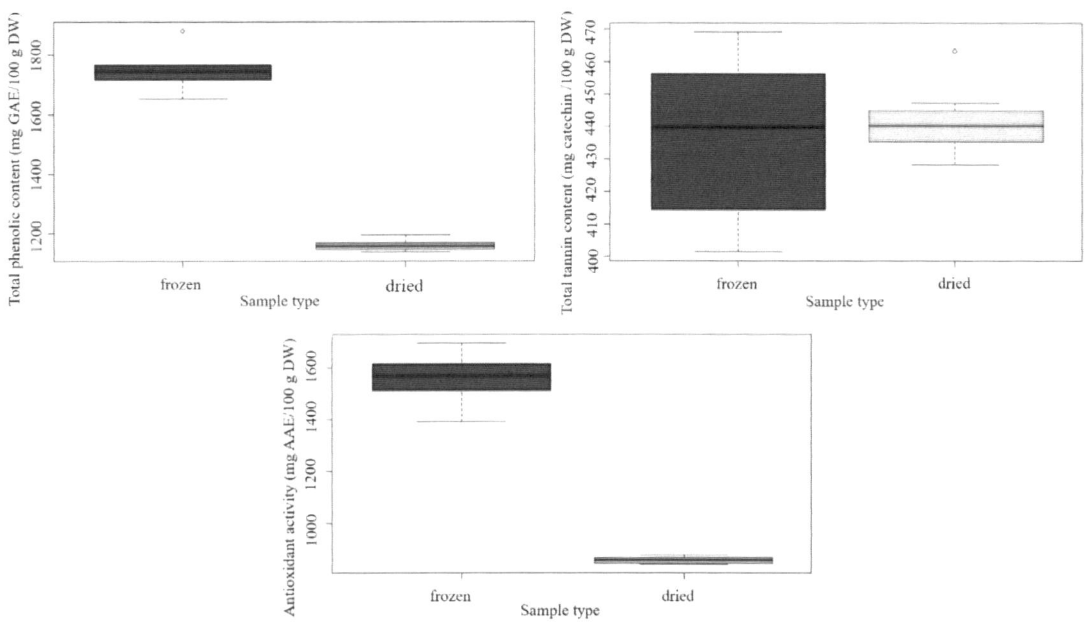

Figure 4. Boxplot representation of the distribution of phenolic/tannin contents and antioxidant activity according to the sample type (frozen, oven-dried).

Regarding the investigation of the MW effect on the antioxidant content (phenolics, tannins) and activity, the efficiency of the extraction of the targeted compounds increased under MW irradiation of solvent mixtures obtained from frozen leaves, in particular at 2.3 GHz of MW. No statistically significant differences between the means of the concerned bioactive characteristics and irradiation parameters (exposure time, frequency, incident electric field strength) were found. By our knowledge, no studies on MAE of phenolic compounds and antioxidant activity have been reported so far. Because literature did not report MAE of phenolic compounds from *Ailanthus altissima* leaves, our results could not be compared with similar in the literature. However, the MW exposure produced higher content of total phenolics than other reported modern extraction techniques, such as ultrasound-assisted extraction which led to a decrease of the phenolic content compared to reflux extraction, as shown by some authors [21]. No relevant MW influence on the extractability of the investigated compounds was noticed in the experimental runs performed with oven-dried *Ailanthus altissima* leaves.

The GLM model analysis performed on the obtained experimental data indicated an important correlation between the total phenolic content and the FRAP antioxidant activity, the first one being a statistically significant predictor variable in the model, its increase leading to augmentation of the antioxidant activity. This observation is in agreement with other reported results, which confirmed the existence of the positive correlation between the two variables [7, 22-23].

The vast majority of work on the current topics [11, 16, 24-25], used irradiation instruments of the type multi-mode or single mode reactors [26] for MAE of chemical compounds. These instruments usually radiate inside a MW cavity high powers, respectively hundreds of watts up to 1500 W, providing significant and fast temperature increase in the samples up to (72 - 94) °C after just a few minutes of irradiation. The generated heat facilitates the solvents' diffusivity into the sample and release of the targeted phytochemicals out of the sample. Such exposure set-ups provide high heating rates in just tens of seconds while the whole irradiation usually takes around 5 minutes. Completely different from such MW exposures, much milder ones as the ones we provided in the present experiments, seem to produce relevant results, as well. With no more that 10-12 W input into a directive antenna with a gain not higher than (9.1 – 9.8) dBi, we irradiated samples which did not exceed 43 °C after 3 h of continuous exposure. A much lower slope of temperature increase, completed with a much longer irradiation time seems to be able to induce an effect similar to the drastic MW heating provided by the spread techniqueFs of MW reactors. We could observe with this different approach that the solvent diffusion and extraction kinetics seem to depend on the total absorbed energy (in Joules) and much less on energy rate deposition (in Joules/second) in the sample.

3.2. ATR-FTIR spectroscopy analysis of Ailanthus altissima leaf extracts

The ATR-FTIR spectra recorded for the ethanol extracts obtained from frozen and oven-dried *Ailanthus altissima* leaf samples, control and MW irradiated are represented in Figure 5. The irradiated samples were obtained by exposure of the extracts at 1.74 and 2.3 GHz at various exposure/ extracting times (0.5, 1 and 3 h).

The absorption peaks in the FTIR spectra of frozen leaf ethanol extracts (control) were assigned based on the literature reports [27, 28, 29], as follows: strong broad overlapping peaks at 3366 cm^{-1} and 3232 cm^{-1} due to H-bonded O-H stretching (phenol/ alcohol) and N-H stretching of amine/ amide; 2949, 2922, 2852 cm^{-1}asym./ sym. alkyl C-H stretching of CH_2, CH_3 and methoxy groups (O-CH_3) of polyphenols; strong sharp peaks at 1646-1636 cm^{-1} due to C=O stretching (lipid ester, phenolic acids, free and associated amide I of peptide bonds), C=C stretching (alkene) and N-H bending of amine; peaks at 1558-1507 cm^{-1} attributed to C=C-C aromatic ring stretching (amino acids, flavonoids, polyphenols), N-H bending and C-N stretching of amide II band; overlapping peaks at 1396-1175 cm^{-1} due to C-OH/ C-O-C streching (carbohydrates, glycosides, ether/ methoxy group), C-N stretching of amine and C-H bending/ rocking of CH_3 group (other possible stretching of S=O sulfate group or P=O of phosphates); strong sharp peaks at 1067 and 1012 cm^{-1} assigned to C=C, C-O stretching (ester, anhydride or sulfoxide); medium peaks at 852-800 cm^{-1} assigned mainly to C-H out-of-plane bending (alkene, aromatic ring substitution). Regarding the FTIR spectra of oven-dried leaf ethanol extracts, in principle, the similar peaks were found as those for frozen leaf extracts, but several differences were noticed, as follow: two peaks better separated and shifted at 3472 and 3150 cm^{-1} due to N-H stretching of amine and intermolecular bonded O-H; overlapping peaks at 2925 cm^{-1}; evidenced peak at 1721 cm^{-1} assigned to C=O stretching (carboxylic acid, aldehyde, ketone); peak of higher intensity at 1359 cm^{-1} (probably C-H rocking of CH_3 group, C-H bending of CH_3 of carbonyl compounds); evidenced medium peaks at 1257 and 1203 cm^{-1} (C-O stretching of ether). Such differences may be attributed to changes in some phenolic compounds as previously shown by their quantitative evaluation (total phenolic content). However, more concentrated samples for spectra collection were involved in case of dried leaf extracts (same solvent/ solid ratio).

No major spectral changes occurred in samples exposed to MW at 1.74 and 2.30 GHz for various time compared to control, with some exception (disappearance of the weak peak at 1747 cm^{-1}, merging of the weak peaks in the region 1558-1507 cm^{-1} in irradiated extracts of frozen samples), denoting that MW exposure of 1.74 and 2.30 GHz did not significantly influence the chemical structure of compounds, but assisted the extraction due to mechanical events on plant cells.

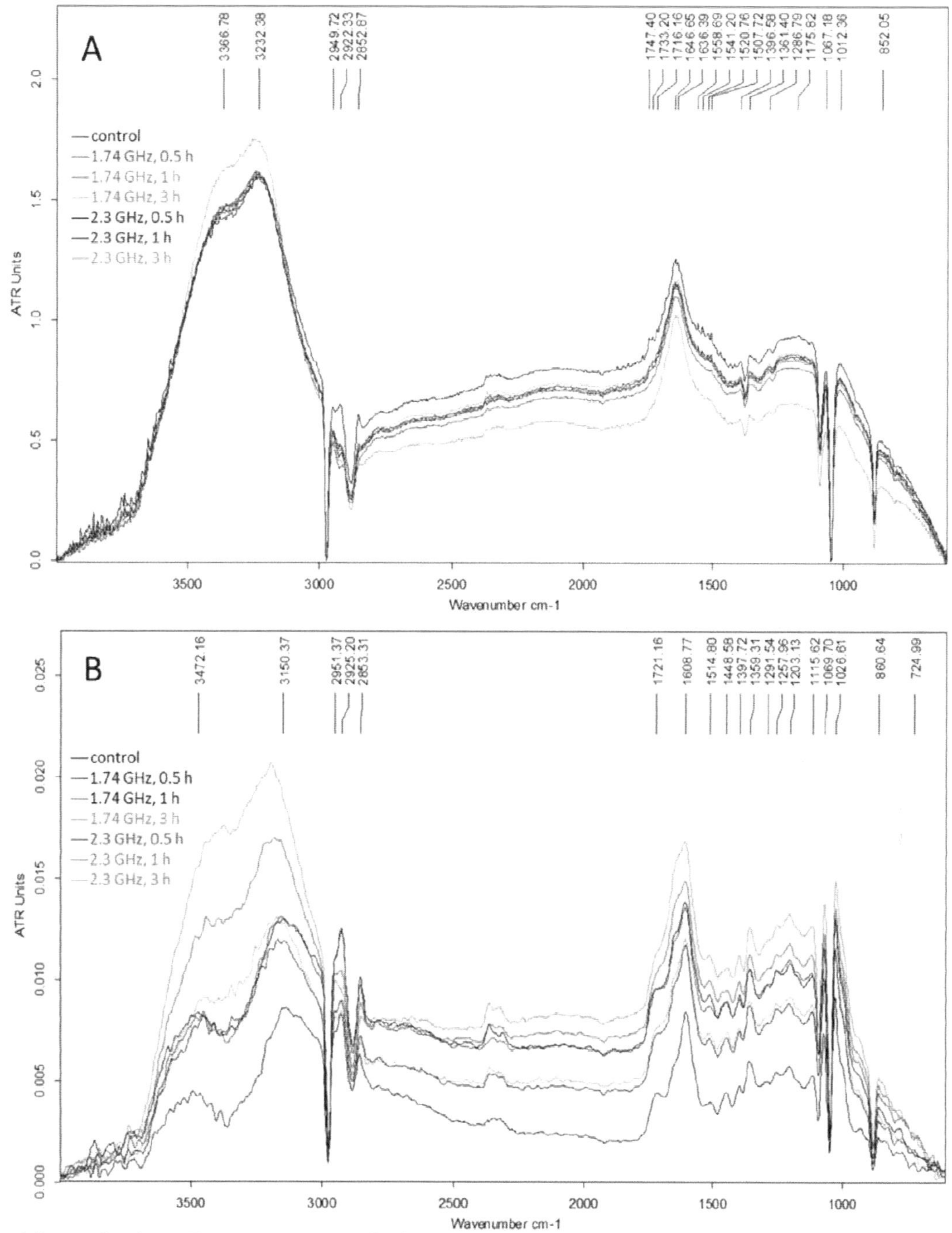

Figure 5. The ATR-FTIR spectra of ethanol extracts of *Ailanthus* leaf prepared from frozen (A) and oven-dried (B) samples (peaks of control sample are given).

4. Conclusions

Increasing the extractability of valuable compounds from different plant sources including invasive ones represents an important task when industrial applications of final products are targeted. The present study investigated for the first time the influence of MW of 1.74 and 2.3 GHz, for two incident electric field strengths and three exposure times on the phenolic/ tannin content and antioxidant activity of *Ailanthus altissima* leaf extracts. Significant differences between the antioxidant content/activity and the type of raw material were found, in favor of frozen *Ailanthus altissima*. Increase of total phenolics, tannins and antioxidant activity after MW irradiation was found in mixtures of ethanol solution-*Ailanthus altissima* frozen leaves, without statistically significant differences in relation to exposure parameter conditions. No significant effect was observed in case of ethanol extracts prepared from oven-dried samples at 60°C. The results underline over again the fact that leaves of the tree of heaven are a valuable resource of bioactive components, which can be highly recovered by MW irradiation.

References

[1] Peter A, Žlabur J Š, Šurić J, Voća S, Purgar D D, Pezo L and Voća N 2021 Invasive plant species biomass—Evaluation of functional value *Molecules* **26** 3814

[2] Rai P K and Singh J S 2020 Invasive alien plant species: Their impact on environment, ecosystem services and human health *Ecol. Indic.* **111** 106020

[3] Saina J K, Li Z Z, Mekbib Y, Gichira A W and Liao Y Y 202 Transcriptome sequencing and microsatellite marker discovery in *Ailanthus altissima* (Mill.) Swingle (Simaroubaceae) *Mol. Biol. Rep.* **48** pp 2007-2023

[4] Petruzzellis F, Nardini A, Savi T, Tonet V, Castello M and Bacaro G 2019 Less safety for more efficiency: water relations and hydraulics of the invasive tree *Ailanthus altissima* (Mill.) Swingle compared with native *Fraxinus ornus* L. *Tree Physiol.* **39** pp 76-87

[5] Walker G A, Gaertner M, Robertson M P and Richardson D M 2017 The prognosis for *Ailanthus altissima* (Simaroubaceae; tree of heaven) as an invasive species in South Africa; insights from its performance elsewhere in the world *S. Afr. J. Bot.* **112** pp 283-289

[6] Rebbeck J, Hutchinson T, Iverson L, Yaussy D and Fox T 2017 Distribution and demographics of *Ailanthus altissima* in an oak forest landscape managed with timber harvesting and prescribed fire *For. Ecol. Manag.* **401** pp 233-241

[7] Poljuha D, Sladonja B, Šola I, Dudaš S, Bilić J, Rusak G, Motlhatlegoe K E and Eloff J N 2017 Phenolic composition of leaf extracts of *Ailanthus altissima* (Simaroubaceae) with antibacterial and antifungal activity equivalent to standard antibiotics *Nat. Prod. Commun.* **12** 1934578X1701201021

[8] Caramelo D, Pedro S I, Marques H, Simão A Y, Rosado T, Barroca C, Gominho J, Anjos O and Gallardo E 2021 Insights into the Bioactivities and Chemical Analysis of *Ailanthus altissima* (Mill.) Swingle *Appl. Sci.* **11** 11331

[9] Mohamed H R, El-Wakil E A, Abdel-Hameed E S S, El-Hashash M M and Shemis M 2021 Evaluation of total phenolics, flavonoids, and antioxidant and cytotoxic potential of *Ailanthus altissima* (Mill.) swingle leaves *J. Rep. Pharm. Sci.* **10** p 130

[10] Daga M, Pizzimenti S, Dianzani C, Cucci M A, Cavalli R, Grattarola M, Ferrara B, Scariot V, Trottac F and Barrera G 2019 Ailanthone inhibits cell growth and migration of cisplatin resistant bladder cancer cells through down-regulation of Nrf2, YAP, and c-Myc expression *Phytomedicine* **56** pp 156-164

[11] Silva A M, Lago J P, Pinto D, Moreira M M, Grosso C, Cruz Fernandes V, Delerue-Matos C and Rodrigues F 2021 Salicornia ramosissima Bioactive Composition and Safety: Eco-Friendly Extractions Approach (Microwave-Assisted Extraction vs. Conventional Maceration) *Appl. Sci.* **11** 4744

[12] Osorio-Tobón J F 2020 Recent advances and comparisons of conventional and alternative extraction techniques of phenolic compounds *J. Food Sci. Technol.* **57** pp 4299-4315

[13] Destandau E, Michel T and Elfakir C 2013 *Microwave-assisted extraction, in Natural Product Extraction. Principles and applications* vol **21** ed M A Rostagno and J M Prado (The Royal Society of Chemistry) p 113

[14] Shukla H S and Pratap A 2017 Comparative studies between conventional and microwave assisted extraction for rice bran oil *J. Oleo Sci.* **66** pp 973-979

[15] Jovanović A A, Lević S M, Pavlović V B, Marković S B, Pjanović R V, Đorđević V B, Nedović V and Bugarski B M 2021 Freeze vs. Spray drying for dry wild thyme (*Thymus serpyllum* L.) extract formulations: The impact of gelatin as a coating material *Molecules* **26** 3933
[16] Mandal V, Mohan Y and Hemalatha S 2007 Microwave assisted extraction—an innovative and promising extraction tool for medicinal plant research *Pharmacogn Rev.* **1** pp 7-18
[17] Singleton V L and Rossi J A 1965 Colorimetry of total phenolics with phosphomolybdic-phosphotungstic acid reagents *Am. J. Enol. Vitic.* **16** pp 144-158
[18] Price M L, Van Scoyoc S and Butler L G 1978 A critical evaluation of the vanillin reaction as an assay for tannin in sorghum grain *J. Agric. Food Chem.* **26** pp 1214-1218
[19] Benzie I F and Strain J J 1996 The ferric reducing ability of plasma (FRAP) as a measure of "antioxidant power": the FRAP assay *Anal. Biochem.* **239** pp 70-76
[20] R Core Development Team 2020 R: A Language and Environment for Statistical Computing, version 4.0.3.; R Foundation for Statistical Computing: Vienna, Austria; Available from: http://www.r-project.org (accessed on 27 November 2020)
[21] Lungu L, Savoiu M R, Manolescu B N, Farcasanu I C and Popa C V 2016 Phytotoxic and Antioxidant activities of leaf extracts of *Ailanthus altissima* swingle. *Rev. Chim.* **67** pp 1928-1931
[22] Luís Â, Gil N, Amaral M E, Domingues F and Duarte A P 2012 *Ailanthus altissima* (Miller) Swingle: a source of bioactive compounds with antioxidant activity *Bioresources* **7** pp 2105-2120
[23] Saifullah M, McCullum R, McCluskey A and Vuong Q 2019 Effects of different drying methods on extractable phenolic compounds and antioxidant properties from lemon myrtle dried leaves *Heliyon* **5** e03044
[24] Alara O R, Abdurahman N H, Ukaegbu C I and Kabbashi N A 2019 Extraction and characterization of bioactive compounds in *Vernonia amygdalina* leaf ethanolic extract comparing Soxhlet and microwave-assisted extraction techniques *J. Taibah Univ. Sci.* **13** pp 414-422
[25] Alara O R, Abdurahman N H and Ukaegbu C I 2021 Extraction of phenolic compounds: A review *Curr. Res. Food Sci.* **4** pp 200-214
[26] Kappe C O and Stadler A 2005 *Microwaves in Organic and Medicinal Chemistry* (Weinheim: Wiley-VCH Verlag GmbH & Co. KGaA) pp 31-54
[27] Deepika S, Harishkumar R, Dinesh M, Abarna R, Anbalagan M, Roopan SM and Selvaraj C I 2017 Photocatalytic degradation of synthetic food dye, sunset yellow FCF (FD&C yellow no. 6) by *Ailanthus excelsa* Roxb. possessing antioxidant and cytotoxic activity *J. Photochem. Photobiol. B: Biol.* **177** pp 44-55
[28] Awwad A M and Amer M W 2020 Biosynthesis of copper oxide nanoparticles using *Ailanthus altissima* leaf extract and antibacterial activity *Chem. Int.* **6** pp 210-217
[29] Nandiyanto A B D, Oktiani R and Ragadhita R 2019 How to read and interpret FTIR spectroscope of organic material *Indones. J. Sci. Technol.* **4** pp 97-118

IOP Conf. Series: Materials Science and Engineering 1254 (2022) 012022 doi:10.1088/1757-899X/1254/1/012022

Limitations of using electro-optical probes for the measurement of the electromagnetic field emitted by the new generations of wireless communication devices

A Sârbu, P Bechet and S Miclăuş

Communications, IT and Cyber Security Department, "Nicolae Bălcescu" Land Forces Academy Sibiu, Romania

E-mail: paljanosanna@yahoo.com

Abstract. The rapid evolution of communication standards caused a paradigm shift in electromagnetic field (EMF) measurement campaigns, as exposure is more likely to take place in the proximity of the device. With the increased spatial variability of EMF exposure, electro-optical (EO) technology stands out as a promising alternative to conventional near field measurement instrumentation due to their miniature sizes and dielectric structure that does not interfere with the measured field. Motivated by the recent progress in the EO sensing technology, it is this article purpose to quantify the limitations of using a set of commercially available EO probes for measurement of EMF emitted by an IEEE 802.11ax router and a VoLTE mobile phone. To our knowledge, this is the first attempt to characterize the capabilities of this emerging sensing technology for EMF exposure assessment in real world communication situations. Present findings suggest that the sensitivity of the investigated EO probes is insufficient for measuring broadband signals at field levels that characterize state-of-the-art communication devices used by the public in common usage scenarios. This work also documents the EO probes suitability and response with respect to frequency, electric field strength and signal bandwidth.

1. Introduction

The performances of modern communication standards are continuously expanding by enabling more advanced modulation techniques combined with increased channel bandwidths, spatial multiplexing, and beamforming. To this extent, human exposure to electromagnetic field (EMF) originated from the state-of-the-art communication devices is characterized by increased temporal and spatial variability. This inherent variability in real-world exposure constitutes the major source of uncertainty in the exposure assessment [1] and must be addressed by extensive in-situ measurement campaigns performed by agile systems that must respond to the challenging demands both in terms of bandwidth and miniaturization.

The advent of 5G communications has attracted considerable amount of research to investigate human exposure to such emitted fields [2-3]. In [2], the authors appreciate that the antenna beamforming and the complexity of 5G technology will require the design of new and effective measurement techniques to achieve the desired accuracy and temporal complexity that characterize new generation signals. The authors of [3] developed a 3D Ray-Launching technique for an in-depth study of the E-field characterization of the personal mobile communications within urban public trams by considering the impact in the use of directive antennas and beam forming techniques, within realistic operation conditions.

The guidelines for limiting human exposure to EMF were recently updated to respond to the requirements of the fifth-generation communication (5G) standards [4-5]. Above 2 GHz the limits are expressed in terms of incident power density which require accurate measurement of the EMF in the proximity of the device, as recent findings suggest that in future networks, EMF exposure will be dominated by the end user device [6]. The near field nature of exposure was also observed for various publicly used technologies like short-range wireless RFID technology [7-8] or occupational sources such as military communication technology [9-10]. When conducted in the near field, EMF measurements are complicated and require sophisticated measurement devices and procedures with several sources of uncertainty [11]. In the near field, evaluating human exposure requires measurements of both electric and magnetic field strengths. These types of measurements become extremely challenging when performed on devices in real operating situations. Traditional EMF probes (dipoles) are built from conductive materials and present several notable performances like high sensitivity and, with recent advances, even miniaturization. Open ended wave guides are used in state-of-the art EMF measurements, like the cutting-edge near-field measurement technique presented in [12]. However, due to their metallic structure their positioning in the near field region of an emitting device produces both scattering and source coupling, and thus alters the real field value to be measured.

EO sensors represent an alternative to EMF field measurements by conventional sensing techniques. They have several advantages like compact size, strong adaptability to harsh environments, and immunity to electromagnetic interference [13] but are also limited by the bandwidth to be measured (mainly for low frequencies) [14] and their sensitivity. Recent progress has been made on EMF measurements based on EO probes [13-15], and their improved characteristics sustain their use for in-situ measurements of EMF originated from communication devices. Developments have been made for measuring soldier exposure in operational environments in the 1-500 MHz frequency range [16]. This type of harsh EMF environment can also be found in medical or other type of occupational hazards. However, up to now the authors did not document any experimental study conducted for measuring EMF originated from public sources by means of EO technology.

To this extent, in this article we present the limitations of using one type of commercially available EO probes, for measurements of EMF emitted by new generation communication devices used by the public. To our knowledge, this is the first attempt to present the results and limitations of using EO probes for in-situ measurements of publicly used communication devices, at the field levels present in everyday usage situations.

2. Materials and methods

We used a measurement system based on EO probes manufactured by Kapteos, presented in Figure 1. The measurement system is composed of an opto-electric (OE) converter (frequency range 1.6-6.4 GHz) and three EO probes that operate based on Pockels effect (laser interaction with the EMF). We use two transversal probes (model ET5) and one longitudinal probe (model EL5) [17], for measuring the electric (E) field strength in all three orthogonal directions. The OE converter is of type HF 1.6-6.4 3T and has 3 channels for synchronous measurements of the E field strength by all three EO probes. The EO probes are made of quartz and alumina, being fully dielectric and EMF transparent. Their size (34.5 x 5.5 mm), working principle and constructive materials recommend them for high spatial resolution measurements and increased accuracy. The sensitivity of the probes is 40 $mV_{RMS}/m\ Hz^{1/2}$ for the longitudinal probe and 200 $mV_{RMS}/m\ Hz^{1/2}$ for the transversal probe. Other EO probes with increased sensitivity were documented [15] but their spatial resolution is much higher than the selected probes.

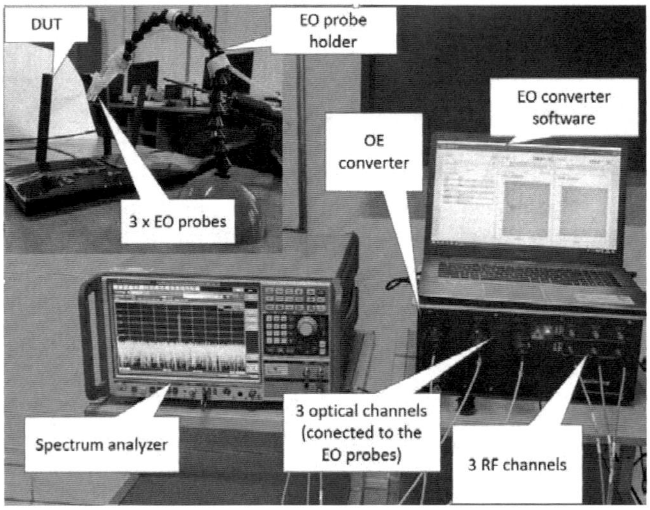

Figure 1. EMF measurement system based on EO probes.

An adiabatic transverse electromagnetic (ATEM) cell model eoCAl HF6 from Kapteos is used to calibrate the EO probes. In the ATEM cell, the field propagation mode between the inner conductor (septum) and the mass plane is predominantly transversal electric (TE) with suppression of first high-order modes. The E field strength value inside the ATEM cell is calculated by applying equation (1) – provided by the manufacturer.

$$E_{ATEM}\left(\frac{V_{RMS}}{m}\right) = 12.85\ x\ 10^{\frac{P_{in\ ATEM}}{20}} \quad (1)$$

The sensitivity axis of the ET5 probe is transversal to the probe and to calibrate it we inserted it into the ATEM at a 90° angle (Figure 2a). On the contrast, the sensitivity axis of the EL5 probe is longitudinal and to measure the TE field in the ATEM cell we insert it at a 60° versus vertical angle (Figure 2b) by a custom-made holder. The sensed E field by the EL5 probe is half the E field strength in the ATEM cell – calculated by equation (1), as cos60°=1/2.

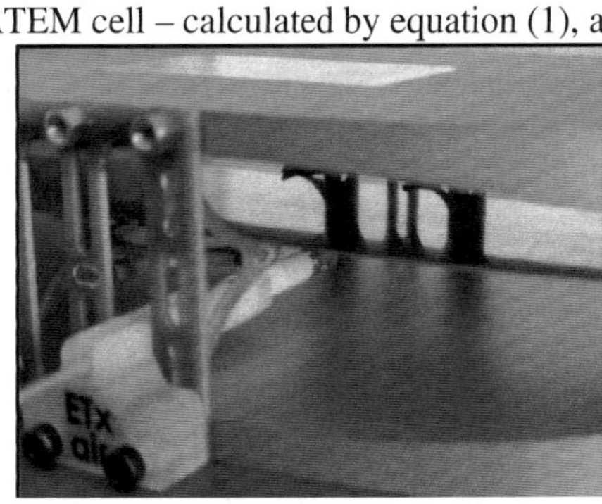

Figure 2(a). ET 5 probe inside the ATEM cell during calibration.

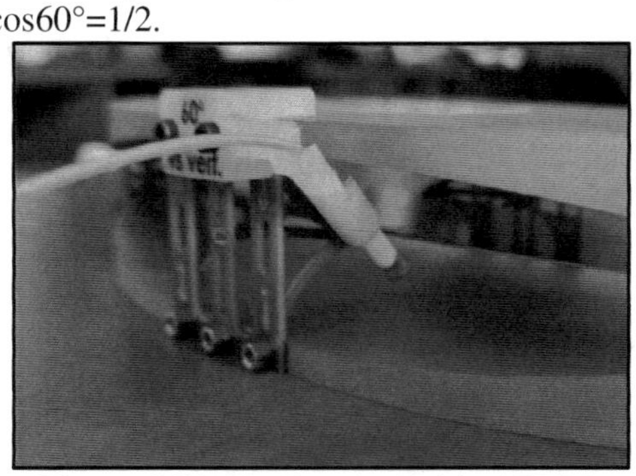

Figure 2(b). EL 5 probe inside ATEM cell during calibration.

To calibrate the probe a continuous signal is inputted to one of the ATEM cell connectors while the other is matched by a 50Ω impedance. The probe is connected to one of the optical inputs of the OE converter and the corresponding radiofrequency (RF) channel is connected to the spectrum analyzer (SA) input. We have used a FSV13 R&S SA. The signal is then measured by the SA and the signal value is recorded. Based on the input power to the ATEM cell and the measured signal power at the SA, the OE converter software calculates an antenna factor (AF). The value of the calculated AF is used in equation (2) to transform the SA power reading (P_{SA}) to field strength value (V/m).

IOP Conf. Series: Materials Science and Engineering 1254 (2022) 012022 doi:10.1088/1757-899X/1254/1/012022

$$E\left(\frac{V_{RMS}}{m}\right) = 10^{\frac{\left(AF\left(\frac{dB}{m}\right)+P_{SA}(dBm)-13,01\right)}{20}} \quad (2)$$

We investigated the EO probes response to the EMF emitted by two types of devices. The first device under test (DUT)is a dual band Wi-Fi router performing download in a 20 MHz channel bandwidth (central frequency =5280 MHz) and the second DUT was a 4G mobile phone with an active VoLTE call (central frequency = 1750 MHz). The DUTs were also measured by a traditional near-field E probe (sniffer - Model Aaronia PBS E1), for comparative purposes. Figure 3 shows the investigated scenarios.

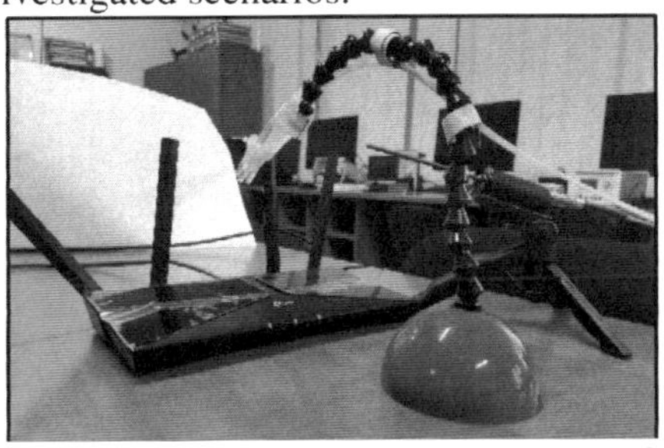

Figure 3(a). *EO probes and PBS E1 probe placed in the proximity of the Wi-Fi router – DUT1.*

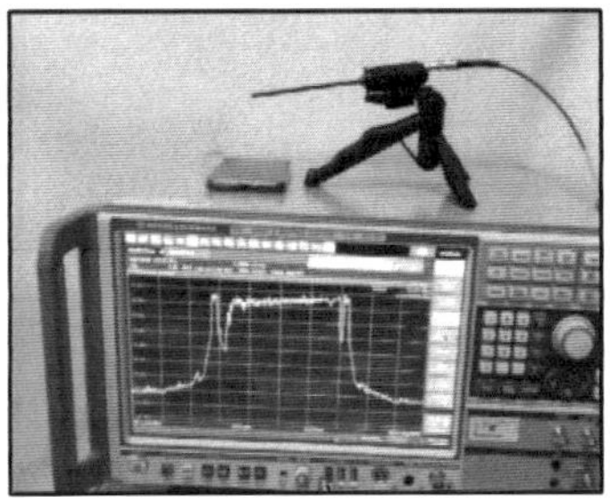

Figure 3(b). *PBS E1 probe measuring the EMF emitted by the mobile phone during VoLTE call - DUT2.*

Figure 3(c). *EO probes measuring the EMF emitted by the mobile phone during VoLTE call - DUT2.*

3. Results and discussion

To perform frequency selective measurements, the RF output of the OE converter must be connected to the RF input of a SA. The signal outputted at the OE channel is characterized by a high noise Figure, leading to an increase in the SA noise floor of about 25 dB. In Figure 4 we present the comparative power measured at the SA terminated by a 50Ω impedance and the power measured with the powered on EO converter (EO probes placed in a shielded box).

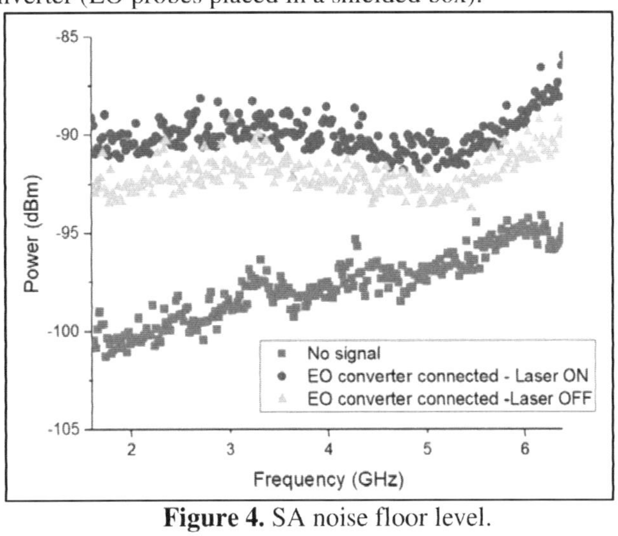

Figure 4. SA noise floor level.

From Figure 4 one can observe that the noise across the entire frequency range increases when the OE converter is connected to the SA. The noise Figure of the OE converter is even higher when the laser is turned on. The sensitivity of the EO probes and the existence of this noise across the entire operating frequency band of the EO converter (1.6-6.4 GHz) requires very low values of the resolution filter bandwidth (RBW) at the SA. RBW values of less than 100 Hz are required to achieve an adequate signal-to-noise ratio (SNR). The small RBW values influence the sweep time (SWT) of the SA, which increases significantly for low RBW values. For the used SA the SWT increased to 2.6 seconds when a RBW of 1Hz was used. This long SWT value is insufficient to capture the signal dynamics of the modern telecommunication devices in 5th or 6th generation communication standards. Higher RBW values can be used if the measured E field strength is of the order of hundreds of V/m.

In Figure 5a we present the power level for the EL 5 and ET 5 EO probes when placed in the ATEM cell, with an applied field strength of 406.3 V/m – continuous wave signal. One can observe that the power readings of the EL 5 probe are about 10 dB higher than for the ET 5. This is because the EL 5 probe is more sensitive than the ET 5 probe. Because the measured power is not very flat on the entire frequency range, AF value must be calculated for every operating frequency. However, for a channel bandwidth of 40/80 MHz, the power readings were stable with a recorded variation of +/-0.8 dB suggesting that a single calibration at the channel central frequency is sufficient when measuring signals with higher bandwidths.

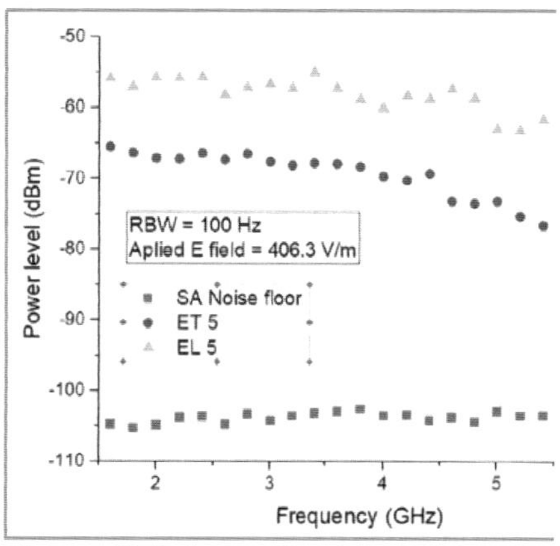

Figure 5(a). Power level for EL5 and ET5 EO probes (Applied field 406.3 V/m, FSV SA, RBW=100 Hz).

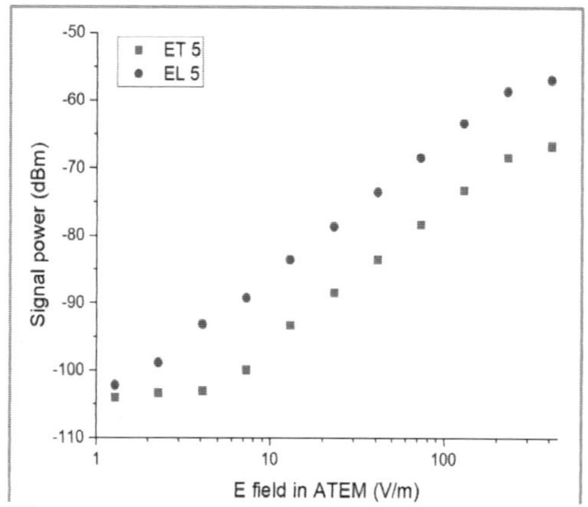

Figure 5(b). Signal power measured by EL5 and ET5 EO probes for different E field values (f=3.5 GHz).

In Figure 5b we present the signal power measured for different field strength values. The power readings were taken for the EO probes placed in the ATEM cell fed by continuous wave of 3.5 GHz with variating power. One can observe that the power levels approach the SA noise floor around the value of 1 V/m. The field applied in the experimental scenario shown in Figure 5a corresponds to a power density (plane wave) of 437 W/m², which is 8.7 times higher than the reference value mentioned for human occupational safety standards and 43.7 times higher than the maximum value in the guidelines for general population exposure [5]. The E field strength values measured in the vicinity of modern communication devices are as low as few V/m, most commonly of the order of tens ofmV/m. These low E field strengths are not measured accurately by EO probes (Figure 5b) because the signal power reading is close to the noise floor of the SA (low SNR).

The results shown above, in Figures 4 and 5, are for continuous wave. Modern communication devices operate on radio channels with bandwidths between 1.4 MHz (LTE minimum) and 160 MHz (IEEE 802.11ax). When using radio channels, the power is spread in the channel so that the maximum in-band power is lower than for continuous wave case. This makes it easier for EO probes to approach their sensitivity threshold. Figure 6 plots the response of an ET5 probe (placed in the ATEM cell) to an LTE signal with different bandwidths. The transmitter power of the generator is 30 dBm, corresponding to a field strength of 406 V/m.

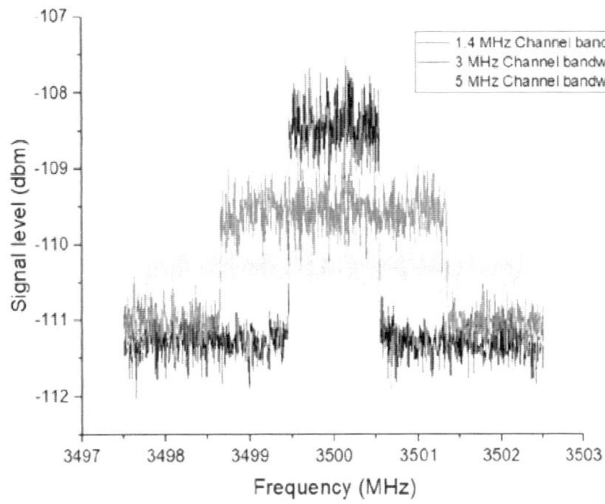

Figure 6. Response of the ET 5 EO probe to LTE signals with different bandwidths (E =406 V/m).

From Figure 6, one can observe that the SNR value is very low, only 3dB for the 1.4 MHz bandwidth, and decreases to 1 dB for the 5 MHz wide channel. These SNR values are not sufficient to achieve accuracy in the measurement campaigns. Moreover, the field level (406 V/m) is far from being realistic so when using communication devices (mobile phones, routers) the EO probes are unable to measure the signal variation.

In Figure 7 we have shown the measurement of the E field emitted by a router (a) in transmission - data download and a mobile phone in a VoLTE call (b). The field strength measurement was performed alternately with the EL 5 EO probe (because of the higher sensitivity) and the PBS E1 field probe.

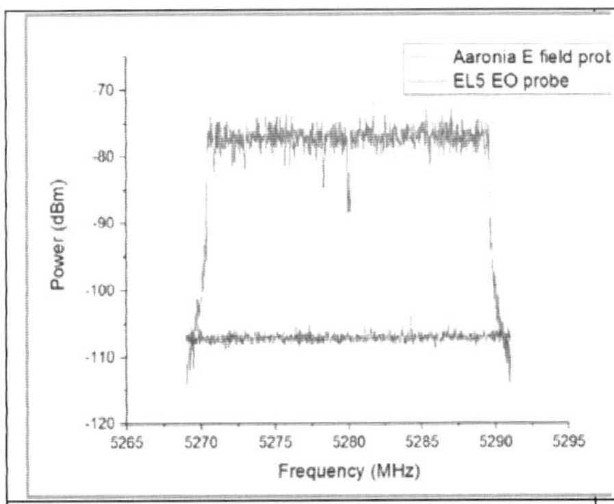

Figure 7(a). Signal level for Wi-Fi router scenario (IEEE 802.11ax) - Figure 3(a).

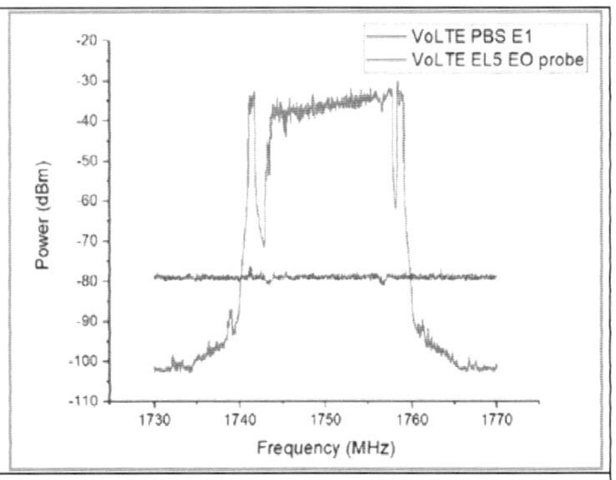

Figure 7(b). Signal level for VoLTE scenario – Figure 3(b),(c).

In none of the experimental situations did the EO probe prove the sufficient sensitivity to detect signal level variation in the channel bandwidth (20 MHz). In contrast, the traditional electric field probe (PBS E1) captured the signal variations in the monitored bandwidth. It can also be observed that the noise floor of the SA increases considerably when the EO measurement system is used.

4. Conclusions

In this work we analysed the current limitations of a commercially available measurement system based on EO probes for evaluating the EMF emitted by nowadays used communication devices. To our knowledge, this is the first attempt to characterize the capabilities of this emerging sensing technology for EMF exposure assessment in real world communication situations. We have demonstrated that the selected EO probes can provide a viable alternative for high accuracy EMF measurements conducted in harsh environments, but only when extremely high field strength levels are emitted. However, even with the recent progress of the EO field sensing technology, their sensitivity remains insufficient for sensing the EMF originated from state-of-the-art communication devices. The sensitivity of the EO probes was documented to be as low as 33mV/m at a RBW of 10 Hz [15] for continuous wave. Even with this improved sensitivity, the RBW value is insufficient for enabling fast SWT at the SA, required to capture the emitted signal time dynamics. Even if EO sensing technology is extremely promising and has already demonstrate several applications in EMF measurements, the limitations in terms of sensitivity makes this technology inadequate for in situ characterization of publicly used communication devices.

Acknowledgments

This work was supported by a grant of the Romanian Ministry of Education and Research, CNCS - UEFISCDI, project number PN-III-P1-1.1-PD-2019-0500, within PNCDI III.

References

[1] Foster K R, Ziskin M C and Balzano Q 2022 Three Quarters of a Century of Research on RF Exposure Assessment and Dosimetry—What Have We Learned? *Int. J. Environ. Res. Public Health* **19** 2067 https://doi.org/10.3390/ijerph19042067

[2] Capriglione D 2021 In-Situ RF Measurements of EMFs for Human Exposure Assessment Due to Modern Cellular Base Stations *IEEE Instrumentation & Measurement Magazine* **24**(8) pp 31-36, doi: 10.1109/MIM.2021.9580794

[3] Celaya-Echarri M, Azpilicueta L, Karpowicz J, Ramos V, Lopez-Iturri P and Falcone F 2020 From 2G to 5G Spatial Modeling of Personal RF-EMF Exposure Within Urban Public Trams *IEEE Access* **8** pp 100930-100947 doi: 10.1109/ACCESS.2020.2997254

[4] IEEE Std. 2019 IEEE Standard for Safety Levels with Respect to Human Exposure to Electric, Magnetic, and Electromagnetic Fields, 0 Hz to 300 GHz; IEEE: New York, NY, USA, pp 1–312

[5] ICNIRP 2020 Guidelines for Limiting Exposure to Time-Varying Electric, Magnetic, and Electromagnetic Fields (up to 300 GHz) *Health Phys.* **118** pp 483–524. Available from: https://www.icnirp.org/cms/upload/publications/ICNIRPrfgdl2020.pdf (accessed on 15 November 2020)

[6] Kuehn S, Pfeifer S and Kuster N 2020 Total Local Dose in Hypothetical 5G Mobile Networks for Varied Topologies and User Scenarios *Appl. Sci.* **10** 5971

[7] de Miguel-Bilbao S *et al* 2021 Near Field Exposure Conditions by UHF-RFID Systems in Smart Healthcare Environments, 2021 IEEE International Joint EMC/SI/PI and EMC Europe Symposium pp 13-18 doi: 10.1109/EMC/SI/PI/EMCEurope52599.2021.9559189

[8] Zradziński P, Karpowicz J, Gryz K, Owczarek G and Ramos V 2021 Modelling and Evaluation of the Absorption of the 866 MHz Electromagnetic Field in Humans Exposed near to Fixed I-RFID Readers Used in Medical RTLS or to Monitor PPE *Sensors* **21** p 4251 https://doi.org/10.3390/s21124251

[9] Colella M, Biscarini M, de Meis M, Patrizi R, Ciallella T, Ferrante D, De Gaetano A, Capuano M, Pellegrino G, Martini E, Cavagnaro M, Apollonio F and Liberti M 2022 Numerical Evaluation of Human Body Near Field Exposure to a Vehicular Antenna for Military Applications *Frontiers in Public Health* (Vol. **9**) Frontiers Media SA. https://doi.org/10.3389/fpubh.2021.794564

[10] Sobiech J, Kieliszek J, Puta R, Bartczak D and Stankiewicz W 2017 Occupational exposure to electromagnetic fields in the Polish Armed Forces *International Journal of Occupational Medicine and Environmental Health* Nofer Institute of Occupational Medicine https://doi.org/10.13075/ijomeh.1896.00696

[11] Dlugosz T and Trzaska H 2010 How to Measure in the Near Field and in the Far Field *Communications and Network* **2**(1) pp 65–68 (Scientific Research Publishing, Inc) https://doi.org/10.4236/cn.2010.21010
[12] Lundgren J, Helander J, Gustafsson M, Sjoberg D, Xu B and Colombi D 2021 A Near-Field Measurement and Calibration Technique: Radio-Frequency Electromagnetic Field Exposure Assessment of Millimeter-Wave 5G Devices *IEEE Antennas and Propagation Magazine* **63**(3) pp 77-88 doi: 10.1109/MAP.2020.2988517
[13] Peng J, Jia S, Bian J, Zhang S, Liu J and Zhou X 2019 Recent Progress on Electromagnetic Field Measurement Based on Optical Sensors *Sensors* (Basel, Switzerland) **19**(13) p 2860 https://doi.org/10.3390/s19132860
[14] Calero V, Suarez M A, Salut R *et al* 2019 An ultra wideband-high spatial resolution-compact electric field sensor based on Lab-on-Fiber technology *Sci Rep* **9** 8058 https://doi.org/10.1038/s41598-019-44644-y
[15] Behague F, Calero V, Coste A, Godet A, Suarez M, Gaborit G, Duvillaret L, Baida F I, Bernal M-P and Courjal N 2021 Minimally invasive optical sensors for microwave-electric-field exposure measurements *Journal of Optical Microsystems* **1**(2) SPIE-Intl Soc Optical Eng. https://doi.org/10.1117/1.jom.1.2.020902
[16] https://en.kapteos.com/products-services, accessed on 09.03.2022
[17] https://en.kapteos.com/eoprobe-datasheet, accessed on 09.03.2022
[18] Zhang J, Zhao Z, Li C and Li Y 2017 Broad-Band Integrated Optical Electric Field Sensor Using Reflection Mach-Zehnder Waveguide Modulator *Fiber and Integrated Optics* **36**(4–5) pp 157–164 (Informa UK Limited) https://doi.org/10.1080/01468030.2017.1357770

IOP Conf. Series: Materials Science and Engineering 1254 (2022) 012023 doi:10.1088/1757-899X/1254/1/012023

Magnetic Hyperthermia with biocompatible coated nanoparticles: A temperature analysis

I Aștefanoaei and A Stancu

Faculty of Physics, "Alexandru Ioan Cuza" University, Iasi, Romania

E-mail: iordana@uaic.ro, alstancu@uaic.ro

Abstract. The magnetic nanoparticles (MNPs) having the special (magnetic and thermal) properties are promising for Magnetic Hyperthermia. To increase their biocompatibility, these MNPs are covered by different organic shells as: chitosan, oleic acid or silica. When an external time - dependent magnetic field is applied, the temperature developed within a malignant cell is strongly influenced by the type of the material which covers the magnetic nanoparticle. This paper studies the temperature field induced by the MNPs covered by an organic shell within a concentric tissues configuration (malignant and healthy tissues). The shell thickness influences significantly the temperature field developed within a malignant tissue. These important effects should be considered in the computation of the specific particle doses corresponding to the hyperthermia treatment.

1. Introduction

In the Magnetic Hyperthermia – the magnetic nanoparticles (MNPs) receive a special attention in the all theoretical and experimental researches. MNPs are the heating sources of the malignant tissues which are destroyed when the temperature increases in the hyperthermic range $(40 \div 46)^{o}$C [1÷8]. These nanosystems are important mediators for the conversion of the magnetic energy in the thermal energy [6÷12]. Their design can be controlled to obtain efficiently the hyperthermic temperature range. Usually, these particles are covered by different organic shells as: chitosan, oleic acid, silica, to improve their biocompatibility when are injected within malignant tissues [7- 8]. A temperature analysis can optimise the design of a MNP used in the therapy. This paper investigates the influence of the organic shellon the temperature field developed by the MNPs injected within a malignant tissue (T) when an external time - dependent magnetic field was applied. The shell thickness and the type of the organic materials used to cover the nanoparticle contribute significantly to the bioheat transport and implicitly to the thermal damage of the malignant tissues. This is a very important temperature effect which helps to elaborate a controlled design of the MNPs used in the Magnetic Hyperthermia.

2. Thermo-Fluid Model

A temperature analysis allows identifying of the temperature effects with the central role in the hyperthermia treatment. Using a temperature model, the best design of the functionalized magnetic particle can be obtained. In this paper, MNPs are composed by a core (having the radiusR_1) and an organic shell with the thickness $\delta = R_2 - R_1$. The incompressible diluted colloidal fluid (with a small concentration $c \leq 4\%$ by volume) is injected at the center of the concentric geometry composed of the malignant tissue (**T**) with radius R_3 surrounded by the healthy tissue (**H**) with radius R_4 (Figure 1a, b).

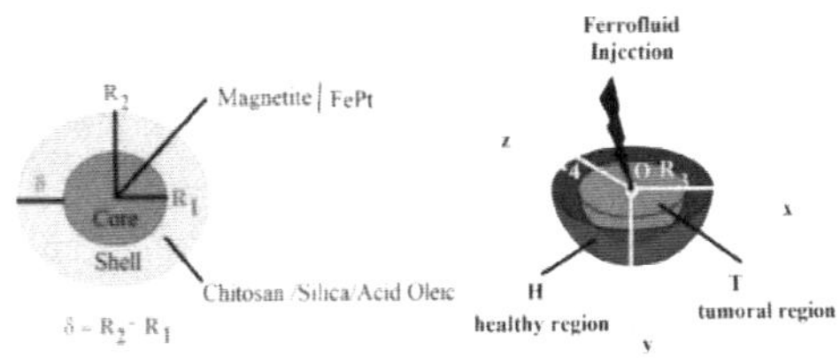

(a) **(b)**

Figure 1 (a) MNP covered by an organic shell and (b) Tumor – Healthy tissues

The ferrofluid flow through tissues is described by Darcy's law [13, 14]:

$$\vec{v}_j = -\,k_j\,\nabla P_j \quad (j = 1,2) \qquad (1)$$

ε_j and K_j are the porosity and permeability of the malignant and healthy tissue; $k_j = K_j/(\mu_j)$. The MNPs concentrations, $C_j = C_j(r,t)\ (j = 1,2)(\text{mg/cm}^3)$ are described by the solutions of the convection-diffusion equation[14-17]:

$$\frac{\partial C_j}{\partial t} + \nabla\cdot(\vec{v}C_j) = \nabla\cdot(D_j^*\nabla C_j) \qquad (2)$$

$D_{j,f}^*$ – MNPs diffusion coefficient in the interstitial fluid. The MNPs volume fractions were given by the relations: $\Phi_j(r,t) = \frac{C_j(r,t)}{\rho_{MNP}}$,where ρ_{MNP} is the mass density of MNPs. The following *boundary conditions were imposed*: i) $C_2 = 0$ on the external boundary of the geometry ($r = R_4$); ii) Neumann boundary condition at the all inner interfaces: $C_1(r = R_3) = C_2(r = R_3)$and $D_1^*\frac{\partial C_1}{\partial r}\Big|_{r=R_3} = D_2^*\frac{\partial C_2}{\partial r}\Big|_{r=R_3}$; iii) The maximum value of the volume fraction was considered: $\Phi_{\max} = \frac{C_{max}}{\rho_{MNP}}$ at the center of the geometry, where the ferrofluid injection within tumor tissue was done.

The space-time temperature distribution within malignant and healthy tissues is described by the solution: $T_j = T_j(x, y, z, t), (j = 1,\ 2)$ of the bioheat transfer equation in the living tissues [13, 17-18]:

$$\rho_j c_j \frac{\partial T_j}{\partial t} = \nabla[k_j\,\nabla T_j] + \rho_b\omega_j^b c_b[T_{art} - T_j] + \frac{Q_{met}^j}{e^{\Omega_j}} + Q_j \qquad (3)$$

Boundary Conditions: i) Dirichlet boundary condition was considered on the external surface of the healthy tissue: $T_2[r = \mathrm{R}_2] = 37°\mathrm{C}$. ii) Neumann boundary conditions are considered at malignant – healthy tissue interface: the heat flux coming from **T** is completely received by **H**:

$$k_1\left[\frac{\partial T_1}{\partial r}\right]_{R_3} = k_2\left[\frac{\partial T_2}{\partial r}\right]_{R_3} \text{ and } T_1[r = \mathrm{R}_3] = T_2[r = \mathrm{R}_3]$$

The index j defines **T** (j = 1) and **H** (j = 2): ρ_j - the mass density, c_j - specific heat capacity, ρ_b - mass density of the blood, c_b- specific heat capacity of the blood, k_j- the thermal conductivity, T_{art} - blood temperature, Q_{met}^j - metabolic heat production. In the equations (3), $Q_j = \alpha\,\Phi_j\,P_1\left(\frac{\text{W}}{\text{m}^3}\right)$ was the volumetric heating rate dissipated by the volume fraction Φ_j within **T**, when the magnetic field was applied. The specific loss power released by the one particle is P_1 and susceptibility χ'' depends strongly on the particle size [17, 19-20]:

$$P_1 = \mu_0\, f\pi\chi''\, H_0^2 \text{ and } \chi'' = \frac{\mu_0 M_s^2 V}{3\,k_B T}\frac{2\pi f\tau}{1+(2\pi f\tau)^2} \qquad (4)$$

H_0 is the intensity of the magnetic field, f - frequency of the magnetic field, $\mu_0 = 4\pi\cdot 10^{-7}$H/m - the permeability, M_s is the saturation magnetization and $k_B = 1.38\cdot 10^{-23} J/K$is the Boltzmann constant. The effective relaxation time contains the Brown relaxation time$\tau_B$and the Néel relaxation time τ_N: $\tau[R] = \frac{\tau_N\,\tau_B}{\tau_N+\tau_B}$; $\tau_B = \frac{3\,\eta V_H}{k_B T}$; $\tau_N = \tau_0\frac{\sqrt{\pi}}{2}\frac{\exp\left[\frac{KV}{k_B T}\right]}{\sqrt{\frac{KV}{k_B T}}}$.τ_0 is the precessional decay time, ηcarrier liquid viscosity, $V_H = V\left(1+\frac{\delta_1}{R}\right)^3$is the hydrodynamic volume of the particles, $\delta_1$is the surfactant layer thickness and K is the anisotropy constant of the magnetic particles. In the relations (3) the coefficient α is a *correction factor* due to the heating rate by the spherical shell of the MNPs. Using Fourier heat conduction equation in the core-shell particle, the temperature in the shell was computed as: $T(r) = T_1 - (T_1 - T_g)\frac{\frac{1}{R_1}-\frac{1}{r}}{\frac{1}{R_1}-\frac{1}{R_2}}$, where $T_1 = T_1(r = R_1)$ and $T_g = T(r = R_2)$. Heat flux, $q = -k_2\frac{\partial T}{\partial r} = k_2(T_1 - T_g)\frac{\frac{1}{r^2}}{\frac{1}{R_1}-\frac{1}{R_2}}$ leads to the heating rate $Q_{shell} = 4\pi r^2 q = \frac{4\pi k_2(T_1-T_g)}{\frac{1}{R_1}-\frac{1}{R_2}}$. Volumetric heating rate of the nanoparticle P_{shell} is:

$$P_{shell} = \frac{Q_{shell}}{V_{shell}} = \frac{3k_2(T_1-T_g)x_1}{R_1^2(x_1^3-1)(x_1-1)} \qquad (5)$$

where: $x_1 = \frac{R_2}{R_1} = \frac{\delta}{R_1} + 1$ and $\delta = R_2 - R_1$ is the shell thickness. An analogous calculus for a spherical MNP (shell free, $\delta = 0$) leads to the following result:

$$P_{sph} = \frac{3k_1(T_1-T_g)}{R_1^2} \qquad (6)$$

The correction factor α from the relation (3) was obtained using the relations (5) and (6):

$$\alpha = \frac{P_{sph}}{P_{sph}+P_{shell}} = \frac{1}{1+\left(\frac{k_1}{k_2}\right)\left(\frac{(x_1^3-1)(x_1-1)}{x_1}\right)} \qquad (7)$$

The correction factor is $\alpha = 1$ in the case of $\delta = 0$(shell free). This coefficient depends on the shell thermal conductivity (k_2) and shell thickness (δ). The thermal damage Ω_j depends on the tissue temperature $T_j = T_j(r,\tau)(j = 1,2)$, according to the Arrhenius relation [17]: $\Omega_j(r,t) = \int_0^t A_j \exp\left(-\frac{E_a^j}{R\,T_j}\right) d\tau$,$E_a^j[J mol^{-1}]$ is the activation energy; A_j [s^{-1}] the frequency factor; R[$J\, mol^{-1}K^{-1}$]–the universal gas constant. The tissue damage: $\theta_j(r,t) = \exp[-\Omega_j(r,t)]$ decreases from the value$\theta_j = 1$, to the value $\theta_j = 0$ in the hyperthermic range. In the equation (3), the blood perfusion rate for tumor-healthy tissue is dependent on the thermal damage $\omega_j^b = \omega_j^0 \exp(-\Omega_j(r,t))$ [13].

3. Results and discussion

The temperature field within liver (T-H regions) was computed for a value of the magnetic field parameters: frequency f = 300 kHz and the magnetic field amplitude H_0: 5 kA/m. The magnetite ($\varepsilon - Fe_2O_3$) and FePt nanoparticles with the radius of 8 nm were covered by an organic shell with various materials as: chitosan, silica and oleic acid. The tumor and healthy regions are two concentric domains having the diameter of 40 mm and 100 mm, respectively (Figure 1). The temperature values depend strongly on the MNP sizes, their shell thickness (δ) and the type of the core-shell materials (Figure 2). In this analysis, the values of the shell thickness were considered in the following range:$10 \div 30$ nm.The thermal conductivity of the organic shell material influences strongly the temperature field within **T** region. The silica shell (with the thermal conductivity, approximately 10 times higher than chitosan), determines a high temperature rise compared by the ones given by chitosan shell, which is a thermal insulator (low thermal conductivity) (Figure 2). Silica shell allows a good transmission of the heat within **T** region. The chitosan shell blocks the heat transmission to **T-H** regions. As a significant characteristic, all these materials reduce MNP heat generation ability. The temperature values were compared with the case when no shell arround MNP ($\delta = 0$). The temperature within T region decreases with the increase of the shell thickness.These important theoretical results are in good agreement with the experimental data [20÷22]. Figure 3 shows the temperature within tumor tissue heated by the FePt nanoparticles covered by silica/chitosan/oleic acid shell succesivelly when the external time - dependent magnetic field was applied.Organic silica shell is a good candidate to be used in the Magnetic Hyperthermia.

4. Conclusion

In the Magnetic Hyperthermia, the shell thickness and type of the organic material which cover the MNP have an important role in the temperature control of the malignant region. The shell thickness of the MNP should be chosen accordingly with the thermal properties of the covered organic material. The shell material with a high thermal conductivity (as silica) allows a better propagation heat within tissues. Low thermal conductivity of the organic shell (as chitosan or oleic acid) block the heat transmission in the malignant tissue. The heat generation of the MNPs within the malignant tissues was reduced due to the presence of organic material coating. The temperature values decrease with the increase in the shell thickness. The temperature effects, concluded by this thermo-fluid analysis are in good agreement with the experimental data.

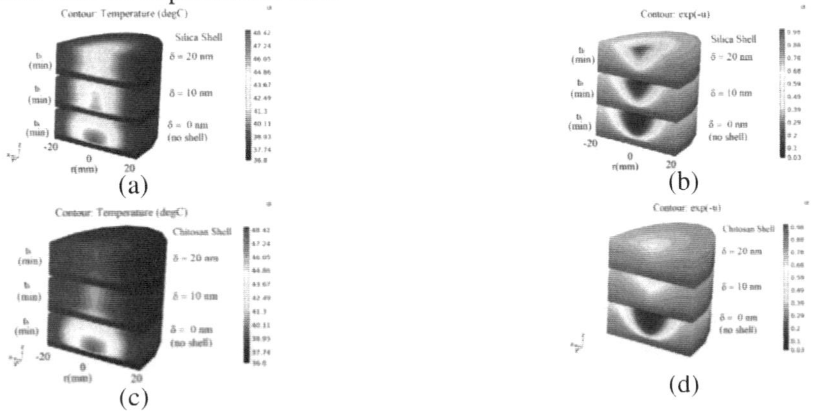

Figure 2. Temperature (a,c) and Tissue damage (b, d) withthe shell thickness.

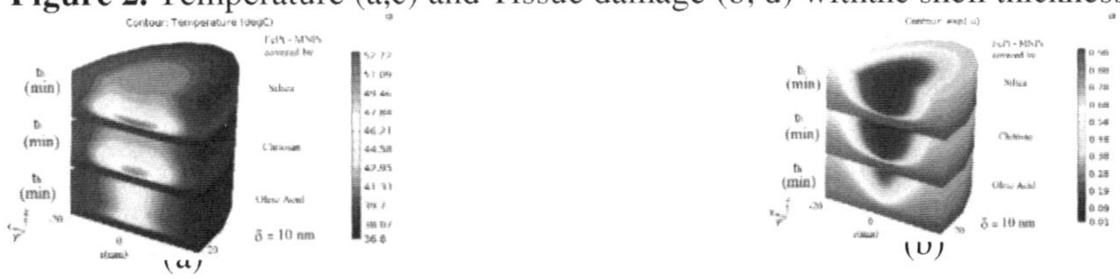

Figure 3. Temperature (a) and Tissue Damage (b) using FePt particles covered by shell materials.

References

[1] Carrey J, Mehdaoui B and Respaud M 2011 Simple models for dynamic hysteresis loop calculations of magnetic single-domain nanoparticles: Application to magnetic hyperthermia optimization *Journal of Applied Physics* **109** 083921

[2] Pankhurst Q A, Connlly J, Jones S K and Dobson J 2003 TOPICAL REVIEW: Applications of magnetic nanoparticles in biomedicine *Journal of Applied Physics* **36**(13) pp 167-181

[3] Tang Y D, Jin T and Flesch R C C 2020 Effect of mass transfer and diffusion of nanofluid on the thermal ablation of malignant cells during magnetic hyperthermia *Applied Mathematical Modelling* **83** pp 122-135

[4] Aştefanoaei I, Stancu A, 2021 Optimal control of the hyperthermic thermal damage within intravascular tumoral regions *Journal of Magnetism and Magnetic Materials* **537** pp 168221-168230

[5] Aştefanoaei I and Stancu A 2021 Thermo-fluid porosity –related effects in magnetic hyperthermia *European Physical Journal Plus*, **136**(12) pp 1216 -1226

[6] Tang YD, Flesch R C C and Jin T 2017 Numerical investigation of temperature field in magnetic hyperthermia considering mass transfer and diffusion in interstitial tissue *Journal of Physics D: Applied Physics* **51**(3) pp 035401-035412

[7] Patil-Sen Y *et al* 2020 Biocompatible superparamagnetic core-shell nanoparticles for potential use in hyperthermia-enabled drug release and as an enhanced contrast agent *Nanotechnology* **31** (37) pp 5102-5114

[8] Wang W, Tang B, Ju B and Zhang S 2015 Size-controlled synthesis of water-dispersible superparamagnetic Fe_3O_4 nanoclusters and their magnetic responsiveness *RSC Adv.* **5** pp 75292–75299

[9] Aştefanoaei I, Stancu A and Chiriac H 2017 Numerical simulation of the temperature field in magnetic hyperthermia with Fe-Cr-Nb-B magnetic particles *The European Physical Journal Plus* **132**(2) pp 1-12
[10] Aştefanoaei I and Stancu A 2017 Thermal performance of Fe-Cr-Nb-B systems in magnetic hyperthermia *Journal of Applied Physics* **121**(10) pp 104701-104711
[11] Aştefanoaei I and Stancu A 2017 A computational study of the bioheat transfer in magnetic hyperthermia cancer therapy *Journal of Applied Physics* **125**(19) pp 4701-4712
[12] Aştefanoaei I and Stancu A 2017 Magnetic Hyperthermia with Fe-Cr-Nb-B magnetic particles *AIP Conference proceedings* **1796**(1) pp 040006
[13] Singh M, Ma R and Zhu L 2019 Theoretical Evaluation of Temperature Elevation, Thermal Damage, Tumor Porosity Enhancement and Magnetic Nanoparticle Migration in tumor during local heating, Sumer Biomechanics , Bioingineering and Biotransport, Conference Paper
[14] Aştefanoaei I and Stancu A 2017 Advanced thermo-mechanical analysis in the magnetic hyperthemia *Journal of Applied Physics* **122**(16) pp 164701-164712
[15] COMSOL Multiphysics, Reference Manual, version 4.4
[16] Aştefanoaei I *et al* 2019 Modelling of working parameters of Gd and FeRh nanoparticles for magnetic hyperthermia *Materials Research Express* **6**(12) pp 125089-125099
[17] Aştefanoaei I, Chiriac H and Stancu A 2016 Investigation of the temperature field in the magnetic hyperthermia using FeCrNbB magnetic particles *European Physical Journal Plus* **131** (9) pp 1-9
[18] Salloum M, Ma R and Zhu L 2008 An in-vivo experimental study of temperature elevations in animal tissue during magnetic nanoparticle hyperthermia *International Journal of Hyperthermia* **7**(24) pp 589-601
[19] Lahonian M and Golneshan A A 2011 Numerical study of temperature using lattice Boltzmann method *IEEE Trans. Nanobiosci.* **10**(4) 262-268
[20] Moorthy M S *et al* 2016 Synthesis of amine-polyglycidol functionalised Fe3O4@SiO2 nanocomposites for magnetic hyperthermia, pH- responsive drug delivery and bioimaging applications *RSC Adv.* 6 pp 110444110459
[21] Gu Y *et al* 2020 Magnetic hyperthermia with ε-Fe_2O_3 nanoparticles *RSC Adv* **10** 28786-28797
[22] Ansari L and Malaeken B 2016 Magnetic silica nanocomposites for magnetic hyperthermia applications *International Journal of Hyperthermia* **33**(3) pp 354-363

On the adapting ability of living organisms to stationary and non-stationary electromagnetic fields

G Roșu[1], S Spandole-Dinu[2], A-M Catrina[2], L Tuță[1], O Baltag[3] and L O Fichte[4]

[1]"Ferdinand I" Military Technical Academy, Bucharest, Romania
[2]"Cantacuzino" National Military-Medical Institute for Research and Development, Bucharest, Romania
[3]"Grigore T. Popa" University of Medicine and Pharmacy, Iasi, Romania
[4]"Helmut Schimdt" University, Hamburg, Germany

E-mail: georgianamarin01@gmail.com

Abstract. The adapting ability of living organisms is the key to survival of the species, and its scope is defined by homeostasis, meaning the tendency of any living system to maintain its biologic parameters. Heat shock proteins are part of a homeostasis system in charge with the regulation of damaged proteins caused by the action of several types of stressors. Among all types of stressors, the electromagnetic radiation has not yet been investigated in terms of heat shock protein response. Recent investigations regarding in vivo experiments using electromagnetic exposure from laboratory sources and commercial mobile phones indicate that the biologic response is more intense when using real life sources. This paper proposes a model for heat shock proteins expression as response to electromagnetic radiation exposure, based on a Hill controller.

1. Introduction

The interaction between living organisms and nonionizing electromagnetic radiation is still an issue of controversy. Still, by applying the precautionary principle, there are several national and international regulations imposing limitations on the exposure to electromagnetic fields (EMFs). These regulations rely on the guidelines of the International Commission on Non-Ionizing Radiation Protection (ICNIRP). One drawback of these guidelines is that they are only focused on the thermal effects of EMF exposure [1]. However, several recent studies indicated serious non-thermal biological effects on humans and animals (rats) from EMFs: affected cognitive functions from mobile phone use [2], DNA damage and antioxidant status in lymphocytes from individuals residing in the vicinity of mobile phone base stations [3], affected reproductive parameters of male rats from long-term exposure Wi-Fi RF-EMF [4].

Most of in vivo experiments rely on the use of laboratory sources, with the emission pattern defined by a continuous wave or a stationary wave [4-7]. A recent review paper comparing the effects of laboratory and commercial off the shelf (COTS) mobile phones [8] indicated that the real (COTS) sources signals are more bioactive than the lower variability ones, implying that the livings organisms have the ability to adapt to EMF exposure when the pattern has low variability. The adapting ability is characteristic to all living organisms and is the key to survival of the species. It is applied through several mechanisms, and its scope is defined by homeostasis, meaning the tendency of any living system to maintain its biologic parameters [9-10].

Extracellular stress proteins are involved in maintaining cellular homeostasis and also in pathophysiological processes induced by different types of stressors. Their expression (increase/decrease of cytosol concentration) is modulated by a double activator-inhibitor mechanism, in order to maintain/adapt the biological parameters of the system [11]. Heat shock proteins are part of a homeostasis system in charge with the regulation of damaged proteins caused by the action of several types of stressors: thermal, chemical, etc. The regulation mechanism have been thoroughly investigated for thermal stressors [12-13], but there have not been any analyses on the impact of EMF to the expression and regulation of heat shock proteins.

This paper proposes a model for the ability of living organisms to adapt to EMF exposure by means of expression of heat shock proteins and is structured as follows:

Part I gives the introduction to the research, Part II deals with the main thermal and non-thermal effects of biologic exposure to non-ionizing radiation [14-17]. The third part describes the characteristics of laboratory sources and commercial sources and their corresponding EMF emissions. Part IV deals with the complexy of biologic systems and the main characteristics that reflect this complex behaviour, and more specifically, the capacity the adapt to changes in the external environment. Part V presents the principles of modeling the ability of biologic systems to adapt to the stress applied by EMF and describes the proposed adaptation model based on a Hill controller. Finally, part VI illustrates the model's behavior under several excitation conditions and discusses the model features. Part VI provides the conclusions

2. Biologic response to electromagnetic fields

Exposure of organisms to external stress factors, such as trauma (injury, infection, surgery), illness (viral, bacterial, or fungal agents), inadequate oxygen supply, hypoglycemia (low blood sugar), hormonal and/or biochemical imbalances, nutritional deficiencies, dehydration, environmental pollution – including electromagnetic radiation, trigger a complex molecular, cellular and systemic response in an attempt to maintain homeostasis. The immune system plays a crucial role in coping with stress factors and its response can be modulated by a wide range of factors, including non-ionizing electromagnetic radiation.

The interplay between electromagnetic fields (EMF) and biological processes is still not fully understood; yet it is clear it depends on biophysical and biochemical properties of exposed organism, as well as EMF parameters (i.e. frequency, intensity, polarization, exposure duration, waveform, variation pattern). Electromagnetic radiation is too weak to break atomic bonds and therefore it doesn't cause ionization in biological systems. However, when a biological system is exposed to EMF, depending on the wavelength, it absorbs different amount of energy. Also, the properties of the biological system, such as water content, can impact the absorption of EM radiation. One of the proven hazards of RF-EMF exposure is related to thermal heating. In general, the biological effects of the RF radiation have been classified as thermal and non-thermal effects.

2.1. *Thermal effects*

Thermal effects of electromagnetic fields are related to the dielectric heating effect. High-frequency electromagnetic radiation stimulates the oscillation of dipolar molecules, such as water molecules, at a very high speed, thereby converting electromagnetic energy into kinetic energy [18]. The heating effect is proportional to the absorption of electromagnetic energy quantifiable by absorption rate (SAR) and expressed in electromagnetic energy in a unit per mass of biological tissue (10.1515/reveh-2016-0011). As a result of dielectric heating due to EM radiation, the main biological thermal effects cascade as a result of the biological system's homeostasis mechanisms: blood flow increases/decreases to keep the body temperature constant (thermoregulation), and metabolic rate increases as a result of increasing the body temperature [19].

Heath shock proteins (HSP) are found in all living organisms and are involved in the cellular and molecular response to stress conditions capable of inducing damaged proteins (e.g. thermal stress, oxidative stress, fever, inflammation, heavy metals, etc.). Particularly, Hsp70 and Hsp90 were associated with the biological response to EMF as several studies showed EMF-exposure can modulate their expression [20-21].

2.2. Non-thermal effects

There are several studies regarding the non-thermal effects induced by RF-EMFs on living organisms, but this topic is still under debate. These biological effects depend on the applied EMF, the model system used and the environment. RF-EMF at 900 MHz has been shown to induce a decrease of cell viability and an activation of apoptotic signaling on cell cultures [22]. A recent study seems to indicate that RF-EMF may strongly influence the ion fluxes through voltage-rectified ion channels, which may inhibit the growth of cancer cells which supra-express such channels [23].

Non-thermal effects of EMF on immunity have been reviewed [17] based on resonant absorption, ion transportation, signaling pathways modulation, magnetic properties of biomolecules, etc. However, to this day, there is no unanimously accepted biological mechanism by which EMF interacts with the immune system. Other hypothesized interactions between EMF exposure and gene expression, heat-shock protein levels, signal transduction pathways, cell morphology and cell membranes cannot be neglected. Studies have shown a synergistic effect of concomitant exposure to EMF and other stimuli, as well as the fact that the immune system's response is strongly dependent on the degree of variability in EMF. Lai, 2021 [16] conducts a comprehensive analysis of studies on the genetic effects of EMF, particularly genotoxicity (DNA damage) and regulation of genes involved in cell cycle arrest, apoptosis, stress response, as well as heat shock proteins.

3. Ambient electromagnetic emission sources

3.1. Stationary and non-stationary sources

A stationary or strict-sense stationary process is a stochastic process whose statistical parameters remain constant in time. A stochastic process can also be wide-sense stationary if its mean is constant in time and its autocorrelation depends only on the time difference $\tau = t_1 - t_2$ and not the individual moments in time t_1 or t_2. Stationarity can also be classified based on the number of moments and which moments in particular are not time dependent. This is called N-th order stationarity and is a weaker form of stationarity. For example, second-order stationarity, which is extremely common, implies that the mean, variance and autocorrelation of the stochastic process are constant and non time-dependent, but any other statistics (high order moments) may vary in time.

Since electromagnetic emissions can be modeled as stochastic or quasi-random processes, any ambient sources that generate these emissions can be classified as stationary or non-stationary based on how the electromagnetic field parameters vary in time. For example, continuous wave radars used for medical imaging and remote monitoring of patients' cardiorespiratory activity [24] can be classified as stationary electromagnetic sources, as the radar signal's variance does not change in time. This means the emitted field's power is constant and not time-dependent. On the other hand, a WiFi router's activity, when both transmitting and receiving WiFi signals, is highly dependent on the data traffic and packet requests sent within a certain time window [25]. This means that the WiFi signal power or variance changes in time. Thus, assuming the signal mean will ideally remain constant in time, we can classify the WiFi router as first-order stationary at best, which is the weakest form of stationarity.

The stationarity of an electromagnetic emission sources can have a significant impact on living organisms. The reason is that non-stationary sources will radiate EM fields with power that varies in time. The organism will be thus be exposed to a stimulus that permanently changes in intensity. This will restrict the organism's ability to adapt to the incoming stimulus. For this reason, some medical applications can use continuous wave radars, as these are stationary electromagnetic emission sources that will not endanger the well-being of the patients. On the other hand, the effects of Wi-Fi routers on living organisms represent a current topic of interest, as such emission sources either have weak-stationarity or are non-stationary and can potentially have negative effects on humans.

3.2. Real sources description

The sources of electromagnetic radiation can be classified as natural and non-natural. Natural sources can be cosmic like celestial bodies such as the Sun, or biological like electrical impulses, currents or potentials, produced by living organisms. However, most of the exposure humans receive from electromagnetic radiation is caused by non-natural sources. These sources are human-made and are generally comprised of technical devices and equipment. Some of most significant ones are: telecommunications systems, power grid transmission lines, high power radars, railroad and tram systems, medical equipment, and even some household appliances.

Non-natural electromagnetic sources account for most of the electromagnetic pollution that has occurred in the past decades. Extensive research is directed nowadays towards studying the effects of electromagnetic emissions on living organisms. One certainty is that high power exposure, even for short duration, can have negative effects on human health. The impact of mobile phones and similar emission sources is a contentious subject as these devices generate significant field levels compared to other everyday devices, and are held in close proximity to the user. Most studies have shown however, that the field levels of such mobile emission sources are generally not enough to be harmful to humans.

4. Biologic systems complexity

4.1. Biologic systems modelling scales

The process of modeling a system starts from defining the objectives of the model and selecting the components that are relevant for the purpose of the study, as well as the level of detail of the results and the degree of generality of the model. Mathematical formulation assumes an adequate structure for the objectives of the model, able to describe the study hypotheses in a quantitative form [26]. Biological systems are characterized by an extremely high degree of complexity, which also lies in the fact that certain processes or mechanisms cannot be observed on a single scale to describe the system [27].

The unit of scale of biological systems is the cell. Starting from this element, the scales of system representation are defined: intracellular, cellular, intercellular and macroscopic. The intracellular scale is usually used to describe the biochemical state of the cell. Depending on the level of scaling, the dynamic behavior of one or more features can be described. In complex natural phenomena, there are aspects observed on several scales - spatial and temporal, for this reason multi-scale models are an appropriate solution in modeling the behavior of biological systems [26].

4.2. Living systems characteristics

The modeling of biological systems must include the basic principles related to the formulation of a survival strategy at the macroscopic level and the spontaneity of the response of the living system to environmental changes [28]. Living systems have flexibility and adaptability; they do not have a purely deterministic behavior. The relationship between the system and the output size depends on the environmental conditions, but it also has a fluctuating character given by the survival strategy.

In the modeling of complex systems, for reasons of saving computing resources, it is practiced to choose an appropriate representation scale (or set of scales) and an associated variable. The mathematical structure selected to describe the dynamics of the system can be divided into functional subsystems that interact with each other.

4.3. Living systems adaptability

Living systems can be represented as networks of elements interacting in a multi-hierarchical / multi-level manner. One of the fundamental characteristics of living systems is formulating a survival strategy, which translates into adaptation to the environment.

Homeostasis can be defined as the tendency of any living system to maintain the internal parameters relatively constant. Through homeostatic mechanisms, essential biological parameters, such as temperature, pH, osmolarity, and the concentrations of metabolites (glucose, ATP, etc.) and ions (K, Ca, Na etc.) are kept within a narrow range of values, in order to ensure the optimal functionality of the organism. Such coping mechanisms to environmental changes take place in every cell, tissue and organ, as well as in the bodily fluids such as blood and lymph, and are essential for the body's overall wellness.

The central nervous system is the command-and-control center for homeostatic processes that involve organs and body fluids that generally consist of a receptor, a control center and an effector bound in a negative feedback loop. When the receptor senses stimulus, it sends a signal to the control center (brain) that determines an appropriate response to an effector (muscle, gland or organ) and thus a negative feedback loop is generated.

For instance, the thermoregulation involves such a mechanism. When body temperature decreases, the heat is conserved through blood vessels constriction, inhibition of fluid secretion from sweat glands and muscular contraction. The control center is the hypothalamus. If the body temperature rises, the regulation center from the brain acts to decrease the body temperature [29].

One of the primary effects of heath shock on cells is the generation of damaged proteins. The misfolding of these proteins could lead to effects similar to neurodegenerative diseases. In order to counteract heat shock effects, cells respond by expressing more heat shock proteins (HSPs). HSP act as molecular chaperones by recognizing and binding to misfolded proteins and reestablishing their native confirmation, or escorting them to proteosomes for degradation [12]. The control of blood glucose is another example of negative feedback mechanism. In this case, the glucose blood level acts as a stimulus for pancreas (the control center). When the blood sugar increases, the pancreatic β-cells are stimulated to secrete insulin in order to decrease the glucose level. On the other hand, when the glucose level decreases, the pancreatic α-cells secrete glucagon which stimulates liver to produce glucose [30].

5. Mathematical formulation of biologic systems adaptation to EMF exposure

5.1. Control systems principles in modelling homeostasis in EMF interaction

Biological systems are represented as a network of elements interacting on several levels, so as to keep their process parameters at certain reasonable intervals. Regardless of the nature of the stimulus, biological systems respond by emitting regulating chemicals. Homeostasis involves maintaining a stable operation by adapting the system parameters, following the changes in the external conditions.

Homeostasis modeling is based on the implementation of process control principles in describing the dynamics of the biological system with multi-scale representation, so as to ensure the stability of system operation. In systems engineering terms, homeostasis means that biological systems compensate for deviations from the reference point, in an attempt to maintain a state of internal constancy [31]. With a broader sense, allostasis is defined as maintaining the stability of the system by adjusting the reference points. Modeling one of the cellular homeostasis mechanisms in interaction with heath shock induced by EMF can involve the expression of HSPs. HSPs are naturally omnipresent in the cell at a baseline level. However, when a stressful stimulus is applied (e.g., heat generated by non-ionizing electromagnetic radiation causing protein denaturation), in order to counteract its effects, a cascading reaction leading to the activation of genes encoding HSPs is put in motion in order to increase the number of HSPs in the cell. Regulation of the HSPs typically invoke a negative feedback loop [32].

For example, Hsp70 is expressed by an inducible gene. Very briefly, the normal level of Hsp70 in the cell is low and this protein is bound to another protein called HSF1 (Heat Shock Factor 1). The purpose of HSF1 is to activate the gene encoding Hsp70. In order for the cell to produce more Hsp70, HSF1 has to be released. When cellular proteins are damaged due to heating, Hsp70 binds to these proteins, releases HSF1, thus enabling the production of more Hsp70. When thermic shock dissipates and all damages proteins are dealt with, Hsp70 becomes free again and binds back to HSF1, stopping the production of more Hsp70 [33].

In literature, there are mathematical models for regulating the expression of heat shock proteins, namely Hsp70, in response to the application of a thermal stimulus through an adaptive process based on the idea that the body has the ability to adjust its process parameters according to environmental conditions. Although the models are rigorous and experimentally validated, they show only the response to a thermal stimulus. To date, there are only a few studies investigating the expression of heat shock proteins following exposure to EMF [20].

The Hill function is generally used to describe the gene expression rate depending on the transcription factor concentration, but it can be extrapolated to describe the response of biologic systems to external stimuli [34]. Two types of Hill functions are used: the activator and the repressor mechanisms, modelling the activation and the reduction of the gene transcription rate. A Hill controller includes the two types of reactions, resulting in [31, 35]:

$$u(t) = \frac{\beta\, y^n}{y^n + \left(y - e(t)\right)^n} \tag{1}$$

Here y represents the reference point of the controller, β is a coefficient modeling the maximum level, e represents the deviation from the reference point, and n is the Hill coefficient indicating the steepness of the controller response.

5.2. EMF adaptation model schematic

Any living organism will adapt when exposed to no-ionizing electromagnetic radiation. This is a homeostatic mechanism that expresses heat shock proteins Hsp70. The mechanism is based on a PI (proportional-integral) response, according to the following equation [31]:

$$\frac{dv}{dt} = v + (k_p + k_c)\cdot\left(e + \int_0^t e(\tau)\,d\tau\right) \tag{2}$$

Here v represents the expression level of Hsp70, e represents the deviation from the refference point, and kp and kc represent the proportional response gain and compensatory response gain. Choosing a PI-type response for the error is based on the premise that an organism's capacity to adapt diminishes as error or deviation from homeostasis accumulates in time. The two gains can be expressed via Hill controller, as follows:

$$k_p = \frac{\beta_p \cdot \gamma_p{}^n}{\gamma_p{}^n + \left|\int_0^t \left|e(\tau)\right| d\tau - \gamma_p - \varepsilon_p\right|^n} \tag{3}$$

$$k_c = \frac{\beta_c \cdot \gamma_c{}^n}{\gamma_c{}^n + \left|\int_0^t \left|e(\tau)\right| d\tau - \gamma_c - \varepsilon_c\right|^n} \tag{4}$$

In the mathematical expressions for the two response types – proportional and compensatory, stress accumulation in time is represented by the integral of the error's absolute value. We assume that the system has a certain stress tolerance, defined by an upper limit γ, and a lower limit ε of the error's time accumulation.

IOP Conf. Series: Materials Science and Engineering 1254 (2022) 012024 doi:10.1088/1757-899X/1254/1/012024

Figure 1 presents a Simulink implementation for the EMF adaptation model. The values and meaning of the parameters are presented in Table 1. These values have been chosen based on the system's response to a constant value A=1, while targeting a fast regulation capacity for the model. In order to highlight the impact of the Hill coefficient n and the lower limit of the compensatory mechanism ε_c, Figures 2-4 show the response yielded by systems with different coefficients, to the same input, as shown by the final oscilloscope in the Simulink schematic.

Table 1. System model parameters.

Parameter	Notation	Value
Hill coeffcient	*n*	6
Multiplication coefficient of the proportional response	*βp (beta_p)*	1
Multiplication coefficient of the compensatory response	*βc (beta_c)*	1
Lower limit of the proportional response	*εp (eps_p)*	1
Upper limit of the proportional response	*γp (gama_p)*	10
Lower limit of the compensatory response	*εc (eps_c)*	1
Upper limit of the compensatory response	*γc (gama_c)*	100

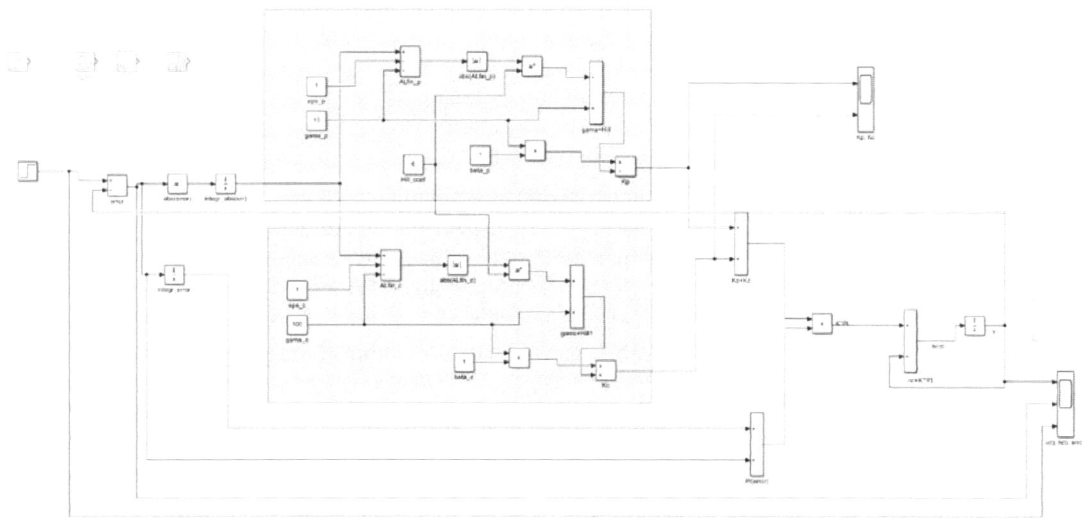

Figure 1. Simulink implementation of the Hill controller modeling the adaptation mechanism.

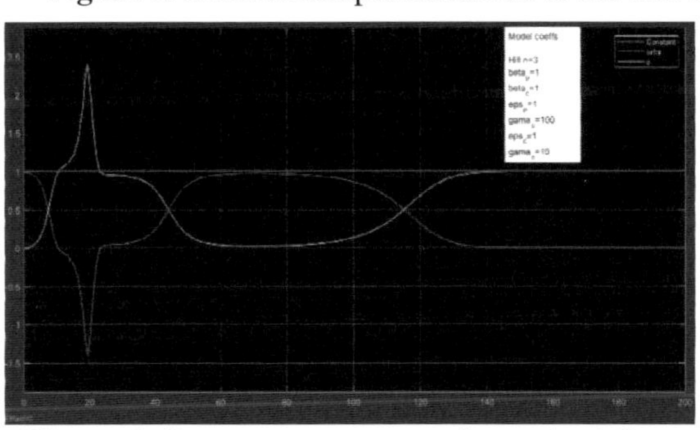

Figure 2. Response to constant input A= 1 for the system defined by Hill coefficient n=3, and ε_c= 1.

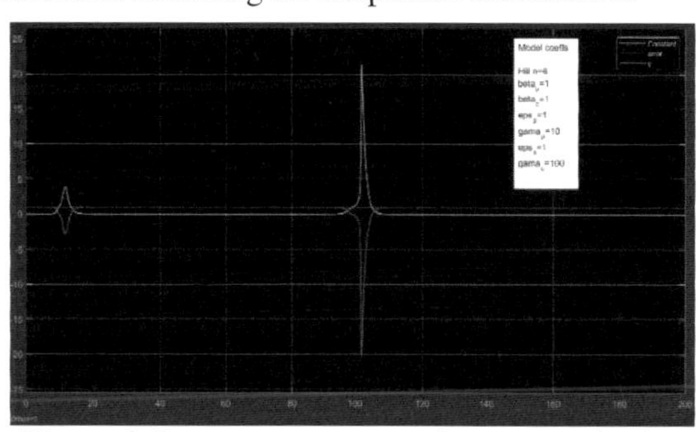

Figure 3. Response to constant input A= 1 for the system defined by Hill coefficient n=6, and ε_c= 1.

The yellow line describes the system response v(t), the red one – constant the excitation, and the blue one – the error. The system coefficients in all three cases are indicated in the textboxes. The higher the system order – meaning the Hill coefficient n, the lower the time constant of the system.

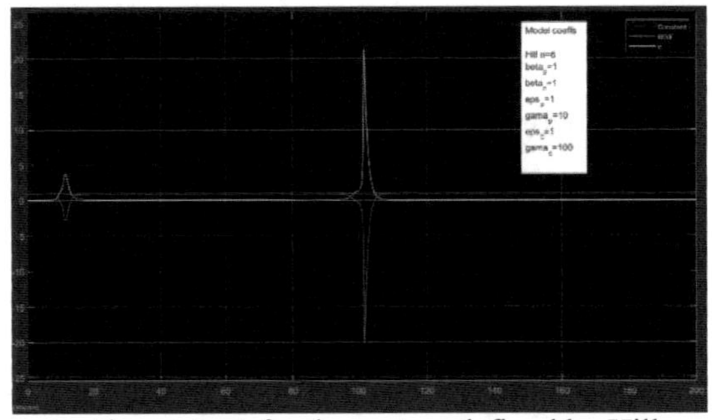

Figure 4. Response to constant input A= 1 for the system defined by Hill coefficient n=6, and ε_c= 10.

It is noted that the lower limit of the compensatory mechanism εc does not have a high impact on the system response, if the system order is high enough (n=6). Thus, a lower value was chosen for εc, so that the compensatory response has a lower threshold level. Therefore, the optimal selection of parameters from all the analyzed configurations is the one indicated in Table I, and the corresponding system response to a constant input in the one illustrated in Figure 3.

6. Model testing

6.1. System response to constant excitation

The first analysis performed upon the system model is to determine its response to a constant excitation, and to find out how the expression level of heat shock proteins varies depending on the excitation level. For this scope, several level of the constant input were selected A=1, 2, 5 and 10, respectively and the system response in all cases is illustrated in Figure 5. The typical response of a Hill controller is illustrated in all cases, having two peak values of the output level, with peak distance and width depending on the stimulus level. It is noted that the higher the input, the closer the two peaks are, but the system tends to adapt to the stimulus in all cases, the output falls to zero after the two transients.

6.2. System response to sinewave excitation

When discussing the system response to a sinewave excitation, two parameters render significance: the excitation level – which was chosen to have an RMS value equal to the constant one (A= 1, 2, 5, and 10, respectively) and the period T. In order to compare the results to the ones in the previous stage, and to highlight the impact of each parameter, Figures 6 and 7 describe the system response to a sinewave, when varying the excitation level, and the period, respectively. Markers were added to the peak values of the output signal in all cases. In Figure 6, it is noted that the higher the level of excitation, the higher and also the closer the two peak levels are, meaning that the system response is more intense both in terms of magnitude and frequency. When applying a sinewave excitation of various frequency, the system response is more acute when the frequency is higher - for period T = 10.

6.3. System response to rectangular excitation

Regarding the response to rectangular signals, an additional quantity – the duty cycle, needs to be discussed, besides the amplitude and the period. The duty cycle is defined as the ratio between the pulse width (for which the signal is equal to its amplitude) and the period. In the time interval outside the pulse width, the signal has zero value. The reason for choosing this type of excitation is that a rectangular signal is comprised of a sum of sinewaves of frequencies equal to multiple values of the fundamental frequency. This represents an initial step for the frequency analysis of the system response. Figure 8 describes the system response to a rectangular signal with various duty cycle values, ranging from 10%, 20%, 50%, and 99%. The final version is almost a constant excitation. It is noted that maximum value is obtained for the highest duty cycle, and its peak values are Vp1 = 6.01, Vp2 = 30.29, separated in time at approximately 10 samples. This can be approximated by the constant excitation A=10 case illustrated in Figure 5 – lower right side, which yielded two peak values Vp1 = 6.02, Vp2 = 30.36, also separated in time by approximately 10 samples. If the duty cycle is decreased, the excitation level is lower, therefore the system response will be weaker, the two peaks having lower values and a higher time separation.

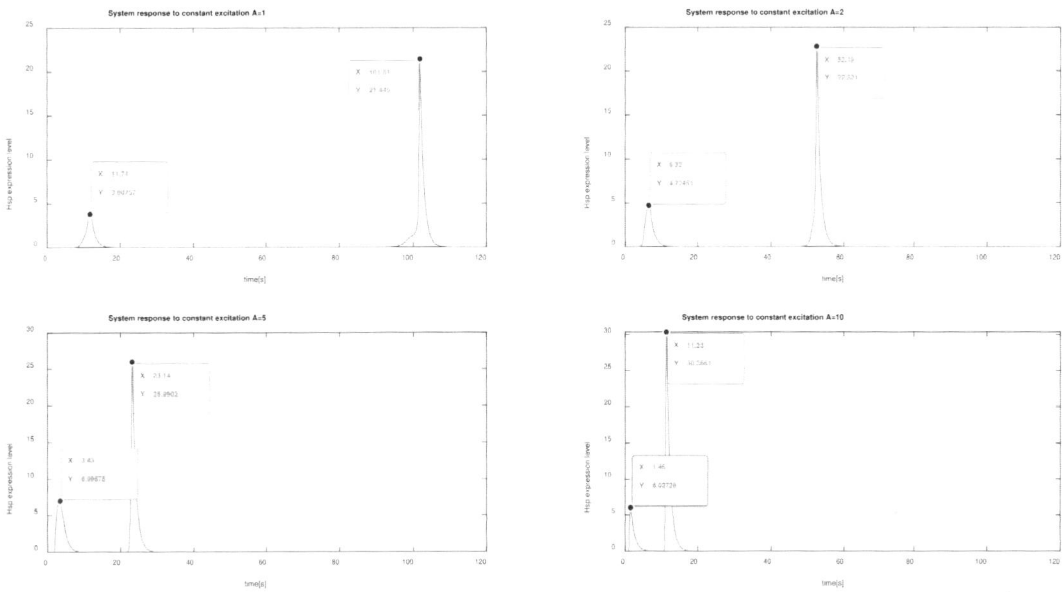

Figure 5. System response to constant input A= 1, 2, 5 and 10, respectively.

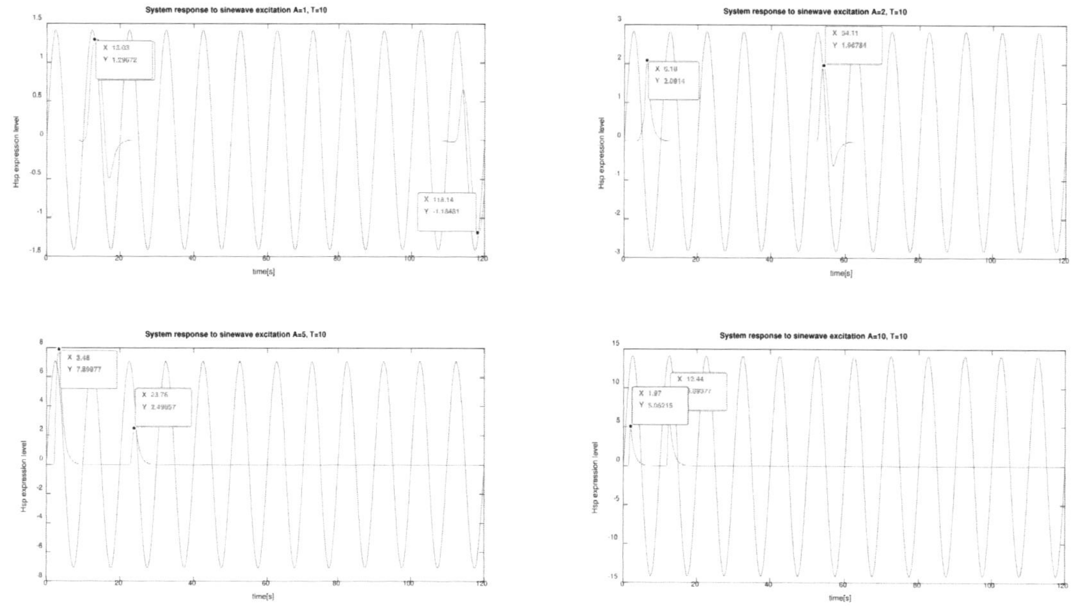

Figure 6. System response to sinewave input with period T = 10, and RMS value of A= 1, 2, 5 and 10, respectively.

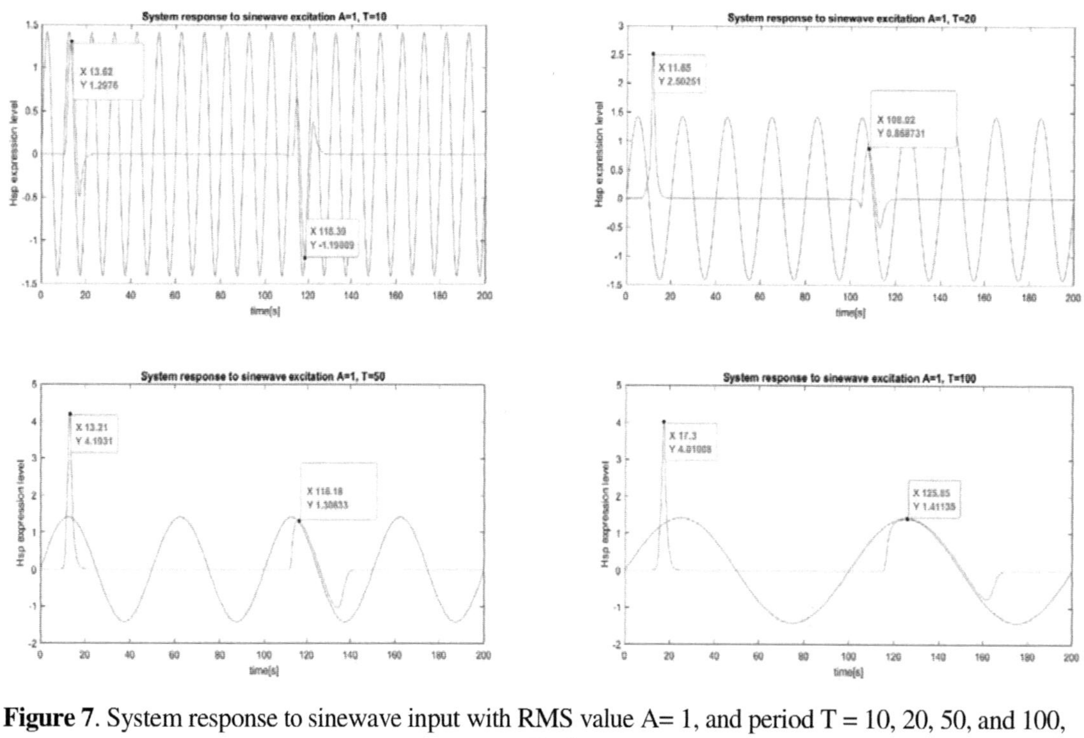

Figure 7. System response to sinewave input with RMS value A= 1, and period T = 10, 20, 50, and 100, respectively.

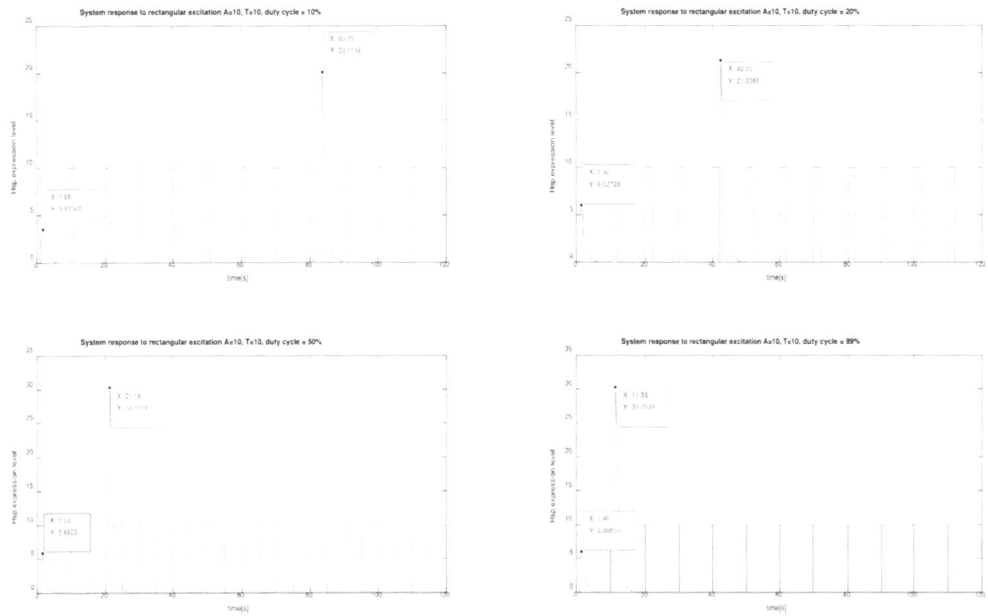

Figure 8. System response to rectangular excitation with amplitude A= 10, period T = 10, and varying duty cycle = 10%, 20%, 50%, and 99%, respectively.

7. Results discussion

The adaptation model of the biological system to electromagnetic fields of several variation patterns yields the expression level of heat shock proteins, through a double activatory-inhibitory mechanism implemented by means of a Hill controller. The system response consists in two peak levels, of various amplitudes, and separated at different time intervals. Table II includes all the data on the system response in the analyzed cases.

It is noted the output amplitude increases with the input level, as well as its infliction duration. For inputs with low value and low periodicity, the inhibitory response becomes more obvious – as in the case of sinewave with amplitude A=1 and period T=10. Also, the two peaks in the system response are closer for higher excitation levels.

Table 2. System response.

Excitation type, parameters	Peak 1 (amplitude)	Peak 2 (amplitude)	Time interval (samples)
Constant excitation A=1	3.8	21.44	90
Constant excitation A=2	4.72	22.82	46
Constant excitation A=5	6.99	25.99	20
Constant excitation A=10	6.02	30.36	10
Sinewave excitation RMS A=1, period T=10	1.29	-1.18	105
Sinewave excitation RMS A=2, period T=10	2.09	1.96	48
Sinewave excitation RMS A=5, period T=10	7.89	2.49	20
Sinewave excitation RMS A=10, period T=10	5.05	6.89	10
Sinewave excitation	2.5	0.86	94

Excitation type, parameters	Peak 1 (amplitude)	Peak 2 (amplitude)	Time interval (samples)
RMS A=1, period T=20			
Sinewave excitation RMS A=1, period T=50	4.19	1.3	103
Sinewave excitation RMS A=1, period T=100	4	1.41	108
Rectangular excitation amplitude A=1, period T=10, duty cycle = 10%	3.53	20.11	82
Rectangular excitation amplitude A=1, period T=10, duty cycle = 20%	6.02	21.32	40
Rectangular excitation amplitude A=1, period T=10, duty cycle = 50%	5.89	30.31	19
Rectangular excitation amplitude A=1, period T=10, duty cycle = 99%	6.00	30.29	10

8. Conclusions

The paper investigates the response of a proposed model for the adaptation of living systems to the exposure to electromagnetic radiation, by means of heat shock proteins expression. The model is based on a Hill controller, comprising a dual respnse - proportional and a compensatory, and it models the variation in the cytosol concentration of heat shock proteins. The model is tested against several types of inputs: constant, sinewave, rectangular, and its response is quantified in terms of peak levels and time duration.

The model is at an early stage, and it lacks experimental validation. However, it is based on a standard controller utilized in modeling biologic systems. It will be further developed by comparing the model results with the ones from biologic tests. A more precise model, developed in a later stage of research, could provide an explanation of the biologic response to non-stationary stimuli.

References

[1] ICNIRP 2010 Guidelines for Limiting Exposure to Time-Varying Electric and Magnetic Fields (1 Hz - 100 kHz) *Health Phys.* **99**(6) pp 818-836

[2] Dasdag S, Taş M, Akdag M Z and Yegin K 2015 Effect of long-term exposure of 2.4 GHz radiofrequency radiation emitted from Wi-Fi equipment on testes functions *Electromagnetic Biology and Medicine* **34**(1) pp 37-42 doi: 10.3109/153683

[3] Zothansiama M, Zosangzuali M, Lalramdinpuii M and Jagetia G C 2017 Impact of radiofrequency radiation on DNA damage and antioxidants in peripheral blood lymphocytes of humans residing in the vicinity of mobile phone base stations *Electromagnetic Biology and Medicine* **36**(3) pp 295-305 doi: 10.1080/15368378.2017.1350584

[4] Wang K et al 2017 Effect of 1.8 GHz radiofrequency electromagnetic radiation on novel object associative recognition memory in mice *Nature - Scientific Reports* **7**(44521) doi:10.1038/srep44521

[5] Hansen J W, Asif S, Singelmann L, Khan M S, Ghosh S, Gustad T, Doetkott C, Braaten B D and Ewert D 2015A far-field radio-frequency experimental exposure system with unrestrained mice*SpringerPlus* **4**(669)

[6] Kivrak E G, Altunkaynak B Z, Alkan I, Yurt K K, Kocaman A and Onger M E 2017 Effects of 900-MHz radiation on the hippocampus and cerebellum of adult rats and attenuation of such effects by folic acid and Boswellia sacra *Journal of microscopy and ultrastructure* **5**(4) pp 216-224

[7] Odaci E, Bas O and Kaplan S 2008 Effects of prenatal exposure to a 900 MHz electromagnetic field on the dentate gyrus of rats: a stereological and histopathological study *Brain Res* **1238** pp 224-229

[8] Panagopoulos D J, Johansson O and Carlo G L 2015 Real versus Simulated Mobile Phone Exposures in Experimental Studies *BioMed Research International* 607053 pp 1-8

[9] Modell H, Cliff W, Michael J, McFarland J, Wenderoth M and Wright A 2015 A physiologist's view of homeostasis *Adv Physiol Educ* **39** pp 259–266 doi:10.1152/advan.00107.2015

[10] Vodovotz Y, An G and Androulakis I 2013 A systems engineering perspective on homeostasis and disease *Frontiers in Bioengineering and Biotechnology* **1**(6) pp 1-8

[11] Pockley A G and Henderson B 2017 Extracellular cell stress (heat shock) proteins-immune responses and disease: an overview *Philos Trans R Soc Lond B Biol Sci* **373**(1738) pp PMC5717522 doi: 10.1098/rstb.2016.0522

[12] Rieger T, Morimoto R I and Hatzimanikatis V 2005 Mathematical Modeling of the Eukaryotic Heat-Shock Response: Dynamics of the hsp70 Promoter *Biophysical Journal* **88** pp 1646-1658 doi: 10.1529/biophysj.104.055301

[13] Peper A, Grimbergen C A, Spaan J A E, Souren J E M and Van Wijk R 1998 A mathematical model of the hsp70 regulation in the cell *International Journal of Hyperthermia* **14**(1) pp 97-124

[14] Funk R and Monsees T 2006 Effects of Electromagnetic Fields on Cells: Physiological and Therapeutical Approaches and Molecular Mechanisms of Interaction. A Review *Cells Tissues Organs* **182** pp 59-78 DOI: 10.1159/000093061

[15] Panagopoulos D, Karabarbounis A, Yakymenko I and Chrousos G 2021Human made electromagnetic fields: Ion forced oscillation and voltage gated ion channel dysfunction, oxidative stress and DNA damage (Review) *Int. J. of Oncology* **59**(92) pp 1-16

[16] Lai H 2021 Genetic effects of non-ionizing electromagnetic fields *Electromagnetic Biology and Medicine* pp 1-10 DOI: 10.1080/15368378.2021.1881866

[17] Piszczek P, Wojcik-Piotrowicz K, Gil K and Kaszuba-Zwoinska J 2021 Immunity and electromagnetic fields *Environmental Research* **200**(111505) pp 1-9

[18] Izza Z, Miller J and Zipes D 2019 Ablation Energy Sources*Clinical Arrhythmology and Electrophysiology* pp 206-237 doi: 10.1016/B978-0-323-52356-1.00007-4

[19] Clarke A and Fraser K 2004 Why does metabolism scale with temperature? *Functional Ecology* **18**(2) pp 243-251doi: 10.1111/j.0269-8463.2004.00841.x

[20] Misa Agustiño M, Leiro J , Jorge Mora M, Rodríguez-González J, Jorge Barreiro F, Ares-Pena F and López-Martín E 2012 Electromagnetic fields at 2.45 GHz trigger changes in heat shock proteins 90 and 70 without altering apoptotic activity in rat thyroid gland *Biology Open* **1**(9) pp 831-838 doi: 10.1242/bio.20121297

[21] Misa Agustiño M, Leiro-Vidal J, Gomez-Amoza J, Jorge Mora M, Jorge Barreiro F, Salas-Sanchez A, Ares-Pena F and López-Martín E 2015 EMF radiation at 2450 MHz triggers changes in the morphology and expression of heat shock proteins and glucocorticoid receptors in rat thymus *Life Sci* **127** pp 1-11 doi: 10.1016/j.lfs.2015.01.027

[22] Grasso R, Pellitteri R, Caravella S, Musumeci F, Raciti G, Scordino A, et al 2020 Non thermal effects of radiofrequency electromagnetic field exposure on neural cells *EPJ Web Conf. Italian National Conference on the Physics of Matter (FisMat 2019)* doi: 10.1051/epjconf/202023000004

[23] Wust P, Kortum B, Strauss U, Nadobny J, Zschaeck S, Beck M, Stein U and Ghadjar P 2020Non-thermal effects of radiofrequency electromagnetic fields *Scientific Reports* **10**(1) p 13488 doi: 10.1038/s41598-020-69561-3

[24] Pisa S, Pittella E and Piuzzi E 2016 A survey of radar systems for medical applications *IEEE Aerospace and Electronic Systems Magazine* **31**(11) pp 64-81 doi: 10.1109/MAES.2016.140167

[25] Roșu G, Tuță L, Halca A, Bicheru S, Hertzog R and Baltag O 2020 Experimental Platform for Studying the Impact of Environmental Electromagnetic Fields on Living Organisms *13th International Conference on Communications (COMM)*, Bucharest

[26] Motta S and Pappalardo F 2012 Mathematical modeling of biological systems *Briefings in Bioinformatics* pp 1-12 doi:10.1093/bib/bbs061
[27] Fischer H 2008 Mathematical Modeling of Complex Biological Systems. from Parts Lists to Understanding Systems Behavior *Alcohol Research & Health* **31**(1) pp 49-59
[28] Coscia V 2011 On the mathematical theory of living systems, I: Complexity analysis and representation *Mathematical and Computer Modelling* **54** pp 1919-1929 doi:10.1016/j.mcm.2011.04.031
[29] Bo Duan X S X 2019 How to Break a Fever: A Feedback Circuit for Body Temperature Control *Neuron* **103**(2) pp 179-181 doi: 10.1016/j.neuron.2019.06.023
[30] Bich Leonardo M M S A M 2020 Glycemia Regulation: From Feedback Loops to Organizational Closure *Frontiers in Physiology* **11**(69) pp 1-26 doi: 10.3389/fphys.2020.00069,
[31] Acevedo A and Androulakis I P 2017 Allostatic breakdown of cascading homeostat systems: A computational approach *Heliyon* **3** pp 1-20 doi: 10.1016/j.heliyon.2017
[32] Krakowiak J, Zheng X, Patel N, Feder Z, Anandhakumar J, Valerius K et al 2018 Hsf1 and Hsp70 constitute a two-component feedback loop that regulates the yeast heat shock response *eLife* **7** pp 1-17 doi: 10.7554/eLife.31668
[33] Morimoto R, Jurivich D, Kroeger P, Mathur S, Murphy S, Nakai Aet al 1994 Regulation of heat shock gene transcription by a family of heat shock factors *The Biology of Heat Shock Proteins and Molecular Chaperones* (New York, USA: Cold Spring Harbor Laboratory Press) pp 417-455 doi: 10.1016/j.jtbi.2009.03.021.
[34] Frank S A 2013 Input-output relations in biological systems: measurement, information and the Hill equation *Biology Direct* **8**(31) pp 1-25
[35] Santillan M 2008 On the Use of the Hill Functions in Mathematical Models of Gene Regulatory Networks *Math. Model. Nat. Phenom.* **3**(2) pp 85-97

IOP Conf. Series: Materials Science and Engineering 1254 (2022) 012025 doi:10.1088/1757-899X/1254/1/012025

Biological neural network model based on a non-linear stochastic system

C Popovici[1], A Pătrașcu Cutaru[1], L Tuță[1], G Roșu[1], Lars Ole Fichte[2] and O Baltag[3]

[1]"Ferdinand I" Military Technical Academy, Bucharest, Romania
[2]"Helmut Schimdt" University, Hamburg, Germany
[3]"Grigore T. Popa" University of Medicine and Pharmacy, Iasi, Romania

E-mail: georgianamarin01@gmail.com

Abstract. As a contribution to recent activities in the field of modelling electro-biological phenomena, we would like to present an investigation on the electrical properties of a network of neuron and propose a simplified model for brain activity, based on a biological neural network represented by a non-linear system with stochastic components. The network's fundamental element is represented by Izhikevich model, and including several types of responses, grouped into excitatory and inhibitory response. The network's electrical activity is compared to a biological signal obtained from an electroencephalogram recording, and it is noted that the correlation between the biological and the simulated signal increases with the number of neurons in the network.

1. Introduction

Theoretical neurosciences include various models for simulating the brain activity, all based on the electrical nature of the neural signal [1-2]. Within the model, a compromise must be made between the level of detail and the computing resources [3-4]. Artificial neural networks (ANNs) are rough approximations of the nervous system, having the perceptron as the fundamental element [5]. Their purpose is to transport information along a predefined path and represent one of the most frequently used artificial intelligence algorithms. Biological Neural Networks (BNNs) transmit information on a different way: each neuron in the network has its extensions (axons and dendrites) which behave like wires [3,6-7]. Synapses stimulate neurons, thus triggering an action potential [8].

Similar to their artificial correspondents, neurons in BNNs sum up the input signals they receive, and if the threshold is reached, then the action potential is triggered. A major difference between ANNs and BNNs is the learning process [9]. ANN uses a global method such as the downward gradient based on back propagation. In the human brain, the learning process is not yet fully understood, but information through the neuron does not flow in one direction. In simulating the human nervous system, BNNs are much more efficient than ANNs, primarily in terms of resource consumption.

Modeling the human brain activity requires huge resources, since the brain contains almost 100 billion neurons, each having almost 10000 synapses with other neurons. Current human brain emulator projects – such as the Human Brain Project, are based on High-Performance Computing tools and they require massive hardware and software resources to perform simulations of the human brain [10]. This paper proposes a simplified model for brain activity, based on a Biological Neural Network represented by a non-linear system with stochastic components.

2. Modelling the neural activity

2.1. Characteristic electrical parameters of neural activity

The neuron is the structural and functional unit of the nervous system – Figure 1[1]. The structure of a nerve cell differs greatly, depending on its type. The common elements corresponding to each structure are represented by the cellular body and the nerve endings (axon and dendrites). The body cell (soma) is delimited by a selectively permeable membrane for ions; thus, it is electrically polarized. Ionic channels are discontinuities of the cell membrane with a role in the transport of ions. They allow the passage of certain ions, for example, potassium (K^+) or sodium (Na^+). These channels are voltage-dependent because they are activated when changes in the membrane potential occur.

Ionic channels play an important role in the generation and transmission of the nerve impulse. If the stimulant transmitted to the neuronal cell exceeds a certain minimum value (named threshold intensity), then an action potential will be generated and transmitted further to the axon. The length of an axon can vary from a few microns to one meter. The axon is wrapped by the myelin sheath that has a very good insulating capacity and which also increases the transmission speed of the nervous flow.

The following terms are used to describe how neural signals are generated and transmitted: membrane potential, action potential and resting potential [8]. The membrane potential (measured in mV) is the electrical potential difference of the inside and outside of the cell.

The outside of the cell is positively polarized due to the high density of Na^+ ions. The inside of the cell is negatively charged due to the high density of anions. The resting potential is the difference between the internal and external potential, i.e. the resting membrane potential. Its value remains constant in the absence of stimuli, which is around -65 mV, but it can vary from -90 mV to -60 mV. The action potential is a local and rapid change in membrane potential, which lasts for a very short time (1 ms). Several stages can be observed in the process of the generation of an action potential and returning to the resting membrane potential, which are illustrated in Figure 2.

The first stage, of slow depolarization, consists of the slow increase of the action potential, by opening the Na^+ channels, disabling the Na^+/K^+ pump and penetrating the Na^+ ions inside the cell. This penetration of ions will cause, in the second stage, a reversal of the membrane polarity, becoming electropositive on the inner face and electro-negative on the outer face.

In the second phase, the action potential spike develops, which can reach up to +30 mV. The threshold value of -55 mV is the minimum value of the resting membrane potential required to open the ionic channels and generate an action potential. In the third stage, the membrane potential quickly returns close to the resting level, by closing the Na^+ channels and opening the K^+ channels.

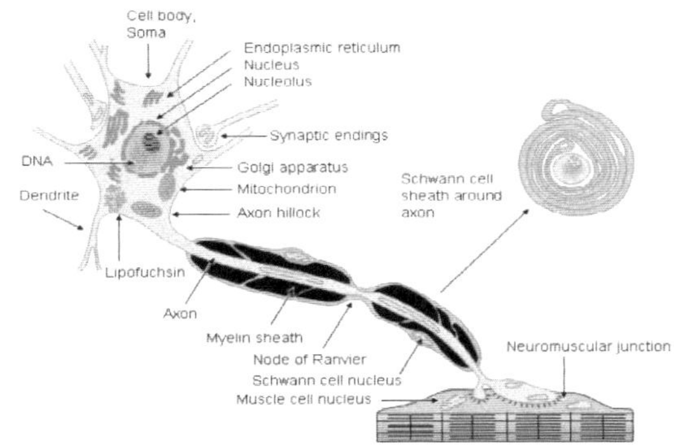

Figure 1. The structure of a neuron [1].

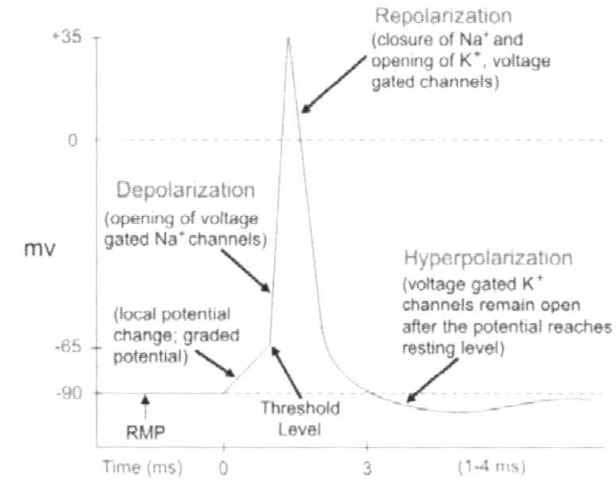

Figure 2. Action potential [9].

In this way, the external face of the membrane quickly becomes electropositive again. The fourth stage, of slow repolarization, manifests itself due to the excess of positive charges (Na^+ ions) left inside the cell due to the closure of the Na^+ channels.

The final stage consists in increasing the potential difference, exceeding the value of the membrane resting potential, by activating the Na^+/K^+ pump, which allows the movement of the Na^+ ions from the intracellular environment to the outside and the K^+ ions in the opposite direction. At the end of this stage, the resting membrane potential will return to the value of -70 mV.

2.2. *Nernst potentials*

Membrane potential changes can be further explained in terms of Nernst potentials [11]. The cell (including the neuron) membrane is a lipid bilayer of 3-4 nm thickness with embedded proteins, which act as channels for specific ions: Na^+, K^+, Cl^-. The cytosol has high K^+ concentration and low concentrations of Na^+ and Cl^- ions, whereas the extracellular medium has reverse ion concentrations, thus yielding the concentration gradients.

Because the membrane is permeable only to specific ions, and the ion concentrations are different in the opposing sides of the membrane, an electric potential occurs. The reversal potential represents the potential that balances a concentration gradient. Nernst equation defines the reversal potential E_{rev} of a particular ion depending on the concentrations of the respective ion in the extracellular $[X]_{out}$ and intracellular environment $[X]_{in}$, respectively [11]:

$$E_{rev} = \frac{RT}{z \cdot F} \ln \frac{[X]_{out}}{[X]_{in}} \quad (1)$$

In (1), R is the universal gas constant R = 8.31 J/mol*K, T is the temperature in Kelvins, z is the number of charges, and F is the Faraday constant F=96485.33 C/mol. Therefore, the Nernst equation describes the electrochemical balance in terms of the reversal potential, which represents the membrane potential value with no net flux of ions. The Nernst potential for Na^+ is +59 mV, K^+=-80 mV, Cl^-=-65 mV.

The resting potential of the membrane represents a weighted average of the Nernst potentials of the ions (E_{Na}^+, E_K^+, E_{Cl}^-), with the weights being the membrane permeability for the respective ions – the conductances g_{Na}^+, g_K^+, g_{Cl}^-:

$$E_{rest} = \frac{(g_{Na^+} * E_{Na^+}) + (g_{K^+} * E_{K^+}) + (g_{Cl^-} * E_{Cl^-})}{(g_{Na^+}) + (g_{K^+}) + (g_{Cl^-})} \quad (2)$$

Since the resting potential mainly depends on the membrane permeability, when the membrane conductance increases for a particular ion, the membrane potential tends toward the Nernst potential of the respective ion. An even more thorough representation of the membrane potential in terms of ions permeability would be the Goldman-Hodgkin-Katz equation [12]. However, a simpler model for the membrane potential variation representation is chosen for the proposed algorithm, which is further described.

2.3. *Hodgkin-Huxley model*

The Hodgkin-Huxley model was developed following the study of the calamari axon and is based on understanding the behavior of the ionic channels of Na^+ and K^+ as the main factors in neuronal signal propagation [1], [11]. The model is synthesized by the differential equation having as variable the membrane potential *V(t)*:

$$C\frac{dV}{dt} = I - \left[g_{Na^+}\left(V - V_{Na^+}\right) + g_{K^+}\left(V - V_{K^+}\right) + g_{Cl^-}\left(V - V_{Cl^-}\right) \right] \tag{3}$$

The quantities in (3) have the following meaning: *C* represents the membrane capacitance (per unit of area) and *I* - the electrical current input to the neuron (per unit area). The sodium and potassium ionic channels are characterized by the conductance g_{Na} and conductance g_K respectively, varying in time, depending on the membrane potential $V(t)$.V_{Na} and V_K represent the reversal potentials of the two ionic channels. The loss channel, which represents the membrane's natural ability to be ion permeable, is modeled by the conductance g_L (per unit of area), which is of constant value, and the reversal potential V_L, respectively.

A set of 3 differential equations that model the behavior of ionic channels is added to the basic Hodgkin-Huxley equation. The Hodgkin-Huxley model is considered to provide high accuracy, but it requires large computing resources.

2.4. *Izhikevich Model*

The Izhikevich model is relatively recent [13], and it is an effective alternative to the Hodgkin-Huxley model, in that it is biologically plausible and computationally effective. The model is based on two differential equations with constant coefficients, being much simpler to solve.

$$\frac{dV}{dt} = 0.04V^2 + 5V + 140 - u + I$$
$$\frac{du}{dt} = a\left(bV - u\right) \tag{4}$$

The quantities in the model are the membrane potential *V(t)* and the recovery variable of the membrane *u(t)*. The parameter *a* [ms^{-1}] represents the time scale of the recovery variable, the parameter *b* [Ω^{-1}] represents the sensitivity of the recovery variable *u* to the threshold fluctuations of *V*, *c*[mV] represents the value at which *V* is reset, and *d* [pA] is the value at which *u* is reset. *I* represents the neuron's excitation current, coming from synapses or other sources. Different parameter values of *a*, *b*, *c*, *d* determine the behavior of the main categories of identified neurons. The neuron types and corresponding Izhikevich parameters are given in Table 1.

Table 1. Types of neurons and values of Izhikevich model parameters.

Neuron type	Izhikevich model parameters			
	a [ms^{-1}]	*b* [Ω^{-1}]	*c*[mV]	*d* [pA]
RS (Regular Spiking)	0.02	0.2	-65	8
IB (Intrinsically Bursting)	0.02	0.2	-55	4
CH (Chattering)	0.02	0.2	-50	2

Neuron type	Izhikevich model parameters			
	a [ms^{-1}]	b [Ω^{-1}]	c[mV]	d [pA]
FS (Fast Spiking)	0.1	0.2	-65	2
LTS (Low Threshold Spiking)	0.02	0.25	-65	2
TC (Thalamo Cortical)	0.02	0.25	-65	0.05
RZ (Resonator)	0.1	0.25	-65	2

The Izhikevich model is used to illustrate the behavior of neurons in the algorithm presented in section III and was chosen due to its simplicity and computational economy [14]. In order to reduce the space allocated to this section, the response of neurons of type RS (Regular Spiking), respectively CH (Chattering), to the same stimulus: the excitation current of constant value I = 10 pA, is represented in Figures 3 and 4.

Izhikevich considers the excitation current injected into the Hodgkin-Huxley model as a bifurcation parameter: when the amplitude of I is below the threshold value, no changes in the response of the cell are observed, but when its amplitude increases, repetitive spikes will occur, depending on the intensity of the stimulus and the neuron type.

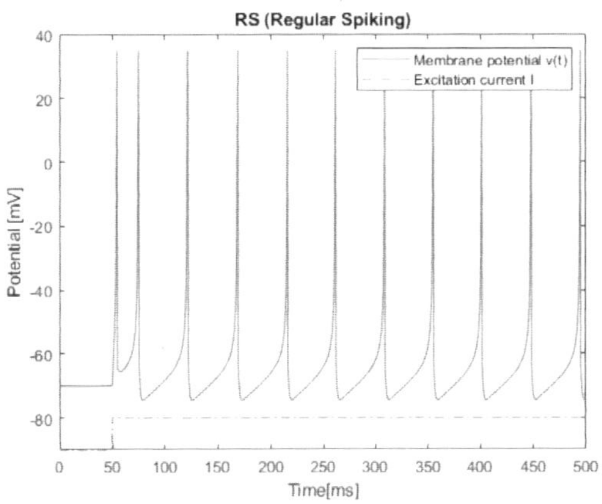

Figure 3.RS (Regular spiking) neuron's response.

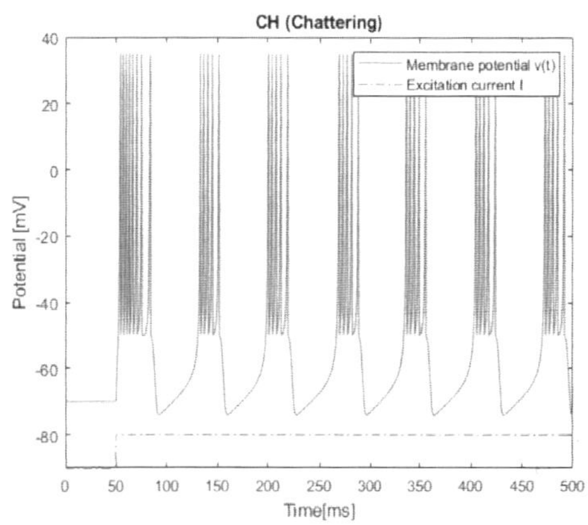

Figure 4. CH (Chattering) neuron's response.

2.5. Spike encoding of the level of the stimulus

Neural encoding refers to the response neurons have to different stimuli, in the sense of how the membrane potential $V(t)$ varies. For example, by applying a constant excitation to a Regular Spiking neuron, its response will change depending on the current's intensity value. Figure 5 and 6 illustrate this answer, for a value of I = 3 pA, I = 5 pA, respectively.

The generation of the action potential begins with the increase in the intensity of the stimulus above the threshold value. The generation frequency increases with the increasing intensity of the stimulus. The form of the action potential propagation in the RS mode does not change, but a train of pulses is maintained as long as there is an impulse strong enough to exceed the threshold value to trigger an action potential.

The response of neurons will also change depending on the variation of the stimulus. If an impulse excitation current or an excitation current with exponential variation is applied, the response of the neuron will vary according to the input variation and the neuron's behavior.

IOP Conf. Series: Materials Science and Engineering 1254 (2022) 012025 doi:10.1088/1757-899X/1254/1/012025

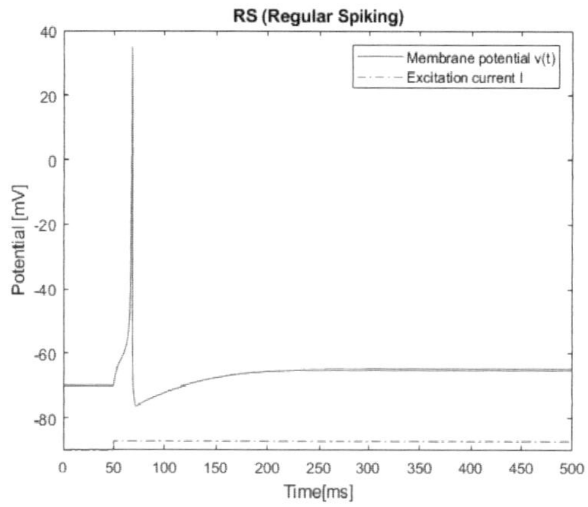

Figure 5. Membrane potential variation for an RS neuron and an excitation current of I = 3 pA.

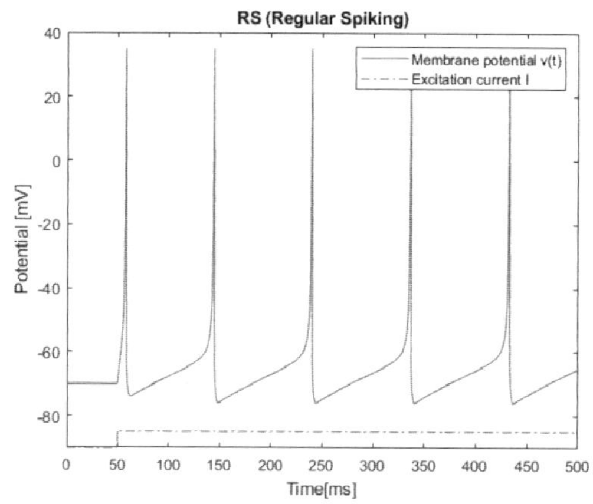

Figure 6. Membrane potential variation for an RS neuron and an excitation current of I = 5 pA.

2.6. Stochastic models in electro-biological neural networks description

In modeling biological neural networks, the neuronal activity of the population is usually characterized by the variable that sums up the membrane potentials of the neurons composing the network.

The transition from the single neuron model to the activity of the entire network is hampered by the following considerations: first of all, by the large number of neurons in the structure of the network; secondly, by the fact that the global properties of the network do not represent a simple multiplication of local properties (of individual neurons). Interactions between neurons (locally) are stochastically described, but from a global perspective, neural activity can be orderly.

In the model proposed by Wilson and Cowan, neural activity is considered as the result of the interaction between two categories of neurons – excitatory and inhibitory [15]. The name of the neuron category reflects the effect they have on other neurons to which they are connected, but also on their own behavior. The Wilson - Cowan model describes the evolution of a biological neural network, divided into two subpopulations: excitatory type E(t) and inhibitory type I(t). Although it is not a detailed biophysical model, it describes the activity of the network using only 2 differential equations, by analyzing how the two subpopulations E(t) and I(t) vary over time.

The variation of the parameters, on which this model is based, generates a multitude of behaviors, explaining the mechanism of transition from normal brain activity to hyper-synchronization of some regions of the brain during epileptic seizures [16]. One of the shortcomings of the model is given by the premise that each neuron receives the same number of excitatory and inhibitory signals.

2.7. Probability distributions used in biological phenomena description

Biological systems are heterogeneous and dynamic, and modeling their behavior assumes knowledge of biological processes, elements of dynamics, as well as the use of appropriate mathematical tools. In modeling the dynamics of neural networks, there is a tendency to use stochastic models over deterministic ones, both at the cell level and at the system or subsystem level. Stochastic models contain at least one element of randomness defined mathematically.

The study of various biological processes indicates a fundamental stochastic character [4], [7]. The choice of the probability distribution is an important step in the accurate modeling of various biological processes. There are two categories of probability distributions – discrete and continuous. The distributions that are used within the biological neural network model are described in section III.

Discrete probability distributions model the probability that a variable will take discrete values. The most commonly used discrete distributions include geometric distribution, binomial distribution, and Poisson distribution.

The Poisson distribution indicates the probability of a series of random and independent events occurring, the individual probability of which is low. Unlike the binomial distribution, which studies the probability of occurrence of n repetitions of an event of probability p, the Poisson distribution is uni-parametric and represents the probability of producing a number of k independent events, knowing the mean value / expected number λ of events recorded in a given time interval:

$$P_{\lambda}(k)=\frac{\lambda^{k}\cdot e^{-\lambda}}{k!} \tag{5}$$

In (5), *e* represents Euler's number, and the other parameters have been defined previously. Thus, the Poisson variable is the variable *k* which represents the total number of successes or repetitions of an event.

The Gamma distribution is a continuous distribution that is used to model positive variables that have asymmetrical distributions. The distribution's parameters are the form parameter *k* and the scale parameter *θ*, and the probability *density* function is described as follows:

$$f(x)=\frac{x^{k-1}\cdot e^{-x/\theta}}{\theta^{k}\cdot\Gamma(k)} \tag{6}$$

Γ(k) is the Gamma function, defined as:

$$\Gamma(k)=\int_{0}^{\infty}t^{k-1}\cdot e^{-t}\,dt \tag{7}$$

For a natural positive integer *k*, the Gamma function is expressed as follows:

$$\Gamma(k)=(k-1)! \tag{8}$$

Several types of probability distributions are used in the Poisson process description, depending on the investigated parameter. Thus, for modeling the number of events *k* desired to occur in the time interval from 0 to t, it is appropriate to use Poisson discrete probability distribution.

If a single event occurs during the time interval 0 to t, then the time moment variable at which the event occurs is evenly distributed between 0 and t, so that the modeling of the time variable of occurrence of the event follows a uniform distribution. On the other hand, an exponential distribution is used for modeling the waiting time until the first event occurs in a considered time interval. The Gamma distribution is used to model the waiting time until the kth event occurs.

3. Biological neural network model

To model the neural activity, first a pattern (deterministic or stochastic) describing the behavior of individual neurons needs to be established, then the model of connection and interaction between the elements of the neural network.

3.1. Basic principles in the model design

The proposed model represents a simplified stochastic version of the Wilson-Cowan model, in the sense that the neural network is made up of individual elements that can have one of two possible states: an active state (excitatory or inhibitory) or a state of rest (quiescent state). The transition between the two states is determined for each neuron, by how it is coupled with the other elements in the network. This coupling is stochastically defined by the probability of existence of synaptic connections, and linearly by a weight that reflects the strength of these connections. The nonlinearity is given by the second-degree equation in (4) describing the variation of the membrane potential.

Thus, the main parameters of the network are: the number of neurons in the network *no_neurons*, the number of external synapses *no_synapses* that connect the network with the outside and on which the input signal (the excitation current) is applied. Each neuron can be of excitatory or inhibitory type, where the determination of inhibitory neurons is carried out randomly, according to a distribution of equiprobable (uniform) probabilities, with the probability *probab_inh*.

The defining stochastic parameter is given by the probability of the existence of an inhibitory neuron, which intervenes in the definition of the network in two ways. First – in establishing the connection between a neuron and a synapse, according to an equivocal distribution, by means of the weight matrix *w_conex* with *no_neurons* lines and *no_synapses* columns, whose elements with value greater than the *probab_inh* probability become zero. Second – in establishing the connection between one neuron with the other neurons in the network, by means of the weight matrix *w* with *no_neurons* lines and *no_neurons* columns. These connections shape the nerve endings of the biological neuron and receive values based on the following principles:

The synapse conductances are weighted by the strength of the connections between the input synapses and the neurons - defined by the weight matrix *w_conex*. 90% of the elements of the *w_conex* array were set to zero, the elements being selected according to a uniform distribution.

The intra-network connections conductances are weighted by the strength of the connections between the neurons in the network –which are defined by the weight matrix *w*. For inhibitory neurons, the weights of neurons' conductances get values according to a Gamma distribution with the shape parameter $k = 2$ and the scale parameter $\theta = 0.003$.

The strength of the connection between an excitatory neuron and an inhibitory neuron is twice as strong as that between neurons of the same type.

Network's dynamics are described based on the time variable, defined by means of the following parameters: the total time allocated to network analysis *t_total*, expressed in milliseconds and the time step set to value $dt = 0.5\ ms$. The network receives an excitation current applied between two moments of time *tinit* and *tfin*, also expressed in milliseconds.

Depending both on the behavior of each neuron given by its excitatory or inhibitory character, and on the way of connecting in the network, the response of each cell will consist in the activation of the neuron (firing), that is, in generating an action potential (spike). The behavior of the neuron is based on the Izhikevich model and is defined by means of the a, b, c, d parameters in (4). Each neuron and each synapse have a specific dynamic, defined by means of a time-varying electrical conductivity, which attenuates with the *tau_g* time constant.

Output variables are given by the variation over time of the membrane potentials of neurons in the network. The network's response is characterized by the neurons' activation map and the moments of time at which the activation is performed, where the character of each neuron (excitatory or inhibitor) is also taken into account.

Another representative form of network's response is the illustration of the electrical activity of the grid that consists in summing up the membrane potentials of all neurons.

3.2. Equations which describe the variations in the membrane potential and the recovery variable

The determination of the neural activity consists of the iterative computation of the membrane potential values $V(t)$ and the recovery variable $u(t)$, at each moment of time t. These variables are determined on the basis of the Izhikevich pattern defined in (4), by writing in matrix form and using matrix notation in bold accented form:

$$\begin{gathered} d\mathbf{V} = \left(0.04 * \mathbf{V}(t) + 5\right) * \mathbf{V}(t) + 140 - \mathbf{u}(t) + \mathbf{I}_{\text{total}}(t) \\ \mathbf{V}(t+1) = \mathbf{V}(t) + d\mathbf{V} * dt \\ d\mathbf{u} = \mathbf{a} \odot \left(\mathbf{b} \odot \mathbf{V}(t) - \mathbf{u}(t)\right) \\ \mathbf{u}(t+1) = \mathbf{u}(t) + d\mathbf{u} * dt \end{gathered} \tag{9}$$

The operator ⊙ suggests the multiplication element by element of two arrays. The matrices have the following dimensions: **V** and **u** have *no_neurons* lines and T columns, where they represent the total number of time samples considered in the analysis, so that **V**(t) and **u**(t) are column vectors containing the values of the two qunatities at time t. The same size (*no_neurons,* 1) is also held by matrices: **dV**, **du**, $\mathbf{I}_{\text{total}}$, **a** and **b**.

$\mathbf{I}_{\text{total}}$ represents the total current in the network, obtained by the sum of the current applied on the input synapses and the current produced by the activation of the neurons in the network $\mathbf{I}_{\text{recurrent}}$, computed as follows:

$$\begin{gathered} \mathbf{I}_{\text{applied}} = \mathbf{w}_{\text{conex}} * \left(\mathbf{g}_{\text{conex}} \odot \mathbf{E}_{\text{conex}}\right) - \left(\mathbf{w}_{\text{conex}} * \mathbf{g}_{\text{conex}}\right) \odot \mathbf{V}(t) \\ \mathbf{I}_{\text{recurent}} = \mathbf{w} * \left(\mathbf{g} \odot \mathbf{E}\right) - \left(\mathbf{w} * \mathbf{g}\right) \odot \mathbf{V}(t) \end{gathered} \tag{10}$$

$\mathbf{w}_{\text{conex}}$ represents the matrix of dimensions (*no_neurons*, *no_synapses*), containing the weights which give the strength of the connections between the network's neurons and the input synapses. The $\mathbf{g}_{\text{conex}}$ and $\mathbf{E}_{\text{conex}}$ matrices have dimensions (*no_synapses*,1) and contain the conductance values and the inhibitory potential of the input synapses, respectively.

Similarly, matrix **w** of (*no_neurons*, *no_neurons*) dimensions contains the weights which express the strength of the connections between the neurons in the network (intra-network connections), and matrixes **g** and **E** of dimensions (*no_neurons*, 1) contain the conductance values and the inhibitory potentials of the neurons in the network, respectively.

Both the input synapses' conductances and neurons' conductances (including their nerve endings) follow a law of attenuation in time, given by the time constant *tau_g* and the step time *dt*:

$$\begin{gathered} \mathbf{g}_{\text{conex}} \leftarrow \left(1 - dt/tau\_g\right) * \mathbf{g}_{\text{conex}} \\ \mathbf{g} \leftarrow \left(1 - dt/tau\_g\right) * \mathbf{g} \end{gathered} \tag{11}$$

3.3. Algorithm's data flow diagram

The algorithm used in simulating the activity of a biological neural network with the parameters described in point III. A and the equations given in point III.B, is structured in the stages illustrated in Figure 7. The algorithm was implemented in the Matlab software, and the input data entry is carried out by means of a graphical interface illustrated in Figure 8. It is noted that the input data supported by the interface is: the number of neurons of the network (parameter *no_neurons*), the number of input synapses into the network (parameter *no_synapses*), the type of excitatory and inhibitory neuron in the network composition– according to the Izhikevich model, and the number of inhibitory cells in the composition of the network. Other input data are related to time: the total time in which the network activity is analyzed, the initial and final time of application of the excitation current – all expressed in milliseconds, as well as the value of the current expressed in picoAmpers.

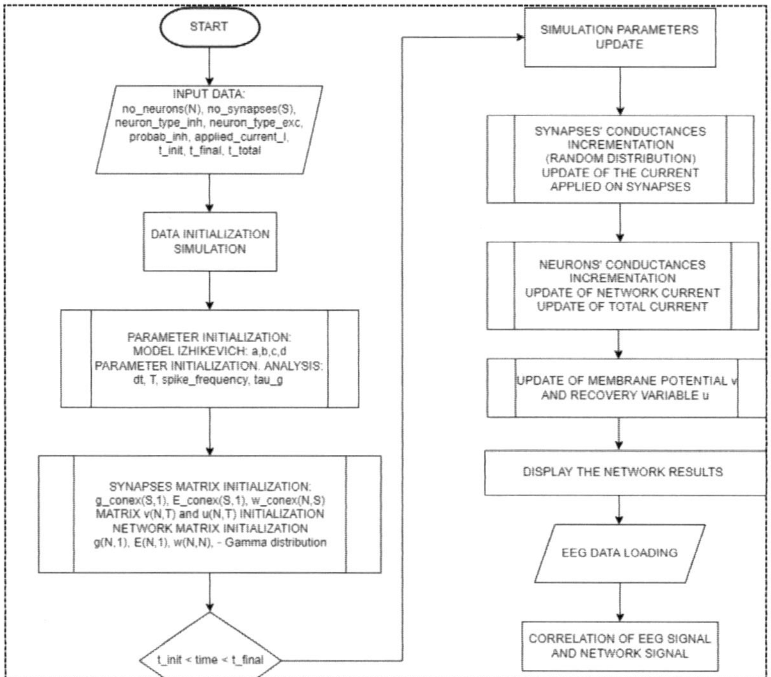

Figure 7. BNN algorithm data flow diagram.

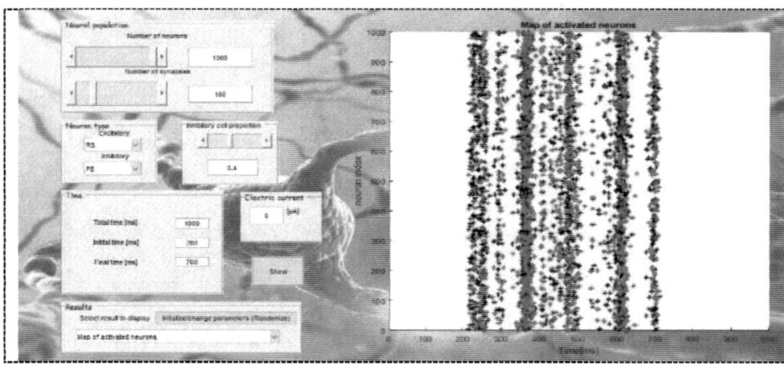

Figure 8. Matlab graphical interface.

The results obtained in the simulation are as follows:

- The electrical activity of the network, i.e. the sum of the membrane potentials of all the neurons;
- The network's electrical current – the sum of all currents, including the ones transmitted by firing neurons, and the currents on the input sysnapses;
- The neurons' map, which represents the neurons' activation moments and the index of the neurons which emit an action potential, where the character of each neuron is taken into account: the excitatory neurons are labeled by black color, the inhibitory ones by red color;
- The action potential of all neurons in the network;
- The membrane potential (PM) of the first and last neuron -with index 1 and *no_neurons* respectively;
- The membrane potential (PM) variation of two randomly selected neurons, one excitatory, the other inhibitory;
- The Gamma distribution histogram of the inhibitory neurons conductances weights *w*;

- A final result is represented by the correlation between the electrical activity of the network and an electroencephalogram record, which is described in section IV.

4. Model validation

The validation method of the BNN model consists of comparing the result regarding the network's activity with an electroencephalogram record.

4.1. Electroencephalogram (EEG)

The electroencephalogram (EEG) is a recording of the macroscopic electrical activity of the cortex layer located in the immediate vicinity of the electrodes through which the signal is collected. In essence, the EEG represents the voltage fluctuations resulting from the ionic currents inside neurons. The EEG measurement method is non-invasive and based on placing a set of electrodes on the scalp, according to a scheme established by the International Federation of Societies for Electroencephalography and Clinical Neurophysiology (IFSECN), called the 10-20 system [17]and illustrated in Figure 9.Each electrode is named in such a way as to indicate the region of the brain on which it is mounted: the prefrontal lobe (Fp), the frontal lobe (F), the temporal (T), the parietal (P), the occipital (O) and the central lobe (C). The electrode name is made by adding the position area, numbered with zero (z) in the sagittal plane, with even numbers for the electrodes mounted on the right side of the skull and with odd numbers for those on the left side.

The data used for the comparison were EEG records from 21 electrodes placed on the subjects' scalp during mental cognitive tasks – arithmetic operations [18]. The data is available through the PhysioBank platform, which is an archive of digital records of various types of biological signals, made available for research purposes [19]. The records of the 21 electrodes of one subject are illustrated in Figure 10.

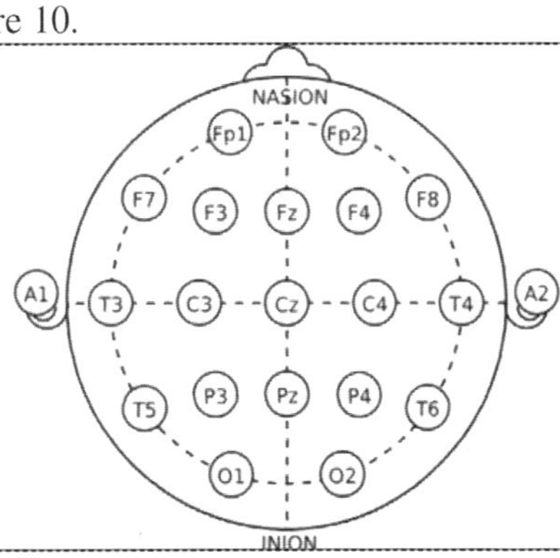

Figure 9. The position of the electrodes according to the 10-20 system [6].

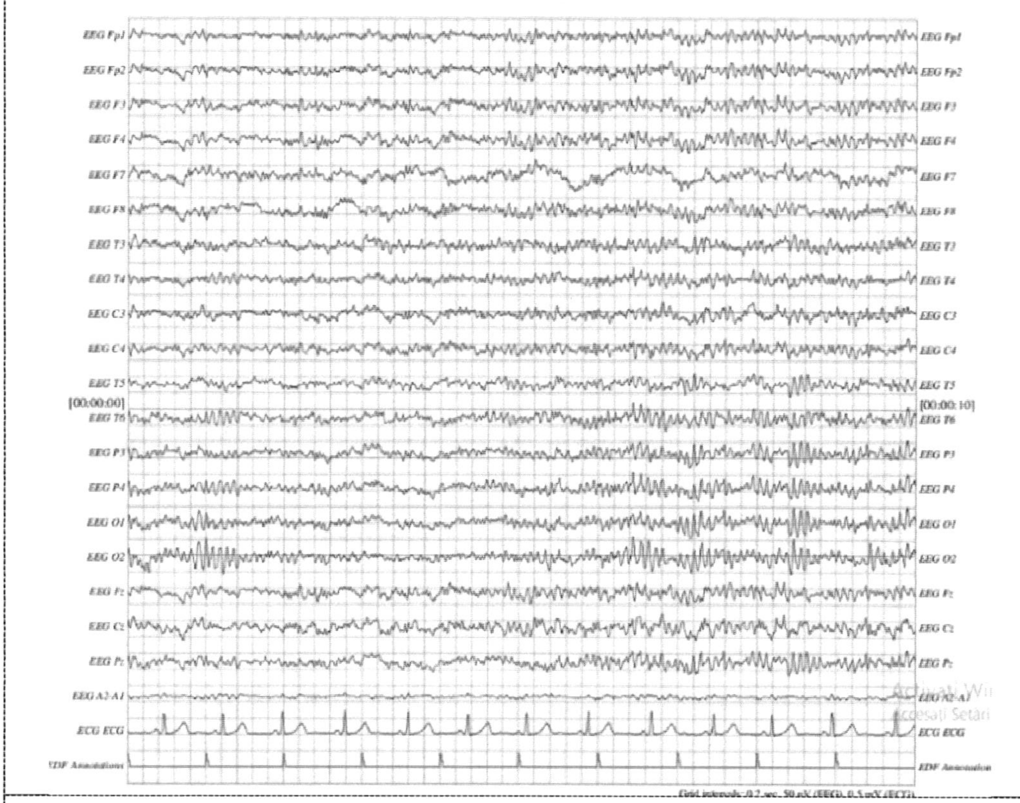

Figure 10. EEG records of those 21 electrodes, while the subject is performing arithmetical operations [18].

4.2. Correlation analysis

In order to perform the correlation between the electrical activity of the modeled network and an EEG signal, there were selected the EEG values from the Fp1 electrode recorded during the time interval of one second. The Fp1 electrode is placed above the frontal lobe - the one responsible with the ability to focus, decision making and intention. In addition to these functions, the frontal lobe supervises almost all brain activity.

Figure 11 illustrates the two signals: the simulated network signals at the top and the one selected from the EEG recording – at the bottom. The parameters of the simulated neural network are: no_neurons = 1000, no_synapses = 100, excitatory neuron type RS, inhibitory neuron type FS, the proportion of inhibitory cells = 0.2, excitation current I= 5 pA, the time interval in which the respective current is applied - from 200 to 700 ms.

The correlation analysis between the two signals consisted in performing a set of signal processing methods. Since the sampling frequencies are different: f_{EEG}= 500 Hz and the $f_{network}$= 2 kHz, the EEG signal was interpolated with a factor of 4. Following interpolation, the frequency of the EEG signal becomes f_{EEG}= 2 kHz. The spectra of the two signals are further investigated, the continuous component is eliminated and the amplitude of the signals is normalized. For the signals thus obtained, the correlation coefficient is computed by using the dedicated Matlab function - Figure 12. It can be seen that there is a maximum near the value of 0, which means that there is a similarity between the signal from the neural network and the EEG signal, with a lag of several samples.

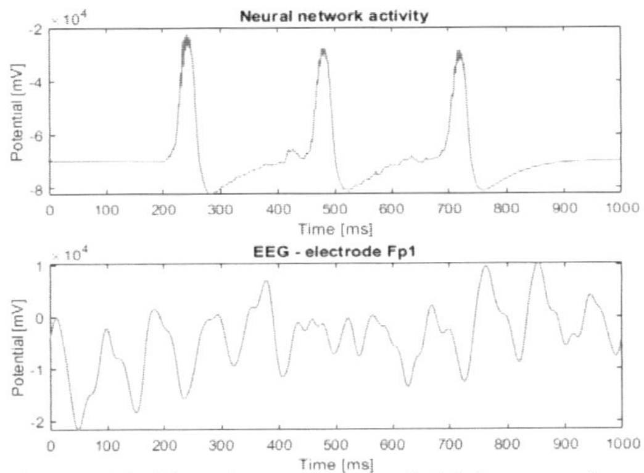

Figure 11. Signals used in the initial stage of comparison: the activity of the modelled neural network (top) and selection from the EEG recording of an electrode (bottom).

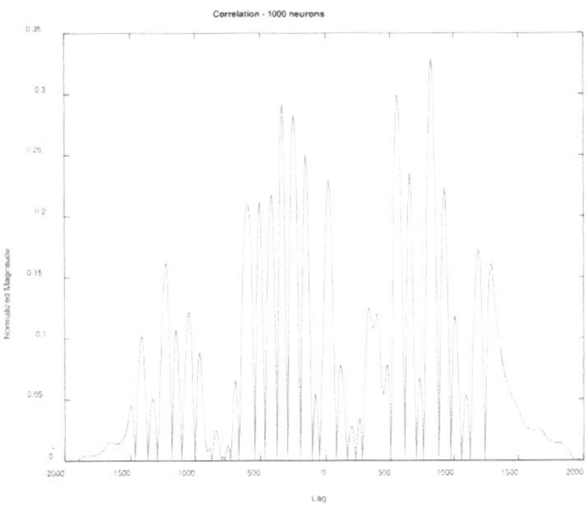

Figure 12. Correlation of initial network signal and EEG signal.

4.3. Results

Since the BNN model is a stochastic one, each run of the program will produce a different result. However, to illustrate that one of the main parameters indicating the accuracy of the model is the number of neurons in the network structure, the correlation coefficient between the network signal and the same EEG signal was calculated, where the parameter *no_neurons* took different values. The comparative results are represented in Figure 13. A nonlinear increase in the maximum value of the correlation coefficient is observed, as the number of neurons increased.

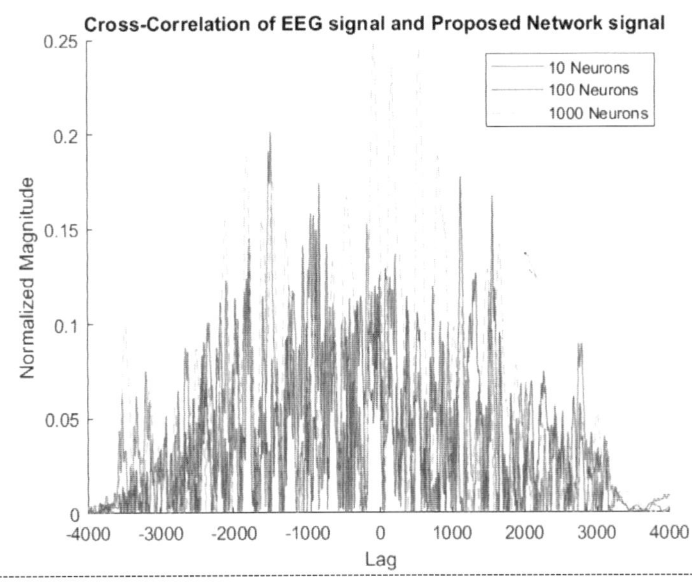

Figure 13. The correlation between the network's signal and the EEG signal: 10 neurons (blue line), 100 neurons (orange line) and 1000 neurons (dashed yellow line).

5. Conclusions

Above, we developed a way to describe the electrical activities of a network of neurons. Starting from an electrical model for a single neuron, we applied a statistical method to combine and express the output of an array of neurons, resulting in a biological neural network based on non-linear system with stochastic components. The non-linearity is given by the equation for the action potential exhibited by the networks' fundamental element – the neuron represented by means of the Izhikevich model. The stochastic components include the Poisson distribution for the synapse's conductance and the Gamma distribution for the initialization of the network's inter-connections.

Being a stochastic model, its results vary from one run to the next one. However, there was noted an increasing similarity to the actual neural activity presented by an electroencephalogram, when the number of neurons within the network was increased, thus improving the model accuracy.

A further development of the model will determine its validity in performing different mental tasks or in simulating different pathologies [16]. Also, the simulated network electrical activity will be analyzed in terms of identifying the signal sources.

References

[1] Malmivuo J and Plonsey R 1995 Bioelectromagnetism. Principles and Applications of Bioelectric and Biomagnetic Fields (New York: Oxford University Press)

[2] Chen H, Yuan B, Baxter D and Byrne J 2005 Study on Simulator for Biological Neural Networks *2005 International Conference on Neural Networks and Brain* pp 1318-1321 doi: 10.1109/ICNNB.2005.1614875

[3] Capone C, Di Volo M, Romagnoni A, Mattia M and Destexhe A 2019 State-dependent mean-field formalism to model different activity states in conductance-based networks of spiking neurons *Phys. Rev. E* **100**(6) pp 1-10 doi: 10.1103/PhysRevE.100.062413

[4] Greenwood P and Ward L 2016 Stochastic Neuron Models (Springer)

[5] Ulnicane I, Knight W, Leach T, Stahl B C and Wanjiku W G 2022 Governance of Artificial Intelligence *The Global Politics of Artificial Intelligence* (eBook) ISBN9780429446726, Chapman and Hall/CRC pp 29-55

[6] Hagen E, Magnusson S H, T V Ness, G Halnes, P N Babu, C Linssen, A Morrison and G T Einevoll 2022 Brain Signal Predictions from Multi-Scale Networks using a Linearized Framework bioRxiv pp 1-41

[7] Wilkinson D 2009 Stochastic modelling for quantitative description of heterogeneous biological systems *Nature Reviews Genetics* **10** pp 122-133 doi: 10.1038/nrg2509,

[8] Muthukrishnan V 2020 Action Potential and Resting Potential *Electrical 4 u*

[9] Shen Y, Wang J and Navlakha S 2021 A Correspondence Between Normalization Strategies in Artificial and Biological Neural Networks *Neural Computation* **33**(12) pp 3179-3202 doi: 10.1162/neco_a_01439

[10] Human Brain Project [Online] Available from: https://www.humanbrainproject.eu/en/. [Accessed 10 03 2022].

[11] Dayan P and Abbott L 2005 Theoretical Neuroscience. Computational and Mathematical Modeling of Neural Systems (Cambridge, Massachusetts: The MIT Press)

[12] Bowman C and Baglioni A 1984 Application of the Goldman-Hodgkin-Katz current equation to membrane current-voltage data *Journal of Theoretical Biology* **108**(1) pp 1-29 doi: 10.1016/S0022-5193(84)80165-4

[13] Izhikevich E M 2003 Simple Model of Spiking Neurons *IEEE Transactions on Neural Networks* **14** pp. 1569-1572 doi:10.1109/tnn.2003.820440

[14] Yang Z, Gandhi V, Karamanoglu M and Graham B 2015 Characterising information correlation in a stochastic Izhikevich neuron 2015 *International Joint Conference on Neural Networks (IJCNN)* pp 1-5 doi: 10.1109/IJCNN.2015.7280534

[15] Wilson H and Cowan J 1972 Excitatory and Inhibitory Interactions in Localized Populations of Model Neurons *Biophysical Journal* **12** pp 1-24 doi:10.1016/s0006-3495(72)86068-5,

[16] Shusterman V and Troy W 2008 From baseline to epileptiform activity: A path to synchronized rhythmicity *Physical Review* DOI: 10.1103/PhysRevE.77.061911

[17] American Electroencephalographic Society guidelines for standard electrode position nomenclature *Journal of Clinical Neurophysiology : Official Publication of the American Electroencephalographic Society* **8**(2) pp 200-202 PMID: 2050819
[18] Zyma I, Seleznov I, Popov A, Chernykh M and Shpenkov O 2018 EEG During Mental Arithmetic Tasks *Physionet* doi.org/10.13026/C2JQ1P
[19] Goldberger A, Amaral L, Glass L, Hausdorff J, Ivanov P, Mark R, Mietus J, Moody G, Peng C and Stanley H 2000 PhysioBank, PhysioToolkit, and PhysioNet: Components of a New Research Resource for Complex Physiologic Signals *Circulation* **101** pp 215-220 doi: 10.1161/01.CIR.101.23.e215

IOP Conf. Series: Materials Science and Engineering 1254 (2022) 012026 doi:10.1088/1757-899X/1254/1/012026

Assessment of human exposure to EMF generated by 5G mobile phone base stations

M V Ursăchianu[1], C Lăzărescu[1], O Bejenaru[1] and A Sălceanu[2]

[1]Romanian "National Authority for Management and Regulation in Communications" (ANCOM), Romania
[2]"Gheorghe Asachi" Technical University of Iasi, Romania

E-mail: marius.ursachianu@ancom.ro, catalin.lazarescu@ancom.ro, ovidiu.bejenaru@ancom.ro, alexandru.salceanu@academic.tuiasi.ro

Abstract. The paper presents a preliminary survey over human exposure to radiofrequency (RF) electromagnetic field (EMFs) levels generated in the vicinity of the new 5G mobile phone base stations recently installed in different major cities in Romania in comparison with other different frequency bands already used for mobile phone base stations. Measurement of electromagnetic field levels has been performed in Iasi, one of the major cities of Romania. Measurements have been performed in different urban locations with a high density of mobile phone base stations (from national operators). The EMF exposure levels for different mobile communications technologies have been measured with a calibrated up to date NARDA SRM 3006 field strength analyser system connected to an isotropic electric field probe. The methodology described by ECC RECOMMENDATION (02)04 has been rigorously respected.

1. Introduction

Our days also mean exponential development of various applications that pollute and agglomerate the electromagnetic environment with permanent emissions (mobile communication, broadcasting services) or non-permanent ones (wireless communication systems, monitoring systems, radar and GPS location detection systems, etc.). This is the main reason for international organizations and national regulatory bodies to seriously and permanently update the topic of human exposure to electromagnetic field radiation both for controlled environments (occupational exposure) and for uncontrolled ones (general public).

Electromagnetic pollution can be classified as an environmental pollution phenomenon and it currently has one of the highest growth rates, also having the widest spread in all populated areas of the globe. The main impact that this type of pollution generates is of psychological nature, especially on the general population. This challenge implies a detailed analysis on the biological effects that this type of electromagnetic radiation could produce on living organisms, mainly on humans. In most cases, human exposure is the outcome of the accumulation of many electromagnetic sources in the same time.

In the context of a permanent dynamics and evolution of electromagnetic radiation sources, the issue of assessing the harmful effects on human health should be raised. It is necessary to avoid any over-exposure by controlling these emissions.

The 5G mobile telephony is one of the worldwide most widespread technologies, with a high rate of growing. In many "corners" the deployment of the 5G cellular networks is followed by strong controversy regarding potentially dangerous health effects. The general concern about the safety of 5G base station installations has been heightened by the ridiculous association with the waves of the Covid-19 pandemic. The only correct answer should be given by rigorously performed measurements and then credibly publicized.

In the world there are two organizations recognized by the WHO (World Health Organization) who carried out different studies on EMF biological effects over the human body: the International Commission on Non-Ionizing Radiation Protection (ICNIRP) and the International Agency on Research on Cancer (IARC). Guidelines produced by the ICNIRP are recommended for the protection of the human body against potentially harmful effects of exposure to different types of EMF.

The EU Council Recommendation 1999/519/EC [1], which is the reference document in the Europe Union, is based on the emission limits set out by the new ICNIRP's 2020 guidelines [2] who is an updated version of ICNIRP 1998 [3] guidelines, aiming to also properly consider the new mobile communication technology 5G NR.

In Romania, the exposure of the population to electromagnetic fields (also including the range of frequencies used for 5G NR technology) is regulated by the Order of the Minister of Public Health no. 1193 from 29.09.2006 [4] . Romania approved norms regarding the limitation of the exposure of the general population to electromagnetic fields from 0 Hz to 300 GHz, described by the Romanian National Ministry Order 1193/2006, being the integral transposition for 519/1999 Recommendation of the Council of the European Union.

According to the European Commission request for conducting a study on the impact on human health, in March 2020 ICNIRP recommendations on exposure to fields with a frequency between 100 kHz and 300 GHz have been updated.

Various complex studies recently carried out by OFCOM (UK) [5], ANFR (France) [6], EMF RATEL (Serbia) [7] have shown that the levels of emissions from the 5G mobile phone base are well below the limits considered safe for human health, as they are specified in Council Recommendation 519/1999 of the European Union.

In Romania, an authorized answer is given for the increased concern of population about the exposure to radiofrequency EMFs sources. The National Authority for Management and Regulation in Communications (ANCOM) monitors the EMF levels. It has been developed an autonomous EMF monitoring system consisting of a number of 150 fixed stations positioned in different cities around the country and an annual mobile measurements program, the results being implemented in an online EMF map with open access for general public [8-9].

Our paper is mainly dealing with 5G NR human exposure levels, using a mobile equipment for EMF level monitoring in the vicinity of such mobile base stations.

We here present the measurement of electromagnetic field levels performed in 2021 in Iasi. For the assessment of human exposure to RF EMF generated by cellular base stations, we have taken into consideration three central locations in the city with high density of mobile phone base stations sites, covering a wide range of mobile phone communications technology types like 2G, 3G, 4G and 5G NR. For each location we investigated the EMF exposure level from all mobile services. For the 5G mobile telephony service we have taken in consideration two situations – with or without generated data traffic.

In the final a comparison between measured fields values data and limits imposed by international organizations and national regulatory bodies has been performed. Also, a compliance check (certification of conformity with respect to simultaneous exposure to different frequency fields) was performed and a set of preliminary conclusions have been drawn.

2. Materials and Methods

According to ICNIRP Guidelines [2-3], the power density reference levels for public exposure, guidelines applicable to mobile phone frequencies bands are presented in Table 1, where f is the frequency expressed in MHz. The measurements have been performed during the peak hours of a working day. The total measurement time was, according to the standard, 6 minutes.

Table 1. ICNIRP Guidelines for public exposure (mobile phone frequencies).

Frequency range	**Power density (S)[W/m^2]**
400 – 2000 MHz	f/200
2 – 300 GHz	10

For determining the EMF levels, we used the Narda SRM 3006 spectrum analyzer. The device operates in 9 kHz to 6 GHz frequency range. Using the SRM-3006 Tools software associated with the device, we have created a special service table called **Mobile Operators**. This service table contains a set of individual frequency bands used for different telephony services between 790 MHz and 3.8 GHz that covers frequency bands of all our national mobile operators. The frequency bands range assigned for Mobile Operators service table is presented in Table 2.

Table 2. Mobile Operators service table created with Narda SRM 3006 Tools.

Index	**Service name**	**Fmin [MHz]**	**Fmax [MHz]**	**Technology**
1	MOBIL 800 DL	791	821	4G
2	MOBIL 900 DL	925	960	2G, 3G, 4G
3	MOBIL 1800 DL	1805	1880	2G, 4G
4	MOBIL 2100 DL	2110	2170	3G, 4G
5	MOBIL 2600 DL	2570	2690	4G
6	MOBIL 3500-3700	3400	3800	5G

The measurement locations of EMF exposure levels generated by the mobile phone base stations in Iasi are presented in Table 3. On a detailed Iasi city map, all three locations can be visualized in Figure 1.

Table 3. The measurement locations

Location number	**Location address**
L1	Boulevard Tudor Vladimirescu nr.111
L2	Street Ștefan cel Mare și Sfânt nr.1
L3	Street Titu Maiorescu nr.4

The EMF exposure levels generated by investigated mobile phone base stations in all chosen location have been measured using a calibrated up to date Narda SRM 3006 field strength analyzer connected to an isotropic probe, an electric field probe [10] for frequency range domain from 420 MHz to 6 GHz to cover all the frequency bands for mobile communications used for mobile phone base stations transmissions, as shown in Figure 2. The measurements were performed with a good visibility to the cellular base station, with no obstacles between measurement system and antennas.

Figure 1. Location of EMF measurement points.

Figure 2. L2 measurement point location and Narda SRM 3006 measurement system.

The measurements have been performed in accordance with ECC Recommendation(02)04 and ICNIRP guidelines, methods for measuring non-ionizing electromagnetic radiation, by integrating signals into specified frequency bands over a 6 (six) minutes time frame (averaged value), in the 100 kHz - 10 GHz frequency range with an electrical probe raised at a ground height of 1,5 m [11].

According to the manufacturer's specifications, the measurement uncertainty of the level measured by the Narda SRM 3006 is + 2.7 / -3.8 dB.

Using an isotropic probe denotes the fact that the measurement results are not affected by the direction of the arrived signal. The result is not also affected by the polarization of the measured field.

The value of the total resultant electric field measured in one location is given by the equation (1):

$$E_{tot} = \sqrt{E_x^2 + E_y^2 + E_z^2}\,, \tag{1}$$

where E_x, E_y and E_z are the electric field components measured on the perpendicular axes of the 3D coordinate system.

After we select the Mobile Operators service table created, we set the Narda SRM 3006 spectrum analyzer connected to the electric field probe to work in spectrum analyzer mode. In this work mode we want to take a short and rapid view over the presence or not of mobile telephony services in locations where the EMF levels have been measured. A synthetic and suggestive image regarding the presence of all mobile telephony services in the spectrum in L2 location is represented by Figure 3. The picture is a screenshot acquired by Narda software.

It is good to know that there is a possibility that not all frequency bands and technologies to be used in a single location (from all three locations investigated in this paper).

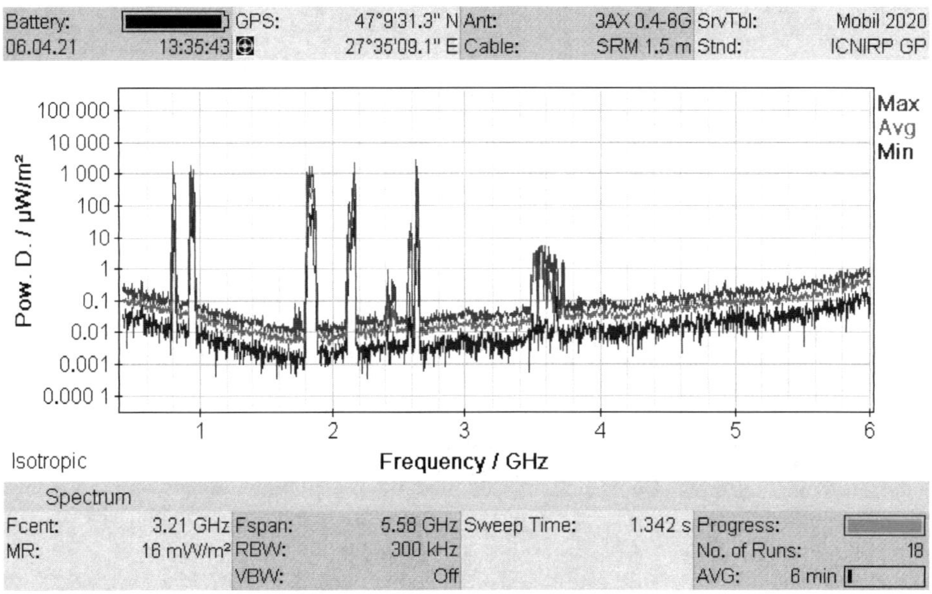

Figure 3. Spectral distribution of mobile telephony services in L2 location.

After we do this short view of presence of telephony services in spectrum, we set the Narda SRM 3006 to work in Safety Evaluation mode.

The Safety Evaluation mode of Narda SRM 3006 device has been especially created for the estimation of an overall exposure in comparison to the permitted limit values imposed by standards and guidelines. It makes possible to do an automatic measurement and evaluation of the field strength exposure in a multi-frequency sources environment. It implements the concept for the measurement of an **instantaneous and maximal exposure at a specific moment of time** for different frequency bands (called services) that can be created and assigned as tables and setups of routines [12].

For each service created, the value of field strength is calculated as a quadratic sum, the device displaying a RMS value. Also, a direct conversion of units can be done, so the final results can also be displayed in terms of power density S[W/m^2] of the plane wave equivalent.

In the Safety Evaluation mode, the RBW (Resolution Bandwidth) can be set to have different values, from a specific value set by user to an individual value set by user when the service table was created. Also, a specific value can be automatically set for RBW by the device.

Aiming to avoid underestimation of the emission, it is important that the RBW not to be set lower than the bandwidth of the signal to be measured. For example, for GSM service, GSM frequency spacing / signal channel width is 200 kHz, this value can be set for RBW for this service [13].

For the study presented in this paper, in the Safety Evaluation mode of Narda SRM 3006, the resolution bandwidth was set manually to a value of 200 kHz for all telephony services created.

The sensitivity of a device in signal detection depends on the input attenuator setting. In Narda SRM 3006 spectrum analyzer this setting is determined by the Measurement Range (MR) parameter, [10]. In case of measuring signals with different power from mobile telephone base station it is important to not overload the device. Also, by choosing a correct MR, the possibility of having false measured results is prevented.

Furthermore, the user can set the device to automatically search the MR before the measurement start in each location. For a safety margin the type of MR search can be done in mode conservative. In that mode of search, the measurement range will be selected so that no saturation will occurs, even in environment with extremely variable signals strength like a multi-frequency environment [10].

For our study, we have defined specific frequency bands for field level strength investigation, thus eliminating other signals that could disrupt the system. In that case we can set a manual value for MR.

After some preliminary tests done in L1 – L3 locations, we conclude that that the value of 16 mW/m^2 for MR is satisfactory.

In each investigated location, a 5G cellular base stations has been present in the area. Because we assumed that the presence of 5G terminals with 5G services active in the area will be very low, the appearance of a traffic beam in the measurement point for 5G telephony service was forced. With a 5G terminal placed next to the measurement device we simulate a downlink, (single user exposure scenario).

To do that we generated a data traffic by using a mobile application called Netograf installed on mobile phone with a 5G active service provided by one of the national operators, as it shown in Figure 4. The Netograf is an application developed internally by ANCOM, [14] both as a desktop or mobile application.

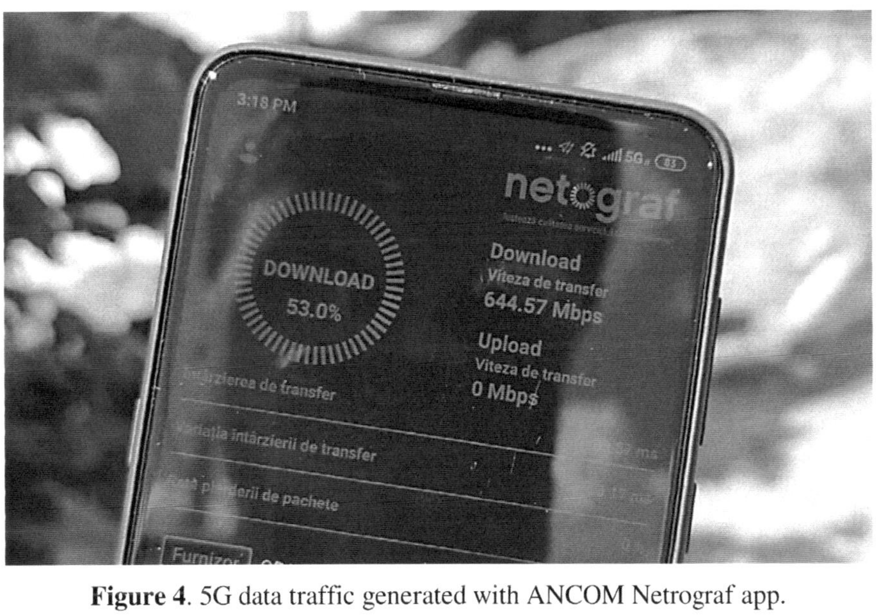

Figure 4. 5G data traffic generated with ANCOM Netrograf app.

For each location where the assessment of human exposure to mobile phone base stations was performed, we measured the instantaneous and maximal possible exposure at a specific time, the moment of time when measurements have been done.

Regarding the types of results that will be displayed, we set the device to display the Maximum (Max) and the Average (Avg) values for each investigated services from the created table. Also, for each service the standard limit value over frequency is set to be displayed.

Avg values refer to the fact that averaging is in term of power. The linear average of power will be automatically calculated by the device and the RMS value will be calculated and displayed for power density of plane wave equivalent. All measurements were performed in the far field region of the source field at an approximate 175 – 200 m distance from mobile phone base station.

For the assessment of human exposure, to display the Avg values will make it possible to determine for example the 6-minute average emission according to ICNIRP guideline.

The Max value is determined by the device in accordance with sweeping method present in EN 50492 standard [15].

With regard to the certification of conformity, an exposure coefficient $(EC)$was calculated. The exposure coefficient can be expressed in terms of power density(S), electric field (E^2) or magnetic field (H^2).

In the case of multi-frequency exposure scenario, a total exposure coefficient $EC_T$was determined according to equation (2).

$$EC_T = \sum_{i=1}^{N} \frac{S_i^{meas}}{S_i^{ref}} = \frac{S_1^{meas}}{S_1^{ref}} + \frac{S_2^{meas}}{S_2^{ref}} + \cdots + \frac{S_n^{meas}}{S_n^{ref}} \qquad (2)$$

where $S_1^{meas}, S_2^{meas}, \ldots\ldots S_n^{meas}$ represents the power density value of the emission measured for frequency $f_1, f_2, \ldots. f_n$while $S_1^{ref}, S_2^{ref}, \ldots\ldots S_n^{ref}$ represents the power density limit values for considered frequencies.

Compliance with the regulations in force shall be demonstrated only if the value of the exposure coefficient is subunit or at most equal to unity.

The exposure coefficient can provide useful information about the exposure caused by multiple types of RF signals (in a wide frequency range) in a multi-frequency exposure scenario, in each investigated location.

According to some studies, it is sufficient to determine only the coefficients corresponding to the ten strongest signals, their contribution to the total exposure being dominant [16].

3. Results and Discussion

The instantaneous and maximal possible exposure in term of field values, expressed as power density (S) (*maximum values and averaged values over a 6-minute time interval*) are presented in Table 4 for L1 – L3 locations.

The values of interest for the assessment of human exposure to electromagnetic fields generated by mobile base stations are S Avg which must then be compared with the power density limit imposed by ICNIRP recommendations, S Std.

From the results presented in Table 4 we can observe that for each location in which field strength levels were measured, the S Avg values are very well below with those limit values imposed by ICNIRP guidelines, especially for the case when data traffic was generated by 5G telephony services.

The highest values of S Avg was recorder in L2 location, one of the busiest areas from city of Iasi, the region being transited by many people every day. This area could be marked like a hot spot region of the city of Iasi, being a representative area in town for mobile phone data traffic.

Graphically, a comparison between S Avg values for all telephony services identified in L2 location, with or without 5G data traffic, is presented in Figure 5.

Table 4. Electromagnetic field power density measured values in L1 – L3 locations.

Location of EMF measurement points	Fmin [Mhz]	Fmax [MHz]	Service name	S Max [W/m^2]	S Avg [W/m^2]	S Std [W/m^2]
L1(no 5G data traffic)	791	821	MOBIL 800 DL	0.0003953	0.0000891	3.95
	925	960	MOBIL 900 DL	0.0005941	0.0004010	4.62
	1805	1880	MOBIL 1800 DL	0.0000584	0.0000211	9.02
	2110	2170	MOBIL 2100 DL	0.0031179	0.0009870	10
	2570	2690	MOBIL 2600 DL	0.0002013	0.0001172	10
	3400	3800	MOBIL 3500-3700	0.0003180	0.0000981	10
L1(with forced5G traffic beam)	3400	3800	MOBIL 3500-3700	0.0198600	0.0025710	10
L2(no 5G data traffic)	791	821	MOBIL 800 DL	0.0014840	0.0007437	3.95
	925	960	MOBIL 900 DL	0.0031660	0.0026000	4.62
	1805	1880	MOBIL 1800 DL	0.0093880	0.0029000	9.02
	2110	2170	MOBIL 2100 DL	0.0036170	0.0015900	10
	2570	2690	MOBIL 2600 DL	0.0024850	0.0008950	10
	3400	3800	MOBIL 3500-3700	0.0003668	0.0001132	10
L2(with forced5G traffic beam)	3400	3800	MOBIL 3500-3700	0.0244900	0.0043140	10
L3(no 5G data traffic)	791	821	MOBIL 800 DL	0.0021490	0.0008201	3.95
	925	960	MOBIL 900 DL	0.0004117	0.0001140	4.62
	1805	1880	MOBIL 1800 DL	0.0007210	0.0005250	9.02
	2110	2170	MOBIL 2100 DL	0.0048520	0.0017540	10
	2570	2690	MOBIL 2600 DL	0.0002130	0.0001150	10
	3400	3800	MOBIL 3500-3700	0.0002100	0.0000980	10
L3(with forced5G traffic beam)	3400	3800	MOBIL 3500-3700	0.0161100	0.0037450	10

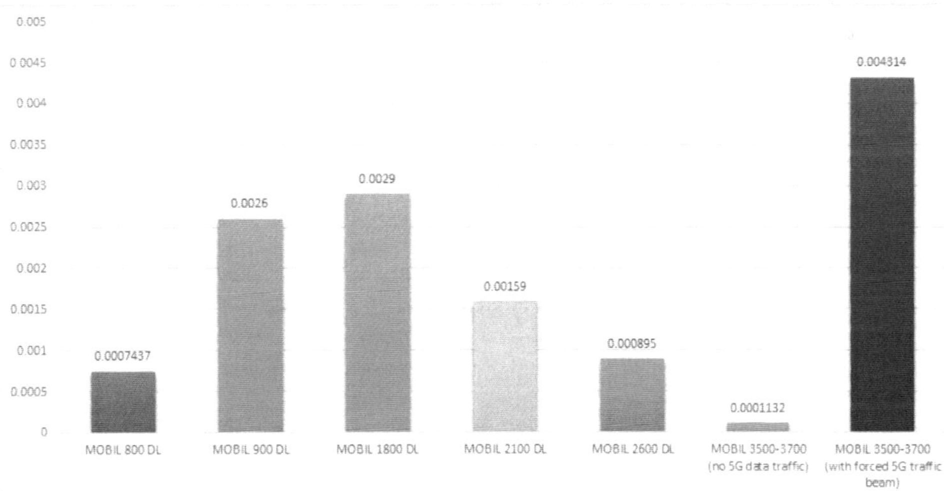

Figure 5. Quantitative and comparative S Avg values, for the identified mobile services in location L2

In the case of conformity assessment for the study here presented, the total exposure coefficient (EC_T) for each location investigated in this paper was calculated according to equation (2) and has the values shown in Table 5. The analysis of the data presented in Table 5 shows that the values of the total exposure coefficient for all investigated mobile communications services are much lower than the unit, checking the compliance with the regulations in force.

Table 5. Values of total exposure quotient for all investigated telephony services.

EMF Measurement Location Points	L1		L2		L3	
	no5G data traffic	**with** forced 5G traffic beam	**no**5G data traffic	**with** forced 5G traffic beam	**no**5G data traffic	**with** forced 5G traffic beam
EC_T	0.000294728	0.000822118	0.000273666	0.00128722	0.001004618	0.001712817

The analysis of the data presented in Table 5 shows that the values of the total exposure coefficient for all investigated mobile communications services are much lower than the unit, checking the compliance with the regulations in force.

4. Conclusions

This paper provides an overview on the assessment of human exposure to RF EMF generated by different types of cellular base stations in three central locations of Iasi-city, Romania. The investigated locations are public, with a high density of cellular base stations sites, covering a wide range of mobile phone communications technology, from 2G to the new 5G NR.

The assessment was realized from the point of view of the instantaneous and maximal possible exposure in term of field values, expressed in maximum and averaged values of power density (S) over a 6-minute measurement time interval.

For each measurement point investigated, the averaged values of power density were very well below with those limit values imposed by international organizations and national regulatory bodies, even for a 5G single user exposure scenario when traffic beam was forced to be generated.

For every investigated location, we have finally noticed that the measurements values present some variations between the exposure levels.

From the point of view of conformity assessment for the case study presented in this paper, the total exposure coefficient EC_T is much smaller than the unit, resulting the compliance with the regulations in force.

From what we know so far, this is the first public scientific study over the assessment of human exposure to EMF fields generated by the 5G cellular base station in Romania, even the traffic beam was forced to be generated in a so called 5G single user exposure scenario. Obviously, the research needs to be extended in a 5G multi-exposure scenario, aiming to be more relevant.

The measurement method proposed in this study can be used successfully when the instantaneous field strength values (averaged values over a 6-minutes period of time) would be enough for an assessment of human exposure to electromagnetic fields generated by various telephony services at a random moment. If the measured averaged field values are well below those impose by limits, the maximum field strength values can be used for a better accurate assessment of the reality of exposure.

From the point of view of limit values for the reference levels, in European Union, twenty states (including Romania) are aligned with the limits imposed by ICNIRP (10 W/m^2 for mobile communications in the frequency range 2 - 300 GHz), while the other eight countries impose much more restrictive limits over the power density of the electromagnetic field value in the same frequency domain. The most restrictive value, 0.1 W/m^2 for power density of plane wave equivalent, is imposed in countries such as Poland, Bulgaria, Lithuania and Italy, which could raise concerns about the development of future 5G infrastructure for frequencies above 2 GHz, unless a relaxation for this limit value will be done.

Without exception, for an accurate and very precise assessment of human exposure to various telephony services, the measured values of the electric field strength must be extrapolated to the maximum load of the cellular network.

Each mobile phone technology has its own extrapolation method. A standardized method of extrapolation for 5G technology is still waiting to be adopted worldwide.

In 2021 Narda developed their own method based on a code selective measurement for evaluating the exposure in case of emissions from a 5G mobile phone base station. They already implement their method in Narda SRM 3006 device as an extra option, suggestively called 5G NR.

Acknowledgment
This paper has been developed in the framework of the collaboration agreement settled between "Gheorghe Asachi" Technical University of Iasi, Faculty of Electrical Engineering and ANCOM, Romania.

References

[1] EU Council Recommendation 1999/519/EC Available from: https://op.europa.eu/en/publication-detail/-/publication/9509b04f-1df0-4221-bfa2-c7af77975556/language-en
[2] International Commission on Non-Ionizing Radiation Protection 2020 Guidelines for Limiting Exposure to Electromagnetic Fields (100 kHz to 300 GHz) *Health Physics* **118**(5) pp 483-524
[3] International Commission on Non-Ionizing Radiation Protection 1998 Guidelines for Limiting Exposure to Time-Varying Electric, Magnetic, and Electromagnetic Fields (up to 300 GHz) *Health Physics* **74** pp 494-521
[4] Ministry Order 1193/2006 Available from: https://www.ancom.ro/uploads/links_files/Odinul_1193_2006_norme.pdf
[5] Ofcom 2020 *Electromagnetic Field (EMF) measurements near 5G mobile phone base stations, Technical Report*, 21 February
[6] Djuric N, Kavecan P, Radosavljevic N and Djuric S 2020 The Wideband Approach of 5G EMF Monitoring *Proc. 12th EAI Int. Conf. AFRICOMM (Ebène City, Mauritius)* pp 86-98
[7] ANFR 2021 Study of the 5G contribution to exposure of general public to electromagnetic waves, Preliminary report (October 2020 to October 2021)
[8] Ursăchianu M V, Bejenaru O, Lăzărescu C, Sălceanu A and Păuleț M 2021 The Assessment of Human Exposure in Iasi-City using Data Provided by The National Autonomous RF-EMF Monitoring System throughout 2020 *Proc. Int. Conf. SIELMEN on Electromechanical and Energy Systems (Iasi, Romania)* pp 225-230
[9] https://monitor-emf.ro/map
[10] https://www.narda-sts.com/en/selective-emf/srm-3006-field-strength-analyzer/
[11] ECC RECOMMENDATION (02)04, Measuring non-ionising electromagnetic radiation (9 kHz – 300 GHz (revised Bratislava 2003, Helsinki 2007)
[12] Wuschek M 2019 *Measuring RF Electromagneti Fieldds at Mobile Communications Base Station and Broadcast Transmitter Sites* (Germany, Narda Safety Test Solution GmbH)
[13] IEC 62232:2017 Determination of RF field strength, power density and SAR in the vicinity of radiocommunication base stations for the purpose of evaluating human exposure
[14] https://www.netograf.ro
[15] CENELEC - EN 50492, Basic standard for the in-situ measurement of electromagnetic field strength related to human exposure in the vicinity of base stations
[16] Luncă E 2015 *Sisteme pentru măsurarea și monitorizarea poluării electromagnetice – Studii teoretice și experimentale* (Iasi: PIM)

Investigation of a radio-frequency temperature transducer with a thermosensitive resistive element based on a complex compound of heterometallic β-diketonate

O V Osadchuk, V S Osadchuk, I O Osadchuk, A O Semenov, V V Martyniuk and M O Prytula

Faculty of Information Electronic Systems, Vinnytsia National Technical University, Khmelnytsky highway 95, Vinnytsya, 21021, Ukraine

E-mail: semenov.a.o@vntu.edu.ua

Abstract. The article considers a new electrical circuit of a microelectronicradio-frequency measuring temperature transducer with a thermosensitive resistive element based on a complex compound of heterometallicβ-diketonate.The main characteristics of the researched radio-frequency temperature transducer with a thermosensitive resistive element based on a complex compound of heterometallic β-diketonate are obtained: the dependences of the active and reactive components of the full impedance of the radio-frequency measuring temperature transducer, the conversion function and the sensitivity equation.

1. Introduction

In recent years, wireless sensor networks (WSNs) and remote sensor systems have been used as part of several important areas, such as social services, military, critical infrastructure monitoring, environmental monitoring, and information collection and processing systems [1-2]. The availability of Internet connections and low production costs have led to a boom in smart facilities, which consist of a CPU, memory and wireless connection. These smart devices consist of sensors that process and transmit data [3].

Leading specialists of the world are developing and creating transducers of physical quantities. The values of physical quantities (such as temperature, humidity, density and others) of a gas mixture (air) significantly affect the well-being of a person and the quality of technological processes in microelectronics and various industries [4-7]. In our work, the emphasis is on temperature.

In the development of new functional materials, an important role belongs to complex compounds in which β-diketonate is a chelating and in some cases a bridging ligand. Nowadays, the field of practical using of functional materials containing β-diketonates of metals is constantly expanding [8]. The hetero-metallic β-diketonates are one of the most interesting complex compounds in this class, because they have semiconductor properties [9].

A promising scientific direction in this field is the creation of frequency measuring devices based on semiconductor structures with negative differential resistance. A great contribution to the development of the theory of negative differential resistance has been made by Ukrainian and foreign scientists [10-13]. The using of the frequency principle of conversion "temperature-frequency" in combination with the technology of microelectronics contributes to the improvement of the metrological and economic performance of temperature transducers, manufactured in the form of integrated circuits [14-15].

Thus, the development and practical using of micro-and nano-electronic transducers of physical quantities based on new nanocomposite materials is an actual task.

2. Theoretical and experimental research

2.1. The research of circuit design of the radiomeasuring frequency transducer of the temperature

The main accent of this work is studying and developing a radio-frequency measuring temperature transducer with frequency output based on a transistor structure with negative differential resistance. This transducer is shown in Figure 1.

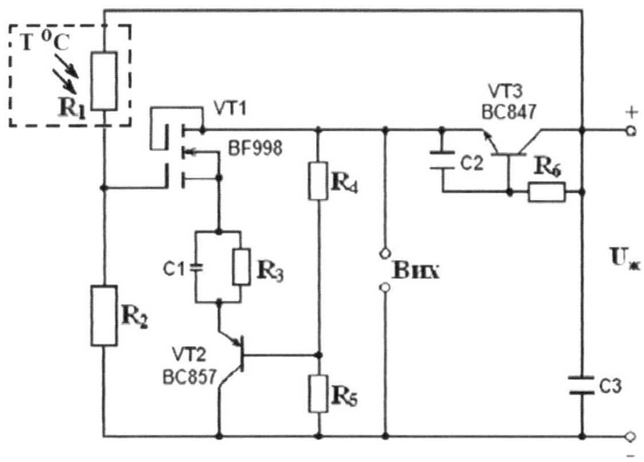

Figure 1. The electric circuit of the radio-frequency measuring temperature transducer.

To determine the behavior of the temperature transducer in the dynamic mode, it is necessary to obtain the dependence of the active and reactive components of the full impedance on the electrodes of the drain-collector of transistor structure. Also it is necessary to obtain the conversion function and sensitivity equation which depend on the temperature. Calculations were based on the nonlinear equivalent circuit of dual-gate MOSFET and bipolar transistors.

2.2. The research of thermosensitive element of the radio-frequency temperature transducer

As a sensitive element, we used the sensitive resistive element produced on a complex compound - tetrakis-μ3-(methoxo) (methanol)-pentaxis (acetylacetonate) (threeicuprum (II), neodymium (III)) methanol (I). The chemical formula of the sensing element is as follow:

$$[Cu_3Nd(AA)_5(OCH_3)_4CH_3OH] \cdot CH_3OH, \text{ whereHAA} = H_3C–C(O)–CH_2–C(O)–CH_3.$$

This chemical structure was obtained by the following method: portions of salts of 3.62 g (15 mmol) of copper (II) nitrate threehydrate and 2.19 g (5 mmol) of neodymium (III) nitrate hexahydrate were dissolved in 130 ml of absolute methyl alcohol containing 40 ml of orthomup ether. Then kept it for 2 hours at room temperature in a tightly closed conical flask. Then 2.7 ml (25 mmol) of acetylacetone was added to the reaction mixture. The conical flask was closed with a reversible water cooler and placed on a heated magnetic stirrer. Then, with continuous stirring and heating (~ 50 °C), piperidine was gradually introduced into the reaction mixture to pH = 8 and continued to heat and stir for one hour. After cooling, a homogeneous blue fine crystalline precipitate precipitated, which was filtered on a glass filter, washed with a small amount of absolute methanol, diethyl ether and dried in a vacuum desiccator over silica gel [16]. The practical yield is 4.07 g, which is 80% of theoretical. The isolated complex compound (I), which is a crystalline powder that is poorly soluble in dimethylformamide, preferably in dimethyl sulfoxide, is practically insoluble in alcohols, chloroform, benzene, acetone [16]. For the synthesized complex compound (I) in the dry powder state, elemental analysis was performed and it was found that it contains:

Cu – 18.65%; Nd – 14.25%; C – 36.68%; H – 5.24%.

As can be seen from the above, the ratio of metals in compound (I) Cu:Nd = 1:3, and its composition corresponds to the following formula: $Cu_3NdO_{16}C_{31}H_{55}$. In addition, magnetochemical, IR spectroscopic and thermogravimetric studies were performed for this isolated heterometallic compound (I) [16]. The analysis of the obtained experimental data of physicochemical research methods allows us to propose the following scheme of arrangement of chemical bonds for complex compound (I) as is shown in Figure 2 [16].

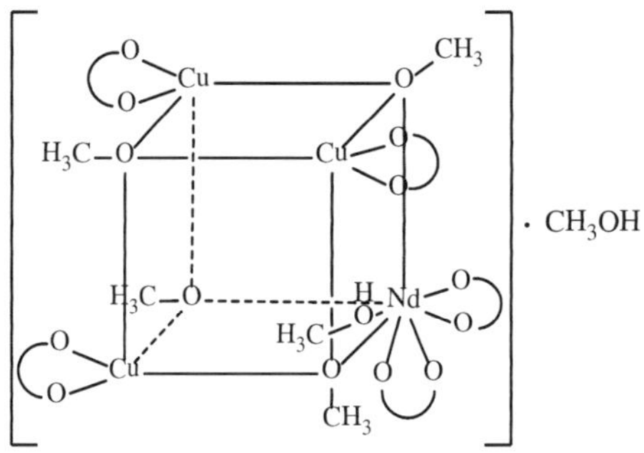

Figure 2. Schematic arrangement of chemical bonds in tetrakis-μ3-(methoxo) (methanol)-pentaxis (acetylacetonate) (threecuprum (II), neodymium (III)) methanol (I).

This sensitive element was used in our radio-frequency temperature transducer.

2.3. The research of the characteristics of the radio-frequency measuring temperature transducer

The active component of full impedance at the electrodes of the drain-collector of transistors VT1 and VT2 has negative value. The reactive component is capacitive. The oscillatory circuit of the device is based on the equivalent capacitance and the active inductance based on the phase-shifting transistor VT3 on R6C2. The equivalent capacitance consists of the capacitance between the drain electrodes of the field-effect transistor VT1 and the emitter of the bipolar transistor VT2[17-18]. Static volt-ampere characteristic is shown in Figure 3(a). Dynamic volt-ampere characteristic is shown in Figure 3(b).

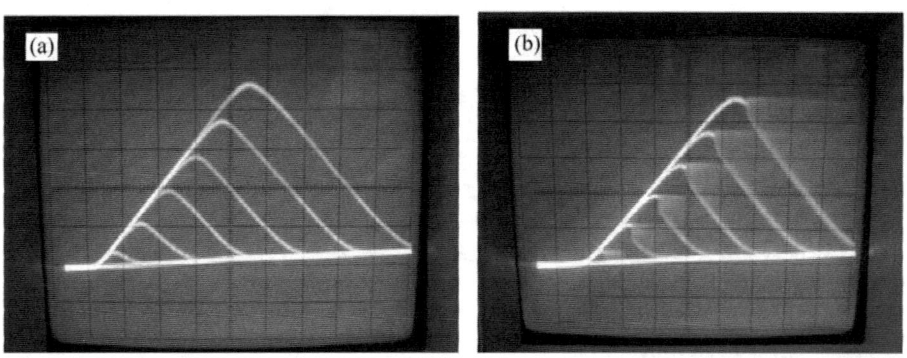

Figure 3. Static (a) and dynamic (b) volt-ampere characteristics of the radio-frequency measuring transducer of the temperature (on vertical axis one division – 2mA, on horizontal axis– 2V).

To determine the conversion function and sensitivity equation, a nonlinear equivalent circuit of the radio-frequency measuring transducer of the temperature is presented in Figure 4. This circuit realizes the dependence of the frequency on the temperature change [19-20].

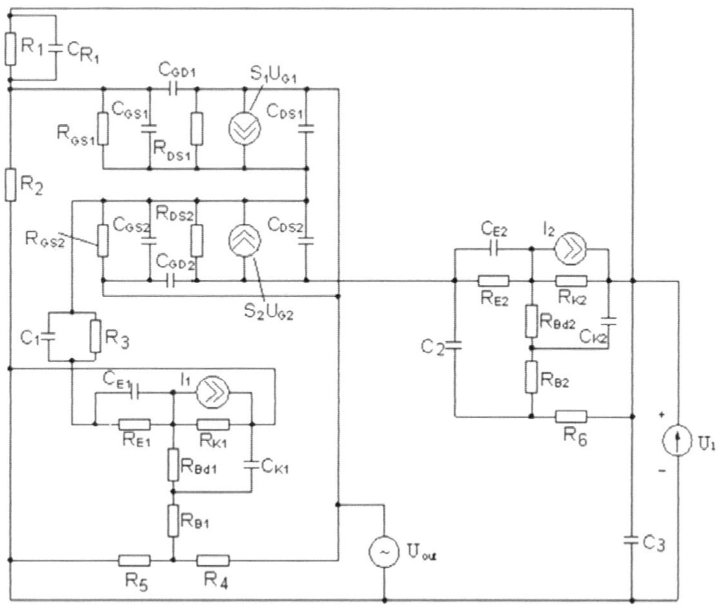

Figure 4. Nonlinear equivalent circuit of the radio-frequency measuring transducer of the temperature.

This circuit takes into account the nonlinear properties of the circuit solution of the radio-frequency measuring temperature transducer based on a transistor structure with negative differential resistance, because the autogenerator of transducer can operate in both linear and nonlinear modes. For the convenience of mathematical modelling, the nonlinear equivalent circuit is transformed into the form is shown in Figure 5.

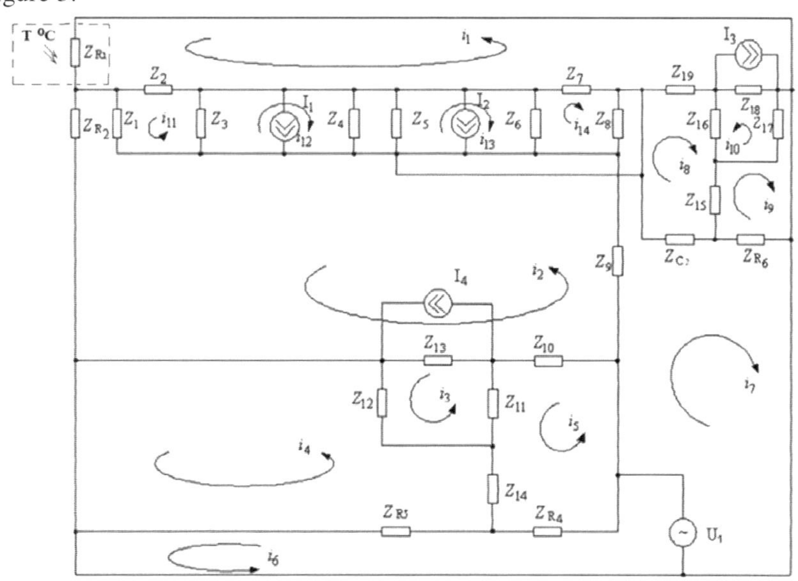

Figure 5. Converted nonlinear equivalent circuit of the radio-frequency measuring transducer of the temperature.

To determine the conversion function of the radio-frequency measuring transducer of the temperature, it is necessary to find the dependence of the generation frequency on the temperature change. This can be done by solving a system of equations that is compiled for alternating current based on a nonlinear equivalent circuit which is shown in Figure 5. Solving the system of equations allows to obtain the value of the total resistance at the electrodes of the collector-collector of the transducer. By decomposing the total resistance into real and imaginary components, it is possible to determine the equivalent capacitance of the oscillatory circuit, which depends on the temperature. The system of equations, according to the directions of the contour currents, has the form

$$\begin{cases}
0=\left(Z_{R_1}+Z_2+Z_7+Z_{19}+Z_{18}\right)i_1+Z_{19}i_8-Z_{18}\left(I_3+i_{10}\right)-Z_2i_{11}+Z_7i_{14};\\
0=\left(Z_{R_2}+Z_1+Z_9+Z_{10}+Z_{13}\right)i_2-Z_{13}i_3-Z_{10}i_5+Z_9i_7-Z_1i_{11};\\
0=\left(Z_{11}+Z_{12}+Z_{13}\right)i_3-Z_{13}\left(I_4+i_2\right)-Z_{12}i_4-Z_{11}i_5;\\
0=\left(Z_{12}+Z_{14}+Z_{R5}\right)i_4-Z_{12}i_3-Z_{14}i_5-Z_{R5}i_6;\\
0=\left(Z_{10}+Z_{11}+Z_{14}+Z_{R4}\right)i_5-Z_{10}i_2-Z_{11}i_3-Z_{14}i_4-Z_{R4}i_6;\\
\dot{U}_1=\left(Z_{R4}+Z_{R5}\right)i_6-Z_{R5}i_4-Z_{R4}i_5;\\
\dot{U}_1=\left(Z_{C_2}+Z_{R6}+Z_8+Z_9\right)i_7+Z_9i_2-Z_{C_2}i_8-Z_{R6}i_9-Z_8i_{14};\\
0=\left(Z_{C_2}+Z_{15}+Z_{16}+Z_{19}\right)i_8+Z_{19}i_2-Z_{C_2}i_7-Z_{15}i_9+Z_{16}i_{10};\\
0=\left(Z_{R6}+Z_{15}+Z_{17}\right)i_9-Z_{R6}i_7-Z_{15}i_8+Z_{17}i_{10};\\
0=\left(Z_{16}+Z_{17}+Z_{18}\right)i_{10}+Z_{18}\left(I_3-i_1\right)+Z_{16}i_8+Z_{17}i_9;\\
0=\left(Z_1+Z_2+Z_3\right)i_{11}-Z_2i_1-Z_1i_2+Z_3(i_{12}+I_1);\\
0=\left(Z_3+Z_4Z_5/(Z_4+Z_5)\right)i_{12}+Z_3i_{13}-(Z_4Z_5/(Z_4+Z_5))i_{13}+Z_3\left(I_1+i_{12}\right)-(Z_4Z_5/(Z_4+Z_5))\left(i_{12}+I_2\right);\\
0=\left(Z_4Z_5/(Z_4+Z_5)+Z_6\right)i_{13}-Z_4Z_5/(Z_4+Z_5)i_{13}-Z_6i_{14}+Z_4Z_5/(Z_4+Z_5)\left(i_{13}+I_2\right);\\
0=\left(Z_6+Z_7+Z_8\right)i_{14}-Z_6(i_{13}+I_2)+Z_7i_1-Z_8i_7,
\end{cases} \tag{1}$$

where

$$Z_1=\frac{R_{GS_1}}{1+\omega^2R_{GS_1}^2C_{GS_1}^2}-j\frac{R_{GS_1}^2\omega C_{GS_1}}{1+\omega R_{GS_1}^2C_{GS_1}^2};\ Z_2=-\frac{j}{\omega C_{GD_1}};\ Z_{R2}=R_2;\ Z_3=R_{DS_1};\ Z_{R5}=R_5;$$

$$Z_4=-\frac{j}{\omega C_{DS_1}};\ Z_5=-\frac{j}{\omega C_{DS_2}};\ Z_6=R_{DS_2};\ Z_7=-\frac{j}{\omega C_{GD_2}};\ Z_{R4}=R_4;\ Z_{R6}=R_6;\ Z_{R3}=R_3;$$

$$Z_8=\frac{R_{GS2}}{1+\omega^2R_{GS2}^2C_{GS2}^2}-j\frac{R_{GS2}^2\omega C_{GS2}}{1+\omega R_{GS2}^2C_{GS2}^2};\ Z_9=\frac{R_3}{1+\omega^2R_3^2C_1^2}-j\frac{R_3^2\omega C_1}{1+\omega R_3^2C_1^2};\ Z_{C2}=-\frac{j}{\omega C_2};$$

$$Z_{10}=\frac{R_{E1}}{1+\omega^2R_{E1}^2C_{E1}^2}-j\frac{R_{E1}^2\omega C_{E1}}{1+\omega R_{E1}^2C_{E1}^2};\ Z_{11}=R_{Bd_1};\ Z_{12}=-\frac{j}{\omega C_{K1}};\ Z_{13}=R_{K1};$$

$$Z_{14}=R_{B_1};\ Z_{15}=R_{B_2};\ Z_{16}=R_{Bd_2};\ Z_{17}=-\frac{j}{\omega C_{K2}};\ Z_{18}=R_{K2};$$

$$Z_{19}=\frac{R_{E2}}{1+\omega^2R_{E2}^2C_{E2}^2}-j\frac{R_{E2}^2\omega C_{E2}}{1+\omega R_{E2}^2C_{E2}^2};\ Z_{R_1}=\frac{R_1(T)}{1+\omega^2R_1^2(T)C_{R_1}^2(T)}-j\frac{\omega R_1^2(T)C_{R_1}(T)}{1+\omega R_1^2(T)C_{R_1}^2(T)}.$$

Elements of the nonlinear equivalent circuit are described by the following values: R_{GS}, C_{GS} are the resistance and capacitance between the gate and source areas of the MOSFET; R_G is the ohmic resistance of the MOSFET gate; R_{DS}, C_{DS} are the drain-source resistance and capacitance of MOSFET; R_S is the resistance of source p-n junction of the field-effect transistor; R_D is the resistance of the p-n junction of the drain of the field-effect transistor; R_n is the resistance of the MOSFET substrate; R_C, R_E, R_B are the resistances of the collector, emitter, base junctions of the bipolar transistor; R_2, R_3, R_4, R_5 are there sistances of the divider of the radio-frequency temperature transducer; I_{DS} is the drain-source current of MOSFET; I_{BS} and I_{BD} are substrate-source and substrate-drain of MOSFET junction current; R_1 is the thermosensitive resistive element based on a complex compound - tetrakis-μ$_3$-(methoxo) (methanol) -pentakis (acetylacetonate) (threecuprum (II), neodymium (III)) methanol (I); C_K, C_E are the capacitances of the collector and emitterjunctions of the bipolar transistor; C_{GD} is the gate-drain capacitance of MOSFET; C_{R_1} is the capacitance of thermosensitive resistive element.

Currents I_{BS} and I_{BD} transitions of substrate-source and substrate-drain MOSFET in linear mode, if the condition $U_{ds} < (U_{gs} - U_T)$ is met, are described by expressions [18]

$$I_{bs} = I_{ss}(\exp U_{bs} / (NU_t) - 1), \tag{2}$$

$$I_{bd} = I_{ss}(\exp U_{bd} / (NU_t) - 1), \tag{3}$$

where U_{bs} is the substrate-source voltage MOSFET; U_{bd} is the voltage substrate-drain MOSFET; I_{ss} is the saturation current p-n junction of the field-effect transistor substrate; N is the coefficient of imperfection of the substrate-drain junction of the field-effect transistor; U_t is the temperature potential p-n junction of the substrate of the field-effect transistor.

The static output characteristic of a MOSFET transistor in linear mode is described by [21]

$$I_{ds} = \frac{\mu \cdot C_0 \cdot W}{L}\left((U_{gs} - U_T) \cdot U_{ds} - \frac{U_{ds}^2}{2}\right), \tag{4}$$

where W is thefield-effect transistor channel width; μ is the media mobility in the MOSFET channel; L is the the length of the channel of the field-effect transistor; U_{gs} is the gate-source voltage; C_0 is the specific capacity of the oxide; U_{ds} is the MOSFET drain-source voltage; U_T is the threshold voltage.

The expression for determining the threshold voltage of the MOSFET transistor for analytical models has the form [20]

$$U_T = \varphi_{SiO_2} + 2\varphi_B - \frac{Q_S + \sqrt{4\varepsilon_S q N_A \varphi_B}}{C_0}, \tag{5}$$

where ε_S is therelative electrical permeability of semiconductor material; Q_S is the specific surface charge; N_A is the concentration of impurities in the semiconductor material.

The Fermi potential included in expression (5) is described by the formula [20-21]

$$\varphi_B = \pm\frac{kT}{q}\ln\left(\frac{N_A}{n_i}\right), \tag{6}$$

where k is the Boltzmann constant; T is the temperature; n_i is the intrinsic carrier concentration in an undoped semiconductor; q is the electron charge.

The drain current of the field-effect transistor in the saturation mode when the condition $U_{ds} \geq (U_{gs} - U_T)$ is met is described by the expression [21]

$$I_{ds_{SAT}} = \frac{\mu \cdot C_0 \cdot W}{6L}((U_{ds_{SAT}} + 2\varphi_B)^2 + U_{gs}(U_{ds_{SAT}} + 2\varphi_B) - 12\varphi_B(U_{gs} - \varphi_B - \frac{4}{3}K\varphi_B^{1/2})), \quad (7)$$

where

$$U_{ds_{SAT}} = U_{gs} - 2\varphi_B + K^2[1 - (1 + 2U_{gs} / K^2)^{1/2}], \quad (8)$$

$$K = (\varepsilon_S q N_A / C_0)^{1/2}. \quad (9)$$

The drain-source resistance of the field-effect transistor R_{ds} in the linear region is determined by the expression [18]

$$R_{ds} = \frac{L}{W\mu C_0}(U_{gs} - U_T), \quad (10)$$

and in the saturation region [18]

$$R_{ds} = \frac{12\left[L(U_d - U_{ds_{SAT}})^{1/2} - 2(\varepsilon_s / qN_a)^{1/2}(U_d - U_{ds_{SAT}})\right]}{W\mu C_0(2\varepsilon_s / qN_A)^{1/2}} \times$$
$$\times \frac{1}{(U_{ds_{SAT}} + 2\varphi_B)^2 + U_{gs}(U_{ds_{SAT}} + 2\varphi_B) - 12\varphi_B(U_{gs} - \varphi_B - 4/3K\varphi_B^{1/2})}, \quad (11)$$

where U_d is the voltage on the drain of the field-effect transistor; $U_{ds_{SAT}}$ is the drain-source voltage of the field-effect transistor in saturation mode.

The solution of the system of equations (1) made it possible to obtain the theoretical dependence of the active and reactive components of full impedance, the frequency of generations on the temperature, and also the sensitivity function of the radio-frequency temperature transducer. The experimental dependences of the active and reactive components of the full impedance of the radio-frequency temperature transducer from temperature change is shown in Figure 6.

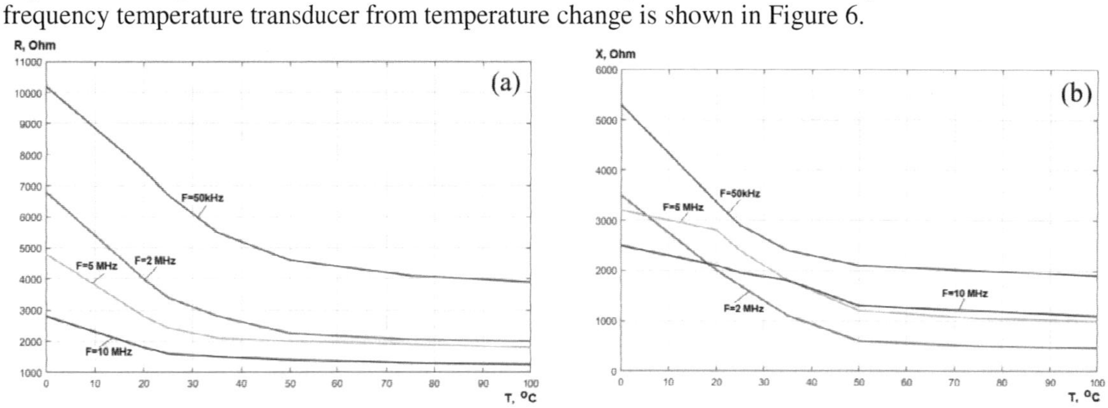

Figure 6. The experimental dependences of the active (a) and reactive (b) components of the full impedance of the radio-frequency measuring transducer of the temperature.

Base on the nonlinear equivalent circuit which is shown in Figure 5, according to Lyapunov's method, the conversion function of the radio-frequency temperature transducer is determined. The conversion function is the dependence of the output frequency of the radio-frequency measuring

IOP Conf. Series: Materials Science and Engineering 1254 (2022) 012027 doi:10.1088/1757-899X/1254/1/012027

transducer on the temperature value. The analytical dependence of the conversion function is described by the expression

$$F = \frac{1}{2\pi}\sqrt{\frac{(C_{R1}^2(T)R_{R1}^2(T) + R_{R1}^2(T)C_{R1}(T)C_{DS2} - L_{екв}C_{DS2}) + D}{2L_{екв}C_{DS2}C_{R1}^2(T)R_{R1}^2(T)}}, \quad (12)$$

where T is the temperature; C_{DS2} is the drain-source capacitance of MOSFET; C_{R1}, R_{R1} are the capacitance and resistance of the thermosensitive resistive element based on a complex compound - tetrakis-μ$_3$-(methoxo) (methanol) -pentakis (acetylacetonate) (threecuprum (II), neodymium (III)) methanol (I); $L_{екв}$ the value of the equivalent inductance based on the bipolar transistor BC847;

$$D = \sqrt{(C_{R1}^2(T)R_{R1}^2(T) + R_{R1}^2(T)C_{R1}(T)C_{DS2} - L_{екв}C_{DS2})^2 + 4L_{екв}C_{DS2}C_{R1}^2(T)R_{R1}^2(T)}.$$

The graphical dependence of the conversion function is presented in Figure 7.

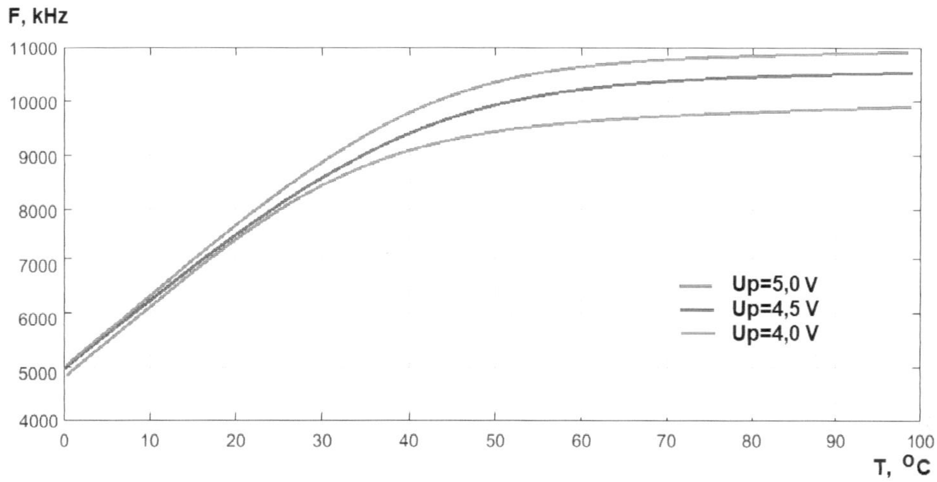

Figure 7. The dependence of the output frequency of the radio-frequency measuring transducer of the temperature.

The sensitivity of the radio-frequency temperature transducer is determined on the basis of the expression (12) and is described by equation (13). The graph of sensitivity dependence is presented in Figure 8.

IOP Conf. Series: Materials Science and Engineering 1254 (2022) 012027 doi:10.1088/1757-899X/1254/1/012027

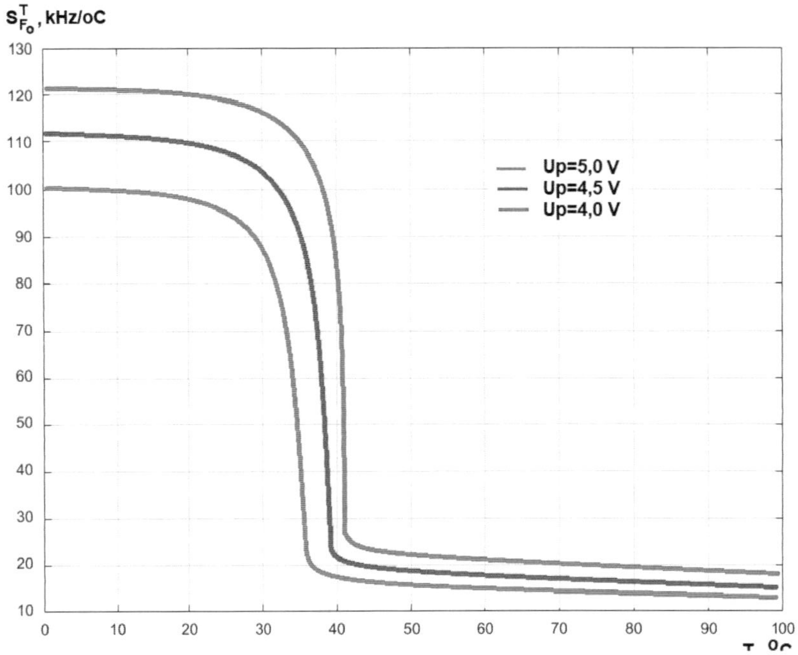

Figure 8. The dependence of the sensitivity of the radio-frequency measuring transducer of the temperature.

$$
\begin{aligned}
S_B^F = & \frac{1}{\sqrt{8}}\Bigg(\Bigg(2C_{R1}(T)R_{R1}^2(T)\left(\frac{\partial C_{R1}(T)}{\partial T}\right)+2C_{R1}^2(T)R_{R1}(T)\left(\frac{\partial R_{R1}(T)}{\partial T}\right)+2R_{R1}(T)C_{R1}(T)C_{DS2}\times \\
& \times\left(\frac{\partial R_{R1}(T)}{\partial T}\right)+R_{R1}^2(T)\left(\frac{\partial C_{R1}(T)}{\partial T}\right)C_{DS2}+\frac{1}{2}\Big(2\big(C_{R1}^2(T)R_{R1}^2(T)+C_{R1}(T)R_{R1}^2(T)C_{DS2}-L_{\text{ekв}}C_{DS2}\big)\times \\
& \times\Bigg(2C_{R1}(T)R_{R1}^2(T)\left(\frac{\partial C_{R1}(T)}{\partial T}\right)+2C_{R1}^2(T)R_{R1}(T)\left(\frac{\partial R_{R1}(T)}{\partial T}\right)+2R_{R1}(T)C_{R1}(T)C_{DS2}\left(\frac{\partial R_{R1}(T)}{\partial T}\right)+ \\
& +R_{R1}^2(T)\left(\frac{\partial C_{R1}(T)}{\partial T}\right)C_{DS2}\Bigg)+8L_{\text{ekв}}C_{DS2}C_{R1}(T)R_{R1}^2(T)\left(\frac{\partial C_{R1}(T)}{\partial T}\right)+8L_{\text{ekв}}C_{DS2}C_{R1}^2(T)\times \\
& \times R_{R1}(T)\left(\frac{\partial R_{R1}(T)}{\partial T}\right)\Bigg)\Bigg/\Big(\big(C_{R1}^2(T)R_{R1}^2(T)+C_{R1}(T)R_{R1}^2(T)C_{DS2}-L_{\text{ekв}}C_{DS2}\big)^2+4L_{\text{ekв}}C_{DS2}\times \\
& \times C_{R1}^2(T)R_{R1}^2(T))^{1/2}\Big)\Big/\Big(L_{\text{ekв}}C_{DS2}C_{R1}^2(T)R_{R1}^2(T))-2(C_{R1}^2(T)R_{R1}^2(T)+C_{DS2}C_{R1}(T)R_{R1}^2(T)- \\
& -L_{\text{ekв}}C_{DS2}+\big((C_{R1}^2(T)R_{R1}^2(T)+C_{DS2}C_{R1}(T)R_{R1}^2(T)-L_{\text{ekв}}C_{DS2}\big)^2+4L_{\text{ekв}}C_{DS2}C_{R1}^2(T)R_{R1}^2(T))^{1/2}\Big)\times \\
& \times\left(\frac{\partial C_{R1}(T)}{\partial T}\right)\Bigg/\Big(L_{\text{ekв}}C_{DS2}C_{R1}^3(T)R_{R1}^2(T))-2(C_{R1}^2(T)R_{R1}^2(T)+C_{DS2}C_{R1}(T)R_{R1}^2(T)-L_{\text{ekв}}C_{DS2}+ \\
& +\Big(C_{R1}^2(T)R_{R1}^2(T)+C_{DS2}C_{R1}(T)R_{R1}^2(T)-L_{\text{ekв}}C_{DS2}\Big)^2+4L_{\text{ekв}}C_{DS2}C_{R1}^2(T)R_{R1}^2(T)\Big)^{1/2}\Bigg)\Bigg/\Big(L_{\text{ekв}}C_{DS2}\times \\
& \times C_{R1}^2(T)R_{R1}^3(T)\Big)\Big)\Bigg/\Bigg(\Bigg(\Bigg(C_{R1}^2(T)R_{R1}^2(T)+C_{DS2}C_{R1}(T)R_{R1}^2(T)-L_{\text{ekв}}C_{DS2}+ \\
& +\Big(\Big(C_{R1}^2(T)R_{R1}^2(T)+C_{DS2}C_{R1}(T)R_{R1}^2(T)-L_{\text{ekв}}C_{DS2}\Big)^2+ \\
& +4L_{\text{ekв}}C_{DS2}C_{R1}^2(T)R_{R1}^2(T))^{1/2}\Bigg)\Bigg/\Big(L_{\text{ekв}}C_{DS2}C_{R1}^2(T)R_{R1}^2(T)\Big)^{1/2},
\end{aligned}
\tag{13}
$$

where T is the temperature; C_{DS2} is the drain-source capacitance of MOSFET; C_{R1}, R_{R1} are the capacitance and resistance of the thermosensitive resistive element based on a complex compound - tetrakis-μ₃-(methoxo) (methanol) -pentakis (acetylacetonate) (threecuprum (II), neodymium (III)) methanol (I); $L_{екв}$ the value of the equivalent inductance based on the bipolar transistor BC847.

As can be seen from the graph, the highest sensitivity of the device is in the range from 0 to 30 °C. Maximal sensitivity 121.85 kHz/°C.

3. Conclusion

The optimal for practical implementation is the frequency method of the temperature measuring based on the dependence of the reactive properties of semiconductor structures with differential resistance on the temperature. Devices based on this method have increasedsensitivity.

The circuit of the radio-frequency measuring transducer of the temperature was proposed in the article. As a thermosensitive element was proposed the sensitive resistive element produced on a complex compound - tetrakis-μ_3-(methoxo) (methanol)-pentaxis (acetylacetonate) (threecuprum (II), neodymium (III)) methanol (I). Static and dynamic volt-ampere characteristics of the radio-frequency measuring transducer of the temperature were obtained. As a result of calculating of the system of equilibrium equations of the equivalent circuit, analytical expressions of the conversion function and the sensitivity equation of the device were obtained. Graphical dependences of the active and reactive components of full impedance on temperature were also presented. It was found that the highest sensitivity of the radio-frequency temperature transducer is in the range from 0 to 30 °C and is 89.05~121.85 kHz/°C.

References

[1] Lepikh J, Romanov V, Gordienko Yu and Dzyadevich S 2011 *Intelligent measuring systems based on new generation microelectronic sensors* (Odessa: Astroprint)

[2] CzubenkoM and Kowalczuk Z 2021 Simple neural network for collision detection of collaborative robots *Sensors* **21** pp 1–20

[3] Nagaraj A 2021 Introduction to Sensors in IoT and Cloud Computing Applications (Bangalore, India)

[4] Neto M, Ribeiro P, Nunes R, Jamone L, Bernardino A and Cardoso S 2021 A Soft Tactile Sensor Based on Magnetics and Hybrid Flexible-Rigid Electronics *Sensors* **21** pp 1–25

[5] Osadchuk A, Osadchuk I and Semenov A 2021 The mathematical model of radio-measuring frequency transducer of optical radiation based on MOS transistor structures with negative differential resistance *Journal of Nano- and Electronic Physics* **13** pp 1–6

[6] Osadchuk A, Osadchuk V, Osadchuk I, Kolimoldayev M, Komada P and Mussabekov K 2017 Optical transducers with frequency output *Proceeding of SPIE* **10445** 104451X

[7] Osadchuk A, Osadchuk V, Osadchuk I, Titova N, Pinaeva O, Kisała P, Rakhmetullina S, Kalizhanova A and Azeshova Z 2020 Optical-frequency gas flow meter on the basis of transistor structures with negative differential resistance *Proceedings of SPIE* **11456** 114560F

[8] Zolotareva N and Semenov V 2013 Diketonates and their derivatives in sol-gel processes *Uspekhi khimii* **80** 964–987

[9] Krisyuk V, Urkasymkyzy S, Baidina I, Romanenko G, Korolkov I, Koretskaya T, Petrova N and Turgambaeva A 2017 Structure and thermal properties of heterometallic complexes for gas-phase deposition of CU-PD films *Journal of Structural Chemistry* **8** pp 1573–1580

[10] Krisyuk V, Tkachev S, Baidina I, Korolkov I, Turgambaeva A and Igumenov I 2015 Volatile Pd–Pb and Cu–Pb heterometallic complexes: structure, properties, and trans-to-cis isomerization under cocrystallization of Pd and Cu β-diketonates with Pb hexafluoroacetylacetonate *Journal of Coordination Chemistry* **68** pp 1890–1902

[11] Shabanova V, Storozhenko T and Zelenov V 2004 Heteronuclear complex compounds of iron(III) and neodymium(III) with hydroxy acids as starting materials for the synthesis of nanomaterials *Ecological Bulletin of Scientific Centers of the BSEC* **3** pp 91–94

[12] Thurston J, Trahan D, Ould-Ely I and Whitmire K 2004 Toward a General Strategy for the Synthesis of Heterobimetallic Coordination Complexes for Use as Precursors to Metal Oxide Materials: Synthesis, Characterization, and Thermal Decomposition of Bi2(Hsal)6·M(Acac)3 (M = Al, Co, V, Fe, Cr) *Inorg. Chem.* **43** pp 3299–3505

[13] Slyusarchuk L, Zheleznova L and Pavlenko T 2018 Synthesis of folding oxides from heteronuclear β-diketonate complexes of 3d-4f-metals *Abstracts of the report at the XX Ukrainian Conference on Inorganic Chemistry (Dnipro)* pp 74–78

[14] Osadchuk O, Martinyuk V, Evseeva M and Seletska O 2019 Magnetically sensitive sensor based on the heterometal complex half *Bulletin of Khmelnitsky National University* **3** pp 97–101

[15] Osadchuk O, Martyniuk V, Yevseeva M and Seletska O 2019 Investigation of the influence of temperature on the physical parameters of the semiconductor μ-methoxo (copper (II), bismuth (III)) acetylacetonate *Bulletin of Vinnytsia Polytechnic Institute* **4** pp 80–86

[16] Samus N, Gandziy M and Tsapkov V 1992 Heteronuclear μ-methoxo(copper, yttrium or lanthanide) acetylacetonates *Journal of General Chemistry* **62** pp 510–515

[17] Osadchuk A, Osadchuk V, Seletska O and Krylik L 2019 Microelectronic transducer of gas concentration based on MOSFET with an active inductive element *Przegląd Elektrotechniczny* **95** pp 237–241

[18] Osadchuk V, Osadchuk O, Baraban S, Zyska T and Zhanpeisova A 2018 Temperature transducer based on metal-pyroelectric-semiconductor structure with negative differential resistance *Proceedings of SPIE* **10808** 108085D

[19] Sze S M and Kwok K N 2007 *Physics of Semiconductor Devices* (Wiley-Interscience: Hoboken, USA)

[20] Vikulin I M and Stafeev V I 1990 *Physics of semiconductor devices* (Moscow: Radio and communication)

[21] SPICE 2020 *Device Models and Simulation Examples* (Oxford University Press)

IOP Conf. Series: Materials Science and Engineering 1254 (2022) 012028 doi:10.1088/1757-899X/1254/1/012028

The benefit of biomedical explorations in elderly patients with respiratory pathology and dental diseases

D Munteanu[1], L I Perțea[2], G Stătescu[3], A Vlase[4], R Ciorap[5] and M R Barbu[6]

[1]Medicine I Department, Faculty of Medicine, Grigore T. Popa University of Medicine and Pharmacy Iasi, Romania
[2]Mother and child Medicine Department, Faculty of Medicine, Grigore T. Popa University of Medicine and Pharmacy Iasi, Romania
[3]Morpho-functional Sciences I Department, Faculty of Medicine, Grigore T. Popa University of Medicine and Pharmacy Iasi, Romania
[4]Surgery Department II, Faculty of Medicine, Grigore T. Popa University of Medicine and Pharmacy Iasi, Romania
[5]Biomedical Sciences Department, Faculty of Medical Bioengineering, Grigore T. Popa University of Medicine and Pharmacy Iasi, Romania
[6]Morpho-functional Sciences II Department, Faculty of Medicine, Grigore T. Popa University of Medicine and Pharmacy Iasi, Romania
*all the authors are quality as first author

E-mail: leo_pertea@yahoo.com

Abstract. Some of the most common chronic respiratory diseases, incapacitating the development and quality of life of patients and directly correlated with oro-dental cavity are represented by asthma and chronic obstructive pulmonary disease (CPOD), a chronic inflammatory disease affecting the airways, in which mast cells, eosinophils and T lymphocytes, play an important role [1]. A better understanding of the diagnosis and treatment of these pathology became possible by accepting that the existence of chronic inflammation, with its variations, is reflected in the clinical condition of the elderly patients, with implications on the dental status. Therefore, the present research, using respiratory biomedical explorations, wants to establish a clinical and functional correlation in the case of elderly patients with respiratory diseases and who associate oro-dental pathologies. In the current conditions, in which the predominant viral damage is SARS-COV 2 infection, it is necessary to develop a clinical-functional algorithm to quickly establish a subsequent evolution and prognosis of elderly patients, which involve an increased predisposition to morbidity and mortality.

1. General considerations

Respiratory pathology is more frequently present in susceptible patients, especially elderly, in such cases infection/inflammation causing recurrent episodes of wheezing, dyspnea (shortness of air) in varying degrees, chest tightness and cough irritating, especially at night and in the morning [2]. These symptoms are associated with the variable but relatively and frequently generalized air flow, which is partially reversible, either spontaneously or under treatment.

Inflammation also causes an increased bronchial reactivity to a variety of stimuli. Exposure to these stimuli causes airway mucosal edema, bronchoconstriction, mucus plugs and airway responsiveness to various stimuli. The result of respiratory flow limitation is reversible (but not always complete), either spontaneously or as a result of treatment. A better understanding of the diagnosis and treatment of asthma was made possible by accepting that the existence of chronic inflammation, with its variations, is reflected in the clinical condition of the elderly patient [3-4].

By improving our understanding on this association with the potential of reducing COPD incidence through evaluation and treatment of the periodontal diseases, known as having important public health and clinical implications [5-6], the present study analyses the effects of the periodontal diseases and systemic biomarkers of inflammation on the occurrence of serious fatal and non-fatal respiratory-related events among subjects with COPD. Thus, chronic inflammation can be assessed and interpreted through routine biomedical examinations, which include spirometry and diffusion through the alveolo-capillary membrane, with results obtained quickly and easily interpreted for the practitioner.

2. Material and methods

To highlight the opportunity, precocity and chronic respiratory effects of oral pathology with reference to asthma and chronic obstructive pulmonary diseases (COPD), a prospective study was conducted on a group of 167 patients hospitalized in the Internal Medicine Clinic, and also on 53 patients consulted in a dental office.

Two aspects have been mainly considered, namely:

- Frequency of dental/oral cavities in patients with chronic respiratory diseases, focusing on the impact of asthma and COPD, since – potentially - all of them will be undergoing dental treatment that can pose risks [7].
- Correlation of respiratory function tests, spirometry and diffusion through the alveolo-capillary membrane, with the frequency and severity of oro-dental diseases.

Respiratory functional examinations were performed with a pre-calibrated type SPIROLAB III spirometer with standardized parameters and were used according to the manufacturer's technical instructions. We use a really complete spirometry system: • FVC, VC with breathing pattern plus MVV tests with real time curves; • over 30 parameters with automatic interpretation and test quality control; • pre/post bronchial response protocol with F/V loop comparison; • predicted values with various authors; • internal temperature sensor for automatic BTPS conversion.

Spirometry assesses the integrated mechanical function of the lung, chest wall, respiratory muscles, and airways by measuring the total volume of air exhaled from a full lung (total lung capacity [TLC]) to maximal expiration (residual volume [RV]) [8]. This volume, the forced vital capacity (FVC) and the forced expiratory volume in the first second of the forceful exhalation (FEV1), should be repeatable to within 0.15 L, upon repeat efforts in the same measurement unless the largest value for either parameter is less than 1 L. The standards used to describe the quality of spirometry measurements are from the Standardization of Spirometry 2019 Update.

Evaluation of lung gas exchange can be performed by measuring the gas transfer factor through the alveolar capillary membrane of carbon dioxide (DLCO), respectively by determining the respiratory gases in the blood arterial (ASTRUP). DLCO provides data on the integrity of the alveolo-capillary membrane and the gases on the fit of the ventilation / infusion ratio in all lung functional units. It often is helpful for evaluating the presence of possible parenchymal lung disease when spirometry and/or lung volume determinations suggest a reduced vital capacity, RV, and/or TLC.

3. Results and discussions

Distribution by sex and age of patients in both groups showed a slightly higher prevalence of males (M/W ratio = 1.3/1), but with a more uniform distribution on age groups (Figure 1 and Table 1).

Table 1. Distribution of patients by age and respiratory disease.

Age (years)	Chronic bronchitis	Emphysema	Asthma	COPD
< 60 years	2 patients	11 patients	6 patients	6 patients
60-70 years	6 patients	3 patients	17 patients	12 patients
70-79 years	15 patients	6 patients	38 patients	34 patients
> 80 years	3 patients	1 patient	9 patients	8 patients

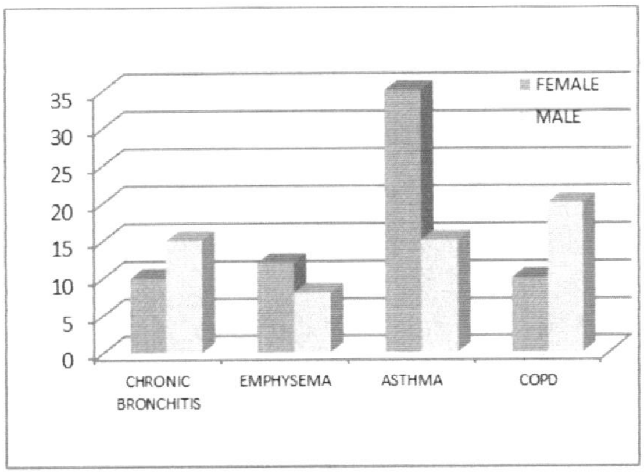

Figure 1. Distribution of patients by age, sex and disease.

Approximately 40% of the patients (67 cases) were discovered accidentally by family doctors or during examination in the ambulatory, being sent to the clinic for investigation and specialized treatment. Statistically, distribution by sex and condition of control group in outpatients with chronic respiratory diseases was relatively uniform.

Other issues investigated in the study were the risk factors for chronic respiratory diseases, first of all smoking (Figure 2), exposure to allergens, the most common being criminalized domestic and digestive, followed by atmospheric and professional conditions, as well as a relatively high ratio of respiratory viral and / or bacterial infections, gastrointestinal reflux or esophageal diseases being frequently involved in a population segment over 60-65 years - and use of anti-inflammatory drugs (NSAID) for osteo-articular inflammatory suffering [9-10].

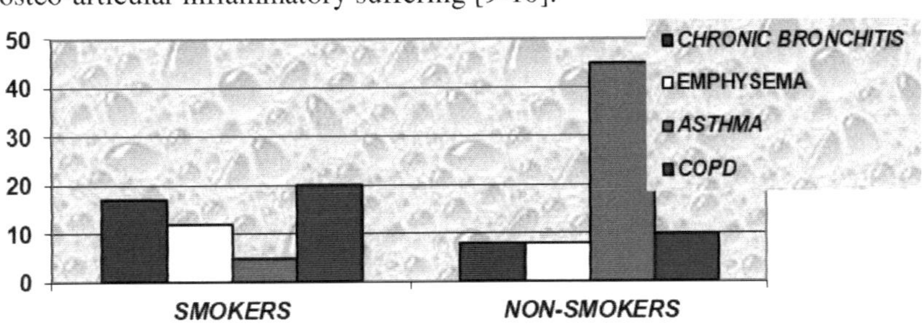

Figure 2. Distribution of smokers and non-smokers in the experimental group.

There is an increased prevalence of smoking patients in the case of chronic obstructive pulmonary disease (COPD), respectively in the case of non-smokers the frequently incriminated condition was asthma, so also an inflammatory and obstructive pathology.

The following investigations referred to diseases of the oral cavity in patients with chronic respiratory pathology, considering the indications of oral hygiene, mainly:

• Gingival indicators: Simplified gingival index, corresponding to the gingival bleeding index (SBI); Papillary bleeding index (PBI) [11].

•Indicators of periodontal inflammation: Periodontal Index (PDI), which evaluate the gingival and periodontium damage [12].

The investigations, conducted in the hospital, respectively in dental offices, showed increased prevalence of gingivitis, periodontitis, chronic marginal superficial, coated tongue, possibly caused by insufficient ventilation in the nose and mouth breathing predominance. Angular cheilitis and stomatitis have been also considered as diseases of the oral cavity in patients with asthma and COPD (figure 3).

In these cases, 69.4% (124 patients) of the subjects with chronic respiratory diseases were diagnosed with superficial chronic periodontitis, respectively chronic bronchitis - 18 cases (14.2%); asthma - 66 cases (53.4%); emphysema - 6 cases (5.1%); COPD - 34 cases (27.3%), the degree of impairment of periodontitis being determined more by the severity of the respiratory disease than by its duration (table 2). Another frequent dental pathology is gingivitis, present in 36.5% (78 patients) of the cases of chronic respiratory disease, respectively - chronic bronchitis 14 cases (17.1%); - asthma 39 cases (49.1%); - emphysema 8 cases (9.4%); - COPD 20 cases (24.4%).

Table 2. Incidence of oral pathology in patients with asthma and COPD

Oro-dental lesions	No of patients	Prevalence	Female (cases)	Male (cases)
Chronic gingivitis	78cases	36.5%	33 42.30%	45 57.69%
Superficial/marginal chronic periodontitis	124cases	69.4%	4637.09%	7862.9%
Coated tongue	5cases	12%	240%	360%
Angular cheilitis	4cases	10%	125%	375%
stomatitis	6cases	12%	2 33.3%	466.7%
Oral candidiasis	3cases	6%	1 33.3%	2 66.7%

The initial results obtained allowed us a first preliminary conclusion regarding the prevalence of obstructive respiratory diseases, respectively the correlation with two oro-dental diseases with disabling potential in elderly patients in the study.

In order to perform a statistically significant analysis with practical potential in the usual medical evaluation, we explored from the point of view of ventilatory function the patients from the category of high-frequency oro-dental diseases (superficial/marginal chronic periodontitis). Thus, spirometry was performed in patients with chronic periodontal disease, considering two ventilatory parameters: the forced vital capacity (FVC) and the forced expiratory volume in the first second of the forceful exhalation (FEV1) (Figure 3)[13-14]. The highest values of FVC and FEV1 will be compared, from the 3 measurements performed. Percentage values between 80-120% are considered normal. In the case of expiratory flows reported to the FVC, default values are used, which delimits normal: FEV1 / CVF (IPB = Bronchial Permeability Index) ≥ 0.7. The assessment of the severity of respiratory dysfunction was made according to the consensus on the strategies for interpreting functional explorations, which recommend the use of FEV1 expressed as a percentage of the predicted value [15].

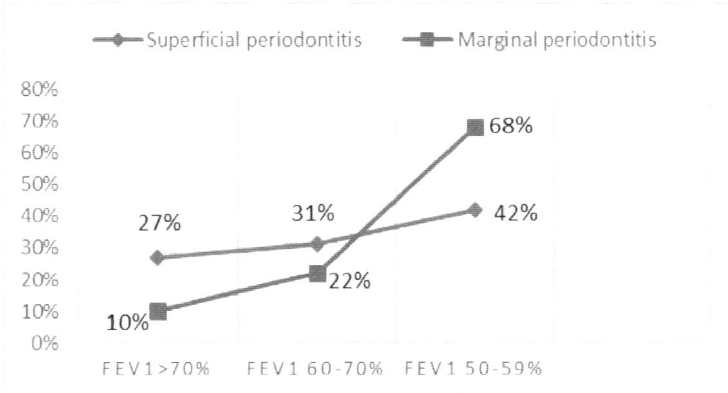

Figure 3. Prevalence of periodontal disease depending on FEV1.

The second respiratory parameter evaluated was diffusion through the alveolo-capillary membrane (DLCO or TLCO, commonly used in Europe) (Figure 4). DLCO provides data on the integrity of the alveolo-capillary membrane and the gases on the fit of the ventilation / infusion ratio in all lung functional units [16-17]. DLCO is a measurement of the conductance or ease of transfer for CO molecules from alveolar gas to the hemoglobin of the red blood cells in the pulmonary circulation.

It often is helpful for evaluating the presence of possible parenchymal lung disease when spirometry and/or lung volume determinations suggest a reduced vital capacity, RV, and/or total lung capacity (TLC) [18-19]. Thus, some patients may experience hypoxemia and a normal DLCO (severe asthma) and other normal oxygen levels in the arterial blood and a low DLCO (pulmonary emphysema).

This parameter is the product of the rate of transfer through the alveolo-capillary membrane and the volume of alveolar volume (VA) and is dependent on • membrane factors: surface area, thickness, functional efficiency; • circulatory factors: volume of blood in the capillaries, respectively hemoglobin value (DLCO value is corrected to the current hemoglobin concentration) [20-21].

The test for acceptance was the respiratory volume≥90% of the total lung capacity (TLC) in the same test session, or an inspiratory volume>85% of the highest TLC in the same test session associated with an alveolar volume (AV) limit 200 ml [22].

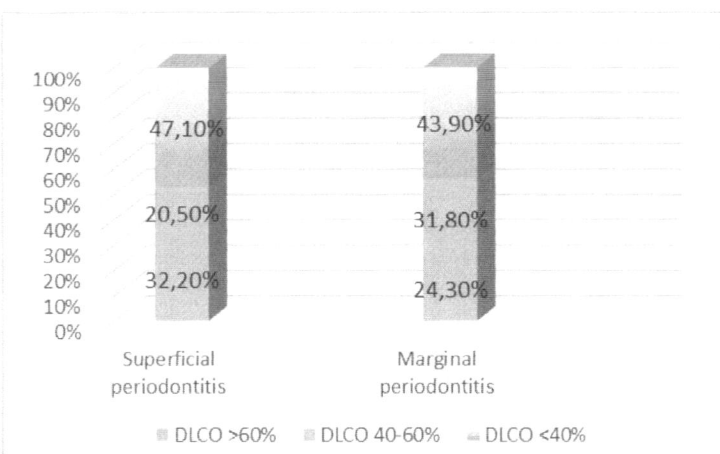

Figure 4. Incidence of DLCO changes in patients with periodontal disease.

The functional respiratory tests performed can guide us to a discussion on the prevalence of periodontal damage in patients with different degrees of severity of obstructive respiratory pathology.

Also compared in patients with asthma and COPD was the incidence of caries lesions of periodontal type, a growing prevalence (49.3% cases) of these disorders being registered in women, the most common being asthma patients, while in those with chronic obstructive pulmonary diseases, untreated odonto-periodontal suffering prevails in males (53.5% cases).

4. Conclusions

Asthma and chronic respiratory pathology are generally an inflammatory disease with multiple implications in dental pathology, thus requiring a precise diagnostic and well-established treatment protocol [23]. The first line therapy to be applied involves inhalation of glucocorticosteroids. The therapeutic consequences are reflected first in the oro-dental condition, inducing pathological transformations of various types and, consequently, improved quality of the primary digestive function and metabolic alterations [24].

Statistical analysis shows that over 60% (176 patients) of the subjects with chronic respiratory diseases were diagnosed with superficial chronic periodontitis: • Chronic Bronchitis - 25 cases (14.2%); • Asthma - 94 cases (53.4%); • Emphysema - 9 cases (5.1%); • COPD - 48 cases (27.3%), the degree of periodontitis impairment being caused by the severity of the respiratory disease rather than by age.

Also, 6.5% (117 patients) of the subjects with chronic respiratory diseases had chronic gingivitis: • Chronic Bronchitis - 20 cases (17.1%); • Asthma - 61 cases (57.3%); • Emphysema - 11 cases (9.4%); • COPD - 25 cases (24.4%). 7.1% of the patients (19 subjects) with chronic respiratory disease, COPD and asthma presents angular cheilitis (3.6% cases), stomatitis (2.3% cases) and candidiasis (1.2% cases).

Functional respiratory examinations performed in patients with chronic periodontal disease, superficial or marginal, have shown a statistically significant increase in the prevalence of FEV1 changes such as moderate-severe impairment (42%, respectively 68%). Thereby, we can discuss and highlight a clinical-functional correlation with practical importance, because the dental clinical examination can be completed by a usual respiratory functional exploration, such as spirometry [11].

From the point of view of the second evaluated respiratory parameter, DLCO, the same increased prevalence of chronic periodontitis is confirmed in elderly patients with moderate-severe impairment (68% of cases) [25].

As a final conclusion, we must emphasize from the research carried out in the hospital and in the dental offices that a more severe change of the studied respiratory parameters (FEV1, DLCO) can be associated with a higher risk of oro-dental damage, like periodontal type [26]. Subsequent research will continue and deepen this studied correlation, with the possibility of developing a clinical-functional algorithm with practical utility.

References

[1] Burney P, Jarvis D and Perez-Padilla R 2015 The global burden of chronic respiratory disease in adults *Int J Tuberc Lung Dis.* **19**(1) pp 10-20

[2] Cortopassi F, Gurung P and Pinto-Plata V 2017 Chronic obstructive pulmonary disease in elderly patients *Clinics in Geriatric Medicine* **33**(4) pp 539-552

[3] Hosseini M, Almasi-Hashiani A and Sepidarkish M 2019 Global prevalence of asthma-COPD overlap (ACO) in the general population: a systematic review and meta-analysis *Respir Res.* **23** p 229

[4] Ciorap R, Corciova C, Ciorap M and Zaharia D 2011 *Proc. Int. Symp. Advanced Topics in Electrical Engineering* (Bucharest: Univ Politehnica Bucharest) pp 143-146

[5] De Souto Barreto P *et al* 2021 The Inspire Research Initiative: A program for geroscience and healthy aging research going from animal models to humans and the healthcare system *The Journal of Frailty & Aging* **10**(2) pp 86-93

[6] Dunn R, Busse P and Wechsler M 2017 Asthma in the elderly and late-onset adult asthma *Allergy* **73** pp 284-294
[7] Mahler D 2017 Evaluation of dyspnea in the elderly *Clin Geriatr Med.* **33**(4) pp 503-521
[8] Medbø A and Melbye H 2007 Lung function testing in the elderly--can we still use FEV1/FVC <70% as a criterion of COPD? *Respir Med.***101**(6) 1097
[9] Bansal M, Khatri M and Taneja V 2013 Potential role of periodontal infection in respiratory diseases - a review *J Med Life.* **6**(3) 244
[10] Gomes-Filho I S, Passos J S and Seixas da Cruz S 2010 Respiratory disease and the role of oral bacteria *J Oral Microbiol.* **21** 2
[11] O'Donnell D E, Deesomchok A, Lam Y M, Guenette J A, Amornputtisathaporn N and Forkert L 2018 Effects of BMI on static lung volumes in patients with airway obstruction *Chest.* **140**(2) pp 461-468
[12] Graham B L *et al* 2019 Standardization of Spirometry 2019 update. An official American Thoracic Society and European Respiratory Society technical statement *Am J Respir Crit Care Med.* **200**(8) pp 70-88
[13] Stanojevic S, Graham B L, Cooper B G, Thompson B R, Carter K W, Francis R W and Hall G L 2017 Official ERS technical standards: Global Lung Function Initiative reference values for the carbon monoxide transfer factor for Caucasians *Eur Respir J.* **50**(3) 1700010
[14] Perez-Padilla R, Wehrmeister F C, de Oca M M, Lopez M V, Jardim J R, Muino A, Valdivia G and Menezes A M B 2018 Outcomes for symptomatic non-obstructed individuals and individuals with mild (GOLD stage 1) COPD in a population based cohort *Int J Chron Obstruct Pulmon Dis.* **13** pp 3549-3561
[15] Coates A L *et al* 2017 ERS technical standard on bronchial challenge testing: general considerations and performance of methacholine challenge tests *Eur Respir J.* **49**(5)
[16] Krishnan J K and Martinez F J 2018 Lung function trajectories and chronic obstructive pulmonary disease: current understanding and knowledge gaps *Curr Opin Pulm Med.* **24**(2) pp 124-129
[17] Gomes-Filho I S, Passos J S and Seixas da Cruz S. 2010 Respiratory disease and the role of oral bacteria *J Oral Microbiol.* **2**
[18] Topalovic M, Laval S, Aerts J-M, Troosters T, Decramer M and Janssens W 2017 Automated interpretation of pulmonary function tests in adults with respiratory complaints *Respiration* **93** pp 170-178
[19] Guder G, Brenner S, Angermann C E, Ertl G, Held M, Sachs A P, Lammers J-W, Zanen P, Hoes A W, Stork S and Rutten F H 2012 GOLD or lower limit of normal definition? a comparison with expert-based diagnosis of chronic obstructive pulmonary disease in a prospective cohort-study *Respir Res.***13**(1) p 13
[20] Graham B L, Brusasco V, Burgos F, Cooper BG, Jensen R, Kendrick A, MacIntyre N R, Thompson B R and Wanger J 2017 2017 ERS/ATS standards for single-breath carbon monoxide uptake in the lung *Eur Respir J.* **49**(1)
[21] Holland AE et al 2014 An official European Respiratory Society/American Thoracic Society technical standard: field walking tests in chronic respiratory disease *Eur Respir J.* **44**(6) pp 1428-1446
[22] Ronaldson SJ *et al* 2018 Determining the optimal approach to identifying individuals with chronic obstructive pulmonary disease: The DOC study *J Eval Clin Pract.* **24**(3) pp 487-495
[23] Ranu H, Wilde M and Madden B 2019 Pulmonary function tests *Ulster Med J.* **80**(2) pp 84-90
[24] Varmaghani M, Dehghani M, Heidari E, Sharifi F, Moghaddam S S and Farzadfar F 2019 Global prevalence of chronic obstructive pulmonary disease: systematic review and meta-analysis *East Mediterr Health J.* **25**(1) pp 47-57
[25] Columbo M, Wong B, Panettieri Jr R A and Rohr A S 2013 Asthma in the elderly: The role of exhaled nitric oxide measurements *Respir Med.* **107**(5) pp 785-787

[26] Du Bois R M, Weycker D, Albera C, Bradford W Z, Costabel U, Kartashov A, King Jr T E, Lancaster L, Noble P W, Sahn S A, Thomeer M, Valeyre D and Wells AU 2011 Forced vital capacity in patients with idiopathic pulmonary fibrosis: test properties and minimal clinically important difference *Am J Respir Crit Care Med.* **184**(12) pp 1382-1389

IOP Conf. Series: Materials Science and Engineering 1254 (2022) 012029 doi:10.1088/1757-899X/1254/1/012029

Electrical properties of the nanocomposite (copper, samarium)-containing complex compound

A O Semenov[1], V V Martyniuk[1,2], M V Evseeva[3], O V Osadchuk[1] and O O Semenova[1]

[1]Faculty of Information Electronic Systems, Vinnytsia National Technical University, Khmelnytsky highway 95, Vinnytsya, 21021, Ukraine
[2] Faculty of Power Engineering and Electromechanics, Vinnytsia National Technical University, Khmelnytsky highway 95, Vinnytsya, 21021, Ukraine
[3]Department of Pharmaceutical Chemistry, National Pirogov Memorial Medical University, Pirogova street 56, Vinnytsya, 21018, Ukraine

E-mail: semenov.a.o@vntu.edu.ua

Abstract. A new semiconductor material tetrakis-μ3-(methoxo)(methanol)-pentakis (acetylacetonate)(tricopper(II), samarium(III)) (I) with composition $[Cu_3Sm(AA)_5(OCH_3)_4CH_3OH]$, where HAA = $H_3C–C(O)–CH_2–C(O)–CH_3$ has been synthesized. By the data of elemental analysis and physicochemical research methods, it was established that the obtained complex compound (I) contained atoms of copper (II) and samarium (III) in ratio Cu:Sm = 3:1, and its composition corresponded to the gross formula $Cu_3SmO_{15}C_{30}H_{51}$. Electrical conductivity of the obtained material in pressed form was measured. For the complex compound (I) such parameters were calculated: the number of valence electrons in one molecule – 272; mass of one molecule – $164.867 \cdot 10^{-20}$ kg; the total number of molecules in the cylindrical sample of a 0.131 g mass and a $18.24 \cdot 10^{-9}$ m^3 volume – $7.946 \cdot 10^{13}$moleculas; the total number of valence electrons – 272. In the 303~413 K temperature range, the resistivity of the pressed sample decreases from $4 \cdot 10^{11}$ to $7 \cdot 10^4$Ohm·cm, which confirms that the isolated compound is a semiconductor with a bandgap ΔE = 1.526 eV. Electrical conductivity properties of the complex compound as thermo- and magnetically sensitive element were studied, for this purpose a test sample of compressed material with 0.5×0.5×1.0 mm geometric dimensions was utilized.

1. Introduction

Automation of technological processes in various fields of industry, research and everyday life requires the development of devices for controlling and regulating various parameters. Such devices are sensors located directly in the control area and designed to measure various physical quantities [1-2]. One of the most common parameters in both industry and research is temperature and magnetic field, their measurement having the largest part among all types of measurements [3-4]. Among other semiconductor devices, thermistors are most often used to measure temperature, they have a nonlinear, noticable temperature dependence of resistance and a negative or positive temperature coefficient of resistance. Thermistors are the simplest semiconductor devices, they are used in automatics and telemechanics, thermometry, electronics, electrical engineering, telephony to solve various technical problems [1-4]. They have a number of advantages over other devices suitable for similar purposes, these advantages are high temperature sensitivity, small size, stability of characteristics over time and almost no need for special care of thermistors during their operation. However, such factors as possible impurities in semiconductor materials, modes of their heat treatment, methods of contacting and others affect electrical parameters that complicates practical application of thermistors. Therefore, despite the fact that many semiconductors have a noticeble temperature dependence of resistivity, not all of them can be used for producing thermistors in large scale. So, the nature of conductivity, the effect of impurities and heat treatment on the electrical conductivity of semiconductor materials should be examined to develop a technology for producing thermistors that would meet requirements of technology.

Therefore, the synthesis of new semiconductor materials with significant conductivity and high values of the temperature coefficient of resistance is relevant, this will provide developing thermistors with high sensitivity to temperature changes.

It is known from [5-10] that heterometallic complex compounds containing atoms of transition metals and lanthanides are widely utilized due to their unique magnetic, electrical, optical, luminescent and other properties. Introducing different complexing metals and chelating ligands into a molecule of complex compound one can obtain compounds of different composition, i.e. to conduct a directed synthesis of compounds with a given range of operating properties. Heterometallicβ-diketonates, which have semiconductor properties, are of particular interest among this class of complex compounds [11-12].

2. Theoretical and experimental research

The aim of this work is studying thermoresistive and magnetic features of a semiconductor material based on heterometallic (copper, samarium)-containing acetylacetonate.

A technique was developed for synthesizing the heterometallic complex compound – tetrakis-μ3-(methoxo)(methanol)-pentakis(acetylacetonate)(tricopper (II), samarium(III)) (I) with such composition:

$$[Cu_3Sm(AA)_5(OCH_3)_4CH_3OH], \text{ where } HAA = H_3C{-}C(O){-}CH_2{-}C(O){-}CH_3.$$

The authors obtained the heterometallic complex compound (I) by the following procedure: salts of 7.23 g (30 mmol) copper(II) nitrate trihydrate and 4.44 g (10 mmol) samarium(III) nitrate hexahydrate were dissolved in 160 ml of absolute methyl alcohol, which contained 70 ml of orthomut ether. The obtained mixture was kept for 2 hours in a hermetically sealed conical flaskat room temperature. Further, 5.4 ml (50 mmol) of acetylacetone was added to the mixture. Then, the conical flask was closed with a reversible water refrigerator and placed on a heated magnetic stirrer. After that, piperidine was introduced in batchesto the mixture to pH = 8bypermanent stirring and heating (~323.15 K). Moreover, it was heated and stirred for an hour more. A color of the derived solution changed and a sediment appeared. Next, the reaction mass was heated (~323.15 K) with continuous stirring for an hour. A homogeneous blue fine crystalline sediment was formedafter cooling. The sedimentwas filtered on a glass filter. Later, it waswashed with a small quantity of absolute methanol, diethyl ether. Then, it was driedover silica gel in a vacuum desiccator. The practical yield was 8.13 g, which was 82% of the theoretically calculated one. So, the isolated complex compound (I) is a fine crystalline powder of blue color, it is soluble in a mixture of dimethylformamide with chloroform (1:1), poorly soluble in alcohols, diethyl ether, better soluble in dimethyl sulfoxide, dimethylformamide, it ruins in water.

Elemental chemical analysis of the isolated complex compound (I) was performed which showed that it contained in dry powder state:

Cu – 19.19%; Sm – 15.11%; C – 36.27%; H – 5.14%.

The above-mentioned data substantiate that the isolated complex compound contains atoms of metals of different nature, and their ratio inside the heterometallic compound (I) Cu is Sm = 3:1.We performed an X-ray phase analysis, magnetochemical and thermogravimetric studies, IR spectroscopic investigations to establish a plausible scheme of chemical bond arrangement in this compound and its structure. On the basis of the performed studies, we established the composition of the selected complex compound (I), which corresponded to a gross formula $Cu_3SmO_{15}C_{30}H_{51}$ and proposed the scheme of chemical bonds arrangement (Figure 1).

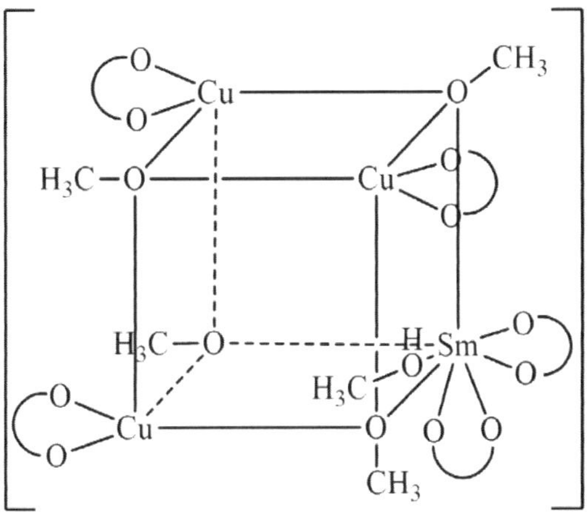

Figure 1. Scheme of chemical bond arrangement in [Cu3Sm(AA)5(OCH3)4CH3OH], where HAA = H3C – C (O)–CH2 – C (O) –CH3.

To conduct measurements of electrical conductivity, the microcrystalline powder of the isolated complex compound $[Cu_3Sm(AA)_5(OCH_3)_4CH_3OH]$ was pressed with a specially designed pressing device. A cylindrical sample of a 0.131 g mass and a $18.24\cdot10^{-9}$ m^3 volume was made and used in experimental studies.

Taking into account these data, the density of the cylindrical sample pressed material was calculated by formula (1):

$$\rho = m / v = 7.182\cdot10^3 kg / m^3, \tag{1}$$

where m is the experimentalsample mass; V is the experimentalsample volume.

The molar mass of the complex compound $[Cu_3Sm(AA)_5(OCH_3)_4CH_3OH]$ was calculated as being 992.5 g/mol, and the number of valence electrons in one molecule was calculated as being 272.

The mass of one molecule of the compound (I) is

$$m_0 = M / N_A = 164.867\cdot10^{-20} kg, \tag{2}$$

where M is the molar mass of the compound (I); N_A is the Avogadro number.

Applying formula (3), we calculated the total number of molecules in the cylindrical sample of a 0.121 g mass and a $18.24\cdot10^{-9}$ m^3 volume:

$$N_{mol} = m / m_0 = 7.946\cdot10^{13} molecules, \tag{3}$$

where N_{mol} is the total number of molecules; m_0 is the mass of one molecule of the compound (I); m is the experimental sample mass.

Total number of valence electrons:

$$N = 272\cdot N_{mol} = 21.6131\cdot10^{15}. \tag{4}$$

The experiment showed that in the 303~413 K temperature range the resistivity of the pressed sample of compound (I) decreases from $4\cdot10^{11}$ to $7\cdot10^4$Ohm·cm, i.e. the isolated compound is a semiconductor.

This experimental investigation contributes to calculations of material'sspecific conductivity at these temperatures and to determination of the bandgap width:

$$\Delta E = \frac{k \ln \dfrac{\sigma_1}{\sigma_2}}{\left(\dfrac{1}{T_2} - \dfrac{1}{T_1}\right)} = 2,4416 \cdot 10^{-19} J = 1,526 eV, \tag{5}$$

where σ is material's specific conductivity; T is the absolute temperature; k is the Boltzmann constant.

The calculations confirmed that this material is a semiconductor.

To use this nanocomposite material as a sensor of non-electric quantities, such as magnetic field induction and temperature, the compound was pressed into an element with geometric sizes of 0.5×0.5×1.0 mm.

Knowing the bandgap width of the studied material and applying the experimental results, we obtained the temperature dependence of specific conductivity. Figure 2 illustrates a received graph of the temperature dependence of specific conductivity.

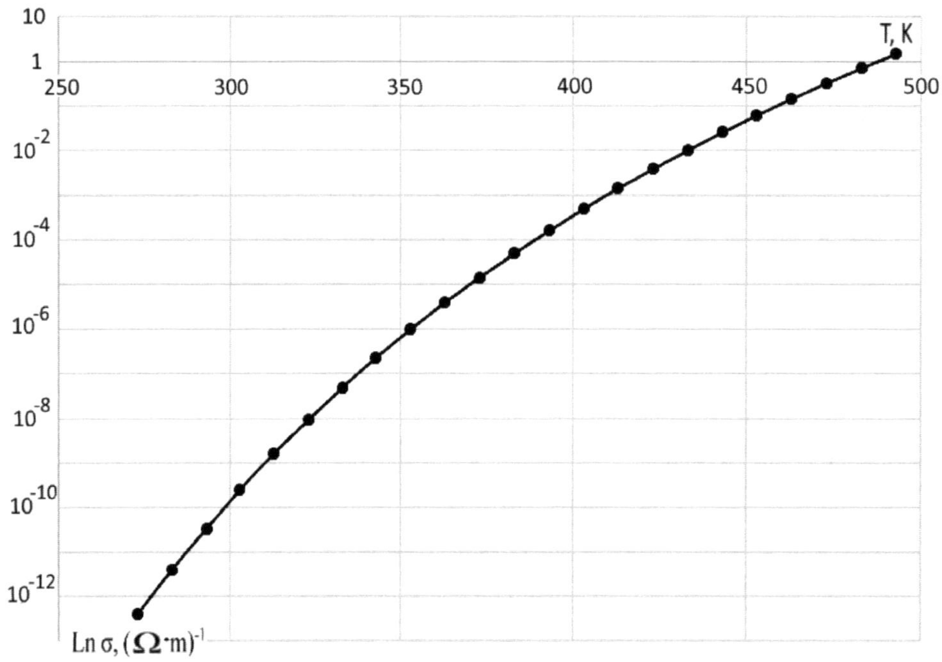

Figure 2. Temperature dependence of specific conductivity of complex compound.

The graph in Figure 2 indicates that the specific conductivity of the studied structure grows from $4.084 \cdot 10^{-13}$ $(\text{Ohm} \cdot \text{m})^{-1}$ to 1.4838 $(\text{Ohm} \cdot \text{m})^{-1}$ when the temperature rises from 273 K to the value of 493 K.

Knowing the behavior of specific conductivity with temperature, we can calculate the inverse value – specific resistivity. The temperature dependence of the specific resistivity of this material is shown in Figure 3.

The graph in Figure 3 shows that the specific resistivity decreases from $2.44 \cdot 10^{12}$ Ohm·m to 0.674 Ohm·m.

Figure 4 presents the temperature dependence of the resistance of heterometallic β-diketonate with sizes 0.5×0.5×1.0 mm.

The dependence in Figure 4 shows that the resistance decreases from $9.79\cdot10^{15}$ Ohms, at a 273 K temperature to 2695.75 Ohms at a 493 K temperature.

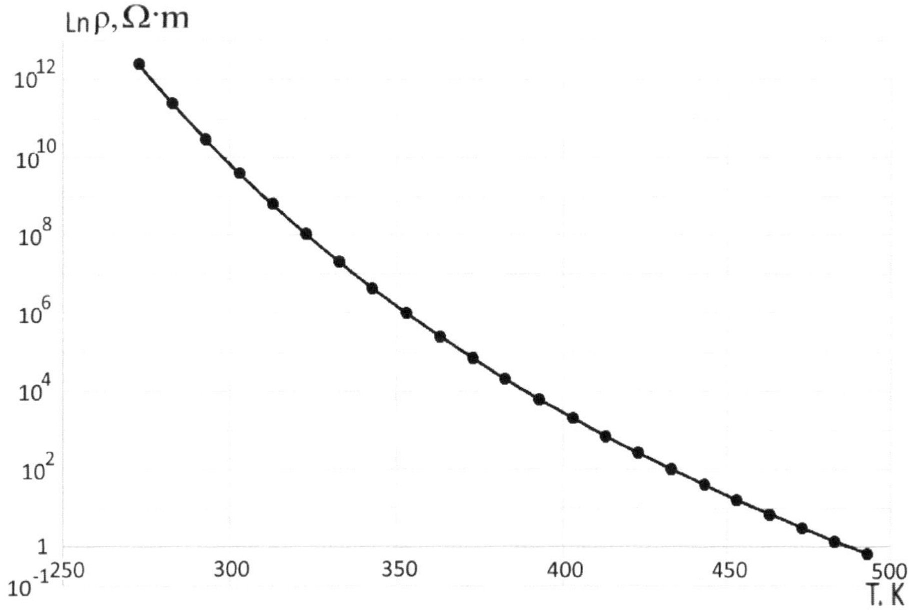

Figure 3. Temperature dependence of the specific resistivity of the material.

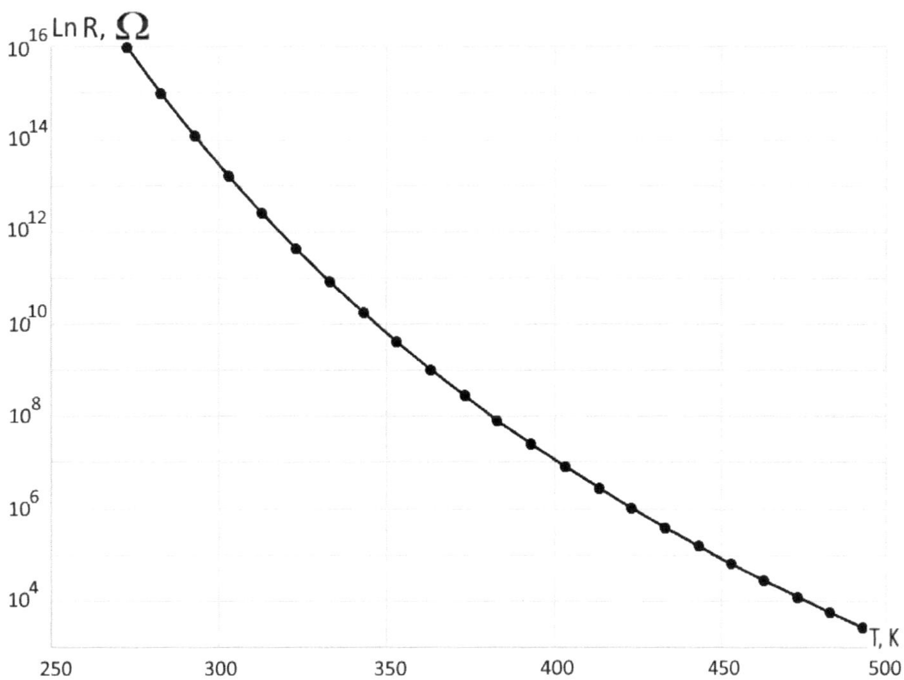

Figure 4. Temperature dependence of resistance of the heterometallic complex compound.

Figure 5 demonstrates the temperature dependence of current density at a supply voltage of 10 volts.

The graph in Figure 5 demonstrates that when the temperature changed by 220 degrees, the current density increased from $4.084 \cdot 10^{-9}$ A/m^2 to $1.48 \cdot 10^4$ A/m^2.

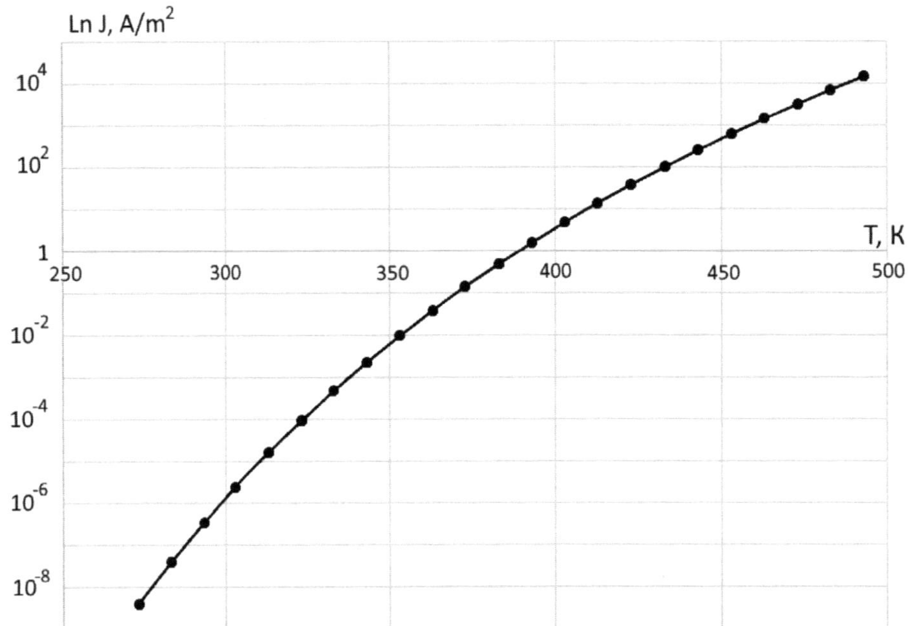

Figure 5. Temperature dependence of current density.

The temperature dependence of concentration can be found from the bandgap width of this material. A graph of the charge carrier concentration change due to the temperature change is shown in Figure 6.

The graph in Figure 6 demonstrates that the charge carrier concentration rises from $2.55 \cdot 10^{13}$ m^{-3} to $9.272 \cdot 10^{25}$ m^{-3} when the temperature grows from 273 K to the value of 493 K.

The quantum Hall constant at 373 K is:

$$R_{qH} = -3\pi / 8nq = 8.2 \cdot 10^{-3} m^3 \cdot C^{-1}, \tag{6}$$

where q is the charge of electron; n is the charge carrier concentration.

Formula (6) yields $R_H < 0$ for the case of electronic conductivity. While, if the current is carried mainly by holes, which are attributed to positive charges, then the Hall voltage changes its sign and $R_H > 0$.

Let us investigate the change in the Hall constant with temperature. For that purpose, we combine the equation of the temperature dependence of the charge carrier concentration and formula (6):

$$R_{qH} = -\frac{3\pi}{8qn_0} \cdot e^{\frac{\Delta E}{kT}}. \tag{7}$$

The temperature dependence of the quantum Hall constant shown in Figure 7was obtained by applying this equation.

The graph in Figure 7 demonstrates that the value of the quantum Hall constant for the compound drops from $2.88 \cdot 10^5$ m$^3 \cdot$C^{-1} to $7.93 \cdot 10^{-8}$m$^3 \cdot$C^{-1} when the temperature grows from 273 K to the value of 493 K.

The charge carrier mobility does not depend on temperature and is a constant value:

$$\mu_n = R_{q\,H} \cdot \sigma \; = \; 1.177 \cdot 10^{-7}\, m^3 \cdot (V \cdot s)^{-1}. \qquad (8)$$

Figure 8 illustrates a dependence of the Hall electric field strength that appears under the action of a magnetic field inside the 0.5×0.5×1.0 mm semiconductor

Figure 8 illustrates the voltage rising from $1.17 \cdot 10^{-2}$ to 1.17 V/m in the 10–1000 mTrange.

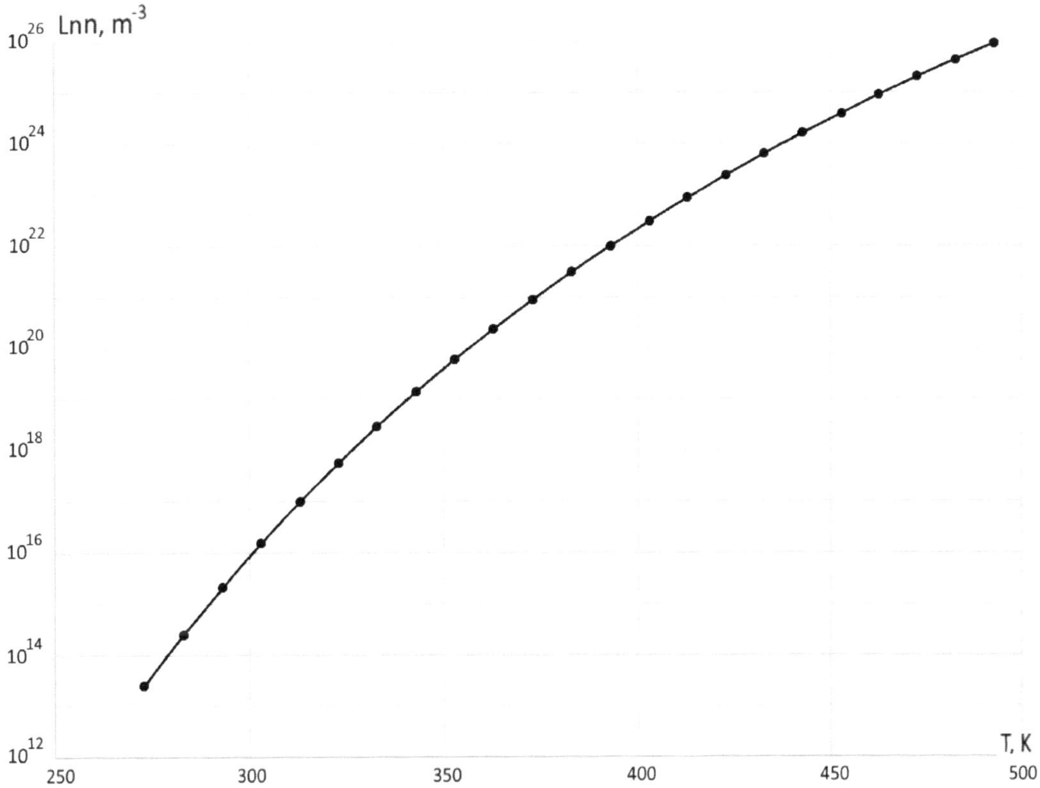

Figure 6. Temperature dependence of charge carrier concentration.

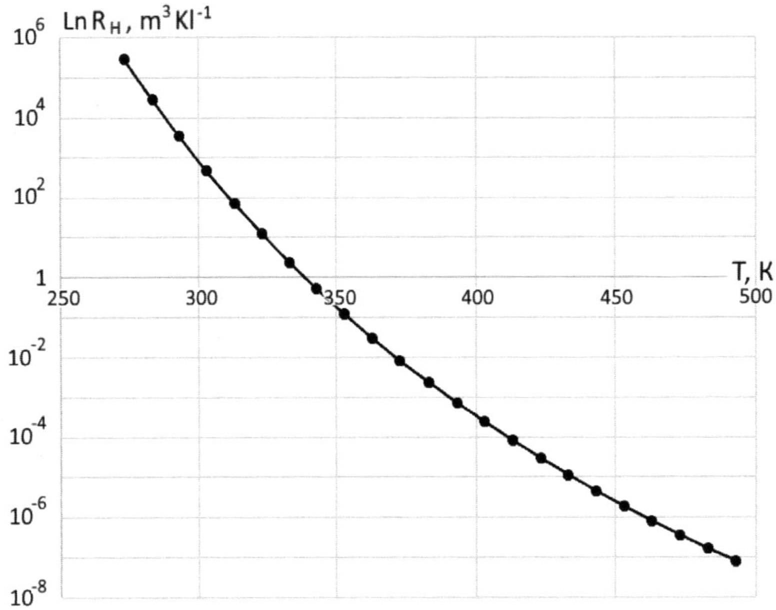

Figure 7. Temperature dependence of the quantum Hall constant of the semiconductor.

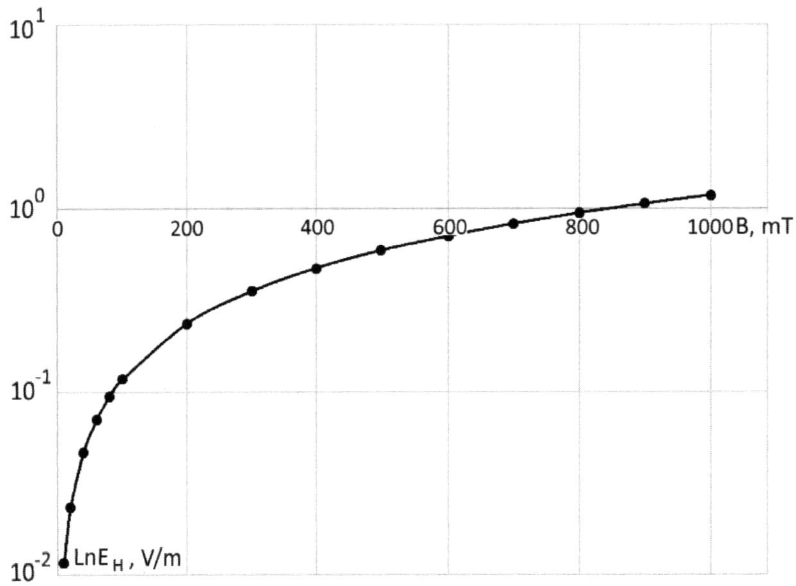

Figure 8. Magnetic field induction dependence of electric field strength in the semiconductor.

Figure 9 depicts the received magnetic field induction dependence of the Hall voltage for the examined compound.

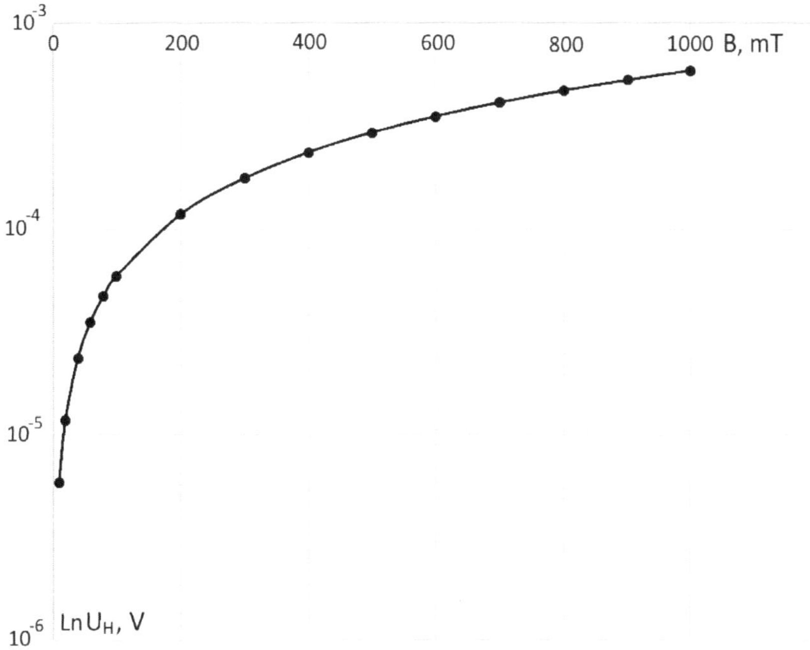

Figure 9. Magnetic field induction dependence of the Hall voltage.

The graph illustrates the Hall voltage growing from $5.88\cdot10^{-6}$ to $5.88\cdot10^{-5}$ V in the 10 to 100 mTrange, from $5.88\cdot10^{-5}$ V to $2.3\cdot10^{-4}$ V in the 100 to 400 mTrange, and from $2.3\cdot10^{-4}$ to $5.88\cdot10^{-4}$ V in the 400 mT to 1000 mTrange.

3. Conclusion

A new sensitive element based on the synthesized complex compound of tetrakis-μ3-(methoxo)(methanol)-pentakis(acetylacetonate)(tricopper (II), samarium(III)) (I) was derived. The technique to synthesize such material is given and the influence of temperature and magnetic field on electrophysical features of such semiconductor is examined.

The study of conductive features of the synthesized complex compound in compressed form in the 273 K ~ 493 K temperature range showed that when temperature increased, its resistivity decreased sharply from $2.44\cdot10^{12}$Ohm·m to 0.674 Ohm·m, and the bandgap was 1.526 eV, which was common for semiconductor materials. The chemical compound decomposes at 523 K, the charge carrier concentration grows from $2.55\cdot10^{13}$ m^{-3} at 273 K to $9.27\cdot10^{25}$ m^{-3} at 493 K, as the quantum Hall constant drops from $2.88\cdot10^{5}$ $m^3\cdot C^{-1}$ to $7.93\cdot10^{-8} m^3\cdot C^{-1}$ when temperature rises from 273 to 493 K, the Hall voltage grows from $5.88\cdot10^{-6}$ to $5.88\cdot10^{-4}$ B in the 0 to 1000 mT magnetic field range.

References

[1] Vuitsik V, Hotra Z Yu, Hryhoriev V V, Kalita V, Melnyk O M and Potentski Ye 2003 *Microelectronic sensors of physical quantities* **1**, ed Hotra Z Yu (Lviv: Liha–Pres)

[2] Vuitsik V, Hotra Z Yu, Hotra O Z, Hryhoriev V V, Kalita V, Melnyk O M and Potentski Ye 2003 *Microelectronic sensors of physical quantities* **2**, ed Hotra Z Yu (Lviv: Liha–Pres)

[3] Hotra Z Yu 2001 *Microelectronic elements and devices for thermometry* (Lviv: Liha–Pres)

[4] Santi W N and Toifur M 2021 Analysis of micro structure and the resistivity of cu / ni thin coat as a low temperature sensor using electroplating method assisted with magnetic field outside of the ion flow *Key Engineering Materials* **885** pp 141–147

[5] Pajuelo-Corral O, Pérez-Yáñez S, Vitorica-Yrezabal I J, Beobide G, Zabala-Lekuona A, Rodríguez-Diéguez A, Seco J M and Cepeda J 2022 A metal-organic framework based on Co(II) and 3-aminoisonicotinate showing specific and reversible colourimetric response to solvent exchange with variable magnet behaviour *Materials Today Chemistry* **24** 100794

[6] Zagaynov I V, Shelepin I V, Fedorov S V, Naumkin A V, Bykov A V and Konovalov A A 2021 Sm(Nd) doped ceria materials for multifunctional application *Ceramics International* (*Article in Press*)

[7] Zhang Q, Wen S, Feng Q and Miao Y 2022 Adsorption characteristics of Cu^{2+} species on cerussite surfaces and implications for sulfidization flotation *Separation and Purification Technology* **2821** 120109

[8] Krisyuk V V, Baidina I A, Romanenko G V, Kryuchkova N A, Shayapov V R, Komarov V Y, Kyzy S U, Benassi E and Igumenov I K 2020 Structural Diversity and Spectral Properties of the Crystals of Heterometallic Complexes Derived from TM(acacen) and Pb(diketonate)2, TM = Cu, Ni, Pd *Crystal Growth and Design* **20**(11) pp 7260–7270

[9] Sokolov V G, Lukina D A, Skatova A A, Moskalev M V, Baranov E V and Fedushkin I L 2021 Samarium and ytterbium complexes based on sterically hindered 1,2-bis(imino)acenaphthene* *Russian Chemical Bulletin* **70**(11) pp 2119–2129

[10] Conradie J 2020 Tris(β-ketoiminato)ruthenium(III) – Structural and electronic data of the neutral, oxidized and reduced forms *Data in Brief* **28** 104833

[11] Devi J and Kumar R 2013 Synthesis and characterization of new heterobimetallic [Ca (II)-Ti (IV)]-μ- oxoisopropoxide and its β-diketonates *International Journal of Chemical Sciences* **11**(2) pp 1075–1085

[12] Mirzaeva I V, Kozlova S G and Krisyuk V V 2021 Quantum chemical study of the stability of copper-palladium complexes in the gas phase *Journal of Structural Chemistry* **62**(1) pp 9–18

IOP Conf. Series: Materials Science and Engineering 1254 (2022) 012030 doi:10.1088/1757-899X/1254/1/012030

A novel approach for power load forecast based on GAN data augmentation

V A Boicea[1], A P Ulmeanu[1] and A Vulpe-Grigorași[2]

[1]Faculty of Energy Engineering, Politehnica University of Bucharest, Romania
[2]Faculty of Electronics, Telecommunications and Information Technology, Politehnica University of Bucharest, Romania

E-mail: valentin.boicea@upb.ro

Abstract. At present, the concept of Smart Grid is intimately related to huge volumes of data generation, known as Big Data. These play a key role as far as the improvement of the network operation is concerned and, above all, when it comes to energy consumption. Hence the load forecast based on Big Data processing represents a reliable alternative in carrying out an accurate load prediction. As such, a new approach for the power load forecast, based on GAN data augmentation, for a hospital, within a Smart Grid, will be presented in this work.

1. Introduction

The concept of Smart Grid (SG) is intrinsically related, above all, with data generation and processing. But what is generating this huge data volume? SG means an important number of sensors, communication devices and PCs. As such, the data coming from these sensors is transmitted to the PCs where it is processed with the ultimate goal of improving the network operation in terms of balance between power generation and consumption. Usually, SGs are fed by renewable sources that have an intermittent character as far as the electricity production is regarded and this makes even more important the balance mentioned above. Thus, also a better economic competitiveness is achieved. Here one should mention the fact that 1% of the mean absolute percentage error (MAPE) in load forecasting is equivalent to several hundred thousand of USD per GW peak in case of a utility company [1]. That is why the load forecast plays a key role when it comes to SGs. Sometimes, these feed strategic objectives like hospitals. Hence, it is the purpose of this paper to introduce a novel approach for the prediction of the energy consumption; the beneficiary will be a hospital located in Chicago.

An important parameter of the forecast is the time horizon for which this is carried out. Paradoxically, there isn't any general accepted separation of the various prediction intervals. Usually, a period of 1 day designates a very short-term load forecast (VSTLF). An interval between 1 day and 3 weeks refers to a short term load forecast (STLF), the medium term load forecast (MTLF) is done for a period between 3 weeks and 3 years while the long term load forecast (LTLF) is carried out for an interval of more than 3 years. In this paper, the energy consumption is predicted for a period of 1 year.

As far as the prediction methods are concerned, these can be classified in two categories: parametric and artificial intelligence (AI) [2].The first category consists of correlating the load demand to the factors that are influencing it (eg. the temperature). The correlation takes place using a mathematical model. This uses in most of the cases statistical techniques [2]. The subcategories of the parametrical forecasting are the Grey Method, the regression methods that are very simple to implement but which are not very accurate, and the time series models. The Grey Method provides better results using a small number of samples but can be sometimes characterized by unacceptable errors. It is used for dynamic forecasting. The time series models are represented by the autoregressive moving average (ARMA), autoregressive integrated moving average (ARIMA) and other similar methods. On the other hand, the AI methods possess also two subclasses: fuzzy logic and artificial neural networks (ANNs). Of course, that the STLF is much more accurate than the LTLF, since the latest is influenced by other various random factors like weather or even social behavior. The STLF being small enough is less influenced by the random factors and permits a forecasting with relatively high precision [2]. However, at present, even the LTLF can be carried out quite accurately using techniques based on ANNs. Another challenge arises when the sample number used to carry out the prediction is not large enough to obtain precise results. As such, the new approach is based on both Generative Adversarial Networks (GANs) and Convolutional Neural Networks (CNNs). The role of the first is to expand the data set based on which the load forecast is carried out while the CNN carries though the prediction. This type of approach is fully compatible with the concept of Big Data and represents a hot topic nowadays when the hospital energy consumption is heavily influenced by the peaks of the pandemic waves. On the other hand, the expansion of the consumption data sets corresponding strictly to the pandemic waves is not very relevant since not enough data is currently available. This will represent an important subject for further study.

2. Generative Adversarial Networks for data augmentation

This kind of ANN is used above all in image processing. As its name already specifies, the GANs consist of a model in which two ANNs compete. One of them has the role of generating artificial samples based on the data set distribution which must be expanded while the other computes the probability that a sample comes from the training data and not from the generated data. Thus, the first is a *generative* model whereas the second is *discriminative* [3]. In other words, the discriminative ANN determines whether a sample comes from the data distribution or from the model distribution. The result of the competition is that the artificial generated data becomes practically almost indistinguishable from the original data set [3]. In the classical implementation, both generative and discriminative models are represented by the *Multilayer Perceptron*. The two are called *adversarial* [3].

In order to determine the distribution of the generator for a data set x, denoted by dst_g, also the noise must be defined, $dst(z)$. The next step is to represent a mapping to data space [3] as $G(z;\pi_g)$ where G is a differentiable function represented by the *Multilayer Perceptron* with the parameter π_g. A second *Multilayer Perceptron*, $D(x; \pi_D)$, is defined and this will output a scalar. On the other hand, $D(x)$ is the probability that x comes from the data and not from the dst_g. At the same time, G is trained to minimize $log(1 - D(G(z)))$ [3]. The competition mentioned above, is represented by the couple D, discriminator network and G, generative networkthat play a zero-sum game, according to (1).

$$min_G max_D V(D, G) = E_{x\sim dstdata(x)} [log D(x)] + E_{z\sim dst(z)} [log(1 - D(G(z)))] \quad (1)$$

Optimizing D completely is not allowed, given the fact that for finite datasets this will result in overfitting. That is why a certain number of steps in optimizing D and one step in optimizing G is carried out. Hence, D will be kept near its global optimum as long as G changes slowly [3]. At the same time, (1) does not result in a sufficient gradient for G to be trained appropriately. During the early stages of the training, when G is even lower from the gradient point of view, D will classify the samples as fake because they will obviously be different from the training data [3]. As such, the following term $log(1 - D(G(z)))$ will saturate. A possible approach could be instead of training G to minimize $log(1 - D(G(z)))$, to train it to maximize D(G(z)). In doing this, $max\ [D(G(z))]$ will provide much stronger gradients during the early stages of the training [3].

Using the GANs, can have advantages and disadvantages. One of the disadvantages is the fact that one does not have a clear representation of dst_g and also that D and G must be perfectly synchronized during the training. That is to be able to avoid a situation in which *G* associates too many values of *z* to the same value of *x* for achieving a good diversity of *dstdata* [3]. An important advantage is that the *Markov* chains are not needed during the expansion of the initial dataset. At the same time, the GANs can capture not only sharp but also degenerate distributions whereas the Markov chains require that distribution must be not clearly defined in order for the chains to be able to pass from one mode to another [3]. Another important advantage that has seen intensive research recently is the use of GAN to augment small datasets with the goal of increasing the classification accuracy. As such, GAN can be used to augment a small training dataset and then the augmented dataset will be used by Convolutional Neural Networks in training [6].

Convolutional neural networks (CNN) have become the de facto accepted standard in Deep Learning applications and usually are trained on big datasets containing millions of points of data, thus ensuring proper training, thus being able to learn complex patterns. The main application for CNN is for image classification, yet in scenarios where signal processing and classification is needed, a 1D-CNN can be used [4,7-8]. 1D-CNN follow the same architecture as a standard CNN, employing convolutional layers with varying kernels for feature extraction, Max Pooling layers to further reduce dimensionality and hidden neurons for classification, the main difference being that the layers are adapted for 1D signals as opposed to 2D signals such as images. Hence, 1D-CNN are well suited, with state-of-the-art performances, for biomedical applications, energy forecast, fault detection and any other application that requires the analysis or forecasting of a time series or signal [4-5, 7-8].

3. Case study and results

For the case study, the electricity consumption data from a hospital in Chicago have been used. These were sampled at 1 value per month from 14.05.2007-14.07.2011 and are available in the free RETScreen software tool, version 2012. The software at hand also offers forecasting tools which we used for comparison. The energy forecast using RETScreen can be seen in Figure 5. The aim of this paper is to improve existing data and energy forecasting through GAN data augmentation. In a first instance, the load forecast in our case has been carried out, without GAN augmentation, for the period 14.06.2010-13.05.2011. In this sense, a CNN was used. On the other hand, RETScreen carries through the load forecast using polynomial regression. The time variation of the actual power consumption of the hospital can be seen in Figure 1.

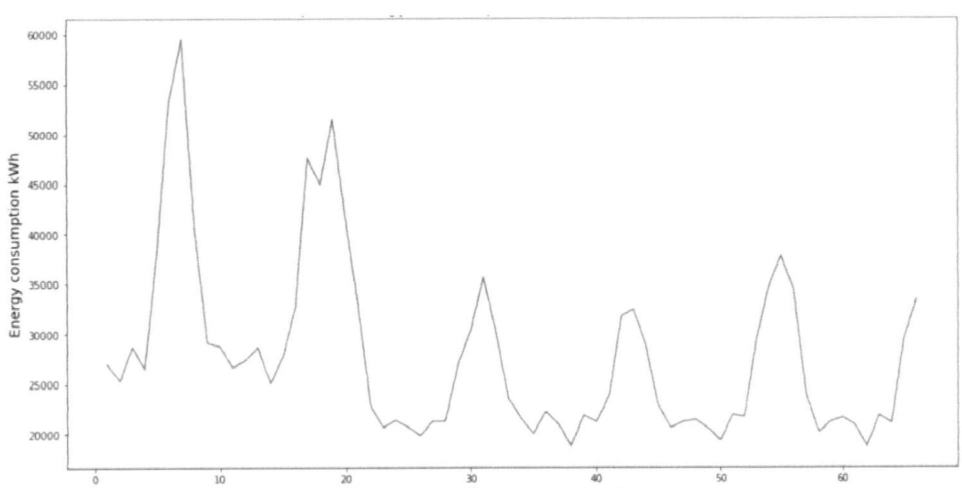

Figure 1. Hospital actual power consumption between 14.05.2007-14.07.2011.

The CNN architecture used for the power load forecast is presented in Figure 2. Given the fact that the total sample number is very low, also the architecture is not complex. One has a number of 62 samples that have been divided as follows: 48 for training and 14 for testing. As such, the CNN architecture contains a single convolutional layer with 64 kernels of size 3, ReLU activation function, stride of 2 and, finally, a hidden layer for classification with 32 neurons of size 3. The training has been done for a batch size of 8 and 100 epochs. The forecast accuracy was 82.51 %.

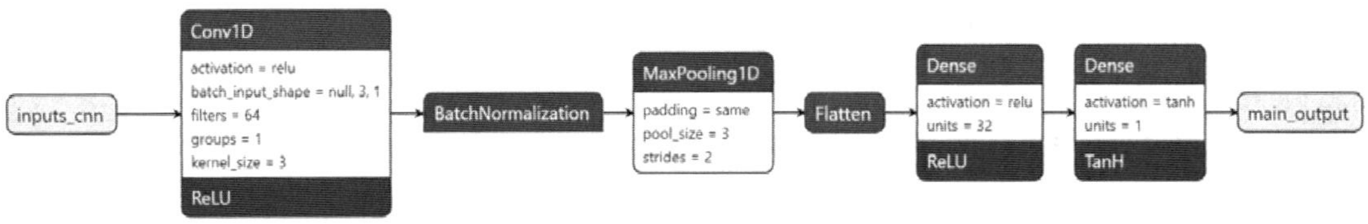

Figure 2. The CNN architecture.

The fact that the network has been trained correctly can be seen in Figure 3 where there is no increase in the loss function either for the training or for the testing. As such, no overfitting takes place.

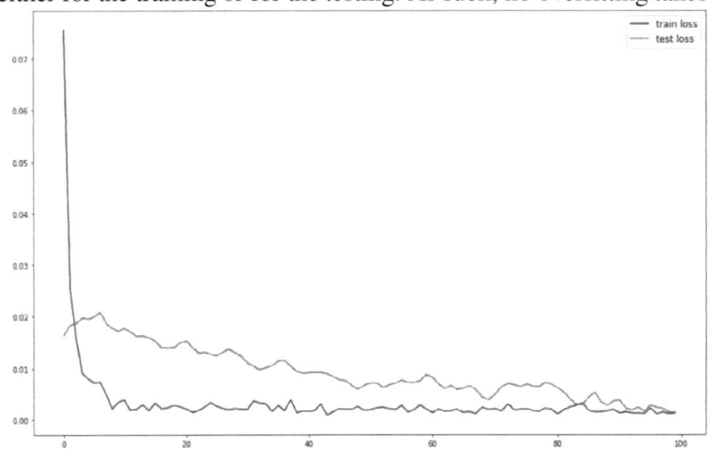

Figure 3. Train and test loss function.

Carrying out the prediction with the CNN, without GAN augmentation, yields the results in Figure 4.

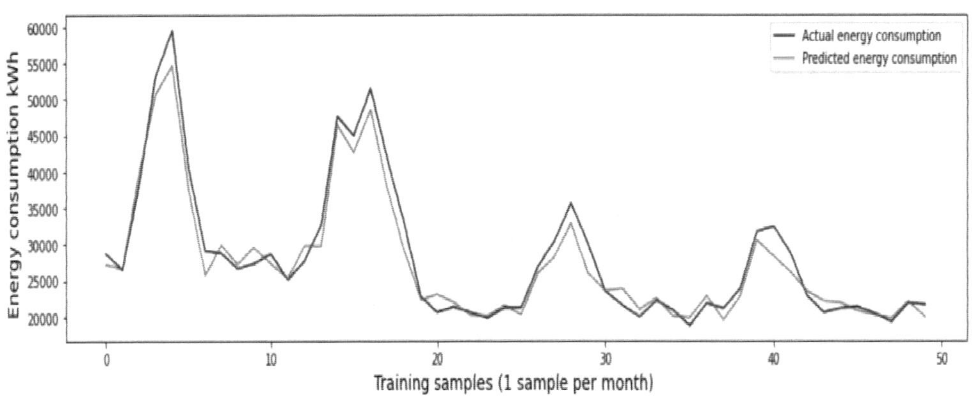

Figure 4. CNN load forecast without GAN augmentation.

Compared to the linear regression offered by RETScreen presented in Figure 5, this prediction is much more accurate.

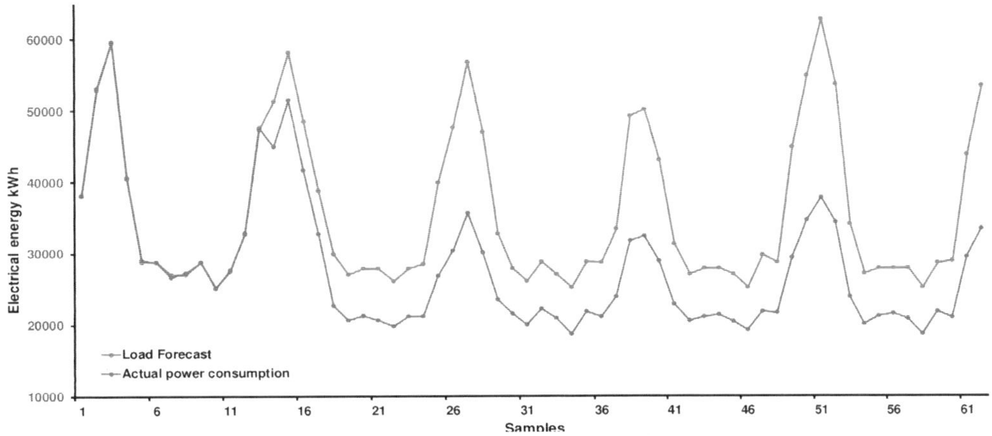

Figure 5. Linear regression load forecast.

But, how is the forecast accuracy when this is GAN augmented? Now, one increases the initial sample number of 62 values to 12000. Figure 6 presents a sample of GAN generated training datasets.

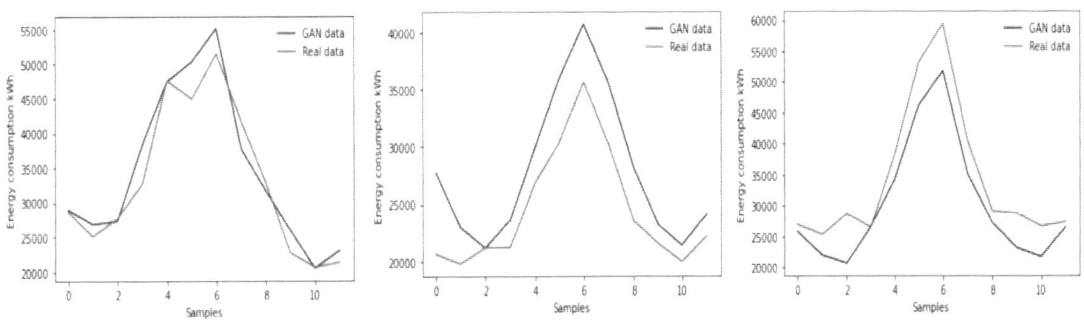

Figure 6. GAN generated data.

If the previous CNN will be used in order to carry out the forecast, the accuracy will be of 78.34 %. That is because the initial architecture is not adapted to 12000 samples. The initial CNN architecture is augmented in order to accommodate for a larger database. As such the second CNN architecture has an additional convolutional layer with 32 kernels of size 3, ReLU activation, a Max Pooling of size 3 and stride 2,and a final hidden layer of 16 neurons for classification. In Figure 7 the modified CNN architecture can be seen. For training, the batch size was 16 and the number of epochs 100. If the initial CNN had 2625 parameters, this one has 7553. In the end, the prediction accuracy will be of 89.93 %.

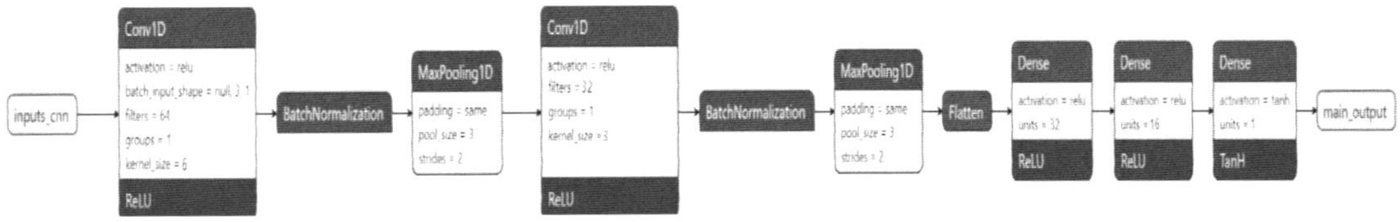

Figure 7. Modified CNN architecture.

The results obtained through CNN and GAN augmentation are presented in Figure 8 while the loss for training and testing is presented in Figure 9. As one can see in Figure 8, there is no increase in the loss function thus the CNN model does not suffer from overfitting.

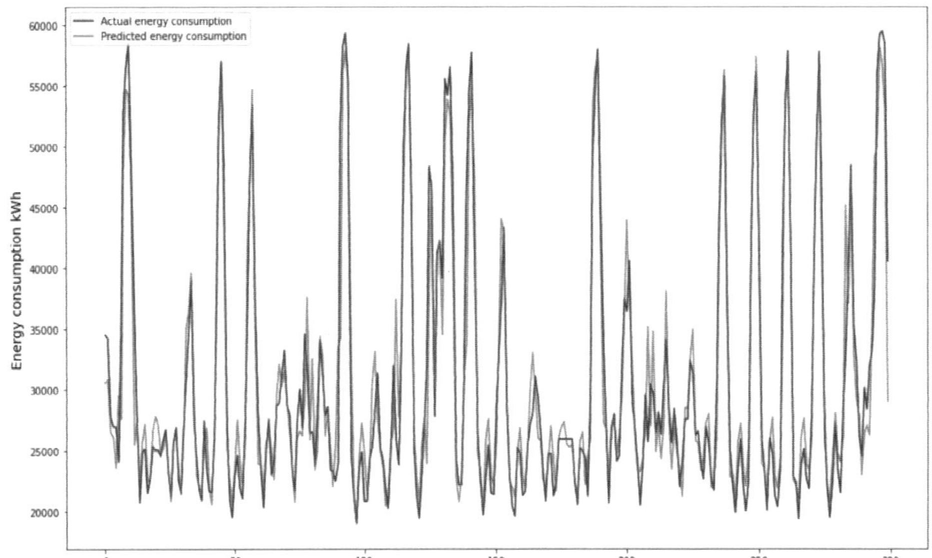

Figure 8. CNN load forecast, augmented with GAN.

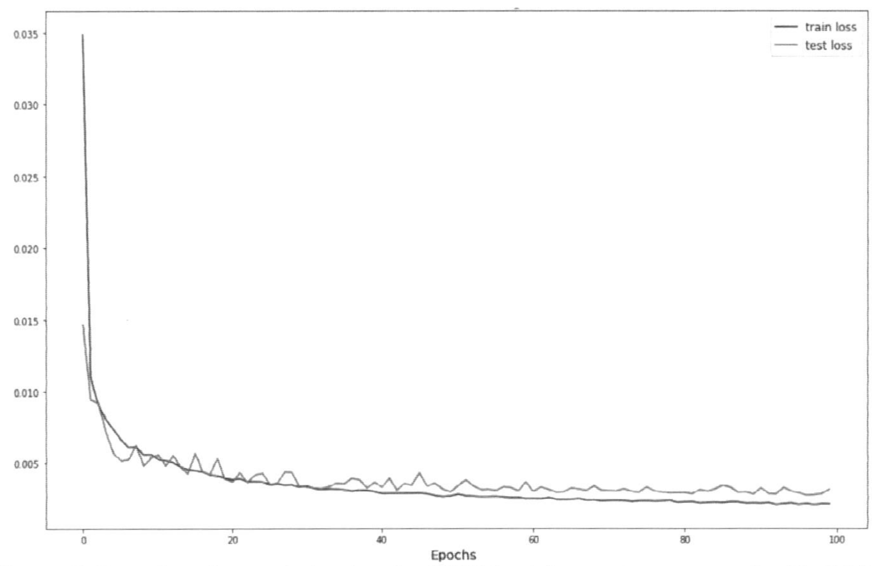

Figure 9. Loss function variation for the CNN load forecast, augmented with GAN.

4. Conclusions

In this work, a novel approach of power load forecast based on GAN augmentation has been presented. In case of a lower number of samples, this method yields very good results. Usually in this case, the consumption characteristic is completely lost and, as such, the forecast is not very accurate in reality. The GAN recaptures this characteristic and leads finally to a prediction that is even more accurate than the initial one. The 7.42 % improvement is obtained also because the complexity of the CNN has been increased in accordance with the augmented data set used for training. The forecast approach presented in this work is fully compatible with the concept of Big Data and helps at determining the load prediction of a hospital under normal conditions. Further study will be needed to conclude if the same type of solution can be used also for expanding the consumption data sets corresponding to pandemic wave peaks using a lower number of samples.

References

[1] Zhang J, Zhang X, Yang J, Wang Z, Zhang Y, Ai Q, Li Z and Sun Z 2020 Deep LSTM and GAN based Short-term Load Forecasting Method at the Zone Level *2020 International Conference on Artificial Intelligence in Information and Communication (ICAIIC)* pp 613-618

[2] Soliman A S, Al-Kandari A M 2010 Dynamic Electric Load Forecasting, ed S A Soliman, A M Al-Kandari, *Electrical Load Forecasting* (Burlington: Butterworth-Heinemann) chapter 8 pp 291-352

[3] Goodfellow I, Pouget-Abadie J, Mirza M, Xu B, Warde-Farley D, Ozair S, Courville A and Bengio Y 2014 Generative adversarial nets *Advances in neural information processing systems* **27**

[4] Kiranyaz S, Avci O, Abdeljaber O, Ince T, Gabbouj M and Inman D J 2021 1D Convolutional Neural Networks and Applications: A Survey *Mechanical Systems and Signal Processing* **151** pp 1-21

[5] Fawaz H I, Forestier G, Weber J, Idoumghar L and Muller P A 2019 Deep learning for time series classification: a review *Data Mining and Knowledge Discovery* **33**

[6] Bowles C, Chen L, Guerrero R, Bentley P, Gunn R, Hammers A, Dickie D A, Hernandez M V, Wardlaw J and Rueckert D 2018 GAN Augmentation: Augmenting Training Data using Generative Adversarial Networks, arXiv: 1810.10863

[7] Sadouk L 2018 CNN Approaches for Time Series Classification *Time Series Analysis - Data, Methods, and Applications* (London: IntechOpen)

[8] Torres J F, Hadjout D, Sebaa A, Martínez-Álvarez F and Troncoso A 2020 Deep Learning for Time Series Forecasting: A Survey *Big Data.* **9** pp 3-21

IOP Conf. Series: Materials Science and Engineering 1254 (2022) 012031 doi:10.1088/1757-899X/1254/1/012031

Digitalizing the life-saving mechanical surgical staplers, for better outcomes in the surgical field

C V Dîrzu and M S Munteanu

Technical University of Cluj-Napoca, Faculty of Electrical Engineering, Cluj-Napoca, Romania

E-mail: cvdirzu@gmail.com

Abstract. This paper attempts to demonstrate the need of implementing data transfer solutions from a mechanical surgical stapler to a display contained in a surgical image chain system, used broadly in any operating theatre, and bring a valid technical solution to fulfil this need. The object of the writing is to present a working electronic system that can be implemented in a circular surgical stapler and bring solid conclusions after testing on animal tissue. The setup contains an Arduino-based system that is capable to detect the pressure exercised on a tissue and display the results on a monitor that is easy to be seen by all the surgical team members, using a wireless connection. The system's role is to provide information to every surgeon in the OR and give them the opportunity to make a collective decision regarding the moment of firing the surgical stapler, in the safest conditions.

1. Introduction

In the last decades, the number of surgical mechanical staplers (SMS) being used in the operating rooms (ORs) is considerably increasing and these types of devices are being used in almost every kind of surgery related to the intestines. Despite the better training of the surgical team members (STMs) and the growing experience in using the SMS, there are still reported incidents of improper handling or improper results on the overall patient treatment and recovery. Some of the incidents are getting to the point that the patients suffer from irreversible conditions, like grade I handicap, sometimes being forced to carry a colostomy bag for the rest of their life, or even including death after severe complications.

In a low anterior resection (LAR), the end-to-end anastomos is made with a circular surgical stapler (CSS) is considered a vital maneuver of the entire surgery. In this step, the proximal and distal end of the colon are being reconnected together [1] and nowadays, most of the times, the procedure is being made using a CSS because of the great medical outcomes it creates [2].

Personal observations made during previous experience and visits made to various ORs show that most of the time, the person in charge of firing the CSS is the 1st Assistant.

Because the steps required to use a CSS in a such procedure are complex, the 1st assistant surgeon will manipulate the instrument between the patient's legs, while the main surgeon remains in the sterile abdominal field [3], without being able to see the status of the maneuver. Because the 1st assistant will be the only one who can see the indicators of the CSS, it will be difficult to take a common team decision regarding the moment of firing the instrument, for creating the proper circular anastomosis.

As it was already stated in various publications, the moment of reaching the force of 8gf/mm² applied on a living tissue is a key factor that influences the results of using a SMS [4], being extremely important to be measured and observed not only by the person in charge of firing the device, but by the entire surgical team, due to the complexity of the maneuver.

As observed in Figure1, the surgical Trendelenburg position of the patient with open legs situation, the sterile surgical drapes and the instruments used to secure the patient on the surgical table [5] that are being used in almost every surgical procedure, creates a sight barrier that blocks the main surgeon's visibility to the hands of his 1st assistant, therefore to the instrument indicators.

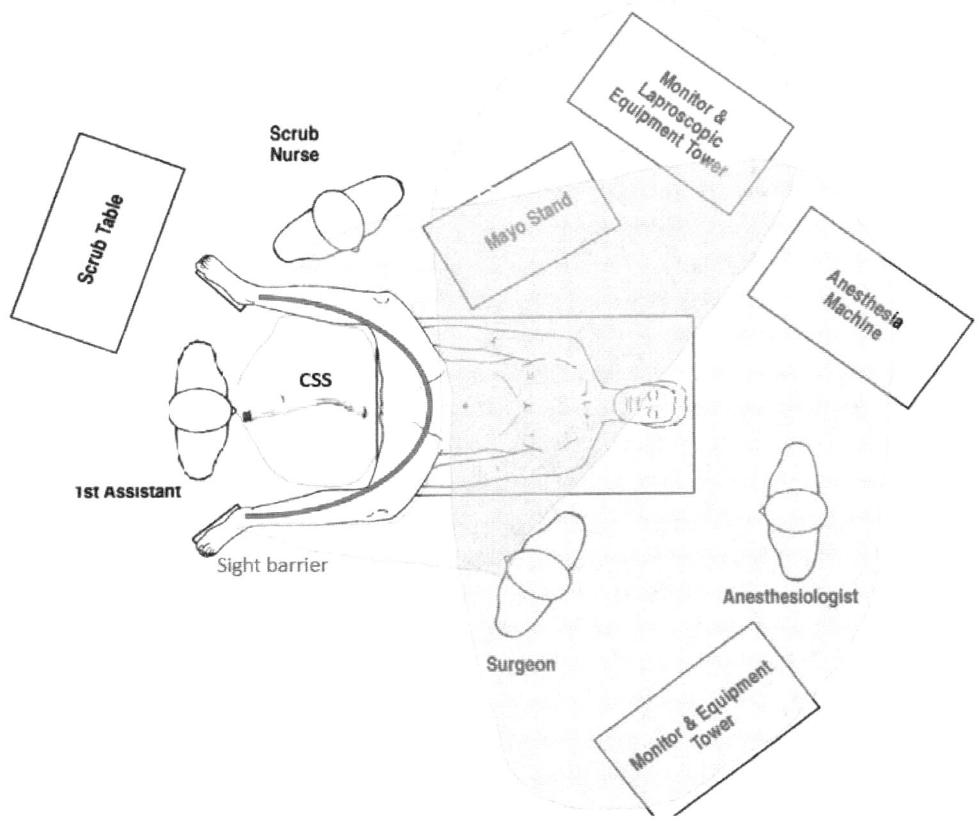

Figure 1. STMs positions during a LAR procedure and visual field when using the CSS.
Visual perimeter of the Nurse, marked with blue
Visual perimeter of the Surgeon, marked with green
Visual perimeter of the 1st Assistant, marked with orange

The main objective of this paper is to solve the problem of some of the STMs that are not able to see the status of CSS, moreover, to allow the main surgeon to support the decision to staple the tissue, by bringing the information in the sight of all STMs, transferring data from the CSS to any screen capable of wireless connectivity that can be found in the OR.

2. Method

The system proposed by this paper is an improvement of previous work made by the author, bringing a functional technological upgrade that will be described in the following chapter.

IOP Conf. Series: Materials Science and Engineering 1254 (2022) 012031 doi:10.1088/1757-899X/1254/1/012031

2.1. Materials

The system includes one traditional SMS and the electronic setup part, being commanded and interconnected by a custom-made operational program in two steps: a measuring hardware rolled by a custom software and a wireless data transfer system (DTS), from the SMS to an external display.

2.1.1. The measuring system

The system used to measure data at the tissue-end is an adaptation of an already existing SMS, created by Johnson&Johnson company: a circular stapler with a 33mm head diameter and two rows of surgical staples, mostly used in LAR procedures. This is a pure mechanical instrument that requires a strict steps-to-use protocol and a comprehensive training in advance, with the rest of the work being made using several electronic parts.

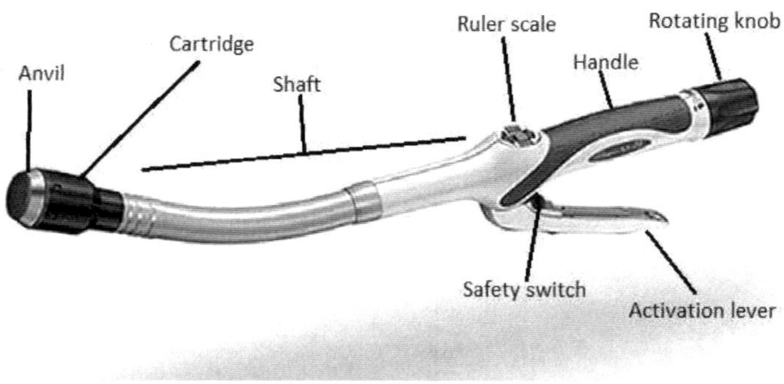

Figure 2. Example of an Ethicon CS and components [1].

To retrieve the data, a round force sensitive resistor (FSR) with a 12.7mm diameter and variable resistance between 100gf-10kgfwas used. The adjustments for balancing the system were made with a multi-turn potentiometer. The FSR was placed between the CSS cartridge and anvil. The FSR role is to vary its resistance accordingly to the pressure applied on its surface, where the tissue is being squeezed. The round shape was selected as it easier to implement it between the CSS jaws and it can be easily designed for later applications [4,6].

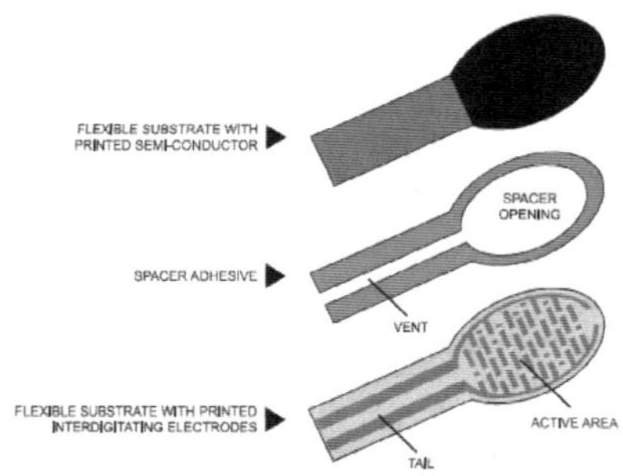

Figure 3. Components of a FSR [6].

For processing the data, a microcontroller development board (MDB) with IEE 802.11 (WiFi) ability was used, model ESP12 MINI V2. The MDB characteristics of 4MB flash memory, 80Mhz/160MHz processor and 11 pins are more than needed to be able to read, process and transfer data to LED module and wireless display in real time and without delays.

For displaying the data at the main CSS user's handle, an RGB LED module was used, to alert the 1st assistant that the threshold of 8gf/mm² of applied force will be reached.

2.1.2. Data transfer system
The DTS is a programming code implemented and run on the MDB, to connect the system to the screen for displaying the resulted data.

For displaying the data accordingly to the other STMs, like the RGB LED used on the CSS, a graphic meter was implemented in the program that will be visible on the wireless display.

To have a wireless display connected to the system, a mobile device with WiFi connection ability was used.

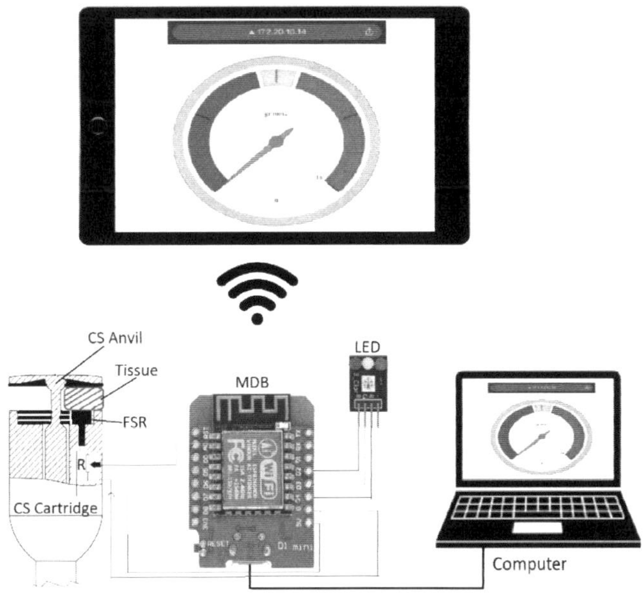

Figure 4. Systemsetup–blockdiagram.

3. Preparation and testing
To better understand the basics of stapling techniques, the ideal force applied to any living tissue must be considered as8gf/mm², described in the medical literature. Also, the tissue dynamics, different thickness of organs or same organ parts and biomechanical properties like the dynamic of tissue fluids must be understood [7-8].

This paper was conducted following the mentioned theories for keeping the testing as close as possible to reality, using pork meat as a target tissue. The tissue used for testing was set aside sufficient time to accommodate the room temperature, to avoid curdled fat for keeping the testing as close as possible to real conditions.

3.1. Connecting the system
To connect the system, the following protocol must be realized, as described in Figure 5:

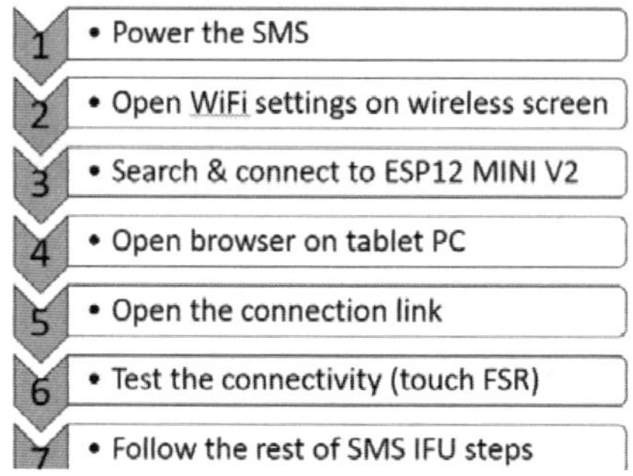

Figure 5. Wireless connection protocol.

The system is meant to be powered by an external battery pack, to be more appropriate for the real-life applications. For testing purpose, both powering methods have been used: battery pack and USB connection to the computer.

4. Results

The connection to the system can be made in less than 2 minutes, as the MDB can be easily found in the wireless connections list of any device. There can be multiple options available for setting a wireless connection, by adding the system to any internet or intranet connections, but the most convenient will be to bridge a connection directly to a wireless device. To maintain a high level of security, a password can be set into the programming code to obtain a safe connection.

In this setup, the connection distance is not an important factor to be aware of, as most of the general surgery OR size is estimated to be around 55m², according to the Guidelines [9], and the surgical table is commonly placed in the middle of the room, with the main equipment around it.

For different connection and display possibilities, multiple browsers were used, like Internet Explorer, Google Chrome and Safari, run on different mobile devices like iPad6, iPhone11, Samsung S7 and a Dell laptop. All the three browsers responded in good conditions on each of the mobile devices, with a small delay being noticed with Safari running on the Samsung S7 mobile phone. The rotating knob was operated multiple times, to observe if the values are transmitted in real time to the display. The same repetition was made using both power methods: battery pack and USB connection to computer.

In this application, a simple digital gauge indicator was used, which respects the same colors as the LED module, showing in the middle-bottom box also the exact value of the pressure exercised on the tissue, like in Figure6.

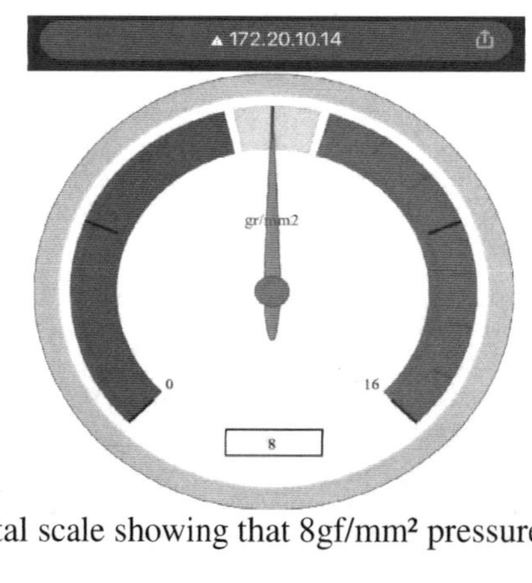

Figure 6. Digital scale showing that 8gf/mm² pressure was achieved.

The digital scale can be adjusted according to the surgeon requirements: it only needs a simple survey on a group of surgeons, for developing the best design.

For each series of tests that were made, the system reacted accordingly, showing the same colour code on both LED module and digital graph on the mobile display.

5. Conclusions

Pure mechanical instruments provide rudimentary feedback to the user if the instrument is not empowered with smart electrical and data systems. A basic SMS that is broadly being used since so many years still bring risks of misfire or unproper placement of staples, even though the steps for use are well documented.

The setup proposed by this paper succeeded to solve a real-life issue, stated at the beginning. It is capable to bring real-time data from the tissue that is being treated, to the sight of STMs, so they can visualize the process of using the SMS during the most important step: while the 1[st]assistant is placing pressure on the tissue at the level of the anastomosis (Figure 7). This is a great step forward in the process of taking the right decision to activate the firing handle and create the anastomosis, when the tissue and the CSS are perfectly engaged to each other, providing the best outcome in the healing process of the patient. Moreover, the system can be also useful for teaching activities in large University Clinics, when the students are observing live surgical procedures inside the OR.

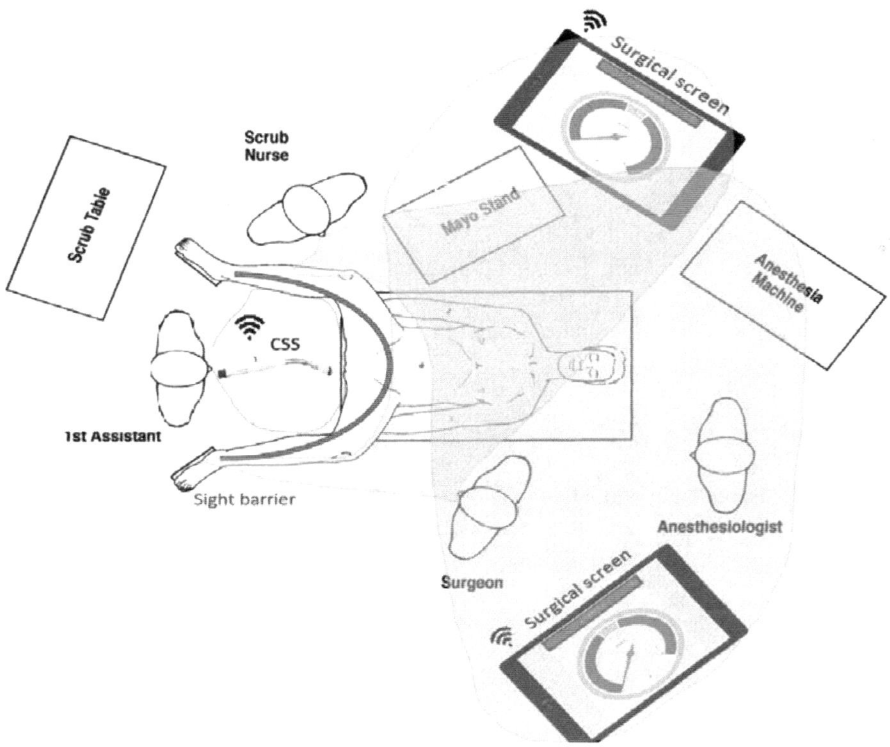

Figure 7. Bringing data in sight of the STMs.
Visual perimeter of the Nurse, marked with blue
Visual perimeter of the Surgeon, marked with green
Visual perimeter of the 1st Assistant, marked with orange

As demonstrated, the data read by the FSR can be easily transmitted to a consumer-goods tablet PC. Because nowadays the ORs are being transformed and are being continuously digitalized, many

WiFi or Bluetooth connections can be set to different equipment found in the respective room. For example, the largest laparoscopic towers manufacturers offer different documentation and audio-video management solutions, which are capable to bring any wired or wireless image source into the image chain matrix and to display it on any interconnected screen [10-11].

To strengthen the results of this work, more testing can be made to check the connectivity possibilities to different equipment, or to observe if the quality of signal can be affected, by placing inside the room different devices that can perturbate the wireless connection.

Acknowledgment
This writing was made based on author's professional education in medical engineering and previous experience in distribution of surgical products, including SMS among others, with continuous follow-up and feedback received from its users, represented by surgeons and nurses.

References
[1] https://medical-dictionary.thefreedictionary.com/end-to-end+anastomosis
[2] Ti T K, Rauff A and Goh H S 1986 Anterior resection using the circular stapling instrument: a Singapore experience. AustNZJSurg. 1986 Dec **56**(12) pp 919-922. doi:10.1111/j.1445-2197.1986.tb01855.x PMID: 3469986
[3] 112615-200325 EMEA Ethicon Circular Stapler ODP Guide
[4] Dîrzu C and Munteanu M 2021 Tissue thickness and pressure measurement solutions in life-saving surgical procedures *9th International Conference on Modern Power Systems (MPS)* |978-1-6654-3382-2/21/
[5] Young-Fadok T M, M.D., M.S., F.A.C.S. *15.1. Colorectal Resections: Patient Positioning and Operating Room Setup* Springer
[6] Force sensing resistor, Integration Guide and Evaluation Parts Catalog, Interlink electronics, Version 1.0, 90-45632 Rev. D
[7] Elariny H, Gonzalez H and Wang B 2005 Tissue thickness of human stomach measured on excised gastric specimens from obese patients. *Surg Technol Int.* XIV **14** pp 119-124
[8] Chekan E and Whelan R L 2014 Surgical stapling device-tissue interactions: what surgeons need to know to improve patient outcomes. *Med Devices (Auckl).* **7** pp 305-318. Published 2014 Sep 12. doi:10.2147/MDER.S67338
[9] Burlingame B 2014 Operating Room Requirements for 2014 and Beyond, 2014 FGI Guidelines Update Series MS, RN, CNOR, September 15
[10] Instructions for use AIDA C WD310, V1.0.0, Karl Storz Endoscope, Tuttlingen
[11] https://www.olympus.co.uk/medical/en/Products-and-solutions/Produts/Product/EASYSUITE.html

IOP Conf. Series: Materials Science and Engineering 1254 (2022) 012032 doi:10.1088/1757-899X/1254/1/012032

Electronic upgrades to a purely mechanical surgical stapler, for better outcomes in a vital surgical maneuver

C V Dirzu and M S Munteanu

Technical University of Cluj-Napoca, Faculty of Electrical Engineering, Cluj-Napoca, Romania

E-mail: cvdirzu@gmail.com

Abstract. This paper attempts to demonstrate the need of implementing an electronic system to collect data, command and control a surgical mechanical stapler, being broadly used in most of the intestinal resections for different pathologies. By implementing this electronic system, we will be able to eliminate some of the most important risk factors in the instrument steps to use, that can bring critical damage to the patient outcomes. The object of the writing is to present a working electronic system that can improve the use of such a device and to bring solid conclusions after testing on animal tissue. The setup contains an Arduino-based system that is capable to sense the pressure exercised on a tissue and automatically control the tension applied to it, before reaching a critical point to irreversibly damage the tissue. The system's role is to eliminate the risks of human errors when using the surgical stapler and automatize the entire process, easing the overloaded and stressful workflow of the surgeons.

1. Introduction

Surgical staplers (SS) and staples are medical devices that are used instead of traditional sutures. Most of the times are used in minimal invasive surgery where removal of organs and reconnecting parts of organs must be executed in narrow places, with maximum efficiency. SS work by compressing two parts of tissue and connecting the two pieces of tissue with staggered rows of B-shaped titanium staples. In some models, the excess of tissue will be automatically cut-out with an included blade.

In 2015, most of the SS found in hospitals, were made in almost 80 percent of the total by only two main manufacturers. The instruments these two companies produce are considered the most performant ones and it literally bring extraordinary improvements in every surgical procedure (SP) where are being used, with documented benefits in the recovery process of the treated patients[1].

Anyway, even with such experience, there are still reported incidents of misfiring or misuse that can damage the healing process or involve errors that can delay the SP or diminish the chances of success.

The U.S.A. Food and Drug Administration (FDA) and other organizations across the Globe are actively monitoring the reported incidents and since the launch of mass-production SSs, they required documented instructions for use (IFU) for every model of SS.

Most of the lawsuits involving SS generally consists of the same named injuries, like organ damage, sepsis, internal bleeding, temporary or permanent handicap, or even death in some cases [2]. For example, one of the main producers recently made a product recall on almost 100.000 SS units, with the concern that they might not fire with enough force to completely form the staples in the correct shape, generally being one of the main reasons for having such post-operative complications [1].

Therefore, even with the well documented IFU and continuous surgeon trainings, we still need to figure further solutions to improve the SSs and eliminate to a maximum the unfortunate incidents where the patients are at risk of suffering.

The concept presented in this paper is meant to take over difficult decisions of the operating surgeon and replace sensitive maneuver steps, by transforming a pure mechanical SS into an automated system that is capable to sense the right moment when the staples must be fired and do so in an automatic manner, that eliminates the risks of mechanical damage when manipulate the stapler by human hand.

2. Method

The system proposed by this paper is an improvement of previous personal work and it will be described in the following lines.

To try and document an eventual IFU for the proposed system and demonstrate the validity of the setup, a series of tests were made.

We used two different DC electrical motor units (EMU) to power the rotating knob and close the anvil to the cartridge, until the signal received from the sensor will give the stop command. A series of repetitions were made for each of both systems, to check the integrity of the components in the setup.

2.1. Materials

2.1.1. Physical components

The main system is an adaptation of an existing circular stapler (CS) manufactured by the company Johnson&Johnson (JNJ). The instrument has a 33mm head diameter and it can apply two rows of staggered staples and realize a circular cut, intended to be used in low anterior resection (LAR) procedures. The largest size was used to facilitate the sensor placement between the jaws of the instrument.

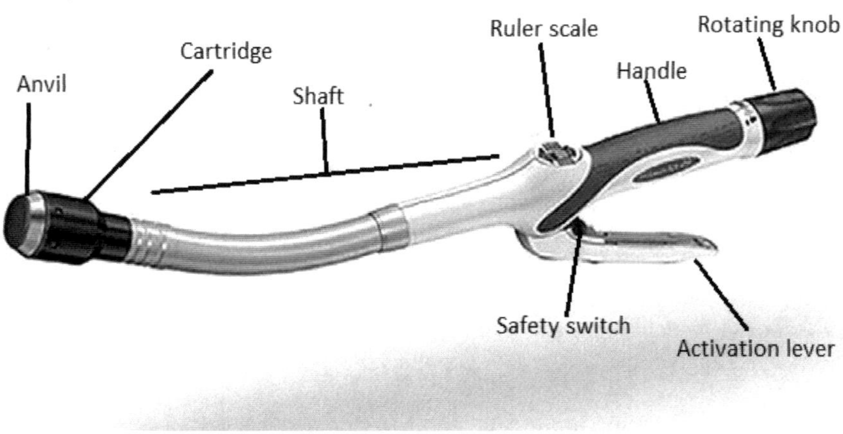

Figure 1. Example of an Ethicon CS and components [1].

To this pure mechanical instrument, a series of electronical components were added, inter-connected and rolled by a custom program code in the open-source Arduino Software (IDE).

To retrieve the data, a round force sensitive resistor (FSR) with a 12.7mm diameter and variable resistance between 100gf-10kgfwas used. The adjustments for balancing the system were made with a multi-turn potentiometer. The FSR was placed between the CSS cartridge and anvil. The FSR role is to vary its resistance accordingly to the pressure applied on its surface, where the tissue is being tightened. AFSR with a round shape sensing area was selected, as it would be easier to implement it between the CSS jaws and benefit of its total contact area [3].

For processing the data, an ESP12 MINI V2 microcontroller development board (MDB) was used. The MDB characteristics of 4MB flash memory, 80Mhz/160MHz processor and 11 pins are more than needed to be able to include all components and execute the program, to read and process the data.

The automatization of the closure system was made with two different EMU: a simple 6V R140 motor and a JGA25-371 geared motor with encoder. The two models were compared, to check if a simple DC EMU can be sufficient to solve the issue or a more complex EMU with reduction and encoder must be considered for such application. The circuits for both systems are described in the following chapter.

The procedural steps were replicated [3] for testing the system in real time, using for checking purpose an electronic digital caliper (EDC), a professional stopwatch (PSW) and the program clock.

For checking the system functionality, an RGB LED module was used to observe the steps carried out by the FSR and EMU when reaching the threshold of 8gf/mm².

2.1.2. Software program

The programming code implemented and run on the MDB is following the logical scheme designed in Figure 2.

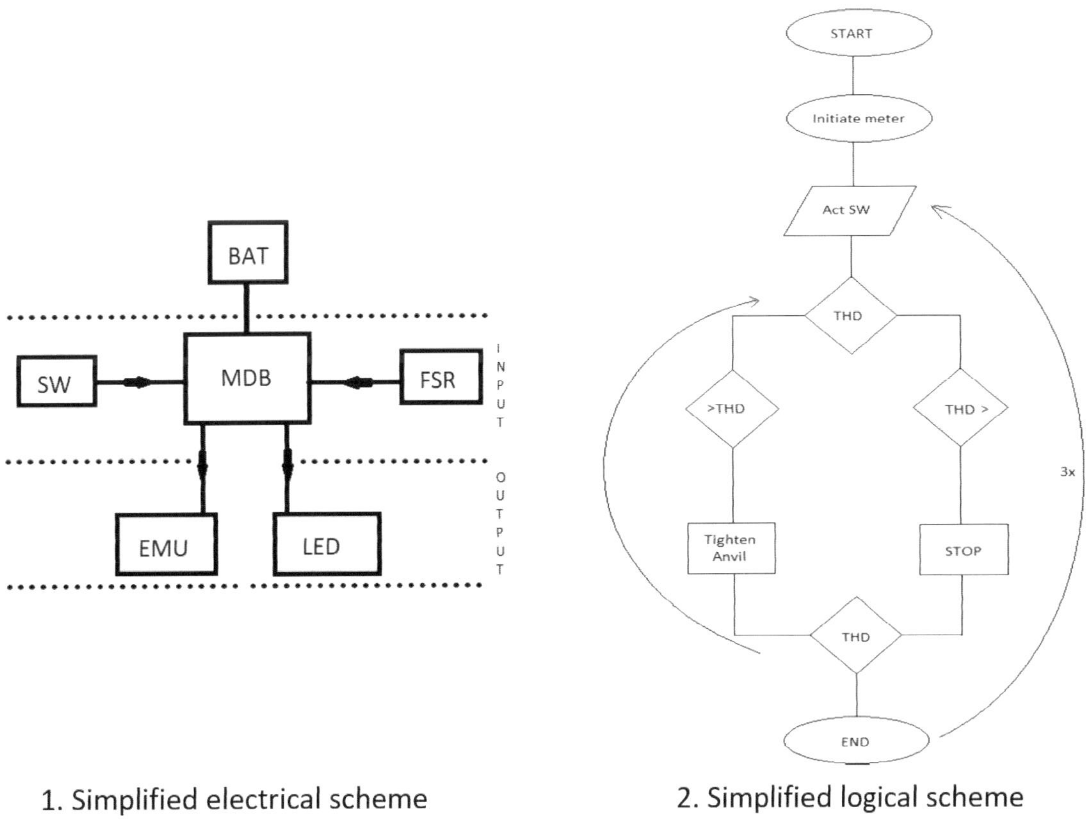

Figure 2. Electrical and logical schemes
Switch (SW), Motherboard (MDB), Battery pack (BAT), Force sensing resistor (FSR), Electrical motor unit (EMU), RGB LED (LED).

After turning the switch (SW) on, the system activates the EMU and start tightening the anvil to the cartridge, until the FSR is sensing the applied force. When reaching the critical value of 8gf/mm², the EMU stops and wait for 5 seconds: this loop starting with the activation of EMU until the stop will be repeated for three times. In total, 15 seconds of optimal applied force must be counted since the threshold was first reached, so the anastomosis will be made with only viable tissue [4, 6],

To display the data accordingly to the steps being performed by the system, the three colours of the LED were gradually added, one for every 5 second loop.

3. Preparation and testing

To better understand the basics of stapling techniques, the ideal force applied to any living tissue should be considered 8gf/mm², as described in the literature. Also, the tissue dynamics, different thickness of organs and same organ parts and biomechanical properties like the contained fluids movements must be understood [4, 6].

This paper was realized following the existing theories and literature for keeping the testing as close as possible to reality. Therefore, multiple tissues were considered for testing, like different pieces of pork, with different fat percentages, but also a disc of soft rubber to check the limitations of the system.

The standard steps for use, noted in the IFU of the modified SS, consist in more than 11 individual and complex steps, each of them with several sub-steps: [5]

Attention is required in all the described procedural elements, with some of them that are challenging the surgeon for an assumed decision. In Figure 3, the sensitive procedural steps to use and decisions to be made by the surgical staff were marked in orange.

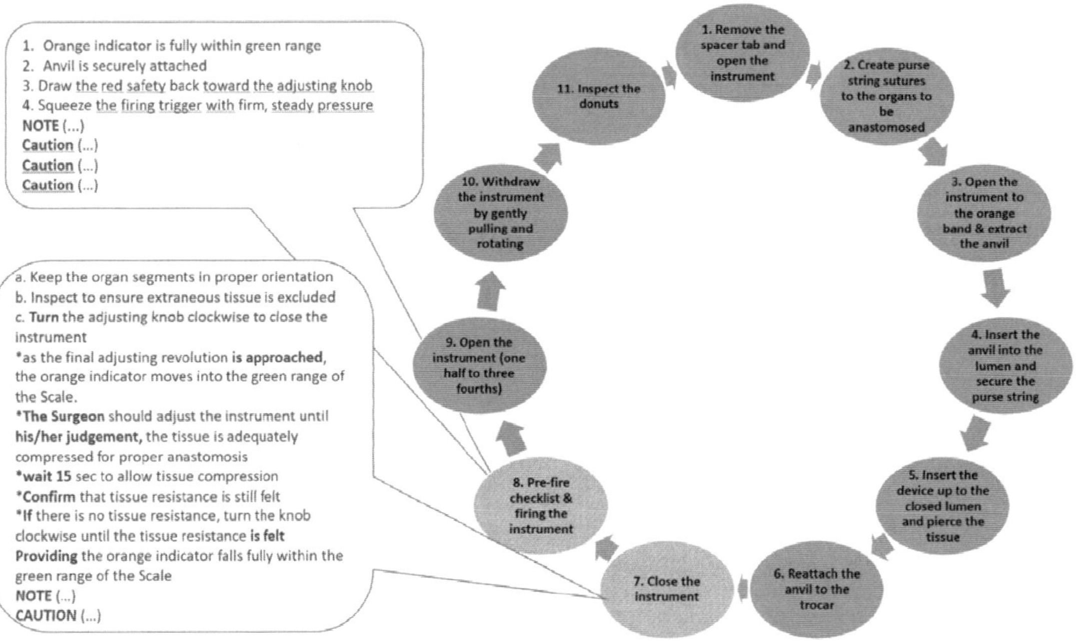

Figure 3. Excerpt of JNJ CS IFU.

3.1. System 1

First setup to be tested was made using a simple 6V R140 EMU. This is a basic circuit selected for its relative reduced costs, and it is described in Figure 4.

Because the EMU will draw more current than MDB can deliver, we had to complete the circuit and include a transistor as well, to protect the MDB from damage. For this, a Darlington TIP122 transistor was used, together with a 270Ω resistor and a diode to protect the MDB from back-currents that might result from the EMU.

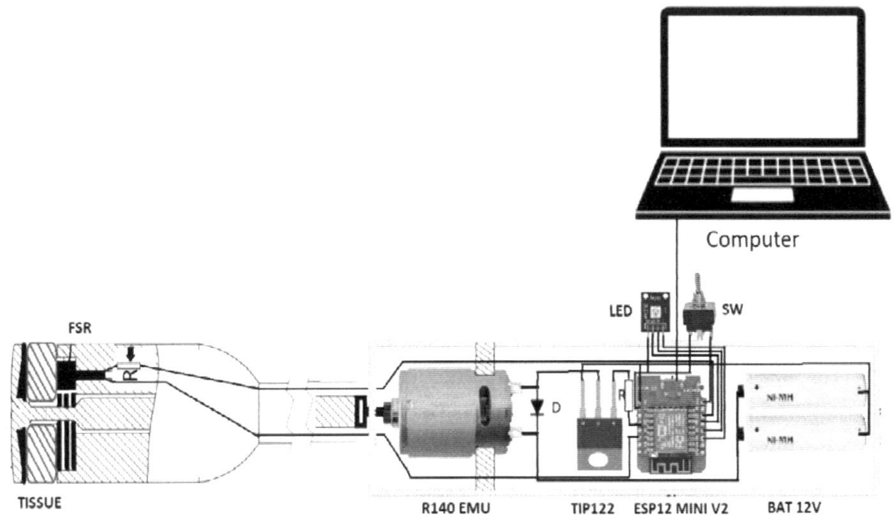

Figure 4. Connection diagram for the setup containing EMU 6V R140.

3.2. System 2

The second setup contains a motor with encoder, type JGA25-371, used in many types of applications, due to its improved quality and upgrades. The gear set contained helps to convert the original high speed and low torque of the motor, to a low speed and high torque state. The encoder role is to monitor and control the speed of the EMU through a motor driver chip model L298N. It is important that when the external load becomes larger, the speed of the ordinary motor will slow down, and it can be adjusted by an algorithm rolled on the MDB. The complete system is described in Figure 5.

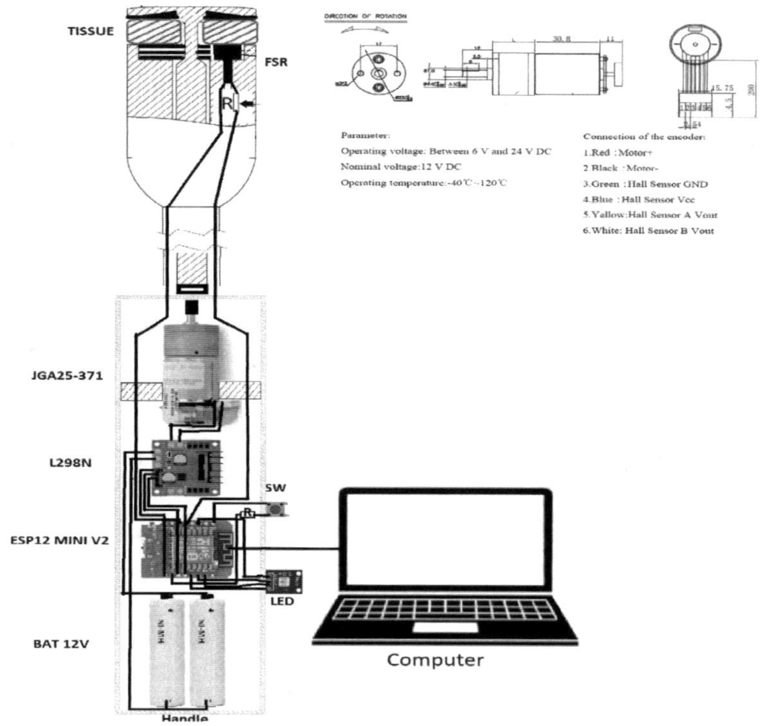

Figure 5. Connection diagram for the setup containing EMU JGA25-371.

Therefore, in this setup we give the EMU a command to reduce to half its speed when reaching 6gf/mm² and continue the rotation until 8gf/mm², when it will operate a full stop. Again, the system will repeat this protocol for three times. This additional function of the system provided by the encoder allows the tissue to better adapt to the pressure applied by the anvil, according to the tissue dynamic properties [6].

Both systems are meant to be powered by an external BAT, to be more appropriate for the real-life applications. For testing and additional checking purpose, they have been powered with a USB data-cable, connected to a laptop.

4. Results

The results of this practical work were to eliminate the most problematic steps to use of a CS, by converting them to automatic protocols made by an electrical system, like the proposed one.

The only unmodified remaining steps for use are the ones related to preparation and positioning in-situ of the SS components, that are practically related to the surgical aspects, as it can be well understood by any surgeon.

The difficulties imposed by the procedural elements that involve surgeon's personal decision were excluded, by implementing the electronic system to the SS: thus, it resulted in an easier IFU protocol, as seen in Figure 6. The only notes that must be known by the surgeon are referring to correct alignment of the intestine segments during SS placement and exclude the other tissues that don't make the object of the anastomosis, marked in the figure with green callouts.

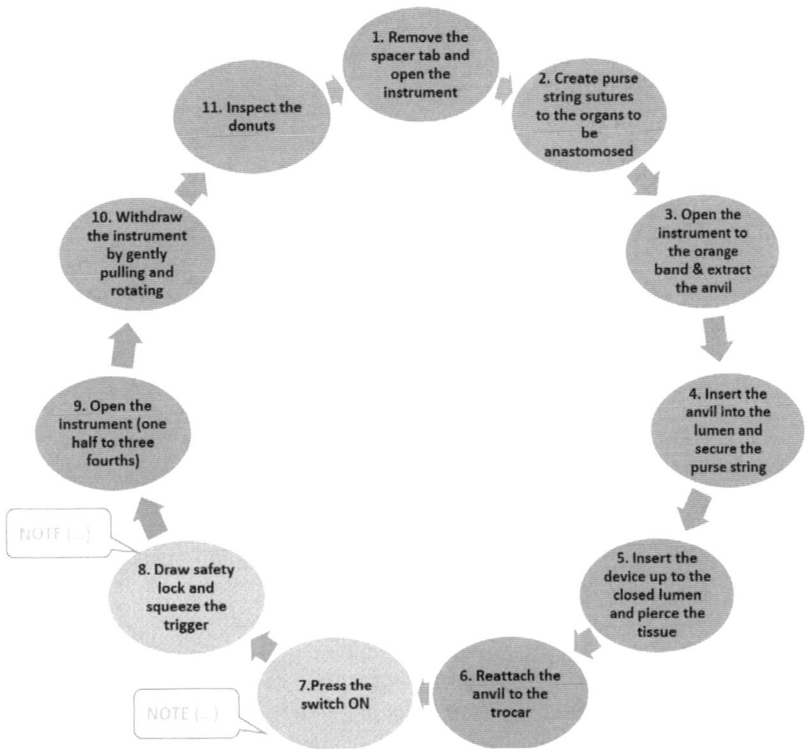

Figure 6. Simplified IFU resulted after SS electrical upgrade.

4.1. Resulted figures

The results obtained after testing the first setup containing the 6V R140 EMU are noted in Table 1. The values were counted and automatically noted in the program code rolled by IDE.

Table 1. System1 resultsTested material	Time to 8gf/mm²**(1)** (in seconds)	Pressure at EMU stop (ingf/mm²)	Pressure after 5sec (ingf/mm²)	Time to 8gf/mm²**(2)** (in seconds)	Pressure after 5sec (ingf/mm²)	Time to 8gf/mm²**(3)** (in seconds)	Pressure after 5sec (ingf/mm²)
10mm pork muscle	10	**8.6**	5.3	1.9	7.1	0.9	**8.2**
5mm pork muscle	11.4	**8.5**	6.4	1.2	7.4	0.6	**8.2**
10mm pork bacon	10.2	**8.3**	7.1	1.6	7.3	0.5	**8.1**
5mm pork bacon	11.7	**8.3**	7.3	1.1	7.7	0.5	**8.0**
10mm pork fat	10.9	**8.2**	7.5	1.2	7.8	0.3	**8.0**
5mm pork fat	12.1	**8.1**	7.6	0.8	7.8	0.3	**8.2**
3mm rubber	12.6	**8.0**	7.9	0.1	8.0	0.1	**8.0**

As it can be observed, this first assembly is a well-working system that is able to control the SS in almost good conditions.

It must be admitted that the simple circuit was not able to provide a sharp stop to the EMU in real time, after receiving the STOP command at 8gf/mm², as seen in the table above. Therefore, this must be considered a risk of over-pressuring the tissue and the setup must be upgraded accordingly. As it turns out from the data in the table, the tissues were constantly placed under greater pressure than the limit of 8gf/mm².

The results obtained after testing the second setup, containing the JGA25-371 EMU, are noted in Table 2. The values were extracted as well from the execution lines of the program, run on the computer connected to the system.

Table 2. System 2 results Tested material	Time to 8gf/mm²(**1**) (in seconds)	Pressure at EMU stop (ingf/mm²)	Pressure after 5sec (ingf/mm²)	Time to 8gf/mm²(**2**) (in seconds)	Pressure after 5sec (ingf/mm²)	Time to 8gf/mm²(**3**) (in seconds)	Pressure after 5sec (ingf/mm²)
10mm pork muscle	22.0	**8.1**	5.6	1.9	7.7	0.5	**8.0**
5mm pork muscle	23.2	**8.0**	6.9	1.2	7.6	0.5	**8.0**
10mm pork bacon	22.4	**8.0**	6.7	1.6	7.6	0.4	**8.0**
5mm pork bacon	22.9	**8.0**	7.2	1.1	7.5	0.5	**8.0**
10mm pork fat	22.9	**8.1**	7.2	1.2	7.9	0.2	**8.0**
5mm pork fat	23.7	**8.0**	7.5	0.8	8.0	0.2	**8.0**
3mm rubber	24.3	**8.0**	8	0.1	8.0	0.0	**8.0**

The geared motor with encoder solved the mentioned issue with success, providing a very good control over the instrument closure, acting in real time after the stop command received from the FSR/MDB. We consider that a similar model can be implemented at larger scale and retake the testing, for documenting the system properly.

4.1.1. General handling of the SS

The SS maneuverability differences after implementing the electrical components were insignificant. Because the EMU automatically runs the entire closing and tightening process, the surgeon only needs to hold and maintain the instrument in the correct position – this means that the involuntary movements and shakes while rotating the knob in a traditional manner, will be eliminated.

The benefits of eliminating this unnecessary and involuntary movements are more than welcome, because they produce fewer malformed staples [7-8] and are even easier to manipulate, especially for surgeons with smaller hands. Also, the advantage of automatic SS has been shown to bring superior performance in clinical practice [9] and even provide an economic benefit [10].

5. Conclusions

Pure mechanical instruments provide rudimentary feedback to the user if the instrument is not empowered with smart electrical and data systems. A basic SS that is broadly being used since so many years in every OR are still bringing risks of misfire, unproper placement of staples or faulty handling of tissue, [2] even though the IFU are well documented.

The system proposed in this paper successfully managed to automatize important steps for use and exclude the risks that can be generated by taking wrong decisions by the surgeon, or by improper handling of the instrument during the SP. The system concept based on a geared EMU has been demonstrated to be very stable and reliable for this application model.

In general, only experienced surgeons are capable to use such SS in a SP. This paper demonstrates that having minimal knowledge of use or fewer technical skills, like the young physicians do, it quickly becomes available for any skill-level surgeon to adopt mechanical suturing in their daily practice.

To strengthen the results of this paperwork, more research can be made on functional tissues in different testing laboratories and could include several categories of tissues, such as fatty / thin tissues, with / without oncologic treatment, male / female or different ages, and observe if the EMU can power in good conditions the SS. Also, a risk analysis and cost analysis can be further worked on to support the idea of mass production.

6. Remarks

Generally, the SS are sterile packed instruments. Because of the properties of the BAT, sterilization can be problematic in developing such a device for real use. Therefore, we can consider that the instrument itself does not need to be sterile, as it will be used in a non-sterile environment, at the back of the patient. We only need to consider the anvil as a sterile part as it goes thru the laparotomy, and include a protocol in the IFU that requires to change the gloves of the 1st assistant after using the SS.

A mention to a further manufacturing protocol must be considered: while the testing phase, we noticed that the system's mechanical characteristics become weak after the fourth repetition. Therefore, to produce an automated stapler that is in perfect conditions to be used, it cannot be pre-fired during the Quality Check more than two or three times, with the fourth one to be realized directly into the SP.

Acknowledgment

This writing was made based on author's professional education in medical engineering and previous experience in distribution of surgical products, including surgical mechanical staplers among others, with continuous follow-up and feedback received from its users, represented by surgeons and nurses.

References

[1] https://www.drugwatch.com/surgical-staplers/#:~:text=Surgical%20staplers%20and%20staples%20are,used%20in%20minimally%20invasive%20surgery
[2] https://www.chaffinluhana.com/risks-and-injuries-associated-with-surgical-staplers/
[3] Dirzu C and Munteanu M 2021 Tissue thickness and pressure measurement solutions in life-saving surgical procedures *9th International Conference on Modern Power Systems (MPS)* |978-1-6654-3382-2/21/
[4] Elariny H, Gonzalez H and Wang B 2005 Tissue thickness of human stomach measured on excised gastric specimens from obese patients. *Surg Technol Int. XIV*; 14: 119-124
[5] Ethicon Endo-Surgery IFU for Curved and Straight Intraluminal Staplers (ILS), REF: CDH21A, CDH25A, CDH29A, CDH33A, SDH21A, SDH25A, SDH29A, SDH33A, Rev.2010-04, P40641P01
[6] Chekan E and Whelan R L 2014 Surgical stapling device-tissue interactions: what surgeons need to know to improve patient outcomes. Med Devices (Auckl) **7** pp 305-318. Published 2014 Sep 12. doi:10.2147/MDER.S67338
[7] Kimura M, Tanaka H and Hato M 2016 Evaluation of a New Stapler with Unique Surface Gripping Technology. *Br J Med Medical Res* 18: 6
[8] Kimura M and Terashita Y 2016 Superior staple formation with powered stapling devices. *Surgery for Obesity and Related Diseases* **12** pp 668-672
[9] Fegelman E, Knippenberg S and Schwiers M 2017 Evaluation of a Powered StaplerSystem with Gripping Surface Technology on Surgical Interventions Required DuringLaparoscopic Sleeve Gastrectomy. *J Laparoendoscopic & Adv Surg Techniq* **27** pp 489-494
[10] Roy S, Yoo A, Yadalam S, Fegelman EJ, Kalsekar I et al. 2017 Comparison of economic and clinical outcomes between patients undergoing laparoscopic bariatric surgery with powered versus manual endoscopic surgical staplers. *J Medical Econom* **20** pp 423-433

Low-cost monitoring system for temperature and pressure in steam sterilizer

D A Ivașcu[1], R Ciubucă[2], A Lodin[3] and M Munteanu[4]

[1]Technical University of Cluj-Napoca, Faculty of Electrical Engineering, Romania
[2]Stefadent SRL Bistrita, Romania
[3]Technical University of Cluj-Napoca, Faculty of Electronics, Telecommunications and Information Technology, Romania
[4]Technical University of Cluj-Napoca, Faculty of Electrical Engineering, Romania

E-mail: danieladrian.ivascu@yahoo.com

Abstract. This article presents the implementation of an electronic system for checking the pressure and temperature of a steam sterilizer, designed for dental offices. As the system was developed on the Arduino platform, the developed prototype has the advantage of a low-cost system (about 120 Euros), but in the same time it has the accuracy of measurements, the results obtained being consistent with the data measured using a reference system (Datalogger produced by Tecnosoft Italy).

1. Introduction

In the actual extremely pressing context of finding optimal solutions and devices for decontamination in general, and for the decontamination of medical instruments in particular [1], the research work in this area is progressing at a rapid pace. This research work is intended not only to develop these sterilizing devices, but also to find sustainable solutions to intervene on them for various operations. Also, the alternatives for intervention and maintenance or verification of these instruments are varied [2], depending on the possible malfunctions as well as on the vision of those who perform these tasks.

According to the Medical Instruments Directive (93/42/EEC) [3], such equipment "means any instrument, machinery, software, material or other type, whether used alone or in combination, used specifically for diagnostic and/or therapeutic purposes for human beings" [4].

It has been debated in the medical literature whether the steam sterilization processes can be monitored and the maintenance of these instruments can be planned [5]. The methods of sterilizing microorganisms by physical factors are diverse, which include a number of very different agencies, such as heat, drying, grinding, pressure and others [6]. Many emerging sterilization methods are based on low temperatures, although steam autoclaving is the most widely used, cheapest and most efficient method available today [7]. A priority condition is ensuring sterile and safe instruments for the daily function of medical offices and hospitals, and this condition is achieved through regulations, best practices, the creation of standard operating procedures, as well as the education of medical staff and technicians [8].

Law regulations on the mandatory use of pressure steam sterilizers in areas with contamination risk have led to a significant increase of their use in various sectors (from dental offices to cosmetic, pharmaceutical or food processing units), because without these devices, companies may not be authorized to carry out their activities. For this reason, the solutions of monitoring the parameters of these equipments have a great dynamic, being part of a continuous process of improvement and modernization.

The aim of the work was to make a practical device for monitoring the pressure and temperature inside a steam sterilizer without coating, with a volume smaller than 24 liters (hereinafter referred to as DMTP-SA). After the making of the prototype, it was mounted on the circuits of a 23 liter sterilizer for measurements, and the values thus obtained were compared with the data obtained from a reference device, a calibrated temperature and pressure datalogger produced by Tecnosoft Italy, but also with those offered by the own monitoring tools of the steam sterilizer.

2. General considerations

The use of steam sterilizers is considered useful in cleaning the surgical instruments from microorganisms and pathogens, and their performance must be properly maintained and monitored; consequently, these instruments require regular maintenance in order to keep them in optimal working order during the sterilization process [9]. The use of an uncalibrated autoclave can lead to failed sterilization, which increases the risk of transmitting various diseases [10]. Therefore, manufacturers recommend the implementation of an appropriate steam sterilizer maintenance program, whose details should be recorded in the sterilizer logbook (including problems, malfunctions, preventive and corrective actions) [11].

The SR EN 13060+A1 standard defines the design, placing on the market, testing of small pressurized steam sterilizers, specifying that "the small steam sterilizer has a chamber volume of less than 60 l and cannot be equipped with a sterilization module". Their importance is due to the fact that they are suitable for sterilizing various instruments that are resistant to moisture and high temperature [12], or when sterilizing materials and equipment that may come into contact with blood or body fluids (tools used by cosmetics, tattoo artists, hairdressers, etc.) [13].

Sterilization cycles are divided into three categories [13]:

- Type B - for sterilization of packaged solid products, porous products and tubular devices (single or multilayer) or unpackaged
- Type N - sterilization of unpackaged solid products
- Type S - Sterilization of products according to the manufacturer's specifications

In terms of efficiency and reliability, it is recommended to reduce water consumption through measures such as the preferential purchase of equipment with waterless pumps, stop the steam generator when the equipment is not in use, and the use of pretreatment systems, these can reduce waste in treatment systems of reversed osmosis water for steam generation [14].

By removing air from a sealed container containing water, a balance is created between liquid water and steam, and the pressure inside the container will correspond exactly to the steam pressure at the existing water temperature. The term "saturated steam" is used to refer to the nature of the steam present in such conditions. Water evaporates as the temperature rises, and so does the pressure until equilibrium is reached again, with a fixed correlation between steam pressure and temperature [15]:

Table 1. Interdependence between steam pressure and temperature [15].

Pressure (mbar absolute)	Temperature (°C)
1208	105
1431	110
2048	121
3043	134
3937	143

By describing the pressure conditioned by temperature, the steam pressure curve is obtained:

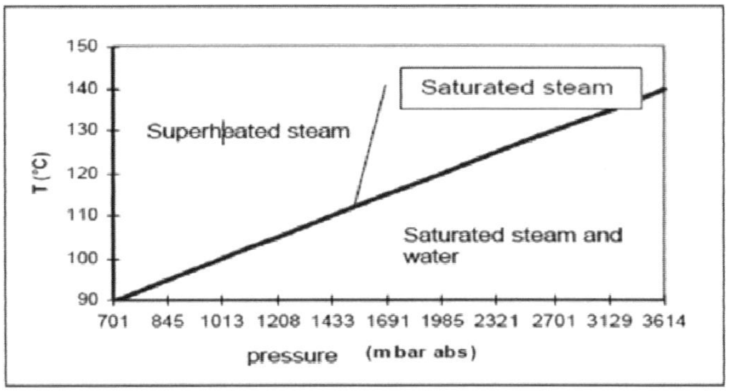

Figure 1. Schematic evolution of saturated steam. Temperature-pressure connection for saturated steam [15].

3. The method and materials used to make the prototype of the DMTP-SA system

The DMTP-SA prototype was developed around the Arduino Uno board, which has the advantage of being an open source platform, defined by its only two components: the ATMEGA 328P-PU physically programmable circuit board, around which are built both the platform and the software (IDE-Integrated Development Environment), that runs on the working computer. It collects data from different sensors connected to the board pins and can act on other devices (servomotors, LEDs, or other types of mechanical devices) based on commands contained in the code loaded in the memory of the microcontroller [6].

The electronic assembly has been inserted into a metal case, and the measured data is displayed on an LCD display (16x2 Blue RC1602 display) with I2C adapter. The data of interest (pressure and temperature) are measured with the help of two dedicated sensors: the pressure sensor MPX4250AP (an absolute pressure sensor for industrial control, with working range 15 ... 700kPa pressure and supply voltage in the range 4.75 ... 5.25 V DC, 20mV per kPa) and Thermocouple K – 400 °C, which measures temperatures from -200 to 1250 °C and can operate in an oxidizing or inert atmosphere, being characterized by a good sensitivity (approximately 41µV/°C) [16]. The measured data is then subjected to an analog-to-digital conversion, via the PmodTC1-MAX31855 circuit, a module powered in the range 3.0V - 3.6V (recommended voltage is 3.3V) and operating in the range 270°C - 1800°C. A temperature conversion takes about 100 ms, generating an output potential (which is then converted to a digital value) by the following equation:

$$V_{out} = (41{,}276\,\mu V/°C) \times (T_R - T_{amb})$$

Where: V_{out} – represents the output voltage of the thermocouple in μV
T_R – represents the reference junction temperature in °C
T_{amb} – represents the surrounding temperature

The wiring diagram and block diagram of the temperature and pressure monitoring device inside a steam sterilizer with a volume smaller or equal to 24 liters are shown in Figures 2 and 3:

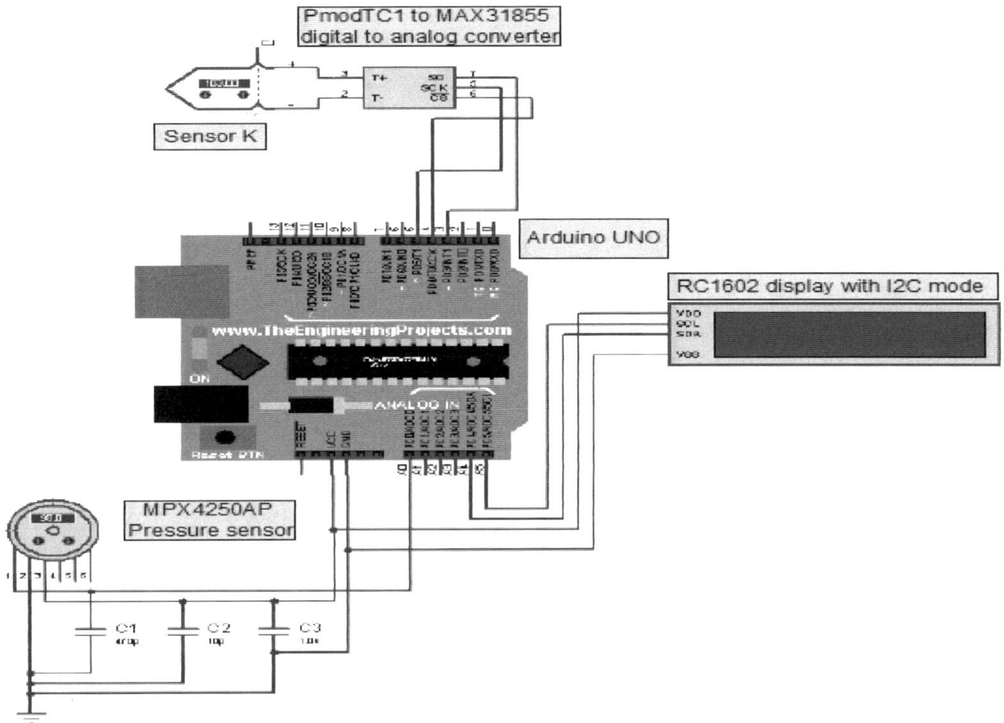

Figure 2. Wiring diagram of the device.

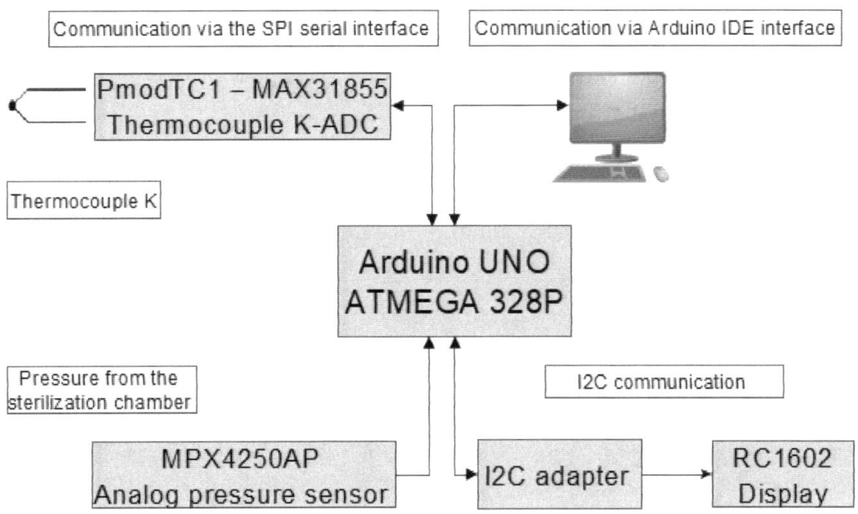

Figure 3. Block diagram of DMTP-SA.

In order to test the performance of the device and validate the measurement set, other equipments were used. Thus, several experiments were performed for different sterilization cycles, using a class B pressure steam sterilizer - Model BTD23-P & T (sterilization volume - 23 L, sterilization temperatures - 121°C/134°C, operating temperature in the range 5°C ~40°C).

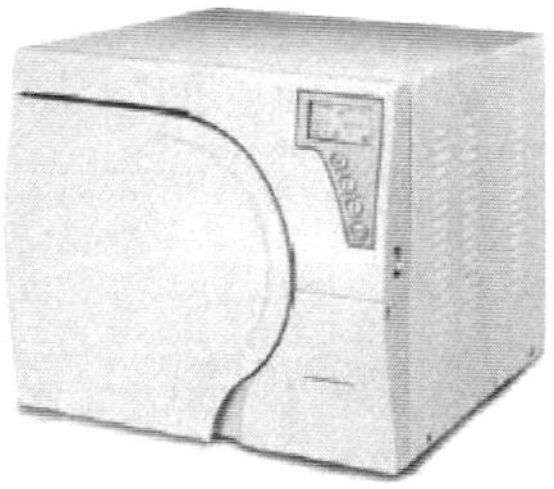

Figure 4. Steam sterilizer under pressure class B - Model BTD23-P & T [17].

This sterilizer used for testing can perform several types of programs (different temperatures and pressures, depending on the load being sterilized), but also provides two test programs, one to assess the level of air leaks in the circuit and the second for assessing the degree of steam penetration.

Also, all the experimental data measured with the made prototype, but also with the reference one (temperature and pressure logger) were then transferred to a digital system (processor frequency 1.1 GHz, Internal RAM - 3.0 GB, SSD - 120 Gb, 64 - Operating System, Windows 10).

4. Making the device and connecting it to the sterilizer enclosure

Starting from the electrical diagram shown in Figure 1, the stages of the actual making of this device are presented in Figure 5.

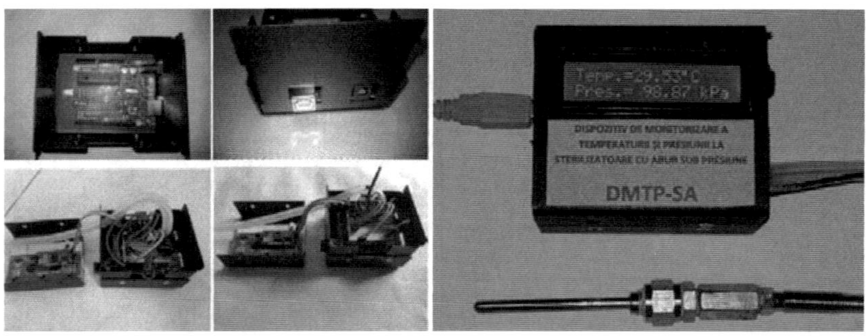

Figure 5. Steps in making the device.

Next, the thermocouple K was connected to the existing mandatory spare connection to the sterilization chamber, as in Figure 6:

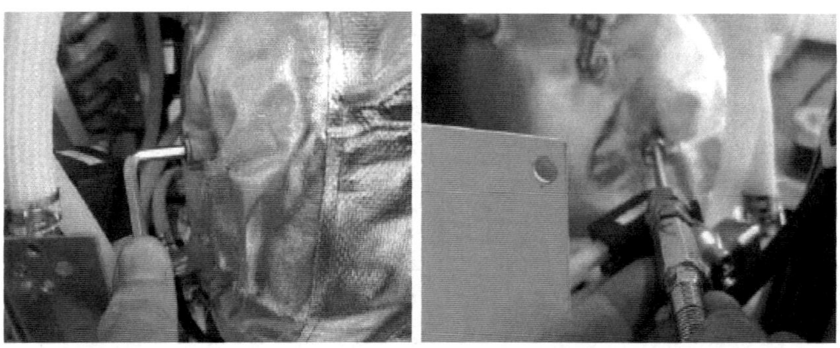

Figure 6. Accessing the spare connection and mounting the K thermocouple on the device.

So, the pressure sensor was connected by a branch to the heat-resistant hose, which ensures the operation of the sterilizer's own pressure sensor, according to Figure 7:

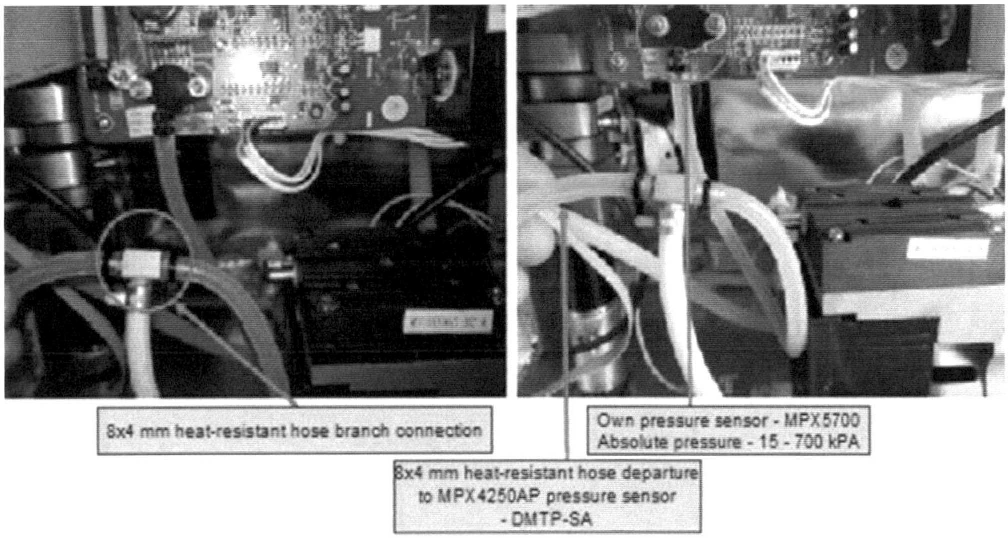

Figure 7. Connecting the MPX4250AP Pressure Sensor.

5. Testing the equipment and the results obtained

To perform measurements on the designated sterilizer, a normal duty cycle was determined by the following parameters: for 20 minutes (19.55 minutes), temperature slightly above 121 C (but not more than 3°C), which represents theoretical temperature depending on the pressure inside. According to the SR EN13060 standard, the theoretical temperature is also considered the measured temperature and the sterilizer must meet this performance criteria.

Figure 8. Parallel verification of sterilizer parameters in the maintenance phase.

The testing of the equipment consisted of performing temperature and pressure measurements and then comparing them with the data obtained by a reference device, namely a calibrated temperature and pressure datalogger (PressureDisk - Tecnosoft Italy).

The theoretical temperature of saturated steam is also a temperature that is considered measured and is calculated as a function of pressure, according to the formula: $T(S) = \left[A * \left(B - \ln\left(\frac{P(S)}{100}\right)\right)^{-1}\right] - C$, (A, B, C) are coefficients, and T (S) and P (S) the temperature respectively the saturated steam pressure [18].

The results of the temperature and pressure measurements performed with the made device and with the reference device are presented in Figures 9 and 10.

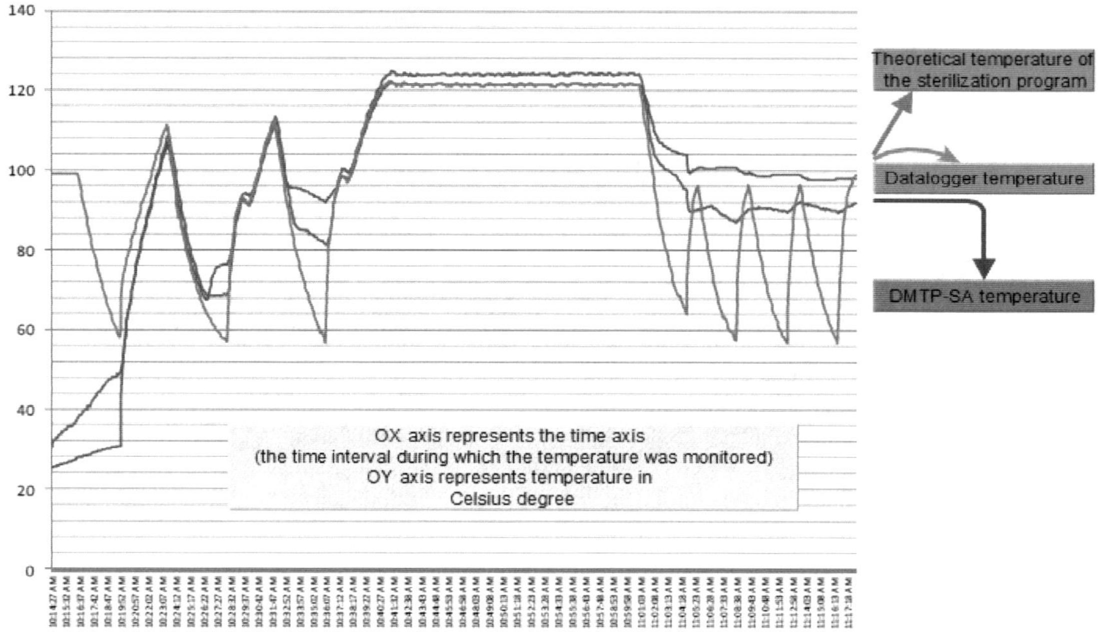

Figure 9. Datalogger and BMTP measured temperatures - GRAPHIC 121-1.

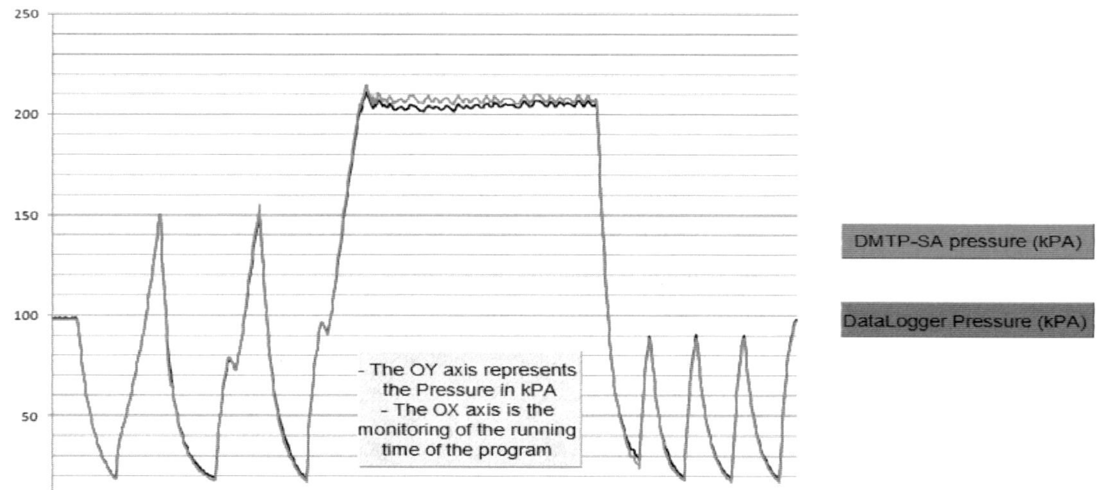

Figure 10. Pressures measured by Datalogger and DMTP - GRAPHIC 121-2.

From the previous graphs it can be seen that the results of the pressure and temperature measurements performed with the designed device are consistent with the measurements made with the system considered as reference.

The analysis of the pressure and temperature data recorded during the sterilization program at 121°C and represented graphically in figures 9 and 10 leads to the formulation of the following observations:

• The internal reference temperature of the PmodTc1 mode for the analogue-to-digital conversion of the temperature provided by the thermocouple K has a linear increase from 28.37°C at the beginning of the program to 32.75°C at the end of the cycle. The variation is 4.38°C for the entire sterilization cycle, being significant. This is not agreed, the reason being not to have any fluctuations in the compensation through the temperature reference junction.

• As in the case of the vacuum test, the pressure recorded from the DMTP via MPX4250AP and Arduino Uno, follows very closely the pressure recorded by PressureDisk on the pressurization slopes, on the vacuum slopes (pre-vacuum or drying vacuum) and during the maintenance period. In the case of the maintenance period, the variations are explained due to the repeated admission of steam. This inlet is controlled by the electronic process control system to maintain pressure and temperature within the standardized sterilization thresholds.

• Regarding the temperature, there are several areas with different behavior of the thermocouple K compared to PressureDisk. Thus, the vacuum stage is accompanied by a decrease of temperature. For three quarters of the sterilization program, the data collected are similar, but in the last part, the thermocouple K provides different data than the PressureDisk. During vacuum 2, at a certain point before reaching -860 mbar, the temperature decreases slowly, after which, once this stage is completed, by the admission of steam for pressurization 2, the temperature starts to increase rapidly. If in pre-vacuum 2 the thermocouple K records approximately the same minimum after which it approaches the data of the other sensor with a delay, in pre-vacuum 3 the minimum recorded by the thermocouple is much lower than that of the calibrated device.

At all pressurization stages there are moments when the data provided by the two temperature sensors coincide, which is valid until the end of the steam intake and the start, as the case may be, of another pre-vacuum or maintenance period.

Then, both the thermocouple K and the PressureDisk stabilize very well on the sterilization maintenance tray, with the mention that the thermocouple K does it at a threshold above about. 2.5°C.

6. Conclusions

The experimental work proposes a flexible low-cost system for monitoring the parameters of a sterilization system. An example of an instrument used to measure pressure and temperature in an autoclave is the High Temperature Pressure Data Logger OM-CP-PR140 which can be purchased at a current price of 1300 euros, the cost difference being significant.

The tests performed with the made equipment were compared with the measurements performed with a reference device, highlighting the fact that the data provided by the developed prototype are comparable to those provided by the calibrated reference system.

Subsequent research aim to perform measurements for different sterilization cycles, at different temperatures and time intervals of this process and with different types of sensors, in order to determine the optimal measurement configuration.

In addition, it is planned to add a wireless data transfer module for the development of a mobile app, so that the verification of the sterilization process, monitoring and possible modification of some parameters is done remotely, without involving the presence of the operator in place, near the sterilizer.

References

[1] Marincaş R *et al* 2017. *Sterilization manual: (practices for decontamination of medical devices)* (Zalău, Romania: Color Print)

[2] Romanian Standardization Association 2015 *SR EN 62353. Electromedical Devices - Recurring inspection and post-repair inspection of an electromedical device*

[3] European Committee for Standardization 1993 *Bundel Steriliseren en Steriliteit. Medical Device Directive 93/42/EEG*

[4] van Doornmalen Gomez Hoyos J P C M 2013 *Surface steam sterilization : steam penetration in narrow channels*(Netherlands: Technische Universiteit Eindhoven) Available from: https://doi.org/10.6100/IR758412
[5] van Doornmalen Gomez Hoyos J P C M 2018 Parametric Release of Loads in Surface Steam Sterilization November 2nd *19th World Sterilization Congress*
[6] Rahn O 1945 *Physical Methods of Sterilization of Microorganisms* (Ithaca, NY: Laboratory of Bacteriology, Cornell University)
[7] Cohen A 2010 *Dental Sterilization Solutions*
[8] Nothum P H 2013 *Managing Steam Sterilization Process*
[9] Romanian Ministry of Health 2016 *ORDER no. 961 of 19 August 2016 for the approval of the Technical Norms on cleaning, disinfection and sterilization in public and private sanitary units, working techniques and interpretation for tests to evaluate the efficiency of the cleaning and disinfection procedure, the recommended procedures for hand disinfection, depending on the level of risk, the methods of application of chemical disinfectants depending on the substrate to be treated and the methods for evaluating the progress and efficiency of the sterilization process.*
[10] Cheng V C *et al* 2013 Management of an incident of failed sterilization of surgical instruments in a dental clinic in Hong Kong *J Formos Med Assoc* **112** 11 pp 666-675
[11] Scottish Dental Clinical Effectiveness Programme Dundee Dental Education Centre. Available from: https://www.scottischdental.org/cep
[12] Dion M and Parker W 2013 Steam Sterilization Principles. Pharmaceutical Engineering *The official Technical Magazine of ISPE* **33**(6) Available from: https://ispe.org/pharmaceutical-engineering
[13] Romanian Standardization Association 2019 *SR EN 13060+A1. Small steam sterilizers*
[14] Laranjeira P R, Bronzatti J A G i, R Queiroz de Souza R and Graziano K U 2017 *Steam Sterilization: Fundamental Aspects and Technical Resources to Reduce Water Consumption*
[15] Miorini T 2008 *Sterilization of Medical Devices* Graz, German: Institut for angewandte Hygiene (Institute for Applied Hygiene)
[16] Thermocouple probe n.d. Available from: https://www.frigotehnie.ro/termostate/1270-sonda-termocuplu-tip-k-6x100mm 011000082710.html
[17] P&T MEDICAL TECHNOLOGY BTD 23 B Autoclava – Bimedis Available from: https://ro.bimedis.com
[18] *ONR 112069-2. Annex 1 to Guideline for validation and routine monitoring of sterilization processes with moist heat for medical devices.* 1.Jänner 2001 Fachnormenausschuss 215 Desinfektion und Sterilisation von Medizinprodukten

IOP Conf. Series: Materials Science and Engineering 1254 (2022) 012034 doi:10.1088/1757-899X/1254/1/012034

Segmentation of panoramic dental radiographies for teeth displacement analysis

O Bunta[1], C M Stăncioi[2], M Munteanu[3], M Abrudean[2] and V Mureșan[2]

[1]Department of Orthodontics, University of Medicine and Pharmacy, Cluj-Napoca, Romania
[2]Automation department, Technical University of Cluj-Napoca, Cluj-Napoca, Romania
[3]Electrotechnics and measurement department, Technical University of Cluj-Napoca, Cluj-Napoca, Romania

E-mail: stancioi_cristina@yahoo.com

Abstract. The primary goal of this work is to analyse the panoramic dental radiography from different perspectives and to obtain the best image after processing it, from which then can be extracted the necessary information. The complete system was built in such a way that a dental specialist may utilize it to better examine the characteristics of a patient's intraoral configuration.The system was developed with the Matlab work environment and includes a graphical user interface that the doctor may easily operate. Image - processing techniques were utilized during the development to process such an image, and the ones that proved successful will be detailed in this article. Furthermore, another aim of this system is to measure the intensity of each element of the denture, permitting a dental specialist to draw additional vital diagnoses. Some diagnoses can be identified automatically even after the raw image has been pre-processed. As a conclusion, each step executed with the software will aid in the analysis of panoramic dental radiography.

1. Introduction

The appearance of the equipment that made it possible to perform panoramic radiography was an innovation in the field of oral imaging, given that, previously, dental radiographs were only intraoral. This radiology procedure was developed in 1946 by Prof. Paatero.

This type of radiography, known as panoramic dental radiography, is called orthopantomography, the name given to the first equipment in the world to perform this type of radiography, Orthopantomograph, produced by Instrumentarium in Finland between 1961 and 1964.

Dental radiographies are used primarily for doctors to see hidden conditions, those that cannot be seen with the naked eye. Thus, at the beginning of any dental treatment, your doctor will ask you to perform a dental x-ray just to be able to make a correct diagnosis and to recommend a treatment plan that is as suitable for one as possible. Dental radiographs are also used during dental treatments to verifythe correctness of their performance.

Dental radiographs are very important elements in establishing certain diagnoses or developing treatment plans. These show many details that are impossible to see without making it.

Many aspects can be followed on a dental x-ray[1]:

- Quantity (height and width) and quality of alveolar bone, possible resorption or bone infections;
- Possible infections of the teeth and bones that support them;
- Possible pathological diseases: fractures, tumours, foreign bodies, decalcifications, eruptiondisorders;
- The position and shape of some important anatomical formations: maxillary sinus, cranial nerves, temporo-mandibular joint;
- Checking the correctness or failure of an endodontic treatment;
- The position and condition of the natural teeth left in the oral cavity;
- Presence and condition of restorations in the oral cavity: fillings, dental work, implants;

There are five types of panoramic dental radiographies (Figure 1) depending on their incidence asfollows:

- standard incidence;
- orthogonal incidence for highlighting interdental spaces, with beam emission positioned perpendicular to the tooth;
- orthogonal incidence for highlighting interdental spaces, with beam emission positioned perpendicular to the tooth;
- frontal incidence without shadows, with increased focus on the front of the teeth;
- pediatric incidence for highlighting permanent eruptive teeth in children [2].

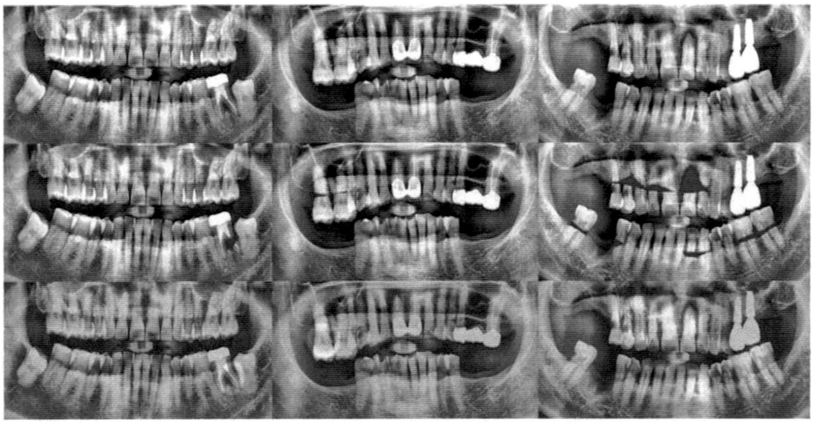

Figure 1. Types of panoramic dental radiographs.

2. Methods

Radiological investigations such as dental radiography should be performed before each dental treatment, regardless of the apparent complexity of the case, as dental problems can occur both in sightand inside the gum or bone.

Advantages of digital dental radiography [3]:

- 90% reduction in irradiation dose compared to classical radiology;
- Decreasing the working time by increasing the speed at which the radiological image is playedback on the monitor, practically in a few seconds;
- Getting high quality images;
- Possibility of image processing and highlighting details that cannot be perceived otherwise bythe human eye;
- Possibility of remote transmission of images over Internet networks;
- Better archiving and management of radiological images.

Because some images are hard to interpret, they need to be processed before the specialist can give a correct and accurate diagnostic.

An application was developed in order to take into consideration all of the above aspects for abetter and much easier evaluation of the panoramic dental radiography.

The application is designed and developed in such a way that it contains all the necessary tools forprocessing an image, after it is done.

Some images are noisy. Digital images are susceptible to various types of noise. The noise is the result of errors in the initial image acquisition, resulting in pixel values that do not reflect the intensity of the true image. Noise can appear in the image in several ways depending on how it is created the image. If the image is scanned from a photo, the grain size of the film or photo paper is a source of noise. The electronic transmission of the image can itself be introduced noise. Noise reduction or favoring of certain components of the image is done by the filtering operation.

Filtering is a technique of modifying or enhancing the image, to favor certain components or to remove other components. Image noise is reduced by replacing the value of each pixel with the medianof neighboring pixel values (considered a circular neighborhood, identified by radius). In the case of a data set, the median is the value number separating the whole set into two halves: lower (values <medianvalue) and upper (values> value) median). The median of a finite data set can be obtained by ascendingthe elements in the set and choosing the one of in the middle. If the set contains an odd number of elements, then there is a single middle value. If the set contains an even number, then the median valueis the arithmetic mean of the two middle elements.

Median filtering described in (1) is also an image smoothing technique (such as Gaussian linear filtration). All image smoothing techniques are effective in terms of removal noise, but it has the disadvantage of affecting the contours of the objects included in that image. The contours are critical elements in an image, so it is very important to eliminate the noise and keep them possibly unaffected. For a moderate level of noise, the median is the ideal way to remove noise, preserving the contours, that iswhy it is a technique often used in image processing [4].

$$(x , y) = \frac{1}{\sigma \sqrt{2 \pi}} * e^{-\frac{(x^2 + y^2)}{2 \sigma^2}} \tag{1}$$

The segmentation process is considered a process of classifying objects in an image, and in a some measure is equivalent to identifying the object itself, as a consequence of segmenting different objects (physical realization of classes or abstract patterns) that are perfectly located within the digital image. The goal segmentation is the simplification of the representation of images in useful information bypartitioning into regions. Segmentation Image is a method used to locate certain objects or outlines inan image. Several segmentation methods have been proposed for medical images. Segmentation obtained using traditional basic image processing techniques, such as threshold or histogram, requires human intervention in order to obtain satisfactory results [5]. Automating these approaches is difficultdue to the complexity of the shape, the shadows and the variability of the individual objects. Moreover,noise and other artifacts of the image may cause incorrect regions and contours, or discontinuities ofobjects obtained by through this method.

Digital image processing biomedical field is an area of independent. This domain has at based on a rigorous mathematical theory, well developed, but implementations are resource - intensive, in especially for real-time use of information extracted from images. The image is a description of the variation light intensity on a surface. A image is therefore a signal two-dimensional; thus, image processing can be also considered a branch of processing digital signal.

Spatial processing can be seen as a transform as in (2) and explained in Figure 2:

$$(x, y) = \{f(x, y)\} \tag{2}$$

Where:

- f is the original image;
- g is the resulted image;
- H is the operator defined for the neighboring of the pixel f(x, y).

Another filter that was applied is the spatial H filter which is described in (3):

$$(x, y) = h \cdot (x, y) \quad (3)$$

Where:

- g is the resulted image;
- h is the filter vector;
- w is the neighboring of the pixel f(x, y).

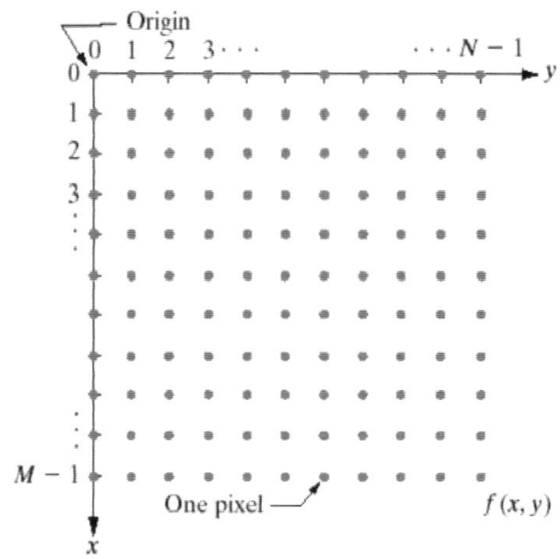

Figure 2. Digital image processing technique [6].

Because of the acquisition of the image, there will be noises identified and some of those noises make the process of identifying the diagnostic very hard. Also, depending on the equipment which is used, the sensors may not capture all the details to be visible. For this reason the pre-processing step is executed, so that the resulted image contains all the necessary information and details needed (Figure 3).

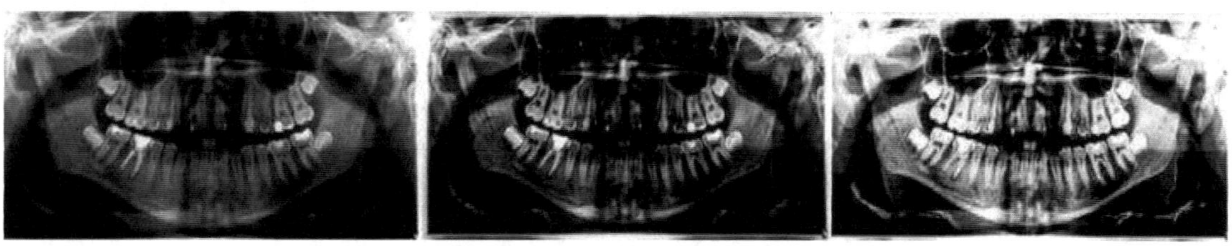

Figure 3. Image processing and filtering.

The distance between two pixels in a digital image is a quantitative quantity. The distance between the coordinate points (i, j) and (k, l) can be defined in the following ways:

- Euclidean distance. The advantage of this way of defining is that it is intuitive and the main thing the disadvantage is the presence of a cost high calculation due to the use of the radical butalso from the cause of the resulting integer value and the necessary interpolation;
- Inter-block distance (Manhattan), is used when only movement is allowed horizontal and vertical;
- Chess distance (Chebychev). This distance is used if in addition to horizontal movements and vertical movements are also allowed [7].

Image contrast adjustment operations are level-oriented transformations gray and represent associations that correlate the gray level of the original image with its new value from the processed image. Linear contrast adjustment linearly extends the original pixel values in a range values so that theentire sensitivity range is used. The method uses two thresholds, one minimum and one maximum, between which the range of sensitivity is defined interest.

3. Results and conclusions

Studies on existing experimental data in the field, but also research the experimental features presentedin this paper were intended to demonstrate that this is possible that by using the methods of improving the quality of the images, to facilitate and to improve the objectivity of the diagnosis and monitoring ofthe cerebral vascular system and a his possible ailments.
The analysis of the proposed methods shows the following conclusions:

- Application of algorithms to improve medical images and eliminate noise that occurs during theacquisition process, as stages of pre-processing a images, is essential.
- The use of different linear or non-linear filtering methods can help doctors in taking of decisionsregarding the diagnosis of different patients.
- Filters applied to different images have the advantage of extracting information from MRI imagesusing grouping algorithms.
- Quality descriptors, targets or hybrids, can be used successfully in quality assessment processedimages, by comparison with the original images or by using the system visually human.

References

[1] Pritchard B, Akbarian Tefaghi F and Makdissi J 2020 Anatomy in panoramic image interpretation *British Dental Journal* **228** 229 10.1038/s41415-020- 1324-1
[2] Bin Ghazali A B 2020 Interpreting panoramic radiographs for beginners, *BDJ Student* **27** 49 10.1038/s41406-020-0144-8
[3] Choi Y J, Young Kim Y H, Han S-S, Jung U-W, Lee C, Lee A and Jeon K J 2020 Alveolar bone height according to the anatomical relationship between the maxillarymolar and sinus *Journal of Periodontal & Implant Science* **50**(1) pp 38-47 10.5051/jpis.2020.50.1.38
[4] de Oliveira Capote T S, de Almeida Gonçalves M, Gonçalves A and Gonçalves M 2015 Panoramic Radiography—Diagnosis of Relevant Structures That Might Compromise Oral and GeneralHealth of the Patient *Emerging Trends in Oral Health Sciences and Dentistry* (InTech: London, UK)
[5] Kamburoglu K, Kolsuz E, Murat S, Yüksel S and Ozen T 2012 Proximal caries detection accuracy using intraoral bitewing radiography, extraoral bitewing radiography and panoramic radiography *Dentomaxillofacial Radiol* **41**(6) pp 450-459
[6] Tikhe S V, Naik A M, Bhide S D, Saravanan T and Kaliyamurthie K P 2016 Algorithm to Identify Enamel Caries and Interproximal Caries Using Dental Digital Radiographs *IEEE 6th International Conference on Advanced Computing (IACC)* pp 226-228
[7] Dinç I and Pusey M L 2015 Emerging Trends in Image Processing *Computer Vision and Pattern Recognition*

Automatic detection of melanoma bydeep learning models-based feature extractionand fine-tuning strategy

S Barburiceanu and R Terebeș

Communications Department, Technical University of Cluj-Napoca, 400114, Romania

E-mail: Stefania.Barburiceanu@com.utcluj.ro, Romulus.Terebes@com.utcluj.ro

Abstract. Skin cancer is one of the most common and lethal cancers in the world. Even though melanoma accounts for a small fraction of all skin cancer types, it is currently the leading cause of skin cancer fatalities. In order to accurately identify the disease inthe initial stages,dermoscopy was designed to help specialists and increase the detection rate. Deep learning has been found in recent research to be particularly beneficial in the detection of skin cancer and medical diagnostics. Within this context, we propose adopting two deep-learning algorithms to detect benign and melanoma skin conditions. The first option entails extracting features from CNN pre-trained networks and applying machine-learning classification to make predictions, while the second explores using a fine-tuning methodology combined with data augmentation. The goal of this work is to compare the two methodologies in terms of classification performance and time efficiency. A public database including dermoscopic images of melanoma and normal skin lesions is used for the experimental evaluation. The experimental evaluation reveals that pre-trained models offer superior performance in terms of accuracy and processing time.

1. Introduction

Melanoma occurs for just approximately 1% of all skin cancers yet accounts for the vast majority of skin cancer mortality [1]. Melanoma is a form of skin cancer that originates in some cells that synthesize the melanin pigment. Melanoma occurs when these cells called melanocytes proliferate uncontrollably. Even though melanoma is far less prevalent than other forms of skin cancer, it is more deadly as the probability of spreading to other regions of the body is higher if not diagnosed at an early stage [1]. Melanoma is a serious public health concern across the world. Over 150.000 new melanoma diagnoses are recorded each year in Europe, whereas approximately 96.000 new patients are diagnosed each year in the United States according to [2-3].

Formerly, the primary means of detection of melanoma was clinical inspection, which has restricted and unpredictable performance, posing considerable hurdles in early cancer identification.Dermoscopy, a high-resolution non-invasive imaging technology that enables the inspection of skin lesions with a dermatoscope, has increased qualified professionals' diagnostic capabilities in the past years [4]. Because specialists' time is restricted when screening a large number of patients for melanoma, automated solutions able to detect cancer might save lives and eliminate wasteful biopsies. Several pieces of research have been conducted to develop methods for processing dermoscopy data that can aid clinicians in making an early diagnosis.

The authors in [5] concentrate on the challenge of melanoma identification and provide a deep-learning-based strategy towards classifying dermoscopic data as normal and cancerous. The presented approach is based on the Vgg model and employs the transfer learning principle. The authors assess their proposed solution on a collection of over 1200 images and conclude that it outperforms the state-of-the-art.

In [6], the authors performed a comparative analysis of the diagnostic made by 145 dermatologists against an automated melanoma classification system. The training was performed on dermoscopic data available online. The authors employed a fine-tuning methodology based on Resnet50. The evaluation was done using clinical images to assess. The authors concluded that the CNN-based technique obtained a lower variance, suggesting superior robustness when compared to human evaluation.

In [7], benign and cancerous dermoscopic images are classified using a decision forest algorithm trained on extracted attributes such as texture, color, and lesion shape. In [8], Generative Adversarial Networks are used to augment the originalInternationalSkin Imaging Collaboration (ISIC) 2019 dataset. The authors employ decision fusion of several pre-trained CNN models to increase the classification accuracy.

Another transfer learning-based technique was described in [9], where the authors used a MobileNet architecture to classify seven forms of skin cancer, including melanoma. Another study [10] concentrated on the development of a deep learning architecture for the classification of different skin lesions from dermoscopic data obtained from several ISIC databases.

The authors of [11] employed two types of textural characteristics generated from Local Binary Patterns and Gray Level Co-Occurrence Matrix to identify melanoma and non-melanoma skin conditions. Color information is also collected by taking into account several color spaces. A Support Vector Machine method is applied to process the joint texture color descriptor.

This paper is focused on the binary classification of benign and melanoma skin lesions on the basis of two deep-learning strategies: performing feature extraction from CNN pre-trained architectures and using a machine-learning classification for predictions and considering a fine-tuning approach along with data augmentation techniques. Several CNN models pre-trained on a very large dataset are analyzed and the most optimal layers for pooling features related to the considered classification task are determined. The objective of this work is to compare the performance of the two strategies in terms of classification accuracy, precision and recall, and time efficiency.

2. Background

2.1. Convolutional Neural Networks

A Convolutional Neural Network (CNN) is a type of Neural Network (NN) that is commonly implied for recognizing patterns in visual data.Neurons, layers, kernel weights, and biases are all essential parts of a neural network. A neuron is a function that produces an output that represents a weighted sum of the received inputs, and a layer is simply a set of neurons that perform the same operation and have the same parameters. In the training step, the network weights and biases are adjusted such that the model is adapted to the considered classification task and database. CNNs make use of a particular type of layer called the convolutional layer, which allows them to learn from image data. CNN's fundamental components are convolution, pooling, non-linearity, and classification operations.

The convolution operation is used to extract features from input data because the learned weights that capture information are found in the convolutional layers. The filters and the feature maps are the two main components of a convolutional layer. The filter, also known as the kernel, is a matrix containing the weights that is able to identify a specific feature by moving over the image's receptive field (a small portion of the input that has the same dimension as the considered filter) or the output of the previous layer. A dot product between the learned filter and the output of the preceding layer is generated. The obtained value is then placed in a matrix calleda feature map. Following that, the kernel shifts, and the procedure is iterated until the filter has gone through the whole image. For each filter, a different feature map is produced. Figure 1 shows an example of convolution where two filters are considered and a stride of 2 (the number of pixels the kernel is moved across the input data). All these weights of the filters are learned by the CNN in the training process, but the number and size of filters must be set by the network designer prior to the training procedure.

By preserving only the most significant information, the objective of pooling layers is to reduce the number of variables of each feature map and thus, the total computation of the network. Max-Pooling is a type of pooling in which a window is slid across the input with a given stride. In this process, only the largest value within that window is extracted. Figure 2 shows an example of a max-pooling operation where a stride of 2 is used.

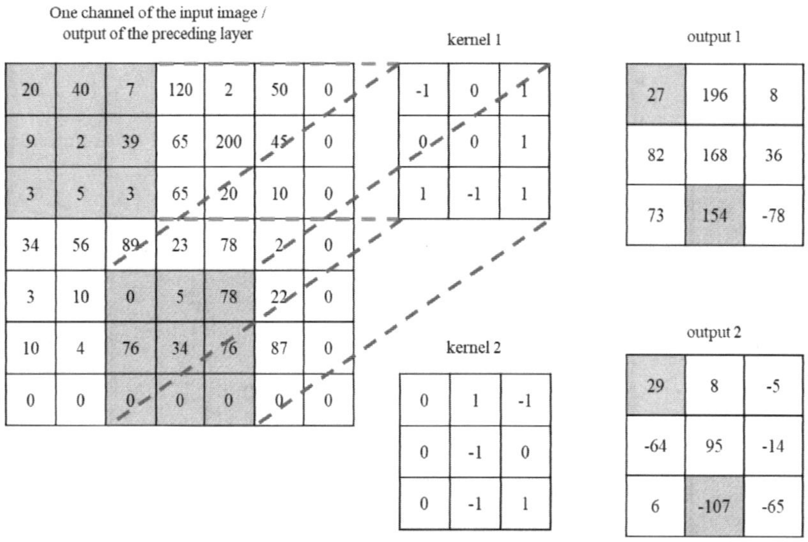

Figure 1. Example of convolution operation with two kernels.

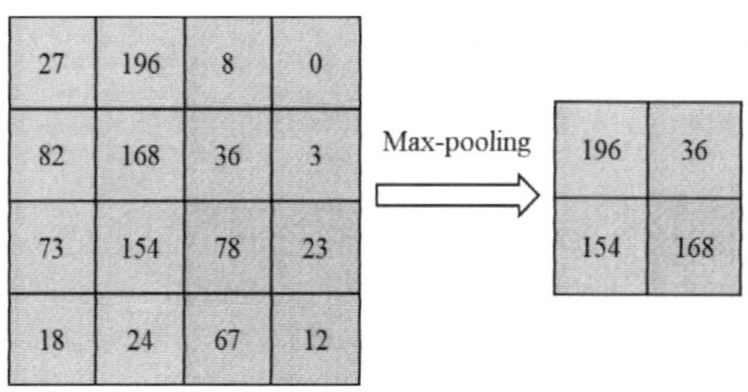

Figure 2. Example of max-pooling operation.

If the CNNs would have lacked a non-linear activation function, they would not have had the impressive results they achieved so far. Since in most cases, the input data is nonlinear, the architecture of CNNs needs nonlinearity. ReLu is one of the most used activations functions that is used to add nonlinearity by converting each negative value to 0 [12].

The convolutional, pooling and nonlinear layers work together to extract relevant features from images, add non-linearity, and minimize dimensionality.

Fully connected layers are standard Multi-Layer Perceptrons (MLP) having a softmax activation function [13] and in this case, every neuron from the previous layer is linked to every neuron from the following layer. The goal of fully connected layers is to categorize the input image into several classes using the features acquired by convolutional layers.

An example of CNN architecture is presented in Figure 3 for the AlexNet model [14].

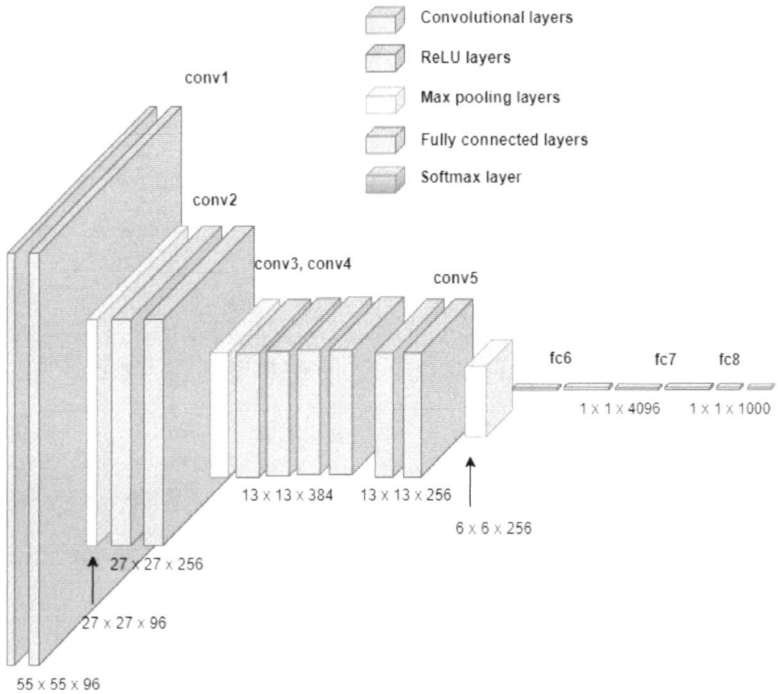

Figure 3. Example of CNN architecture.

2.2. Strategies to use CNNs

2.2.1. End-to-end approaches. In this type of solution, CNN performs both tasks: feature extraction and classification. The network is trained from scratch in order to learn the optimal weights of the network. The forward stage of training consists of passing the input entirely through the network, while the backward stage consists of backpropagating gradients and updating weights. The backpropagation technique works on the premise of modelling a function by changing the weights of the inputs to create the desired output. The training is done by providing to the network the difference between its actual response and the correct response and employing it to change its filter weights. The main disadvantage of this approach is the long time that is needed to design the network architecture including the optimal number of filters and their size, the number and type of layers in the network, and also, to set various hyperparameters. Moreover, the training process is very time-intensive.

2.2.2. Transfer learning in image classification. This technique implies using the information acquired in an initial task as a base for another classification problem [15-16]. The advantage lies in the possibility to obtain promising results without the need to handcraft complex CNN architectures and to use very advanced hardware to process data in a reasonable amount of time. The most common approach is to use models that are already pre-trained on large amounts of data classified in a considerable number of classes (ImageNet database [17]).

Using the capabilities of deep learning approachescan be easily done by performing feature extraction from pre-trained CNN models. This solution is particularly helpful if the hardware resources are not very powerful because it simply requires only one pass of the training data through the network. A pretrained network is used for deriving already acquired image features that are later employed to build a classification model by considering a traditional machine-learning technique such as the Support Vector Machine (SVM). If the new dataset is related to the initial one, the features can be retrieved further down the network because they are specific also to the new classification purpose.

On the other hand, if they are considerably different, deeper features are less valuable and earlier, more general information can be collected from shallower layers.

3. The proposed approaches

In this paper, two transfer learning strategies are considered: using pre-trained models for feature extraction and fine-tuning. A comparison is made between the two approaches for observing which one is better for the classification of dermoscopic images.

3.1. The use of pretrained models as feature extractors

This approach involves the utilization of a CNN architecture that is already trained. The training has been performed on a dataset with a high number of samples, namely ImageNet [17]. The weights of the network have exactly the same values as in training since the training is not performed again in this approach. There are several advantages to using pre-trained networks. First and foremost, because the dataonly travel the network once, the time required for feature synthesis is substantially decreased. Second, even with a small number of training samples, encouraging performance can be achieved.

An already trained CNN is employed as a feature extractor in the proposed methodology. The extraction of information can be done by considering any layer from the convolutional base. In our research, we examine a number of such models and for each of them, we extract information from all layers, and we keep the layer that achieves the most significant performance. Table 1 illustrates the explored models and the associated layers that offer the highest classification accuracy.

Table 1. Employed CNN arhitectures.

CNN model	Considered layer	Size reqiurement
ResNet18 [18]	res3b_branch2a	224×224×3
ResNet50 [18]	activation_48_relu	224×224×3
AlexNet [14]	conv4	227×227×3
Inception-v3 [19]	conv2d_50	299×299×3
Vgg16 [20]	conv5_2	224×224×3
InceptionResnet-v2 [21]	block35_1	299×299×3

The first step in the process of feature extraction is to adjust the dimension of the input images in order to fit the size requirements of each CNN model under consideration. The next stage consists in passing the image through the network and pooling the features from a certain layer. The output in this case is a series of feature maps. In order to obtain the corresponding feature vector, for each feature map, the average value is computed and stored in the output vector. The obtained feature vector has a number of features equal to the number of feature maps. The next step consists in performing a Z score normalization of the data as in equation (1). Normalization is critical in machine learning techniques because it ensures that all of the dataset's attributes are assessed along the same scale. After this step, each feature (column) will have a zero mean and unitary standard deviation.

$$z(x) = \frac{(x-\bar{X})}{S} \quad (1),$$

where z is the result after applying the Z score normalization, x is the input data, $\bar{X}$ is the sample mean and S is the sample standard deviation.

The standardization is performed on columns which means that the mean value and standard deviation are computed separately for each feature type.

The normalized feature vectors are used as input for the Support Vector Machine classifier with Radial Basis Function kernel whose parameters are chosen by using a grid search approach to support

IOP Conf. Series: Materials Science and Engineering 1254 (2022) 012035 doi:10.1088/1757-899X/1254/1/012035

the best classification performance. Figure 4 presents the block scheme that describes the considered strategy.

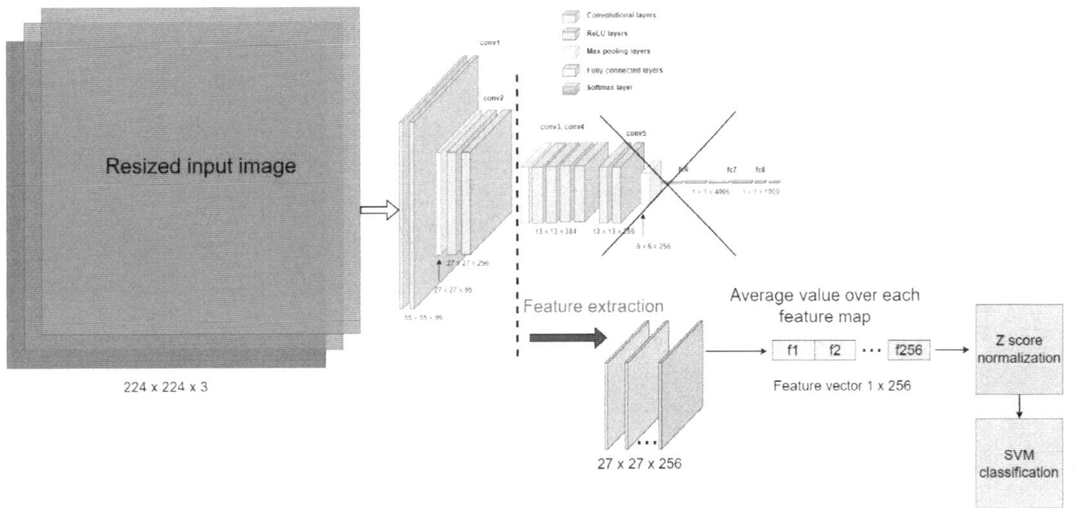

Figure 4. Proposed strategy for feature extraction from pretrained CNN models.

3.2. Fine-tuning strategy

A model that is already trained on ImageNet is employed as an initial point for a new classification problem. Transfer learning makes fine-tuning considerably faster than performing training from scratch with random initialization of parameters.The pretrained model's output layers are set up for 1000 classes corresponding to the classes of ImageNet. The output layers of the pretrained model are substituted by new layers appropriate for training new data. The number of categories in which the new dataset is divided gives the number of outputs for the new layers.

The learning rate for the weights and biases in the output layers is adjusted to obtain faster learning when compared to the other layers that are kept. A high learning rate helps the model to obtain faster learning, but there is possible to achieve weights that are not optimal. On the other hand, if the learning rate has a lower value, the training is longer, but in this way, the weights may be more appropriate. Because the initial layers have already been trained on ImageNet, fine-tuning requires only a small value for the learning rate (the considered initial learning rate is 0.0004). Because the output layers 'weights are randomly initialized, training from scratch needs a higher learning rate (0.004).

The dimension of the input images (both training and testimages) is adjusted exactly as in the previous method in order to fit the size requirements of each model. The training images are also augmented by considering the following transformations: random translation, flipping, rotation, and scaling. This procedure is useful for reducing the risk of over-fitting (a model memorizes the specifics in the training data to the point that it cannot generalize for new data). The augmentation is applied only to training images.

The training is then performed by considering the following parameters: the stochastic gradient descent with momentum, size of the mini-batch (set of training data that is used to calculate the gradient and adjust parameters in the network) of 10, 6 epochs (an epoch is a complete training session over all of the training samples). For fine-tuning, a smaller number of epochs is usually considered when compared to training from scratch. Figure 5 presents the fine-tuning strategy.

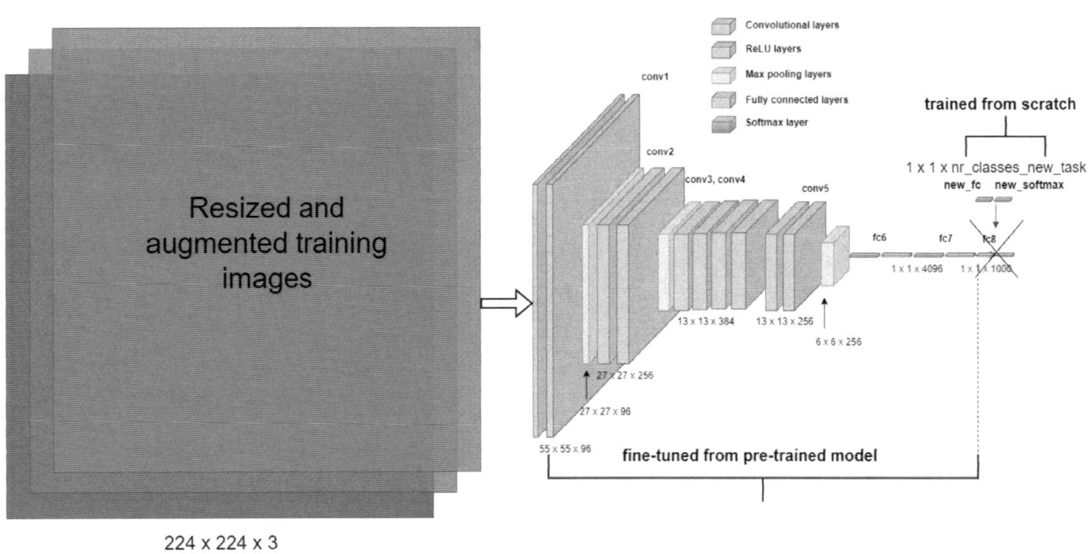

Figure 5. Proposed strategy for fine-tuning.

4. Experimental setup and results

200 RGB dermoscopic images of melanocytic lesions acquired with a mole analyzer are used in the experimental assessment. The dataset is called PH^2 and is acquired by [22]. The considered configuration is presented in table 2. Figure 6 shows some sample images from this dataset.

Table 2. Dataset configuration.

Image category	Number of samples	Train/Test partition
Nevus	160	120/40
Melanoma	40	30/10

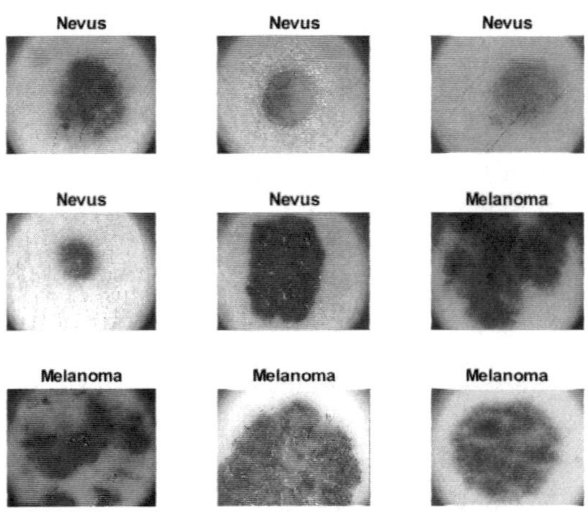

Figure 6. Images from the PH^2 dataset [22].

For performance evaluation of the proposed classification strategies, the following metrics are considered: overall accuracy, macro-averaging precision, and macro-averaging recall. They can be computed by using the confusion matrix. A confusion matrix is a tool for analyzing an algorithm's efficiency. The samples from the real class are shown on each row of the matrix, while the samples from the predicted category are displayed on each column. The confusion matrix for the binary classification for the considered dataset is presented in Figure 7.

Confusion matrix		Predicted class	
		Melanoma (cancer)	Nevus (no cancer)
Actual class	Melanoma (cancer)	**True positive (TP)**	**False negative (FN)**
	Nevus (no cancer)	**False positive (FP)**	**True negative (TN)**

Figure 7. Confusion matrix for binary classification.

The macro-averaging precision and macro-averaging recall are computed for each class separately and then, the final result is the average value of the results obtained for each class. They are given in equation (2) and equation (3) respectively. The overall accuracy is computed by using equation (4).

$$Precision = \frac{TP/(TP+FP)+TN/(TN+FN)}{2} \tag{2}$$

$$Recall = \frac{TP/(TP+FN)+TN/(TN+FP)}{2} \tag{3}$$

$$Overall\ Accuracy = \frac{TP+TN}{TP+FN+FP+TN} \tag{4}$$

The macro-averaging metrics are considered in this approach because of the disproportionate distribution of samples per class in the considered dataset. The overall accuracy cannot describe on its own appropriately such datasets since the obtained value can be very high even if all samples from one of the classes are predicted wrong. This happens because the performance of a class containing a larger number of images has a greater influence on the outcome than a small class's. Thus, good accuracy can be achieved by depending solely on the majority class, which is misleading. For the macro-averaging metrics, in the sense that the final result is the mean value of the results obtained for each class, the imbalance is ignored because every class is assigned the same weight regardless of the number of samples in that class.

The experimental evaluation consists of the comparison of the two strategies for the classification of dermoscopic images. The classification results achieved for the considered dataset by using the fine-tuning approach are given in table 3. Table 4 presents the performance achieved by pooling features from pre-trained models and inputting them into the SVM algorithm for classification. All metrics from tables 3 and 4 are computed for the same random partition of images into training and testing subsets.

Table 3. Classification results [%] for the PH^2 dataset by considering the fine-tuning approach.

	ResNet18	ResNet50	AlexNet	Inception-v3	Vgg16	InceptionResnet-v2
Accuracy	96	96	80	88	86	86
Precision	93.75	91.67	90	81.25	92.55	83.33
Recall	93.75	97.5	50	92.5	65	68.75

Table 4. Classification results [%] for the PH^2 dataset by considering pre-trained CNN models for feature extraction and the SVM technique for classification.

	ResNet18 (res3b_branch2a)	ResNet50 (activation_48_relu)	AlexNet (conv4)	Inception-v3 (conv2d_50)	Vgg16 (conv5_2)	InceptionResnet-v2 (block35_1)
Accuracy	98	98	96	96	96	98
Precision	95.45	95.45	91.67	93.75	97.62	95.45
Recall	98.75	98.78	97.5	93.75	90	98.75

The reported findings show that the feature extraction technique outperforms the fine-tuning strategy for all CNN models. This occurs because the explored dataset has a limited number of images, and through fine-tuning, the network cannot acquire sufficient knowledge as there is not enough data for learning. Fine-tuning is typically more successful than feature extraction granted that a large dataset is available for the CNN to learn additional features. Aside from the classification performance improvement, the training procedure for feature extraction from deep learning models is much faster since SVM trainingis more time-efficient than that of CNNs. The results show promising performance for the detection of melanoma even for a relatively small dataset.

In the fine-tuning approach, by updating the weights for the final layers, the network is adapted to the new task. By considering a higher learning rate, the learning is faster in these layers. However, even if for the other layers the learning rate is smaller and the fine-tuning is done on a small number of epochs, the layers are not frozen entirely. This means that even the weights related to more general layers are modified a little bit during the fine-tuning. Since the considered dataset is very small, the update of the weights can be counterproductive in this situation because the network does not have enough data to learn from. This can be the reason for poorer results for the fine-tuning approach compared to feature extraction. For feature extraction, the weights are kept the same and in this way, the general features are learned correctly since the training was done on a very large dataset.

Based on this preliminary work, we intend to dive deeper into the fine-tuning approach and investigate various scenarios of freezing and unfreezing layers of the considered models. By performing this analysis, we expect to determine that point in the network where the optimal balance between generality and specificity is obtained. After that point, all deeper layers being fine-tuned, the model can learn more specific features related to the new classification problem. This is particularly important for increasing the accuracy of small datasets that are relatively different from the initial task. Additionally, by freezing all layers before that optimal point, the general weights already learned are kept the same which is important for small datasets as by retraining them the performance is reduced (based on the obtained results in the current research).

For the pre-trained network, the obtained results are promising. Extracting features from earlier layers in the network implies using general features that are not specific to the original dataset and can be useful for the new task of melanoma detectionwhere more general features such as texture, color, and shape are relevant. We can also compare our results with the findings obtained in [23] for the same dataset. By extracting features from the last layers of AlexNet, Resnet18, and Vgg16, the authors obtain recall values ranging from 89% to 91%, precision metrics from 93% to 97%, and accuracies from 87% to 88%. By incorporating a feature selection approach and by combining both CNN features and other handcrafted descriptors, the authors improve their results achieving a value of 98% accuracy. Also, an accuracy of 97.5% is obtained in [24-25] by using handcrafted features, 96.5% accuracy by fusing handcrafted and deep-learning-based features in [26], while feature extraction from pre-trained models achieves 93% in [27].

5. Conclusion

In this paper, we propose using two deep-learning algorithms to detect benign and melanoma skin lesions. The first methodology allows extracting features from CNN pre-trained nets and employing machine-learning classification to provide predictions, while the second investigates using a fine-tuning procedure in conjunction with data augmentation. The purpose of this study is to contrast the two techniques in terms of classification performance and time efficiency. The experimental evaluation makes use of a public database that contains dermoscopic data of melanoma and benign skin lesions. The results reveal that pre-trained models outperform the fine-tuning strategy for the considered dataset due to the reduced number of training samples. The obtained findings reveal that fine-tuning more general weights in the network can be detrimental to classification performance. Further investigation is needed to analyze various solutions of freezing and unfreezing layers in the CNN models to determine an optimal point where the balance of generality and specificity is achieved.

As future work, we plan to explore various scenarios for the fine-tuning approach to see how performance changes when the layers are frozen and unfrozen. These findings on refining fine-tuning to deliver better model performance will enable the development of improved transfer learning-based methods for datasets containing a limited number of images.

References

[1] What Is Melanoma Skin Cancer? Available from: https://www.cancer.org/cancer/melanoma-skin-cancer/about/what-is-melanoma.html (02.04.2022)
[2] Global Cancer Observatory, Available from: https://gco.iarc.fr/ (02.04.2022)
[3] Sung H, Ferlay J, Siegel R L, Laversanne M, Soerjomataram I, Jemal A and Bray F 2021 Global Cancer Statistics 2020: GLOBOCAN Estimates of Incidence and Mortality Worldwide for 36 Cancers in 185 Countries *CA. Cancer J. Clin.* **71** pp 209–249
[4] Vestergaard M E, Macaskill P, Holt P E and Menzies S W 2008 Dermoscopy compared with naked eye examination for the diagnosis of primary melanoma: a meta-analysis of studies performed in a clinical setting *Br. J. Dermatol.* **159** pp 669–676
[5] Lopez R A, Giro-i-Nieto X, Burdick J and Marques O 2017 Skin lesion classification from dermoscopic images using deep learning techniques *Proc. 13th IASTED Int. Conf. Biomed. Eng. BioMed 2017* (Innsbruck, Austria: IEEE) pp 49–54
[6] Brinker T J, Hekler A, Enk A H, Klode J, Hauschild A, Berking C, Schilling B, Haferkamp S, Schadendorf D, Fröhling S, Utikal J S, et al 2019 A convolutional neural network trained with dermoscopic images performed on par with 145 dermatologists in a clinical melanoma image classification task *Eur. J. Cancer* **111** pp 148–154
[7] Ferris L K, Harkes J A, Gilbert B, Winger D G, Golubets K, Akilov O and Satyanarayanan M 2015 Computer-aided classification of melanocytic lesions using dermoscopic images *J. Am. Acad. Dermatol.* **73** pp 769–776
[8] Gong A, Yao X and Lin W 2020 Dermoscopy Image Classification Based on StyleGANs and Decision Fusion *IEEE Access* **8** 70640–50
[9] Younis H, Bhatti M H and Azeem M 2019 Classification of Skin Cancer Dermoscopy Images using Transfer Learning *2019 15th International Conference on Emerging Technologies (ICET)* (Peshawar, Pakistan: IEEE) pp 1–4

[10] Iqbal I, Younus M, Walayat K, Kakar M U and Ma J 2021 Automated multi-class classification of skin lesions through deep convolutional neural network with dermoscopic images *Comput. Med. Imaging Graph.* **88** 101843

[11] Al-Mansour E and Jaffar M 2016 Classification of Dermoscopic Skin Cancer Images Using Color and Hybrid Texture Features *Int. J. Comput. Sci. Netw. Secur.* **16** pp 135–139

[12] Nair V and Hinton G E 2010 Rectified linear units improve restricted boltzmann machines *Proceedings of the 27th International Conference on International Conference on Machine Learning* ICML'10 (Madison, WI, USA: Omnipress) pp 807–814

[13] Nwankpa C, Ijomah W, Gachagan A and Marshall S 2018 Activation Functions: Comparison of trends in Practice and Research for Deep Learning *ArXiv181103378 Cs*

[14] Krizhevsky A, Sutskever I and Hinton G E 2012 ImageNet classification with deep convolutional neural networks *Proceedings of the 25th International Conference on Neural Information Processing Systems - Volume 1* NIPS'12 (Red Hook, NY, USA: Curran Associates Inc.) pp 1097–1105

[15] Han D, Liu Q and Fan W 2018 A new image classification method using CNN transfer learning and web data augmentation *Expert Syst. Appl.* **95** pp 43–56

[16] Shaha M and Pawar M 2018 Transfer Learning for Image Classification *Proceedings of the Second International Conference on Electronics, Communication and Aerospace Technology (ICECA 2018)* (Piscataway, New Jersey: IEEE) pp 656–660

[17] Russakovsky O, Deng J, Su H, Krause J, Satheesh S, Ma S, Huang Z, Karpathy A, Khosla A, Bernstein M, Berg A C and Fei-Fei L 2015 ImageNet Large Scale Visual Recognition Challenge *Int. J. Comput. Vis.* **115** pp 211–252

[18] He K, Zhang X, Ren S and Sun J 2016 Deep Residual Learning for Image Recognition *2016 IEEE Conference on Computer Vision and Pattern Recognition (CVPR)* (Las Vegas, NV, USA: IEEE) pp 770–8

[19] Szegedy C, Vanhoucke V, Ioffe S, Shlens J and Wojna Z 2016 Rethinking the Inception Architecture for Computer Vision *2016 IEEE Conference on Computer Vision and Pattern Recognition (CVPR)* (Las Vegas, NV, USA: IEEE) pp 2818–2826

[20] Simonyan K and Zisserman A 2015 Very Deep Convolutional Networks for Large-Scale Image Recognition *ArXiv14091556 Cs*

[21] Szegedy C, Ioffe S, Vanhoucke V and Alemi A 2016 Inception-v4, Inception-ResNet and the Impact of Residual Connections on Learning *ArXiv160207261 Cs*

[22] Mendonça T, Ferreira P M, Marques J S, Marcal A R S and Rozeira J 2013 PH2 - A dermoscopic image database for research and benchmarking *2013 35th Annual International Conference of the IEEE Engineering in Medicine and Biology Society (EMBC)* (Osaka, Japan: IEEE) pp 5437–5440

[23] Filali Y, EL Khoukhi H, Sabri M A and Aarab A 2020 Efficient fusion of handcrafted and pre-trained CNNs features to classify melanoma skin cancer *Multimed. Tools Appl.* **79** 31219–38

[24] Akram T, Khan M A, Sharif M and Yasmin M 2018 Skin lesion segmentation and recognition using multichannel saliency estimation and M-SVM on selected serially fused features *J. Ambient Intell. Humaniz. Comput.*

[25] Nasir M, Attique Khan M, Sharif M, Lali I U, Saba T and Iqbal T 2018 An improved strategy for skin lesion detection and classification using uniform segmentation and feature selection based approach *Microsc. Res. Tech.* **81** pp 528–543

[26] Moura N, Veras R, Aires K, Machado V, Silva R, Araújo F and Claro M 2018 Combining ABCD Rule, Texture Features and Transfer Learning in Automatic Diagnosis of Melanoma *2018 IEEE Symposium on Computers and Communications (ISCC)* (Natal, Brazil: IEEE) pp 00508–13

[27] Salido J A A and Jr. C R 2018 Using Deep Learning for Melanoma Detection in Dermoscopy Images *Int. J. Mach. Learn. Comput.* **8** pp 61–68

Acknowledgments
This paper was financially supported by the Project "Entrepreneurial competences and excellence research in doctoral and postdoctoral programs - ANTREDOC", project co-funded by the European Social Fund financing agreement no. 56437/24.07.2019.

IOP Conf. Series: Materials Science and Engineering 1254 (2022) 012036 doi:10.1088/1757-899X/1254/1/012036

Expending the power of artificial intelligence in preclinical research: an overview

A Diaconu[1], F D Cojocaru[1], I Gardikiotis[1], L Agrigoroaie[1], D M Furcea[1], A Pasat[2], G Suciu[2], C Rezuș[1] and G Dodi[1]

[1]Grigore T. Popa University of Medicine and Pharmacy of Iasi, 16 University Street, 700115 Iasi, Romania
[2]Beia International Consult SRL, 12-22 Peroni Street, 041386 Bucuresti, Romania

E-mail: gianina.dodi@umfiasi.ro

Abstract. Artificial intelligence (AI) is described as the joint set of data entry, able to receive inputs, interpret and learn from such feedbacks, and display related and flexible independent actions that help the entity reach a specific aim over a period of time. By extending its health-care applications continuously, the ultimate AI target is to use machine simulation of human intelligence processes such as learning, reasoning, and self-correction, to mimic human behaviour. AI is extensively used in diverse sectors of medicine, including clinical trials, drug discovery and development, understanding of target-disease associations, disease prediction, imaging, and precision medicine. In this review, we firstly describe the limitations and challenges of the AI tools and techniques utilized in medicine, followed by current uses and applications of AI in the translational field, highlighting the cardio-renal preclinical models with potential to contribute to future clinical research.

1. Introduction

Advances in technology during the last decade have enlightened our knowledge regarding challenging evidence-based medicine supremacy through artificial intelligence (AI). AI refers to computer systems based on mathematical algorithms that are able to simulate human intelligence. The concept of AI dates back to the mid 20th century, but only recently major technological advances generated enthusiasm in the medical field for its potential to help physicians, from establishing a diagnosis to predicting risk of major adverse events and diseases, drug discovery and enabling a path to personalized medicine [1].

AI covers a variety of methods: machine learning (ML), deep learning (DL), natural language processing (NLP), and optical character recognition (OCR) [2].

Though often used interchangeably with AI, ML refers to a group of algorithms that autonomously learn data and identify complex patterns in order to perform multiple tasks. By using ML, AI can be used to study complex relationships that express heterogeneous associations about the analysed medical data. In medicine, ML practice increased rapidly over the last 10 years, being often compared to a black box, due to its ability to generate data that cannot easily be explained [3].

Conventional ML techniques used in the medical field include decision trees, random forests, support vector machines (SVMs), Bayesian classifiers, k-nearest neighbors, k-means, linear discriminant analysis and neural networks. The conventional ML techniques used in the AI medical field are presented in Table 1.

Table 1. Overview of conventional ML used in the medical field.

ML tool	Features	Main advantage	Main disadvantage
decision trees	- based on a tree-like architecture; - mainly used in solving classification and regression problems by using rules to split data in binary values;	- useful when aggregating different types of data to make accurate predictions [4];	- the data samples can over- or under-fit the model particularly when using a small data set [5];
random forests	- based on multiple decision trees used for classification; - the ultimate prediction it is done by majority voting;	- versatile because the technique can be used for both regression and classification tasks, and it is also easy to view the relative importance it assigns to the input feature [6-7] ;	- it does not give precise continuous nature prediction;
SVMs	- built on the idea to design a hyperplane that separates positive and negative examples to minimize the separation error;	- when there is a clear margin of separation between classes it performs well;	- in case of big data sets the technique is not suitable;
Bayesian classifiers	- based on probability theory; - create a category with maximum posterior probability, considering the given prior probability, cost function, and category conditional density;	- they explicitly provide the conditional probability distributions of the values of the class attribute given the values of the other input attributes [4];	- it assumes that the predictors are independent which is not true in real life data;
k-nearest neighbors	- based on searching for k-nearest training instances and classifying a new instance related to the most frequent class of these k instances;	- performs well regarding the search space;	- the prediction accuracy is based on the quality of the data;
k-means	- based on partitioning n samples into k clusters and each new sample is assigned to the cluster with the nearest mean;	- it does not require prior training and it guarantees convergence [8];	- the convergence can be to a local optimum, because the initialization prototype is not the best one [9];
linear discriminant analysis	- based on building functions for prediction that maximize the discrimination between previous built categories;	- it avoids the issue of dimensionality and it reduces the required resources and dimensional costs;	- it does not perform well if the modelling is not balanced: the size of the set of objects in various classes (highly) varies differently;
neural networks	- based on collections of neurons with the purpose of processing signals; - signals are exchanged between neurons that can be organized in layers.	- they can fit complex nonlinear models that cannot be necessarily specified in advance [10].	- they can function in a wrong manner because of over fitting and overtraining phenomena.

Deep learning (DL), a subset of ML, uses artificial neural networks that mimic the electrical impulses of the human brain to allow for complex interactions between data sets [11].

ML approaches add value across the spectrum of mainly clinical trials and observational research, but also improves the success and efficiency of basic science and translational research, being ideal from planning, to study execution, data management and analysis. AI systems are being routinely used in our day-to-day life, but this digitalization encounters various challenges of procuring, examining, and applying that paradigm to solve multifaceted clinical problems. However, its applicability to the healthcare system is still in its early stages. This may be in part due to financial and ethical limitations, but also a lack of large-scale studies to validate their accuracy. In this review, we highlight the use of AI in translational research, with a special focus on cardio-renal disease prediction models that have the potential to contribute to future clinical research.

2. ML in the medical field

Using artificial intelligence in medicine is of great importance, as it can predict different illnesses. AI techniques have been utilized in the clinical setting to diagnose diseases such as hypertension, chronic heart disease, stroke, diabetes, cancer, Alzheimer, skin and liver disease [12].

Besides diagnosis ML-based applications involve early disease prediction, treatment, outcome prediction and prognosis evaluation, [13] personalized medicine, behaviour variation, drug discovery, manufacturing, clinical trial research, radiology and radiotherapy, smart electronic health records, and epidemic outbreak prediction [14]. In the light of the novel Sars-COV-2 pandemic, AI was successfully used for infection disease surveillance [15].

In 2021, Karekarand Vazifdar [16] reviewed the status of research in AI in health care up to September 2019 and included all the clinical studies conducted in the field of AI registered on the ClinicalTrials.gov website in a registry-based audit. According the authors out of a total of 331,536 registered studies, 156 (65 studies registered in 2019) were related to AI, divided in 84 interventional and 72 observational trials. Of these, 26.3% studies were oncology related, followed by cardiology, ophthalmology, psychiatry, and neurology. Almost 34% of the studies comprised of AI-based medical devices intervention testing, followed by 28% of diagnostic studies. It is important to mention that this study does not evaluate the quality of studies.

Another search on deep learning models and medical imaging was performed using Ovid-MEDLINE, Embase, Science Citation Index, and Conference Proceedings Citation Index databases. The systematic review and meta-analysis published by Liu et al. [17] and registered with PROSPERO, id, included 82 studies from Jan 1, 2012, to June 6, 2019, from which only 69 studies provided enough data to construct contingency tables.

The recent review of Weissler et al. [3] details the role of ML in clinical research in transforming the future of evidence generation, based on an overview of the conference attendees proceedings that included biomedical and ML researchers, representatives from the US Food and Drug Administration (FDA), AI technology companies, non-profit organizations, patient advocacy groups, and pharmaceutical companies. The authors highlighted the potential of ML techniques to improve the success and efficiency of clinical research, increasing its positive impact for all stakeholders.

As an example, deep learning techniques have been utilized to analyse large datasets of 12-lead ECGs for detecting asymptomatic cardiovascular disease, such as left ventricular systolic dysfunction, hypertrophic cardiomyopathy or asymptomatic atrial fibrillation from a sinus-rhythm ECG [18]. Deep learning was also investigated as a potential predictor of hemodynamic parameters. A recent study proposed a DL model for estimating on a standard electrocardiogram an elevated mean pulmonary capillary wedge pressure, a parameter obtained invasively via pulmonary artery catheterization [19]. These results come from ECG features not available to the clinician's eye, making of ML a very promising disease-screening tool.

As observed, AI techniques have the potential to offer real-life practicability in order to assist human decision-makers, but as mentioned in Harrer et al. [14] review, is not a "magic bullet that

will make the success rates of clinical trials skyrocket overnight". In order to reach the proposed goal, AI practices must be tested as a complement or replacement alongside the existing technology, and the added value must be validated in a scalable way to users and regulatory bodies.
Although these requirements have started to be explored, we need to acknowledge the role of AI in preclinical research, since successful clinical trials involve significant preclinical examination and development approved by the regulatory bodies.

Preclinical research has evaluated the use of ML algorithms in multiple areas (see Figure 1), from drug discovery [1,20], animal movement data [21] to analysis of histological or radiology images and screening and evaluating diseases [22].

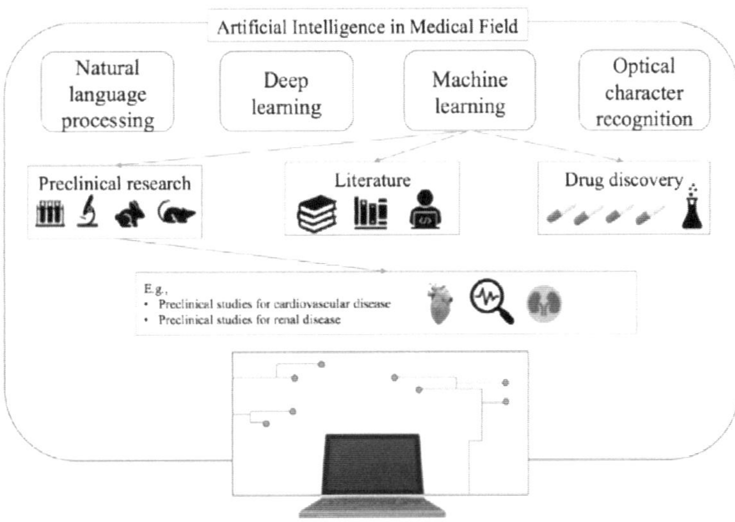

Figure 1. AI/ML in preclinical field.

2.1 AI in renal diseases
AI algorithms, though underutilized in the field of nephrology [23], have been successfully used in the field of renal pathology (see Table 2). Therefore, researchers have developed various ML algorithms to help in the rearrangement care of chronic diseases [24] and identify patients at higher risk of chronic disease.

Ginley et al. [25] used machine learning to analyse digitized renal biopsy samples of patients with diabetic nephropathy and validated it for a streptozotocin mouse model of diabetes mellitus (DM).They used an automatic computational pipeline to quantify glomerular structures, previously simplified into three main components – nuclei, capillary lumina and Bowman space, and later applied the data to the classification of diabetic nephropathy. Simon et al. [26] successfully used a SVM model alone and in tandem with a neural network for glomerular detection in histopathological whole-slide images (WSIs) of mice and rat kidney tissue sections and biopsies. Diabetes-related kidney damages were obtained by using a standard streptozotocin animal model, which involved administration of streptozotocin, followed by induction of mild-moderate diabetic nephropathy after 25 weeks. Kidney tissues were then prepared, stained by different methods (haematoxylin and eosin, Periodic acid-Schiff, Jones silver, Gömöri'strichrome and Congo Red) and the images were registered by a whole-slide bright-field microscope. Other studies validated the use of SVMs [27] or convolutional neural networks (CNNs) [28-29] for glomerular segmentation in WSIs of murine kidneys. Bouteldja et al. [30] trained a CNN for multiclass histological kidney segmentation to evaluate renal disease in multiple murine models. They also evaluated it on human kidney samples, with favourable results, thus holding promise of translating the findings from preclinical to clinical research. Generative adversarial networks (GANs) have also gained popularity for analysing pathological image datasets and virtual staining of histopathological images of animal kidney sections. Govind et al. [31] used a GAN-based pipeline to identify Wilms' Tumor 1 positive cells by translating kidney sections from a streptozotocin treated mouse model stained with Periodic acid-Schiff into Wilms' Tumor 1 labelled immunofluorescence images.

2.2 Cardiovascular diseases

Perhaps the most studied application of ML in the preclinical cardiovascular setting is the detection of various arrhythmias, as large ECG datasets are available. Atrial fibrosis and remodelling might be responsible of the onset of atrial fibrillation, so Fu et al. [32] proposed a novel 11-layer CNN to identify fibrosis in cardiac tissue histologic images of murine hearts. The experiment was conducted on left atrial section from alloxan monohydrate induced diabetic Japanese rabbits stained with Masson's trichrome. A recent study tested a supervised ML tool, FIBER-ML, to quantify fibrosis in histological sections from two models –acute stress-induced cardiomyopathy in rats and HIV-induce nephropathy in mice [33]. Tokarz et al. [34] reported a study evaluating a fully convolutional network deep learning technique for detecting rodent progressive cardiomyopathy lesions (cardiomyocyte degeneration or necrosis, fibrosis, mononuclear cell infiltration and mineralization) in histologic sections of the heart. CNNs can also provide accurate quantification of vascular pathology in a mouse choroidal neovascularization model. [35] Park et al. [36] proposed a deep learning-based ultrasound imaging consisting of an integrated neural network for monitoring flow-vessel dynamics, such as mean flow velocity, wall shear stress and vascular strain, in murine carotid arteries. The proposed DL-based method better assesses flow-vessel dynamics than the conventional ultrasound-based techniques.

Table 2. The use of ML in preclinical research.

Study	Animal	ML technique	Observations
			Renal
Ginley et al. (2019) [25]	Mice	CNN RNN	- glomeruli detection from WSIs: - glomerular boundaries: 0.93±0.04 balanced accuracy; - glomerular nuclei: 0.94 sensitivity and 0.93 specificity; - glomerular structural components: 0.95 sensitivity and 0.99 specificity;
Simon et al. (2018) [26]	Mice Rats	SVM CNN	glomeruli detection by SVM model from WSIs: - precision in mice ranged from 0.83 to 0.98; - in rats, Jones and hematoxylin and eosin stains had the highest precision, above 90%, followed by trichrome-stained WSIs (88.7% precision); by combining the SVM model with a CNN, an overall precision of 97.5% was obtained: - the highest precision stain was Periodic acid-Schiff (99.1%), followed by hematoxylin eosin (98.4%), trichrome (98.2%), Congo red (96.6%) and lastly, Jones (95.3%);
Gadermayr et al. (2017) [27]	Mice	SVM	- glomeruli detection from WSIs: - F_1-score of 0.68, recall of 0.87, precision of 0.58 - 42% of the detections = false positives; - glomeruli segmentation from WSIs: - 51% of glomeruli were accurately segmented

			(DSC > 0.8); - DSC of 0.71 for 80% of the objects; - 40% of the objects were segmented with a DSC of 0.85;
Gadermayr et al. (2018) [28]	Mice	CNN	glomeruli segmentation from WSIs: - DSC of 0.9 (precision of 0.89, recall of 0.92); - 80% of the objects obtained a DSC > 0.9;
Gupta et al. (2019) [29]	Mice	CNN	- glomeruli detection and segmentation: - mean F-score of 0.79 when analyzing eight WSIs;
Govind et al (2021) [31]	Mice	GAN architecture	- detection of Wilms' Tumor 1 – positive cells: - sensitivity of 0.87; - specificity of 0.92;
		Cardiovascular	
Fu et al. (2018) [32]	Rabbits	CNN	- segmentation of histological images, particularly fibrosis identification: - dice similarity coefficient of 0.947;
Facchin et al. (2022) [33]	Rats and mice	Supervised ML	fibrosis quantification in heart and kidney tissue sections;
Tokarz et al. (2021) [34]	Rats	CNN	consensus was reached by both veterinary pathologists and the CNN on 80% the cohort slides;
Jarutis et al. [35]	Mice	CNN	predicting choroidal neovascularization lesion: - dice coefficient of 0.88, comparing to manual annotations;
Park et al. (2022) [36]	Rats	CNN	DL-UVF improves measurements of flow-vessel dynamics- mean flow velocity, wall shear stress and vascular strain with up to 4.6-fold, 15.1- fold, and 22.2-fold respectively.

Abbreviations: RNN - recurrent neural network; DSC - dice similarity coefficient; DL-UVF - deep learning-based ultrasound imaging for flow–vessel dynamics.

2.3. Drug discovery

AI became over the last years an integral part of the pharmaceutical industry due to its potential to cover the entire drug discovery and development value chain, starting from target identification and up to clinical development.

The use of AI-driven solutions in pre-clinical drug discovery is described several review papers, the most recent one focusing on enhancing preclinical drug discovery with AI. Vijayan et al. [20] keynote offers an overview of current AI tools used in preclinical drug discovery with real impact examples, along with opportunities and challenges in implementing AI in drug discovery. First, the authors present the schematic representation of the preclinical drug discovery process highlighting opportunities for AI to be applied across the drug discovery continuum, followed by an illustration on how AI could increase the effectiveness of the DMTA (design, make, test and analyse) cycle. The second part of the keynote summarizes AI algorithms and their application in preclinical drug discovery, namely ML/supervised for binding affinity/pose prediction, PhysChem property; ML/unsupervised for data visualization, dimensionality reduction and molecular pattern recognition, ML/reinforcement learning for de novo molecular design, virtual screening, docking and multiparameter optimization (MPO) and DL for protein structure prediction, binding affinity/pose prediction, de novo molecular design, focused library generation, PhysChem property prediction, absorption, distribution, metabolism, excretion, and toxicity (ADMET) endpoints, MPO and synthesis planning. In the challenges in the application of AI to preclinical drug discovery section, the authors identify several points of AI essential for the growing investment within the pharmaceutical industry. Briefly, large amounts of high-quality training data, standardization and integration of the available drug discovery data, the lack of transparency, are just a few of the many challenges in implementing AI into drug discovery.

3. Conclusion

Advances in technology during the last decade have enlightened our knowledge regarding challenging evidence-based medicine supremacy through AI. Researchers have pursued different versions of AI based on fidelity to human behaviour and rationality in both thought and action, considering that the technology works when data is reliable, success and failure are recognizable, rules are clear and tasks are cyclic. Although AI might not be a solution for every problem in medicine, it is clearly a valuable tool if applied in the correct setting and with the accurate data. According to the findings of this review paper, AI/ML tools have been widely used in drug discovery, translational research and the pre-clinical phase with increasing complexity over the years. All the mentioned examples demonstrate a novel paradigm used to optimize the diagnostic and treatment of diseases, inviting clinicians and researchers to envision and expect for more from AI. It becomes clearer that in a near future randomized controlled trials will no longer be needed and the paradigm will change: the medical community will be offered self-improving decision tools, based not on randomized controlled trials but on deep learning algorithms.

Acknowledgement: This work was supported by a grant of the Ministry of Research, Innovation and Digitization, CNCS/CCCDI-UEFISCDI, project number PN-III-P2-2.1-PED-2019-5446, within PNCDI III.

References

[1] Paul D, Sanap G, Shenoy S, Kalyane D, Kalia K and Tekade R K 2021 Artificial intelligence in drug discovery and development *Drug Discov Today* **26**(1) pp 80-93
[2] Bhatt A 2021 Artificial intelligence in managing clinical trial design and conduct: Man and machine still on the learning curve? *Perspect Clin Res.* **12**(1) pp 1-3
[3] Weissler E H, Naumann T, Andersson T *et al* 2021 Correction to: The role of machine learning in clinical research: transforming the future of evidence generation *Trials* **22** pp 537
[4] Fakhr M and Elsayad A 2015 Diagnosis of cardiovascular diseases with bayesian classifiers *J. Comput. Sci.* **11**(2) pp 274-282
[5] Yan-Yan S and Ying L U 2015 Decision tree methods: applications for classification and prediction *Shanghai archives of psychiatry* **27**(2) pp 130-135
[6] Biau G and Erwan S 2016 A random forest guided tour *Test* **25**(2) pp 197-227
[7] Linardatos P, Papastefanopoulos V, Kotsiantis S 2021 Explainable AI: A review of machine learning interpretability methods *Entropy* **23**(1) pp 18
[8] Sobiech J and Wolfgang D 2013 Observing lake- and river-ice decay with SAR: advantages and limitations of the unsupervised k-means classification approach *Annals of Glaciology* **54**(62) pp 65-72
[9] Eltibi M F and Wesam A 2011 Initializing k-means clustering algorithm using statistical information *Int. J. Comput. Appl.* **29**(7) pp 51-55
[10] Livingstone D J, Manallack D T and Tetko I V 1997 Data modelling with neural networks: Advantages and limitations *J. Comput. Aided Mol. Des.* **11** pp 135-142
[11] Beneke F, Mackenrodt M O 2019 Artificial intelligence and collusion *IIC Int. Rev. Intellectual Property Competition Law.* **50**(1) pp 109–134
[12] Kumar Y, Koul A, Singla R and Ijaz M F 2022 Artificial intelligence in disease diagnosis: a systematic literature review, synthesizing framework and future research agenda *J Ambient Intell Human Comput* pp 1-28

[13] Jiang F, Jiang Y, Zhi H, Dong Y, Li H, Ma S *et al* 2017 Artificial intelligence in healthcare: past, present and future *Stroke Vasc Neurol.* **2**(4) pp 230–243
[14] Harrer S, Shah P, Antony B and Hu J 2019 Artificial intelligence for clinical trial design *Trends Pharmacol Sci.* **40**(8) pp 577–591
[15] Komolafe T E, Cao Y, Nguchu B A, Monkam P, Olaniyi E O, Sun H *et al* 2021 Diagnostic test accuracy of deep learning detection of COVID-19: A systematic review and meta-analysis *Academic Radiology.* **28**(11) pp 1507–1523
[16] Karekar S R and Vazifdar A K 2021 Current status of clinical research using artificial intelligence techniques: A registry-based audit *Perspect Clin Res.* **12**(1) pp 48-52
[17] Liu X, Faes L, Kale A U, Wagner S K, Fu D J, Bruynseels A, Mahendiran T, Moraes G, Shamdas M, Kern C, Ledsam J R, Schmid M K, Balaskas K, Topol E J, Bachmann L M, Keane P A and Denniston A K 2019 A comparison of deep learning performance against health-care professionals in detecting diseases from medical imaging: a systematic review and meta-analysis *Lancet Digit Health.* **1** pp e271-e297
[18] Siontis K C, Noseworthy P A, Attia Z I and Friedman P A 2021 Artificial intelligence-enhanced electrocardiography in cardiovascular disease management *Nat. Rev. Cardiol.* **18**(7) pp 465–478
[19] Schlesinger D E, Diamant N, Raghu A, Reinertsen E, Young K, Batra P *et al* 2022 A deep learning model for inferring elevated pulmonary capillary wedge pressures from the 12-lead electrocardiogram *JACC: Advances* **1**(1) pp 1-11
[20] Vijayan R S K, Kihlberg J, Cross J, Poongavanam V 2022 Enhancing preclinical drug discovery with artificial intelligence *Drug Discov.* **27**(4) pp 967-984
[21] Wijeyakulasuriya D A, Eisenhauer E W, Shaby B A, Hanks E M 2020 Machine learning for modeling animal movement *PLoSOne.* **15** e0235750
[22] Xie Q, Liu Y, Huang H, Hong B, Wang J, Han H *et al* 2020 An innovative method for screening and evaluating the degree of diabetic retinopathy and drug treatment based on artificial intelligence algorithms *Pharmacol. Res.* **159**
[23] Verma A, Chitalia V C, Waikar S S and Kolachalama V B 2021 Machine earning applications in nephrology: a bibliometric analysis comparing kidney studies to other medicine subspecialities *Kidney Medicine.* **3**(5) pp 762–777
[24] Jain D and Singh V 2018 Feature selection and classification systems for chronic disease prediction: A review *Egypt Inform. J.* **19**(3) pp 179–189
[25] Ginley B, Lutnick B, Jen K Y, Fogo A B, Jain S, Rosenberg A *et al* 2019 Computational segmentation and classification of diabetic glomerulosclerosis *J. Am. Soc. Nephrol.* **30**(10) pp 1953–1967
[26] Simon O, Yacoub R, Jain S, Tomaszewski J E and Sarder P 2018 Multi-radial LBP Features as a tool for rapid glomerular detection and assessment in whole slide histopathology images *Sci. Rep.* **8**(1) p 2032
[27] Gadermayr M, Eschweiler D, Jeevanesan A, Klinkhammer B M, Boor P, Merhof D 2017 Segmenting renal whole slide images virtually without training data *Comput. Biol. Med.* **90** pp 88–97
[28] Gadermayr M, Dombrowski A K, Klinkhammer B M, Boor P and Merhof D 2018 CNN cascades for segmenting sparse objects in gigapixel whole slide images *Comput Med Imaging Graph.* **71** pp 40-48
[29] Gupta L, Klinkhammer B M, Boor P, Merhof D, Gadermayr M 2019 Iterative learning to make the most of unlabeled and quickly obtained labeled data in histology *Proceedings of Machine Learning Research* **102** pp 215-224
[30] Bouteldja N, Klinkhammer B M, Bülow R D, Droste P, Otten S W, von Stillfried S F *et al* 2021 Deep learning-based segmentation and quantification in experimental kidney histopathology *J. Am. Soc. Nephrol.* **32**(1) pp 52–68

[31] Govind D, Santo B A, Ginley B, Yacoub R, Rosenberg A Z, Jen K Y *et al* 2021 Automated detection and quantification of Wilms' Tumor 1-positive cells in murine diabetic kidney disease *SPIE-Intl Soc Optical Eng*. 11603

[32] Fu X, Liu T, Xiong Z, Smaill B H, Stiles M K and Zhao J 2018 Segmentation of histological images and fibrosis identification with a convolutional neural network *Comput. Biol. Med.* **98** pp 147–158

[33] Facchin C, Certain A, Yoganathan T, Delacroix C, Garcia A A, Gaillard F *et al* 2022 FIBER-ML, an Open-Source Supervised Machine Learning Tool for Quantification of Fibrosis in Tissue Sections *Am. J. Clin. Pathol* **192** pp 783-793

[34] Tokarz D A, Steinbach T J, Lokhande A, Srivastava G, Ugalmugle R, Co C A *et al* 2021 Using artificial intelligence to detect, classify, and objectively score severity of rodent cardiomyopathy *Toxicologic Pathology*. **49** pp 888–896

[35] Jarutis J, Bijeikis S, Ragauskas S, Mickevičius K, Kalesnykas G, Cerrada-Gimenez M *et al* 2020 Validation of a novel artificial intelligence (AI) machine learning algorithm for quantification of vascular pathology in preclinical models *Information*. **11** p 125

[36] Park J H, Seo E, Choi W, Lee S J *et al* 2022 Ultrasound deep learning for monitoring of flow-vessel dynamics in murine carotid artery *Ultrasonics* **120**

IOP Conf. Series: Materials Science and Engineering 1254 (2022) 012037 doi:10.1088/1757-899X/1254/1/012037

Game-based virtual reality solution for post-stroke balance rehabilitation

R Ciorap[1], D Andrițoi[1], A Casuță[2], M Ciorap[3] and D Munteanu[4]

[1]Biomedical Sciences Department, Faculty of Medical Bioengineering, Grigore T. Popa University of Medicine and Pharmacy Iasi, Romania
[2]Faculty of Electrical Engineering, Gheorghe Asachi Technical University of Iasi
[3]Faculty of Machine Manufacturing and Industrial Management, Gheorghe Asachi Technical University of Iasi
[4]Medicine I Department, Faculty of Medicine, Grigore T. Popa University of Medicine and Pharmacy Iasi, Romania

E-mail: doru.andritoi@umfiasi.ro, radu.ciorap@umfiasi.ro

Abstract. In recent years, there has been a remarkable decline in stroke-related mortality. However, it remains one of the leading causes of death and the most prominent cause of lifelong disability in adults. This positive development results from increased accessibility to various stroke treatment options. In addition, the use of virtual reality in rehabilitation offers a possible alternative to traditional methods of medical rehabilitation of the upper limbs, improving muscle strength or balance. An early diagnosis of equilibrium deficits brings an incredible contribution to recovery. Analysis tests using stabilometric platforms have begun to be used more recently due to the advantages offered. This assessment method highlights and analyzes balance disorders and determines the body's center of gravity position through several sensors or arrays of sensors that transmit information to the processing software. This work aimed to develop a low-cost, stabilometric platform that could be used to build a recovery system using virtual reality-based games. The developed platform uses an Arduino UNO microcontroller that takes data from 4 force sensors. Two applications have been created. The first one was for balance maintenance and the second for execute balance exercises in the anteroposterior and mediolateral direction.

1. Introduction

It is incredible how much technology has been used to solve various problems. For example, game-based virtual reality methods offer computer-generated artificial environments that produce a simulation of reality so successful that the user can get the impression of an almost real physical presence, both in certain real places and in imaginary places [1].

Motivating the patient to perform the recovery exercises is very important, and virtual reality games are beneficial. Furthermore, the possibility to perform repetitive tasks with different levels of complexity, adapted to the patient's performances, supports the effectiveness of virtual reality games for the rehabilitation of chronic stroke patients [2]. Recently, numerous studies have been published on virtual reality in the recovery of post-stroke patients [3-8].

The main challenge in any rehabilitation physiotherapy program is no real objective feedback regarding its efficiency and effect [9]. Therefore, during the rehabilitation program, the physiotherapy specialist follows an already established protocol, making changes to accommodate the patient's current state and health history. In addition, it is essential to monitor a number of the patient's vital parameters during the recovery exercises so that the patient's effort to perform the tasks does not increase the risk of another health incident [10-12].

Balance rehabilitation is an innovative method of improving equilibrium based on advanced knowledge of human biomechanics and motor coordination. It can also be considered a beneficial, progressive, and active method, which can be applied to patients of any age, respecting each individual's ability.

The state of equilibrium depends on the kinesthetic sensations, the visual perception, and the vestibular analyzer's mechanisms. Among the different types of balance formulated in the literature, two major categories are static and dynamic equilibrium. Static equilibrium refers to maintaining a stationary position, while dynamic equilibrium refers to maintaining balance while moving (walking). There are several methods used to evaluate balance disorders [13]. They are divided into two main classes according to their particularities. The wide variety of tests used in medical practice is explainable due to the complex processes that maintain stability.

Balance disturbance can occur in multiple situations: various diseases of the central and peripheral nervous system [14], musculoskeletal disorders [15], sensory disturbances [16], vision problems, various vestibular disorders or stroke [17-18] are just as many factors that can cause imbalance, suspending one or more elements that ensure body position control.

2. System design

The significant advantage in the case of stabilometric tests is that it eliminates the human factor in the measurement chain and, therefore, the source of errors from the subjectivity of the evaluator. Furthermore, in the diagnosis and treatment process, the use of modern transducers [19], which transform the forces that appear at the support surface of the subject (see Figure 1) into electrical signals, measurable, processable, and comparable, is a real advantage.

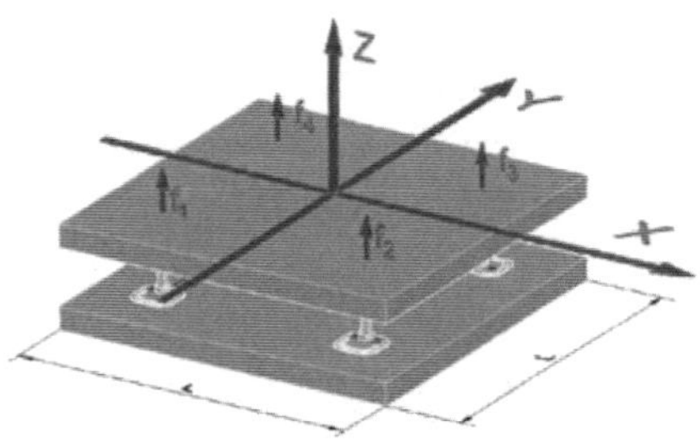

Figure 1. Force distribution on the stabilometric platform.

To determine the position of the center of pressure (CoP), we establish the origin of the axes in the center of the platform (Figure 1) and write the following equations:

$$F_z = f_1 + f_2 + f_3 + f_4$$
$$M_x = F_z \cdot y_0$$
$$M_y = F_z \cdot x_0$$

where F_z is the z-axis resultant of forces, M_x and M_y are the moments on the x and y-axis, respectively. We can write an expression only for the Mx and My moments generated by the limited motions:

$$M_y = (f_2 + f_3) \cdot \frac{L}{2} - (f_1 + f_3) \cdot \frac{L}{2} = (f_2 + f_3 - f_1 - f_4) \cdot \frac{L}{2}$$
$$M_x = (f_3 + f_4) \cdot \frac{L}{2} - (f_1 + f_2) \cdot \frac{L}{2} = (f_3 + f_4 - f_1 - f_2) \cdot \frac{L}{2}$$

The explicit expression of the coordinates of the center of pressure is obtained:

$$x_0 = \frac{(f_2 + f_3 - f_1 - f_4)}{f_1 + f_2 + f_3 + f_4} \cdot \frac{L}{2}$$
$$y_0 = \frac{(f_3 + f_4 - f_1 - f_2)}{f_1 + f_2 + f_3 + f_4} \cdot \frac{L}{2}$$

A series of tests are used for evaluation because balance is a combination of sensations from several neurological systems: the proprioceptive, vestibular, and visual. If any of these three systems are functioning normally, the patient can maintain a fair balance. The key to the test is to remove the visual sensation, if possible, from the equation (by keeping the eyes closed).

The design of such a platform (see Figure 2) seems quite simple but involves many sensitive issues. In the case of the project developed by us, we wanted to use a platform with four sensors. Therefore, we used a load-cell that supports up to 50Kg and offers outstanding technical characteristics as a technical solution.

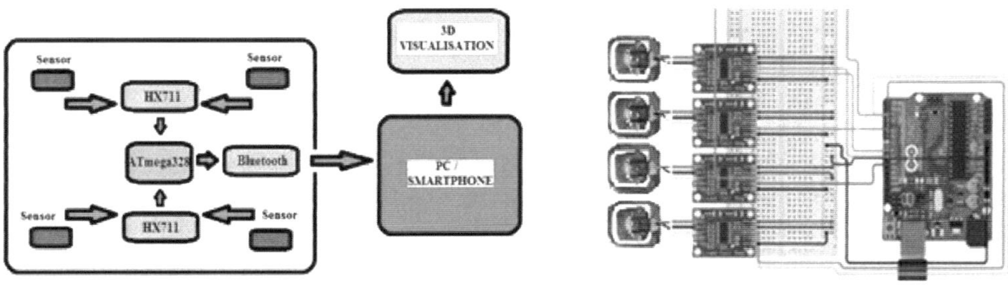

Figure 2. System block and connexion diagram.

The 4 component cells of the system were coupled through two HX711 modules that transfer the information to the decision element represented by an AtMega328 microcontroller integrated on an Arduino UNO type platform [20].

In this way, the imbalances that appear in different situations or positions can be evaluated: when the patient is sitting with his eyes closed, with his eyes open, with his head to the left or right, or in other positions adopted according to the pathology of each patient. Also, with the help of stabilometric platforms, it is possible to determine the projection of the center of gravity inside the supporting polygon, information with a great therapeutic value in analyzing a patient's postural attitude [21].

The software allows the patient's medical data storage following posturographic tests. With all this information available, it is easy to follow patients' progress, make comparative analyses, and evaluate the effectiveness of recommended therapies. The software application that manages the evaluation using the stabilometric platform has been developed in a programming language similar to C++, which allows the programming of the Arduino UNO platform. The interface calculates body weight distribution in real-time and provides visual feedback on the subject's movement performance. It also allows the control of this distribution through the 3D system for rehabilitation exercises.

The stabilometric platform was designed using Autodesk FUSION 360.

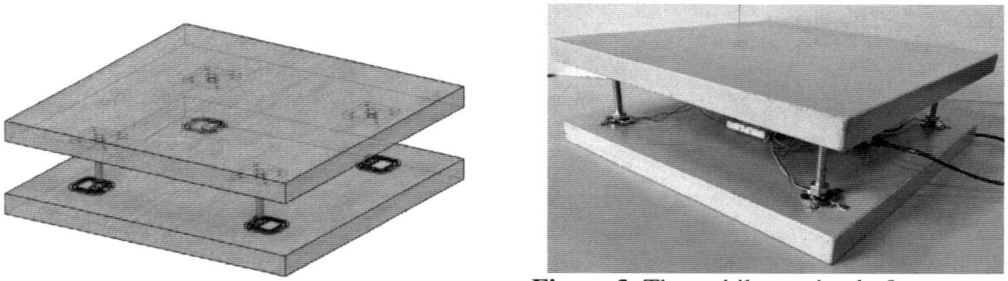

Figure 3. The stabilometric platform.

A pair of MYRYA MY9802 VR glasses were used as the viewing medium. The MYRYA MY9802 glasses are a wearable 3D cinema type device that allows viewing movies in Virtual Reality and 3D format. In addition, this Head Mounted Display (HMD) has built-in headphones that connect to your phone via an audio cable (jack type) and provide auditory feedback. A Bluetooth HC 05 module was used to connect to the mobile phone inserted in the MYRYA glasses.

3. Results

The patient is instructed on how the device works and what to do during the therapy session. Once the patient understands the actions he has to perform, the glasses are mounted on the patients' head, the app is run, and the Bluetooth connection between the Arduino-HX711 module and the android app on the smartphone is made. To develop the game interface that the patient activates, we used a game engine with an integrated development platform created by Unity Technologies. This game engine can make 2D, 3D, virtual reality, and augmented reality games, simulations, and other experiments. The game development platform is user-friendly, easy to use, and does not require high resources.

The application developed consists of moving several balls placed in a box. First, when the patient must maintain the equilibrium, the balls must be kept in the top position. The balls move from one position to the other if the CoP opposition exceeds a specific area for the anterior-posterior displacement. The selection of a particular ball to be moved is done by a mediolateral displacement.

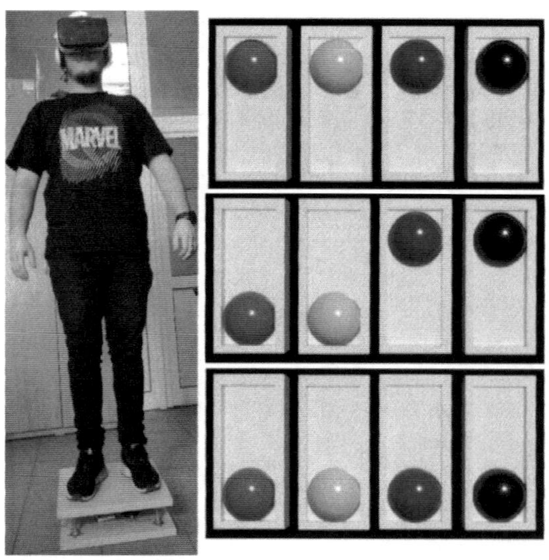

Figure 4. The patient on the stabilometric platform and some screens of the application developed.

We have also developed another mobile phone application in which the patient has to make movements in the anteroposterior and mediolateral direction without losing balance. The application consists of moving a ball from the center of the screen to lateral positions along a predefined path (see Figure 5). The complexity of the trail and, therefore, the degree of difficulty is set from the application menu according to the therapeutic purpose. The patient can select a certain level of complexity and repeat the exercise as many times as they wish. Finally, reaching an obstacle on the route is signaled. After completing a level, the patient is congratulated and can proceed to the next level of complexity.

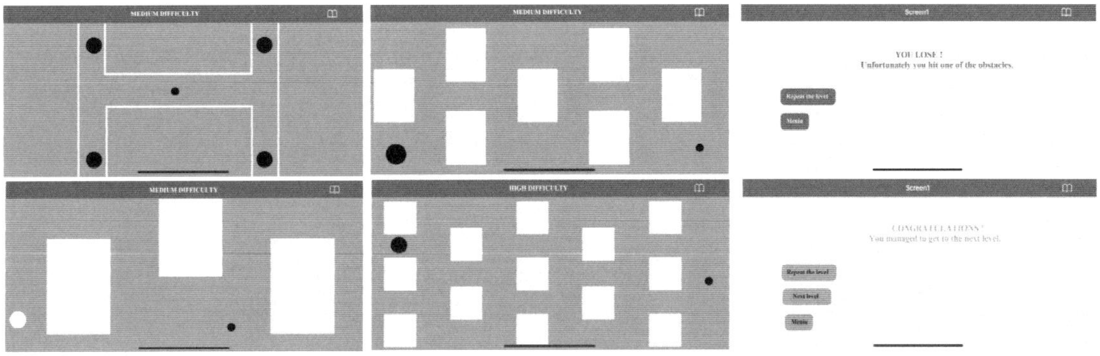

Figure 5. Application for balance training.

4. Conclusion

After several rehabilitation sessions using this device, the affected parts of the patient's neuro-locomotor system begin to regain function. Through this training, the patient begins to perform more and more movements.

The system is low cost and easy to use, which is an advantage and makes it possible for the patient to be trained by the physiotherapist and perform the whole procedure themselves at home.

References

[1] Cano-Manas M J, Collado-Vazquez S, Hernandez J R, Villena A J M and Cano-de-la-Cuerda R 2020 Effects of video-game based therapy on balance, postural control, functionality, and quality of life of patients with subacute stroke: a randomized controlled trial *J Healthc Eng* **2020**

[2] Shahmoradi L, Almasi S, Ahmadi H, Bashiri A, Azadi T, Mirbagherie A *et al* 2021 Virtual reality games for rehabilitation of upper extremities in stroke patients *J Bodyw Mov Ther* **26** pp 113-122

[3] Darekar A, McFadyen B J, Lamontagne A and Fung J 2015 Efficacy of virtual reality-based intervention on balance and mobility disorders post-stroke: a scoping review *J Neuroeng Rehabil* **12**

[4] de Rooij I J M, de Port I G L V and Meijer J W G 2016 Effect of virtual reality training on balance and gait ability in patients with stroke: systematic review and meta-analysis *Phys Ther* **96**(12) 1905-1918

[5] In T, Lee K, and Song C 2016 Virtual reality reflection therapy improves balance and gait in patients with chronic stroke: randomized controlled trials *Med Sci Monitor* **22** 4046-4053

[6] Ibrahim M S, Mattar A G and Elhafez S M 2016 Efficacy of virtual reality-based balance training versus the Biodex balance system training on the body balance of adults *J Phys Ther Sci* **28**(1) p 20

[7] Garay-Sanchez A, Suarez-Serrano C, Ferrando-Margeli M, Jimenez-Rejano J J and Marcen-Roman Y 2021 Effects of immersive and non-immersive virtual reality on the static and dynamic balance of stroke patients: a systematic review and meta-analysis *J Clin Med* **10**(19)

[8] Donath L, Rossler R and Faude O 2016 Effects of virtual reality training (Exergaming) compared to alternative exercise training and passive control on standing balance and functional mobility in healthy community-dwelling seniors: A meta-analytical review *Sports Med* **46**(9) p 1293

[9] Janatová M, Pětioký J, Hoidekrová K, Veselý T, Hána K, Smrčka P *et al* 2022 System for game-like therapy in balance issues using audiovisual feedback and force platform *Electronics-Switz* **11**(8) p 1179

[10] Ciorap R, Andritoi D, Luca C and Corciova C 2019 Monitoring of cardiovascular parameters during rehabilitation after stroke event *IFMBE Proc*, edited by S. Vlad and N. M. Roman (Cluj Napoca, Romania: Springer) vol **71** p 103-107

[11] Andritoi D, Matei D, Luca C, Corciova C and Ciorap R 2015 Preliminary study of HR analysis on patients recovering after stroke *9th International Symposium on Advanced Topics in Electrical Engineering* Bucharest p 23

[12] Ciorap R, Andritoi D, Ciorap M and Munteanu M 2018 Monitoring system for home-based hand rehabilitation *International Conference and Exposition on Electrical and Power Engineering*, edited by M. Gavrilas, C. Fosalau, C. G. Haba *et al* p 836

[13] Jin H Y, Xie L, Xiao Z L and Zhou T 2019 Classification for human balance capacity based on visual stimulation under a virtual reality environment *Sensors-Basel* **19**(12)

[14] Canning C G, Allen N E, Nackaerts E, Paul S S, Nieuwboer A and Gilat M 2020 Virtual reality in research and rehabilitation of gait and balance in Parkinson disease *Nat Rev Neurol* **16**(8) p 409

[15] Dalmazane M, Gallou-Guyot M, Compagnat M, Magy L, Montcuquet A, Billot M, Daviet J-C and Perrochon A 2021 Effects on gait and balance of home-based active video game interventions in persons with multiple sclerosis: A systematic review *Mult Scler Relat Dis* **51**

[16] Babadi S Y and Daneshmandi H 2021 Effects of virtual reality versus conventional balance training on balance of the elderly *Exp Gerontol* **153**

[17] Fishbein P, Hutzler Y, Ratmansky M, Treger I and Dunsky A 2019 A preliminary study of dual-task training using virtual reality: influence on walking and balance in chronic poststroke survivors *J Stroke Cerebrovasc* **28**(11)

[18] Kim J H, Jang S H, Kim C S, Jung J H and You J H 2009 Use of virtual reality to enhance balance and ambulation in chronic stroke: a double-blind, randomized controlled study *Am J Phys Med Rehab* **88**(9) p 693

[19] Veselý T, Janatová M, Smrčka P, Vítězník M, Kliment R and Hána K 2022 Measuring of the energy expenditure during balance training using wearable electronics *Electronics-Switz* **11**(7) p 1096

[20] de Macedo D V, Lustosa E B S, Nogueira M B, Peixoto M E B, do Amaral C M S and Rodrigues M A F 2019 A virtual reality based platform to balance disorders diagnosis and vestibular rehabilitation *Symp Virtual Augment* pp 186-195

[21] Chiari L, Rocchi L and Cappello A 2002 Stabilometric parameters are affected by anthropometry and foot placement *Clin Biomech* **17**(9-10) p 666

IOP Conf. Series: Materials Science and Engineering 1254 (2022) 012038 doi:10.1088/1757-899X/1254/1/012038

Observing the different optimizations in communication protocols for more efficient algorithms in Wireless Sensor Networks

B Zanaj[1], A Peci[1] and E Zanaj[2]

[1]Department of Mathematics and Informatics, Faculty of Economic and Agribusiness, Agriculture University of Tirana, Tirana, Albania

[2]Department of Electronics and Telecommunications, Faculty of Information Technology, Polytechnic University of Tirana, Tirana, Albania

E-mail: bzanaj@ubt.edu.al, apeci@ubt.edu.al, ezanaj@fti.edu.al

Abstract. The different optimization performed for data exchange between the nodes of a wireless sensor network is necessary to reduce as much as possible the energy consumption. The battery constrain is the main challenge in this kind of network implementation. The protocol used for data communication from the source the sensing node toward the sink needs to bring an improvement. It will allow a longer lifetime duration for the network by keeping the nodes alive for a longer time. A longer the lifetime of the network allows its usability in different environment where the monitoring of different conditions maybe a necessity. But lately the Wireless Sensor Networks are used as well for tracking purposes like finding the nurses and doctors in a hospital centre. The data exchange and choices of the next sensor node to pass the data in the new algorithms apply the social behaviour of the birds, bees, ants in their search for food where those are in bunches or colonies. Also, within Wireless Sensor Networks application the nodes communicate by mimicking their social behaviour within the algorithm used for the data exchange, with the only purpose that data transmission and communication with the other nodes might not drain the battery.

1. Introduction

Wireless Sensor Network are widely used for monitoring the environment conditions that keep changing over time. A continuous monitoring asks for a frequent data exchange between the nodes. This data exchange will be held in such a way as to preserve the nodes energy and by doing so we keep the nodes alive for as long as possible. Such a constrain as the remaining energy in the network nodes demands for trying new algorithms. The application of the new algorithm aims to bring some improvements into the node communications for data exchange and to achieve a longer lifetime duration of the network. The network must be connected, where all the nodes can communicate their data in order to send them toward the base station. In this work is discussed the network settling by constructing the graph by using the algorithm of Minimum Spanning Tree. And then the work proceeds with the finding of a routing path with less cost for energy conservation in the network nodes.

2. Network architecture in WSN

The WSN architecture includes the five primary layers and three cross layers' planes. These layers make it possible the receiving of the messages and the sensors collaboration with each other to increase the network efficiency. The architecture applied in WSN is used for different use cases like in hospitals, schools, streets, buildings and for security management, for disaster management, and for crises management. There are two main types of architecture applied for wireless sensor network: layered network architecture and the cluster architecture.

3. Layered network architecture

This kind of network uses hundreds of sensors and applies the usage of a base station as well. The nodes establishment in the network can be performed through concentric layers. There are five primary layers and three cross layers' planes, those are listed below and are as well shown in Figure 1.

1. Application layer
2. Transport layer
3. Network layer
4. Data link layer
5. Physical layer

The three vertical cross layers' planes include:

1. Energy management plan
2. Displacement management plan
3. Task management plan

These three cross layers are used mainly for network controlling and to enable the sensor functioning to increase network efficiency.

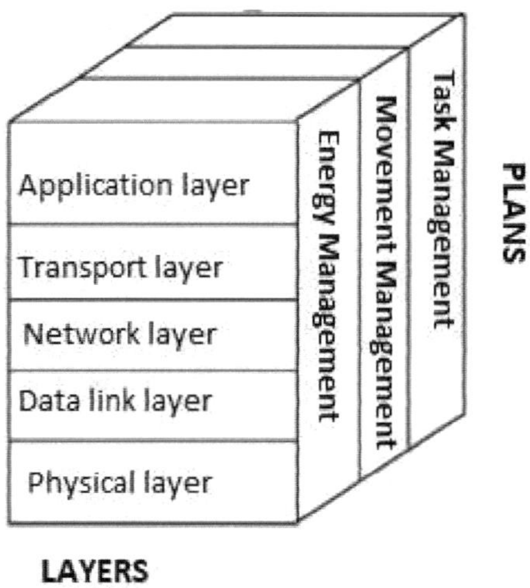

Figure 1. Layered architecture and the cross layers planes of WSN.

3.1 Cluster architecture of the network

In this second type of architecture, the sensor nodes are divided into groups and it is used the Leach protocol for internode communication. The term 'Leach Protocol' stands for 'Low Hierarchy of Adaption for Energy Gathering'. The main characteristics of this protocol are shown as in Figure 2 below.

It is a hierarchical architecture with two levels for data gathering. In cluster architecture is applied a distributed algorithm for node communication for the information exchange where the nodes are arranged into groups known as clusters. There is applied the concept of data fusion in order that the network is efficient for energy consumption. This type of network architecture is extremely used for data fusion characteristic. In every group each of the nodes communicates with the cluster head and only through him it can share its sensed data. All the groups will share their gathered data with the base station through their cluster head node. It will be the cluster head that will communicate the data gathered from its cluster to the base station. The group creation and the cluster head selection for each group is done in complete autonomy as a distributed and independent way.

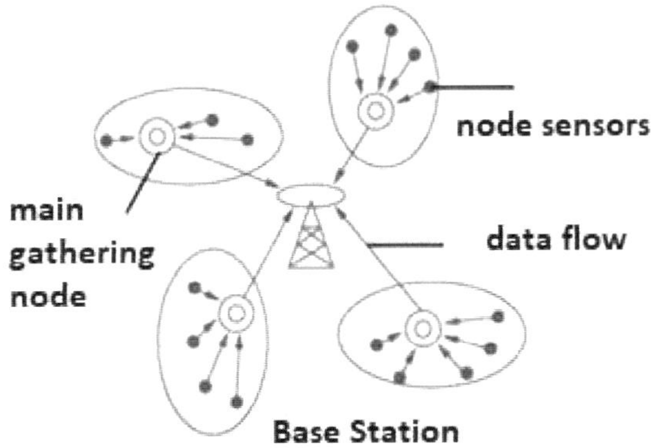

Figure 2. Cluster Architecture with the Base Station and Leach protocol communication protocol.

3.2 Different problems in the network design

The main issues in the architecture designing of a wireless sensor network are like following.

- Energy consumation
- Localization
- Coverage
- Clock
- Computation
- Production cost
- Hardware designing
- Quality of service

Energy consumption: in WSN the energy consumption is the main concern and issue. The battery is the only energy source that is feeding it and it is incorporated in the sensor node. The sensor network may be settled in a site where the battery change is a very complicated operation often not even possible to do it. Energy consumption depends mainly on the node's operation kind like it is transmitting, or it is sensing, or it can perform data elaboration. During the time the node is transmitting the energy consumption is very high. So, the energy consumption can be reduced as much as possible in each layer by using efficient routing protocols.

Localization: for network interoperability the main problem which is critical is the sensor localization. The nodes are arranged randomly in an ad-hoc way so those do not know beforehand their own position. The main difficulty is defining the physical location of the sensors after those are spread in the surface that will be monitored, which is known as node localization. This issue can be solved through the usage of GPS with which the nodes can be equipped, or the localization based on closeness to the other nodes that already know their geographical position.

Coverage: sensor nodes in wireless sensor network use a coverage algorithm for data gathering, and those use a routing algorithm for data transmission. For a total coverage of the whole network there need to be selected the sensor nodes that will be involved in data transmission toward the base station. It is recommended that the choice of most efficient shortest path is selected for the modelling of the coverage protocol.

Clock: in WSN clock synchronization is a serious question. The main function of synchronization is offered a unique scale of clock for all the nodes that belong to the same network. Those clocks might be synchronized for certain application like the monitoring and tracking.

Computation: computation can be defined as the sum of all the data passing through every node of the network. The main demand is that the computation might be performed with the less possible consumption of the node resources. If the life duration of the base station is questionable, then the elaboration of data will be completed in the nodes before transmitting the data toward the base station. In the case when every node has different resources, then the computation will be done in the sink node (gathering node).

Cost of production: in WSN can include a huge number of sensor nodes. If the cost of a single node is high, then the overall cost of the network will be high as well. It is a necessity that the single node cost is kept low. The cost of the sensor node is still a problem for the overall cost of network implementation.

Hardware design: when we design a device like a sensor node, we might take good care on controlling the energy consumption by the microcontroller and by the communication unit to be efficient in energy usage. The overall node design might include a reduced usage of the energy during its operation.

Quality of service: the quality of service or QoS concerns the data distribution on time because some applications that works on real time the data sensing depends too much on the time when it is done. If the data are not distributed on time toward the receiver, these data will become useless.

4. Problems encountered in WSN

Coverage is the main basic issue in WSN, and it addresses on how to put a sensor node in the geographical area in order that each point of the area under observation is sensed by a sensor node. This is a problem of optimization that requires to find an optimal position for the sensor nodes of the network in order that we achieve the maximum coverage of the area. This is addressed as network coverage. For the most efficient setting of a wireless sensor network, it is required to calculate the sensor coverage area. The sensor coverage area is a circle with a ray of Rs meters, where Rs is the sensing zone. The overall coverage of WSN is coloured in grey in Figure 3 and is calculated by the formula 1 as below:

$$Coverage = \sum_{i=1}^{n}(C_i - C_i^0)/C_R \qquad (1)$$

here C_i is a sensor coverage this is variable depending on i, C_i^0 is the overlapped coverage between the i sensor and the other, C_R is the area of the zone of interest. In this formula C_R is used to normalize the values between 0 and 1.

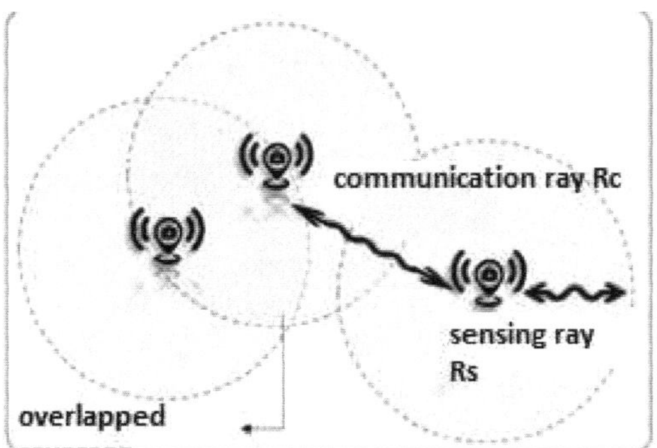

Figure 3. Sensing ray Rs and the overlapped coverage of Rc, normalized by Rc values.

4.1 Life duration

Energy consumption of the sensor nodes where it is fed from a battery is one of the main constrains because it affects the network life duration. It affects directly the WSN applicability in different use cases for environmental observation purposes or for that of health care purposes. Life duration is the main challenge that requires an optimal solution when we discuss the operability of wireless sensor networks for environmental or health care purposes. The life duration will depend on operations on how to decrease the energy consumption of the sensor nodes by increasing the lifetime of the network.

This time duration is estimated by the time it takes for the first node of network to die. Coverage and lifetime influence each other because we want a good coverage and that requires more energy consumption but from the other side a longer lifetime and a less consumption of the energy from the node might not afford a very good coverage. Because energy is fundamental source that we take count of when we construct a WSN, lifetime is considered the main challenge in wireless sensor network implementation. Lifetime duration is given in the formula below:

$$Lifetime = \frac{min\{T_{failure}, i\}_{i=1,2,3,...,n}}{T_{max}} \quad (2)$$

4.2 Minimum Spanning Tree

MST (Minimum Spanning Tree) is an algorithm that has as purpose to find a new path which is less expensive that sends information from the sensing sensor to the source sensor. We can mention its usage in a case when there is a telecommunication company that want to put the cable for a new neighbour. If is obliged to put the cable only through predefined paths like the streets, then the graph will be made of points that can be the houses that connects those paths. Some of the paths can be more expensive because those are the longest, or those might require that the cable to be set too deep in the ground. Those paths will hold a weight which might be the biggest.

Minimum Spanning Tree in this case will find a route which is cheaper in cost that represents the path with less possible cost for the setting of the networkgraph. Laying Tree algorithm will be applied to connect the initial node with the last node. The connection between the source and the sink in Laying Tree algorithm is done through the branches. The branches of the tree has a certain cost and by including the different branches it is calculated the overall cost for each path that the information has to travel to reach the sink node starting from the source node of WSN. The cost of the Spanning Tree is the sum of the weights for every branch in the tree. There can be different spanning trees. The minimum spanning tree is the tree where the cost is minimal through all the other spanning tree. The minimal spanning tree is applied in network designing. Other practical use cases of it are like the following:

-group analysis
-hand writing recognition
-image segmentation

There are two most recognized algorithms for finding the MST that of Kruskal and Prim.

4.2.1 Kruskal Algorithm

Kruskal algorithm is used to create a spanning tree by adding the edges one after the other in the spanning tree. Kruskal algorithm has a greedy approach because after each repetition it finds another edge that has a smaller weight and it adds it into the spanning tree.

Algorithm steps are like the following:

1. We list each vertex of the graph from their weight
2. We start to add the vertexes in MST starting from the edge with the smallest weight till we reach to the other vertex that with the biggest weight.
3. We add only the vertexes that does not create a loop, only those edges that are apart from others.

In Kruskal algorithm in each iteration we need to select the edge with smaller weight. So, it starts with the edge with the smallest weight first, for example the one with weight 1 as is shown in Figure 4. Then it is selected the second edge with the second smallest weight is selected like the edge with weight 2. We can see that the two edges are not connected meaning that those does not have any connection at all with the graph. It continues by finding the edge which has the third smallest weight, this one need to connect the two disconnected part of the graph. We can not continue with the selection of vertex 4 because it will create a loop. The way that the algorithm works does not allow the creation of the loops inside the graph. So, it will continue by selecting the edge with weight equal to 5, the one with the fifth smallest weight. The other two remaining edges creates loops so they will be ignored. What we get at the end is the minimal spanning tree with a total cost of 11 (1+2+3+5).

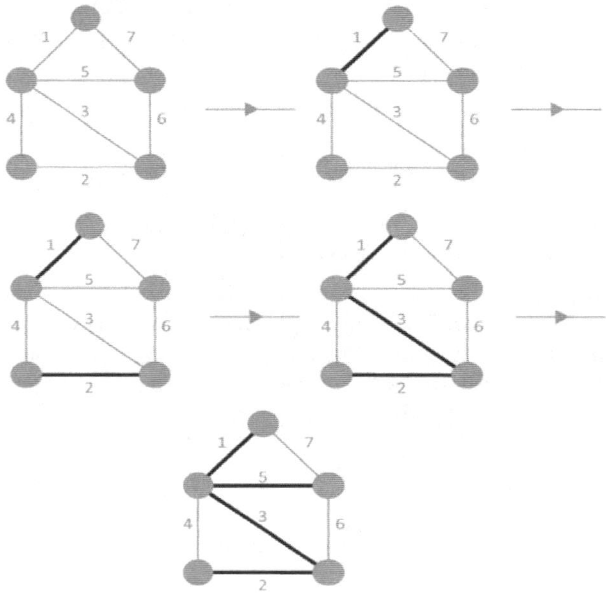

Figure 4. Computing MST with Kruskal algorithm.

4.2.2 Prim Algorithm

Prim algorithm uses a similar way for finding the minimum spanning tree. In Prim algorithm the tree construction starts from a starting point. Differently from Kruskal algorithm where the edges are used to construct the minimal tree, in this case for constructing the minimal tree are used the vertexes. The algorithm steps are as below:

There are divided into two groups the vertices those that are disconnected. In one group are the vertices that are in the tree and in the other groups are the vertices that actually are not in the tree that we are constructing.

There are selected two vertices that has the cheapest edge, then we continue by selecting the vertex that is connected to the cheapest edge that at the moment is not in the tree that we are constructing. Also it is then connected to the tree and it will become part of it. The algorithm continues like this with the other vertices as well by following this priority order, but what we need to take count of is to avoid the loops.

In Prim algorithm we will start with a random node, and we note it down. In each repetition we need to denote a new vertex that is attached to the other already denoted. Like a greedy algorithm, Prim algorithm it will select the cheapest edge and will denote it. So, it will start with the selection of the edge with weight 1 as shown in Figure 5. In other iteration there are three different choices the vertices with weight 2, 3 and 4. So it will choose the vertices with weight 2 and it will be denoted. Again, there are three different options, the vertices with weight 3,4, and 5. But it is not possible to select the vertex with weight 3 as it creates a loop. So, it will select the vertex with weight 4 and we will construct the minimal spanning tree with a total cost of 7 (1+2+4).

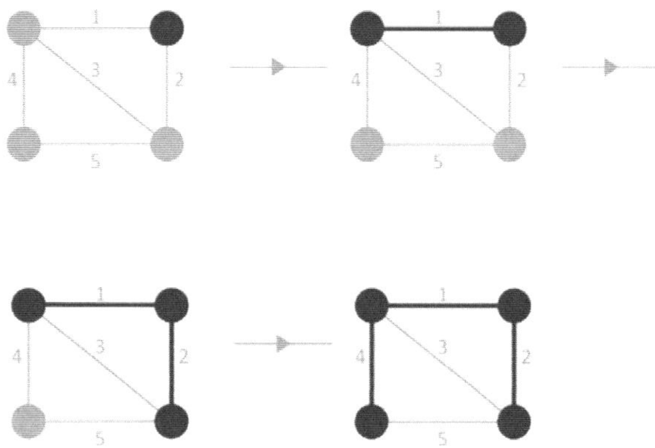

Figure 5. Prim's algorithm and MSP computation.

5. Network setting with Minimum Spanning Tree

Routing in WSN is a very important task that defines how to find the new paths and how to pass the data in the most efficient way from the source to the destination even that the paths can be disconnected. All that matters is to establish and to keep the communication between the network nodes. Energy consumption, selection of vertices, network establishment and scaling, the connectivity and the coverage, also security are the different challenges for the routing protocols in WSN. Like in the Figure 6 below is shown the general structure of the routing protocol in WSN.

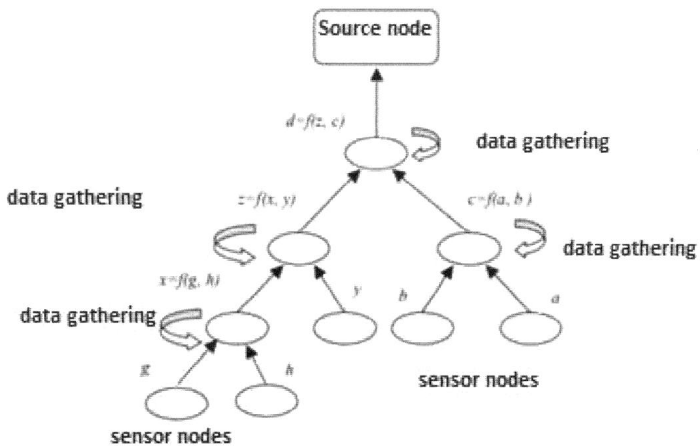

Figure 6. Routing Protocol MST in a wireless sensor network. The arrows show the path direction and f is the function of data gathering.

In this work is proposed the some improvements for the routing algorithm applied in WSN. It will effectively improve the lifetime duration of the network and it guarantees a minimal usage of energy consumption. The proposed protocol will use a "a minimal hoping distance to reach the destination" by calculating a path. The best path is found by following the MST algorithm. The selection of the most efficient path is used in the routing protocol; in this case it is MST which will help in finding the best possible path between all the sensor nodes of the network. The route calculated will start from the source node and it will end to the destination node that is the base station. The route selection criteria are mainly based on the weight; it can be the sum of the costs for the distance along the communication path or it can be the number of hops that are needed to reach the destination node. In this example we have the demonstration of the routing algorithm performance for MST routing by comparing the two different criteria for the path selection in order to improve the efficiency of sensor energy usage. There are two different routing kinds that will be used:

1. Hop based routing
2. Distance based routing

5.1 Data Communication in WSN and the routing protocols

Wireless sensor nodes can communicate in two ways: single-hop communication or multi-hop as shown in Figure 7. In multi-hop communication a package must pass through different nodes in order to reach to the last node for reaching the destination. In Single-hop all the sensors that are circle coloured in blue at Figure 7 can send the gathered information directly to the base station the circle coloured in red in the Figure. In the case of multi-hop communication sensor send their data toward one of its neighbour nodes in order to reach the base station.

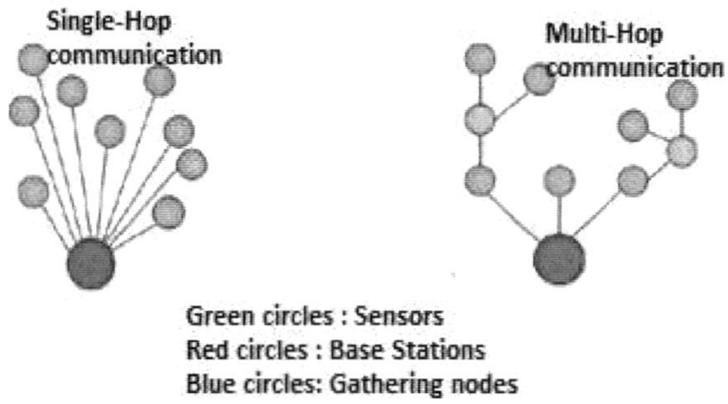

Figure 7. Single-Hop and Multi-Hop communication.

Routing based on MST protocol is used frequently from the researchers for the demand of an efficient usage of energy. This case will show a new approach for constructing a spanning tree with energy efficiency based on minimum hops to direct the data from the sensor node toward the main gathering node (base station) and the maximum remaining energy in each node.

5.2 Cases of MST routing protocol application

In this case is applied with MST the algorithm of Prim, it is made a comparison between the two routing methods: Hop-based routing and Distance-based routing. Firstly, the experiment is done with the setting of 30 sensors and then with the settling of 100 sensors like in Figure 8 and Figure 9.

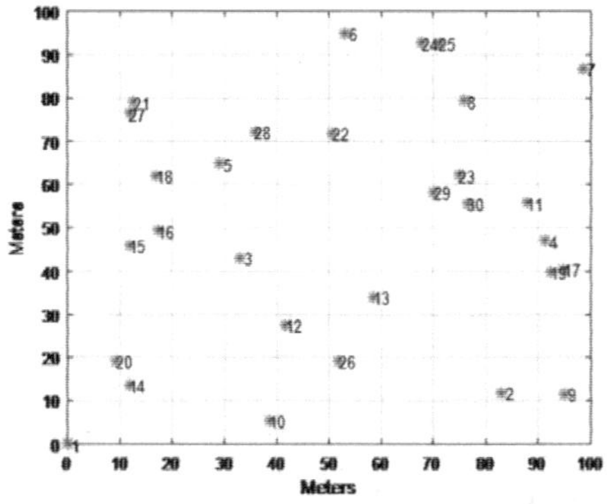

Figure 8. Settling 30 sensors in the environment

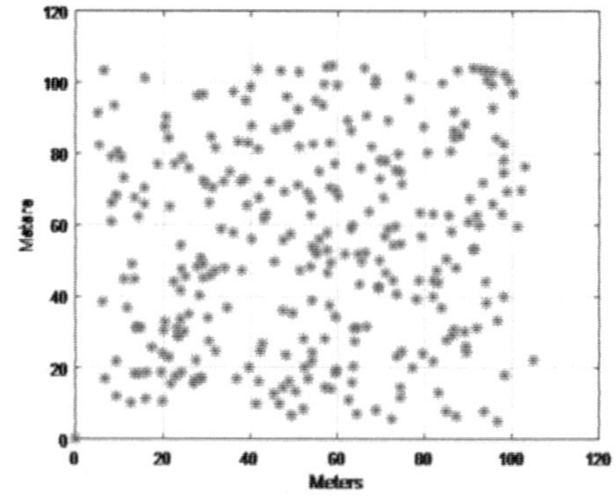

Figure 9. Settling 100 nodes in the environment.

In Figure 10, 11 are shown the path construction based on hoping for 30 and 100 sensor nodes. Instead in Figure 12 and Figure 13 is done the same procedure for the path construction based on distance.

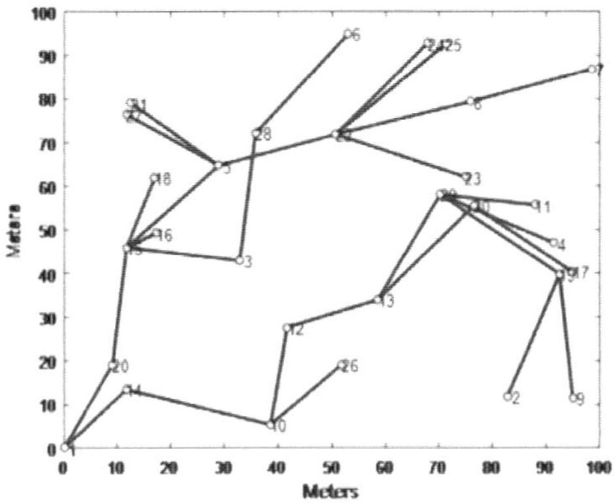

Figure 10. Routing path construction hop based for 30 sensor nodes.

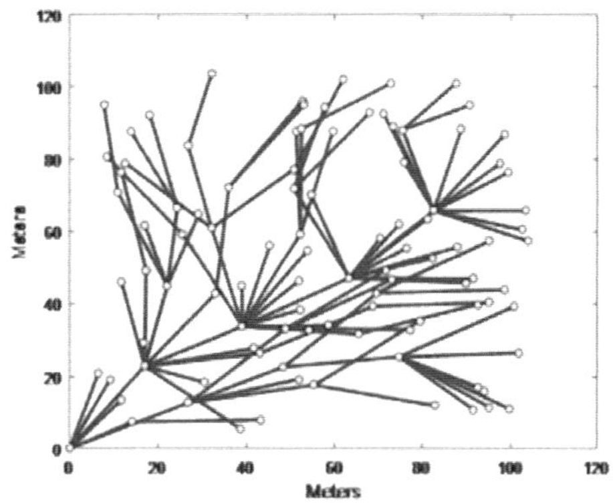

Figure 11. Routing path construction hop based with 100 sensor nodes.

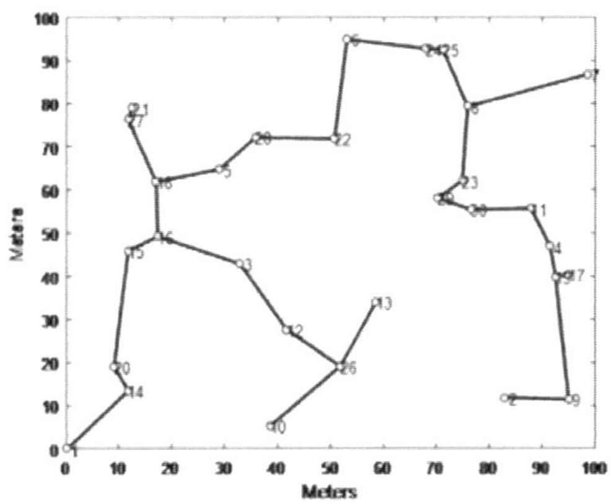

Figure 12. Computing the routing following distance based protocol for 30 sensor nodes.

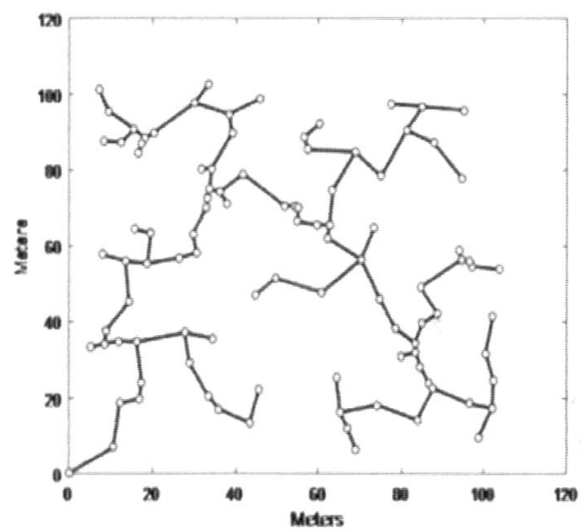

Figure 13. Computing the routing following distance-based protocol for 100 sensor nodes.

Performance estimation is done by comparing the routing protocols based on distance and based on hop. For the approach followed by us the performance will be observed on the average remaining energy in every sensor node, the period duration of stability (the time interval or the number of iteration before the dying of the first node) this is found for the distance. In Figure 14 shows the path construction with minimum hops.

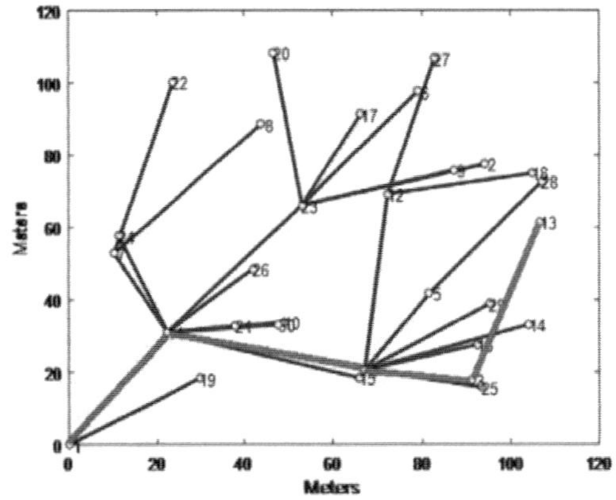

Figure 14. Routing path construction with minimum hops.

In Table 1 will be shown the starting node Id, and the node included in path construction for both protocol used hop-based and distance-based.

Table 1. Routing construction and nodes included for hop-based and distance-based protocol.

Id of sensor node	Routing based on hops	Routing based on distance
17	{ 30; 13; 12; 10; 14; 1 }	{ 19; 4; 11; 30; 29; 23; 8; 25; 24; 6; 22; 28; 5; 18; 16; 15; 20; 14;1 }
26	{ 10; 14; 1 }	{ 12; 3; 16; 15; 20; 14; 1; }
30	{ 13; 12; 10; 14 }	{ 29; 23; 8; 25; 24; 6; 22; 28; 5; 18; 16; 15; 20; 14; 1}

From the data of the above Table no.1 we can understand the routing method hop based, where there are only a few intermediate nodes along the way till it reaches the base station. If we compare it with the distance based routing as shown in Table 2. In hop-based protocol it has less intermediate nodes included so there are fewer hops and fewer will be the sent messages, and all it means that there is less energy spent.

Table 2. Maximal number of hops.

Maximum of hops / **Number of sensors**	**Distance-base routing**	**Hop-based routing**
30	11	4
100	68	4

Table 2 compares the number of hops that the node does from node 13 that were shown in the above graphics to reach the base station where it is used the routing method based on distance and the hop based routing. By looking at the Tables values it is clearly observed that the routing method based on hop tries to minimize the steps to reach the main node (the base station). This will bring to less messages transmitted, and it will bring to an increase in efficiency where there is need to conserve the energy in sensor network.

Figure 15 and 16 shows the energy evolution consumed per each round for distance based protocols and for the hop based one for different network densities. In the Figures15 and 16 are shown the total 1000 rounds of simulations. The starting energy for each sensor node is equal to 0.02nJ. From the graphics from Figures 15 and 16 we can understand that when we do the time test duration of the network till the first nodes dies (when its battery dies), the network applying a hop based protocol those resists more because the first sensor of this network runs out of battery and it happens later if we compare it to the first sensor node for the distance based protocol.

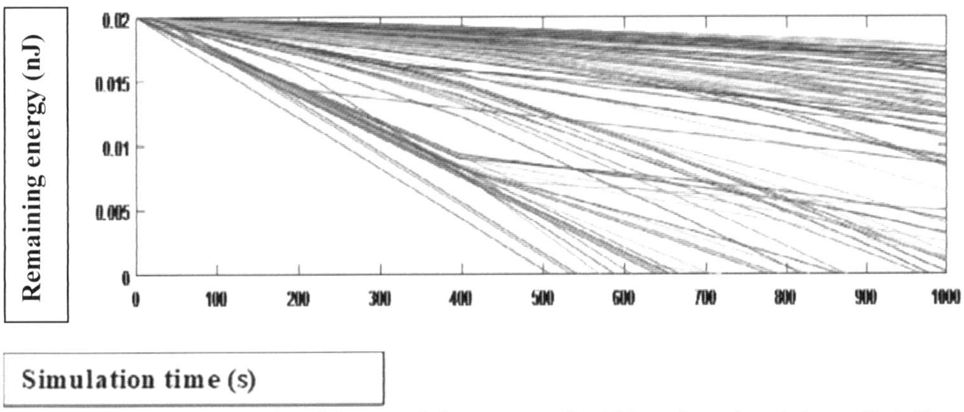

Figure 15. The graphic of the remaining energy for 100 nodes when it is applied distance-based protocol.

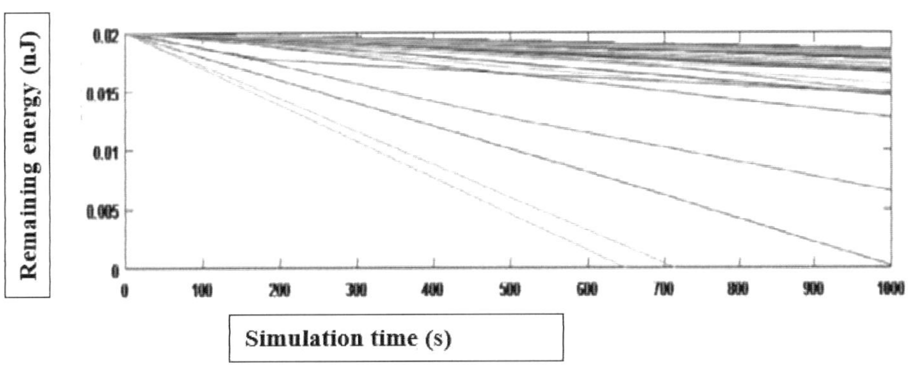

Figure 16. The graphic of the remaining energy for 100 nodes when it is applied the hop based protocol.

As conclusion the routing protocol hop-based conserves more the energy in a longer period of time compared to the routing protocol of type distance based. Consequently, this protocol based on hop achieves to ensure a longer lifetime for the wireless sensor network as shown in Table 3.

Table 3. Lifetime duration for the sensor network by using distance-based and hop-based protocol.

Time of "death" / Sum of sensor nodes	Distance-base routing	Hop-based routing
100	520	647
300	568	641

5.3 Cluster organization and data passing in WSN

In WSN the sensors are set randomly in a zone where they meant to be. If the zone is quite large the sensor needs to observe the area for a longer time, and it requires more energy. In WSN the sensors are small objects those have the battery, and it is not rechargeable or to substitute the battery when the node dies. So, the energy consummation in sensor node is a big challenge in the WSN field of applications. At the beginning every sensor before it is settled its battery is fully charged. But because of transmission process the energy starts to get consumed. If the node has only a few of energy, then it can not participate in the communication. In WSN it is required to have a continuous monitoring of the remaining energy in each node. Every node communicates with the base station when it is within its transmission range, and it transmits in its time interval. If the node can not communicate with the base station when this one it is not within its range of transmission, then the node transmits its data through an intermediary node. So, the communication is made through multi hop. In wireless network with multi-hop, the node near to the base station spends more energy if this one is compared to the other nodes that are far from the base station, because every other node will transmit its data toward the base station through this intermediary node. To avoid this problem, it is used the technique of gathering (Clustering), where the nodes are organized in bunches (groups) where each bunch has a main node known with the term Cluster Head (CH). The nodes transmit the data only toward the main node (CH), and then CH transmits them toward the base station. Routing is another technique that consumes a considerable part of energy. If the route to send the data is not the best from the energy point of view, then consequently more energy is going to be consumed from the node. The same route might not be used for multiple transmissions. Another scenario is that in order that it is achieved the right way of communication it is needed to localize the node correctly. Without knowing the node location, the data can not be transmitted. So, the node localization plays an important role in wireless sensor network. The problems mentioned can not be solved through the usage of deterministic algorithms. These problems can be solved by using high order procedures, that is why there are developed optimizing techniques. The hard problems can be solved easily by using the optimization techniques that gives best solution to the problem.

6. Optimization through new algorithm implementation

Optimization is the process of maximization or minimization of objective function in order to find a solution to complex or difficult problems. The difficult and complex problems of optimization are such kind of problems that can be solved by using deterministic method within a predefined time. Optimization can also be referred to as metaheuristic optimization because it solves the optimization problems by using metaheuristic algorithms. Meta means “the highest level” and heuristic means “solution or procedure”. This means procedure of the highest level for the problems that does not have any deterministic solution. These problems can be single or multiple objectives. The better is the objective function it will become better the optimal solution. With single objective is that all the particles will converge in a single point and therefore that point will be the optimal solution. For the multi-objective case the particles will converge in two or more points and the best solution will be found between them. The metaheuristic algorithm searches for a solution in a research area.

Some algorithms choses the local search others chose the global search. The objective function is formulated by taking count different parameters that relate to the problem statement. There are two types of metaheuristics: metaheuristic based on single solution and metaheuristic based on population. Metaheuristic based on single solution is the oriented usage that means that this type of metaheuristic is limited to local search (to improve the solution). Instead, metaheuristic based on population is oriented toward the search, it means that is limited to global search (to find the new better solution). The techniques based on population are of two types: evolution algorithms and particle intelligence. The evolution algorithms include the genetic algorithm and those differential evolutional. Some optimization algorithms are like follows:

- PSO (Particle Swarm Optimization) algorithm
- ACO (Ant Colony Optimization) algorithm

IOP Conf. Series: Materials Science and Engineering 1254 (2022) 012038 doi:10.1088/1757-899X/1254/1/012038

- ABC (Artificial Bee Colony) algorithm
- BFO (Bacterial Foraging Optimization) algorithm
- FA (Firefly Algorithm)
- COA (Cuckoo Optimization Algorithm)

6.2 PSO algorithm (Particle Swarm Optimization)

The optimization algorithm for particle gathering (PSO-Particle Swarm Optimization) is stochastic optimization technique (probability random based) based on bunches it was first proposed from Kennedy and Eberhart in 1995 [1]. The PSO algorithm simulates the social behaviour of the animals, where are included: insects, birds, and fishes. The bunches converge somehow when they collaborate between them in order to find food, and each bunch member continues to change the searching model by relying on its learning experience or that of the other bunch members. The main idea of the PSO algorithm design is tightly connected with two researches directions: one is the evolutional algorithm. Like the evolutional algorithm also PSO uses a bunch way that makes it searches on the same time the big area in the solution area of the optimized objective function. The other one is the artificial life, so it studies the artificial systems that has some characteristics of natural life. In the study of the social life of the animals in the theory of artificial life, where based on it is constructed the artificial life of the bunch where those are collaborative behaviour of systems. It was Millonas who proposed these main ideas [2].

The main principles are like follows:

1. Vicinity: the bunch might be able to do the computation of area and time
2. Quality: the bunch might be able to detect the difference of the environment quality and to react to this change
3. Diversity in response: the bunch should not be limited; quite the oposite it needs to reach the resources even in a narrow field
4. Stability: the bunch needs not change its behavioural manner indipedently of the environment changes
5. Compatibility: the bunch needs to change its bahavioural way and it is worthy to do it.

Principles number 4 and 5 are the opposite to each other like the two sides of the coin. In PSO the particles might update its positions and its velocity depending on the environment changes, approximately those do fulfil the proximity demand and quality. Instead of it the bunch in PSO does not limit its movement instead it continuously searches for an optimal solution in the pool of all possibilities. The particles in PSO might keep its displacement sTable in the search area, while they keep changing the way they move to adapt it to the environmental change. So the particle gathering systems do comply the above five principles.

PSO searches for the optimal value by updating iterations. The term Swarm means moving particles population. PSO algorithm solves the problems by having a population (that are known as particles) of the possible solutions (that are the particles). The movement of each particle is influenced from its best known position (Local Best Known Position).

With optimization it is understood an optimal point where the conditions are the best and most favourite. We use optimization so that we can find the best solution between some possible solutions. The objective function is used to minimize or maximize the values of the variables we need to optimize. In the case of our work we need the minimal value that is the optimal value that we are searching. The research space in PSO is the environment where the algorithm can calculate the optimal value. If an optimal value overpasses the environment where the algorithm performs the computation, then this value will be reinitialized. How does PSO solve the problems, PSO itself does not guarantee the optimal solution can be found in each case.

IOP Conf. Series: Materials Science and Engineering 1254 (2022) 012038 doi:10.1088/1757-899X/1254/1/012038

6.2 Cases of PSO algorithm

Let us suppose that a group of birds that is flying in a random way and is searching for the food in specific zone [3]. PSO algorithm is inspired from the social behaviour of the birds that flies together in a bunch. In the zone is searched only a piece of food for the bird's bunch as shown in Figure 17.

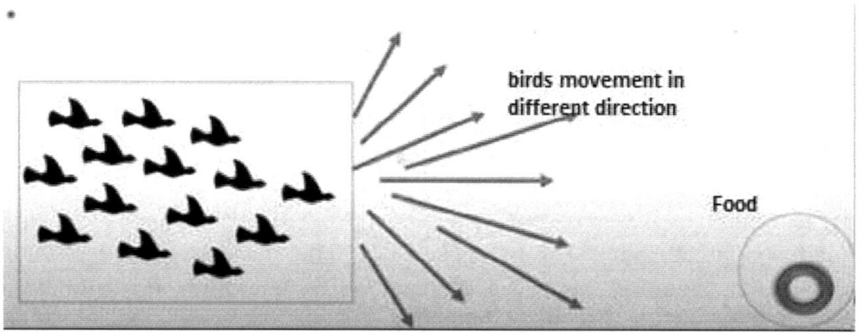

Figure 17. The strategy for searching food.

None of the birds does not know where is the food but they know how far they are in each iteration as in Figure 18.

Figure 18. 5th iteration and the birds are nearer to the food.

The best strategy to find the food is like this to follow the bird that is nearer to the food. The birds do not know with precision where is the food, but after each iteration they know how fare they are from the food. If one of the bunch member finds a favourite path that brings to the food, then the rest of the bunch will follow him.

In PSO every bird is called a particle and the all the birds that are within the bunch is known with the term swarm. Starting from a population randomly started and by moving in random directions, each particle moves in the search area and remembers its previous positions of itself and that of its neighbours [4-6]. The neighbour in this case is the bird that needs to be followed, known as the Global Best (Gbest). So the particles know their previous position and the best position in the bunch, Global Best.

Each particle has velocity where it gives the flight direction of the particles as shown in Figure 19. So the term velocity is the particle velocity in a certain direction.

Figure 19. Birds searching the food with a certain velocity.

PSO initialization

1. PSO algorithm is initialized from a random particles group (where each particle is a solution).
2. Every particle searches for the optimal value by updating iteration
3. In every iteration each particle is updated (with the two best consecutive values):
   a- The fist better value is the value of a better fitness
   b- The second better value is found out from PSO algorithm
4. After the nodes are updated with 2 best values, the particle will update its velocity and the position. The particles can update it position with the formula 3:

$$x_i^{t+1} = x_i^t + v_i^t * t \tag{3}$$

5. the particle velocity is as in the formula 4 below:

$$v_{k+1}^i = wv_k^i + c_1 r_1 (xBest_i^t - x_i^t) + c_2 r_2 (gBest_i^t - x_i^t) \tag{4}$$

Where: - $xBest_i^t$ is the best location for the particle
- $gBest_i^t$ is the best group position
- w is the inertia weight
-c_1,c_2 are two positive constants
- r_1, r_2 are two random parameters that belong to the interval [0 1].

For the update of the position is used the formula
Now= previous position + Velocity (v) where v= Particle velocity and Presence = Actual position

Steps of PSO Algorithm

Step 1. Initialization

It has to be accounted that the research strategy for PSO algorithm is based on the following of bird that is nearer to the food.
.Parameter initialization
.Population initialization
.Position initialization (X_i) randomly for each particle
.Velocity initialization (V_i) randomly for each particle

Step 2. Fitness Evaluation f(X_t^i)

Calculate the fitness value for each particle.
If the fitness value is better than the best fitness value (gBest), then we set the new value like the new gBest. It means that we are going to select the particle with the best fitness value as gBest, this one is the minimal value.

Step 3. For each particle it is updated the velocity and position of each particle by calculating them.
.Calculate the position of particle like in formula 4.4
$x_i^{t+1} = x_i^t + v_i^t * t$ (4.4)
.Calculate the velocity with formula 4.5 as follow
$v_{k+1}^i = wv_k^i + c_1r_1(xBest_i^t - x_i^t) + c_2r_2(gBest_i^t - x_i^t)$ (4.5)

Step 4. Evaluation of Fitness f(X_t^i)
We find the best current value for [gBest]. This process is repeated till we reach to the objective (optimal value).

Step 5. Update of t=t+1
In this step we increment the counter and repeat the cycle till we reach to step 6 (**gBest &x_i^t**)

Step 6. Draw conclusions (outputs) of gBest and x_t^i

In Figure 20 is given the PSO algorithm steps in flow chart diagram.

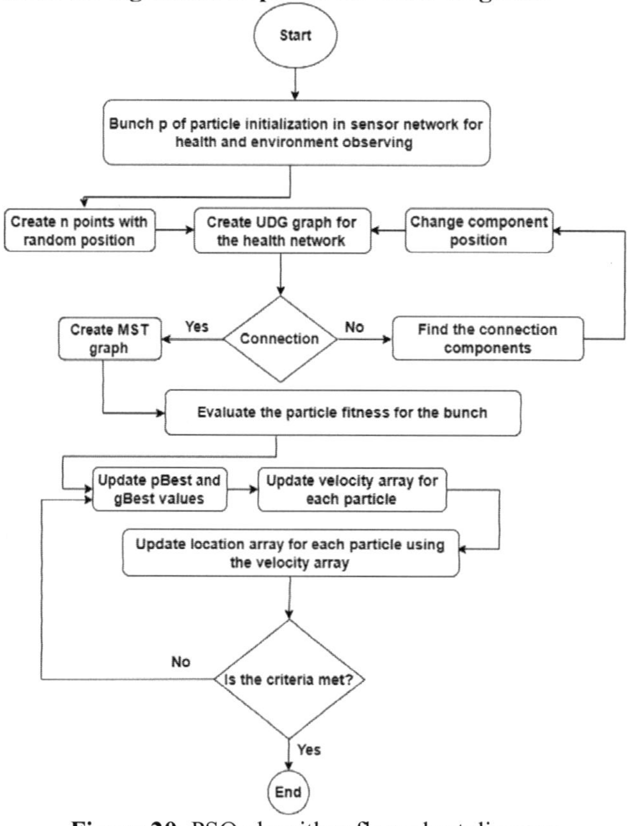

Figure 20. PSO algorithm flow chart diagram.

7. Conclusions
In this work is simulated the wireless sensor network for a certain number of nodes using the Prim algorithm that uses MST as for graph construction and topology setting for the network. The routing protocols that are tried are two one is hop-based and the other is distance-based for route calculation in the network. The protocol for the routing that performs faster in finding the shortest and less expensive path is the best due to energy constrain that this kind of network suffer from. It was observed that the

lifetime duration of the network with the same number of nodes where longer when we applied hop-based protocol. Also, the remaining energy is higher in the case of hop-based one. It was introduced also the algorithm for data communication between the nodes PSO, where in such algorithm takes count and is based on the social behaviour between the birds for finding of food.

References

[1] Wang D, Tan D and Liu L 2017 *Particle swarm optimization algorithm: an overview* © (Springer-Verlag Berlin Heidelberg)

[2] van den Bergh F 2005 *An Analysis of Particle Swarm Optimizers* (University of Pretoria etd)

[3] Saeidian B, Mesgari M S, Pradhan B and Ghodousi M 2018 Optimized Location-Allocation of Earthquake Relief Centers Using PSO and ACO, Complemented by GIS, Clustering, and TOPSIS *ISPRS Int. J. Geo-Inf.* **7** p 292

[4] Jones K O and Bouet A 2008 Comparison of bees algorithm, ant colony optimization and particle swarm optimization for PID controller tuning *Proceedings of the 9th International Conference on Computer Systems and Technologies and Workshop for PhD Students in Computing*, Gabrovo, Bulgaria, 12–13 June p IIIA-9

[5] Luo G, Zhao H and Song C 2008 Convergence analysis of a dynamic discrete PSO algorithm *First International Conference on Intelligent Networks and Intelligent Systems* (Wuhan, China) 1–3 November pp 89–92

[6] AL-Samarrie A K, Alyasiri H and AL-Nakkash A H 2016 Proposed Multi-Stage PSO Scheme for LTE Network Planning and Operation *Int. J. Appl. Eng. Res.* **11** pp 10199–10210

Quality of 3D reconstruction based on photogrammetry for small objects, a case study

I Enesi [1], A Kuqi [2], E Zanaj [3]

[1]Electronic and Telecommunication Department, Polytechnic University of Tirana, Tirana, Albania
[2]Electronic and Telecommunication Department, Polytechnic University of Tirana, Tirana, Albania
[3]Electronic and Telecommunication Department, Polytechnic University of Tirana, Tirana, Albania

E-mail: ienesi@fti.edu.al, anduel.kuqi@fti.edu.al, ezanaj@fti.edu.al

Abstract. 3D reconstructions are widely used, the main challenge of them is the accuracy for small and detailed objects. Various software exist for 3D reconstruction, free and paid ones with various performances. In this paper the performance of 3D object reconstruction will be evaluated in terms of size accuracy regarding number of source photos. The aim of the paper is analysing the size accuracy of the reconstructed 3D models based on photogrammetry for small objects by configuring pipeline in Meshroom. Meshroom is used for 3D photogrammetry reconstruction and MeshLab is used for measurement, both are free software. Experimental results show a high accuracy for objects sizes. Although the tests are done for small objects, the results are promising for cloning body parts used for prosthetics.

1. Introduction

Undoubtedly that the uses of 3D computer vision are growing at an extraordinary rate in different fields of industry, among which we can mention 3D printing, archeology and medical, etc. As the fields for 3D application possibilities are constantly increasing, so are the various solutions for providing these technologies. Our work consists exactly in a study of software that provides 3D reconstruction from images taken with any camera (whether taken from mobile phones or professional cameras). The used software is Meshroom [1]. It is a free, open-source 3D Reconstruction Software based on the AliceVision framework [5]. AliceVision is a Photogrametric Computer Vision Framework which provides 3D reconstruction and camera tracking algorithms [5]. Meshroom is designed around a nodal engine [6]. This is a very special feature of meshroom as it is a software created on the basis of nodes and parameters can be changed very easily, as well as in Meshroom you can add other nodes besides those provided by default, the parameters wanted for the added nodes can be changed easily. Meshroom is developed in Python while the AliceVision framework is developed in C ++ [6]. It is a very good place for 3D reconstruction based on photogrammetry, for analysis and for measurements [1-6]. It is widely used by researchers. Measurements are performed in MeshLab [12].
Different sets of photos of the same object are used as input, reasonable changes are made in the Meshroom workflow in order to increase performance in terms of quality. On each case measurement results are extracted, the appearance of the obtained 3D object is shown. Dimensional results of the

reconstructed object are compared with the real ones. Experiments are performed for small objects with complex shapes, since the prothesis have complex shapes and in some times measuring the body parts is difficult. The proposal in the paper is to clone objects as accurate as possible based on their 3D photogrammetry reconstruction.

part and its analysis of results, concluding with the conclusions of the paper.

2. Methodology

2.1. Photogrammetry

Photogrammetry is a technique that enables the creation of 3D models from photos taken from real objects in different positions, possibly keeping the object static [6-7]. It works by extracting 2D data and superimposing them. Since objects are of different sizes, photogrammetry is used in various fields and applications such as topographic maps or points clouds [8]. The process of obtaining 3D models would be much more complex if we did not use modern software today.

For this article the software we will use are Meshroom and MeshLab.

2.2. Meshroom software

The inputs (in our case the photos) are placed below the left-top part, the right-top part serves to display the output of the photo processing (more specifically it serves to display the final output, the 3D object created but here can also be displayed the result of special nodes when executed). The bottom part is undoubtedly the most important part, more specifically in the left-bottom part is given the graph editor which shows the nodes that participate in the execution workflow. As we mentioned, the nodal environment makes Meshroom Software very special because each node is performed individually. While the bottom-right section shows the characteristics for each particular node, more specifically the outputs, statistics and status are displayed as shown in Figure 1. The photos were taken through usual cameras.

The individual tasks are represented by nodes combined into directed acyclic dependency graphs that are named pipelines [9].

Figure 1. Interface of Meshroom.

The default nodes of the Meshroom workflows shown in Figure 2 are: camera initialization, feature extraction, image matching, feature matching, structure from motion, depth map, depth map filter, meshing, mesh filtering and texturing.

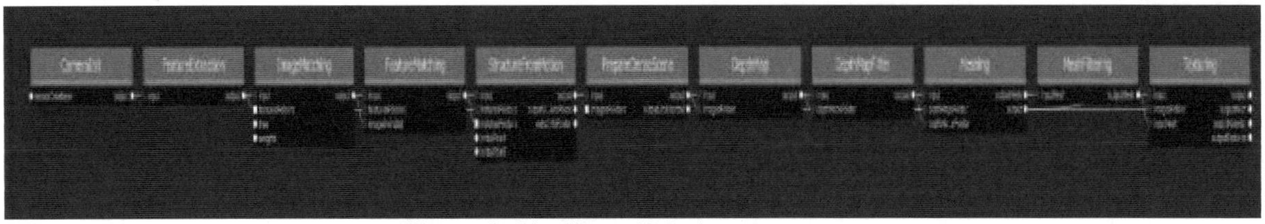

Figure 2. Pipeline of Meshroom.

CameraInit loads image metadata, sensor information and generates viewpoints.sfmcameraInit.sfm. FeatureExtraction extracts features from the images as well as descriptors for those features [10]. ImageMatching is a processing step which figures out which images make sense to match each other. FeatureMatching finds the corrrenspondence between the images using feature descriptors. StructureFromMotion will reconstruct 3D points from the inpute images. DepthMapretrieves the depth value of each pixel for all cameras that have been resolved by SFM. Certain depth maps will claim to see areas that are occluded by other depth maps. The DepthMapFilter step isolates these areas and forces depth consistency. Meshing generates mesh from SfM point cloud or depthmap. MeshFiltering filter out unwanted elements of yor mesh. Texturing projects the texture change quality and size/file type of texture [10-11].

2.3. Meshlab software

Meshlab is an open source systems for 3D image processing and for preparing models for 3D printing. It works based on point clouds or in meshes. A set of tools are provided from Meshlab software as rendering, meshes, texturing, measurement of distances, cleaning, healing etc. [12].

2.4. Small objects reconstruction

Small objects are difficult to be reconstructed, they must be well distinguished from the background, especially if they have details on them and not a small number of photos is required. Measuring the sizes from the reconstructed objects helps us in the assessment of cloned ones using a 3D printer. Small objects with complex shape are difficult to be reproduced and the accuracy of the sizes obtained from the 3D reconstruction plays an important role.

3D reconstruction of objects is realized in Meshroom. Its reconstruction is very sensitive from the input images, usually a considerable overlap between images is suggested to have a better reconstruction.

The reconstructed object is obtained as a scaled version of the real one. To reconstruct it with real sizes, a known size element is needed. Rescaling the reconstructed object using the known dimension, enables the acquisition of the object in real size.

3. Experimental analysis

As mentioned above Meshroom offers a nodal environment and each node will perform specific functions. Our work will consist in analyzing the size accuracy of 3D reconstruction using MeshLab. The photos were taken through the Xiaomi mobile phone Readme Note 8 pro.One method to enable the elimination of unnecessary parts is to increase the "Minimum Observations Angle constraint Space Estimation"" from 10 which was by default to 40 (we create a new meshing2 node and evaluate the output) as in the Figure 3.

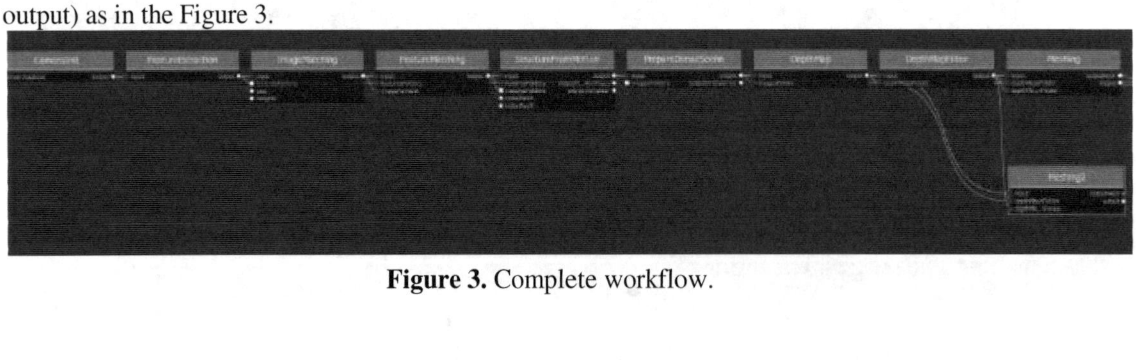

Figure 3. Complete workflow.

If we compare it with mesh case 1 (Mesh gives the result with Minimum Observations Angle constraint Space Estimation at value 10, the value by default), we notice that the background has some improvement (improvement in this case consists in removing as many parts as possible from the background) as shown in Figure 9. This is the case of comparison from visual presentation. In terms of time processing, comparing with the first case, processing time in the meshing node is reduced by 437 seconds (about 7 minutes). The number of Mesh Triangles and Mesh Vertices is also reduced. As expected, adding MeshFiltering and Texturing nodes, the total processing time will be smaller comparing with the first case. The added nodes MeshFiltering and Texturing will be left with the configurations as in the first case. The processing time in each of the two added nodes is reduced comparing with the same node of the first case. Changing the meshing node and with the two added nodes MeshFiltering and Texturing, the execution workflow is shown in Figure 4.

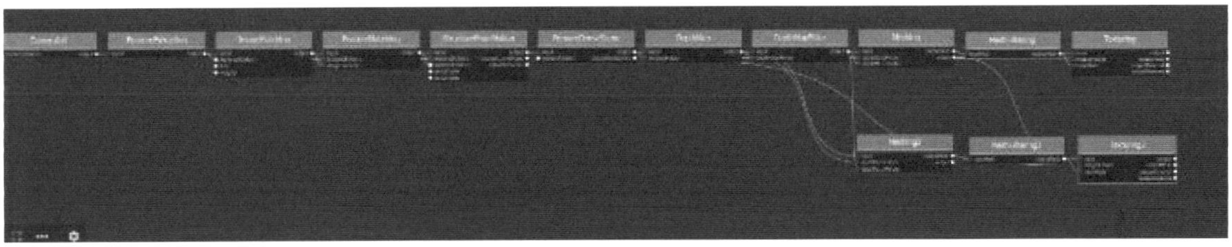

Figure 4. The new pipeline.

As shown in Figure 5, if we compare it in terms of quality with the 3D object obtained from the case 1 we have an improvement of the object in focus. In conclusion we say that both in terms of quality and in terms of time processing case 2 gives a better result.

Figure 5. Reconstruction with new parameters

If we take the case again when we increase the "Minimum Observations Angle constraint Space Estimation" equal to 40 we will try to make another case study. The meshing node in this case is applied directly after structure from motion, eliminating the PrepareDenseScene, DepthMap and DepthMapFilter nodes where in the figure below it is given by the StructureFromMotion - Meshing3 connection as shown in Figure 6.

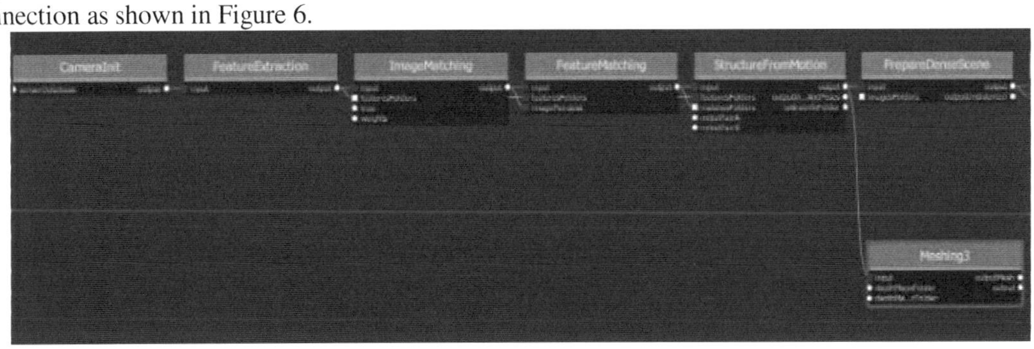

Figure 6. New pipeline with meshing

Many elements of the background are eliminated if we compare case 3 with the two experiments performed above but if we look at the object in focus there are shortcomings of some components compared with both experiments above.

A small detailed object is used for 3D reconstruction. As an object of study an ordinary office key is used and reconstruction is done on Meshroom software. We will try to determine the dimensions of this reconstructed object and we will compare these measurement results with those that belong to the real object. We will do this measurement analysis using MeshLab software.

Initially the experiment was performed starting from a set of 40 photos that I randomly selected as the object number we specified above (keys). But in this case a white surface was used as the background. After I uploaded the photos to Meshroom (which I left with the default settings) and ran the program, the process of executing the nodes went to the node structure from motion and did not continue. This is due to the white background I use, which stops the processing of photos in the structure from motion node.

Therefore we continued the experiment by changing the background but again holding the same object for review (key) as shown in Figure7.

Figure 7. Target object for reconstruction.

The reconstructed object using Meshroom is shown in Figure 8.

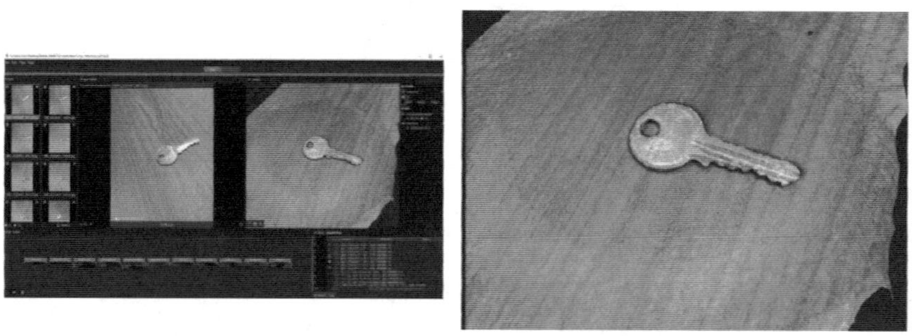

Figure 8. Reconstructed object from photos.

From above we notice that the set with 31 photos as input, was all passed for further processing. And the reconstructed 3D object is of a very good quality, as even the most complex part of the key, which is the part of the teeth, has been completely reconstructed and is easily distinguishable.

To determine the dimensions of the reconstructed object as specified above we will use MeshLab. Initially we will import meshing that was generated by Meshroom in MeshLab and the view as shown in Figure 9.

Figure 9. Object in Meshlab for measurements.

As can be seen from the photo, the part of the teeth is not very clear visually in terms of quality, but the shape of the key is clearly visible (while in the final object obtained in Meshroom, the teeth are visually very clear). With this final result so far that leads MeshLab to try to determine the dimensions of the object in focus. After substituting the new values at x, y and z, now the distances to be measured will be of the same nature as those of the real object.

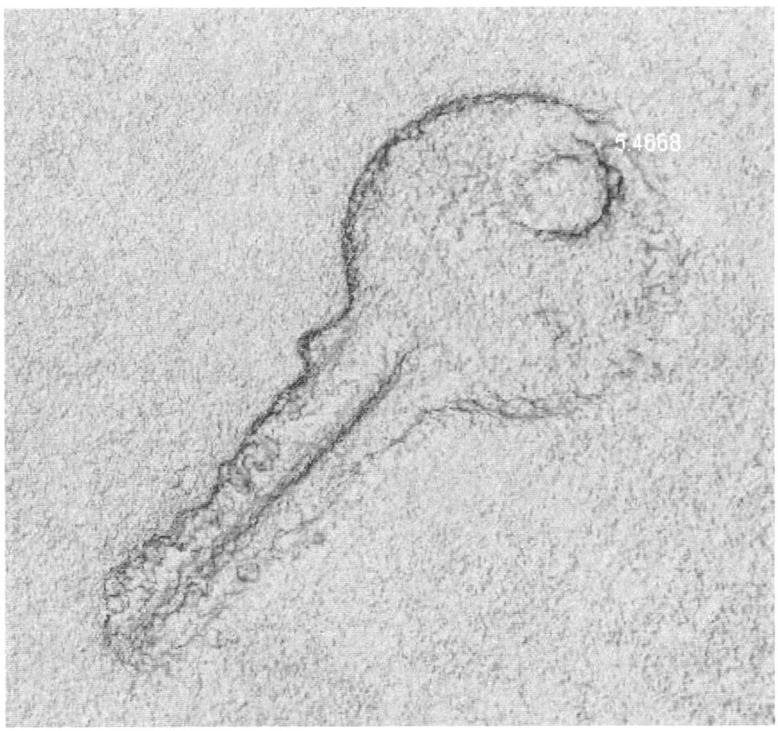

Figure 10. Measuring the key length.

The total length of the key measured by MeshLab results in 5.4668 cm while the real length of the same distance is 5.5 cm, so we have a measurement error of only 0.0332 cm as shown in Figure 10.

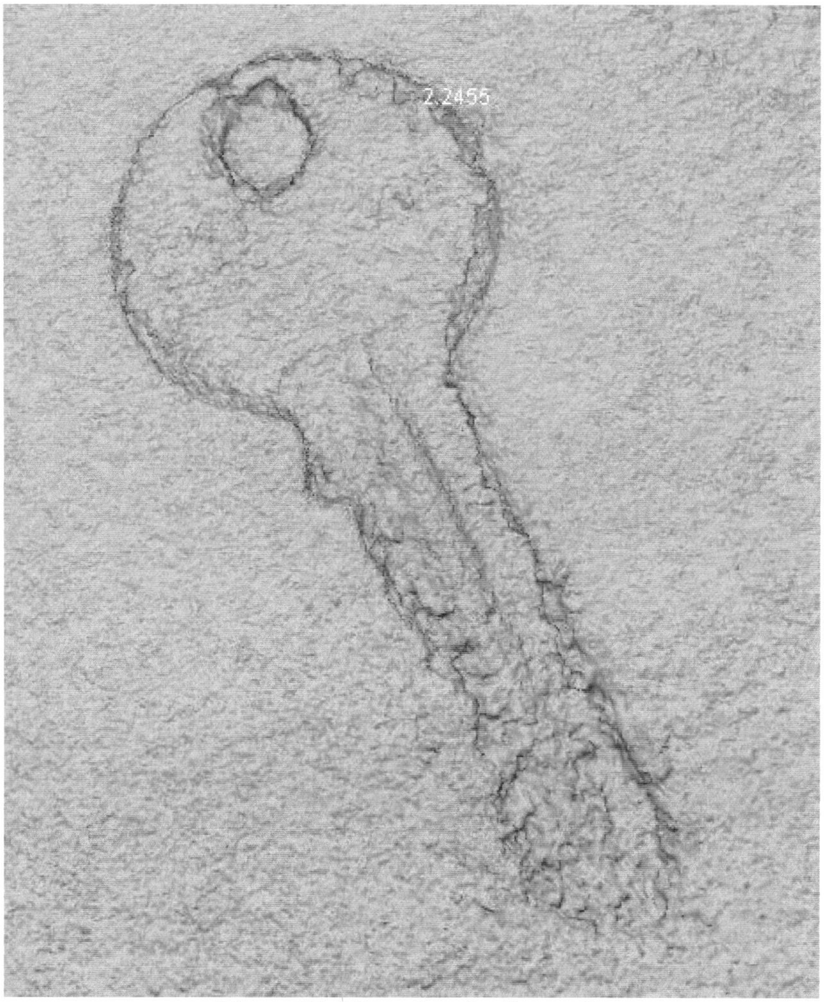

Figure 11. Measuring the diagonal.

The distance of the two key points as in the picture above measured by MeshLab results in 2.2455 cm while the real length of the same distance is 2.3 cm, so we have a measurement error of only 0.0545 cm as shown in Figure 11.

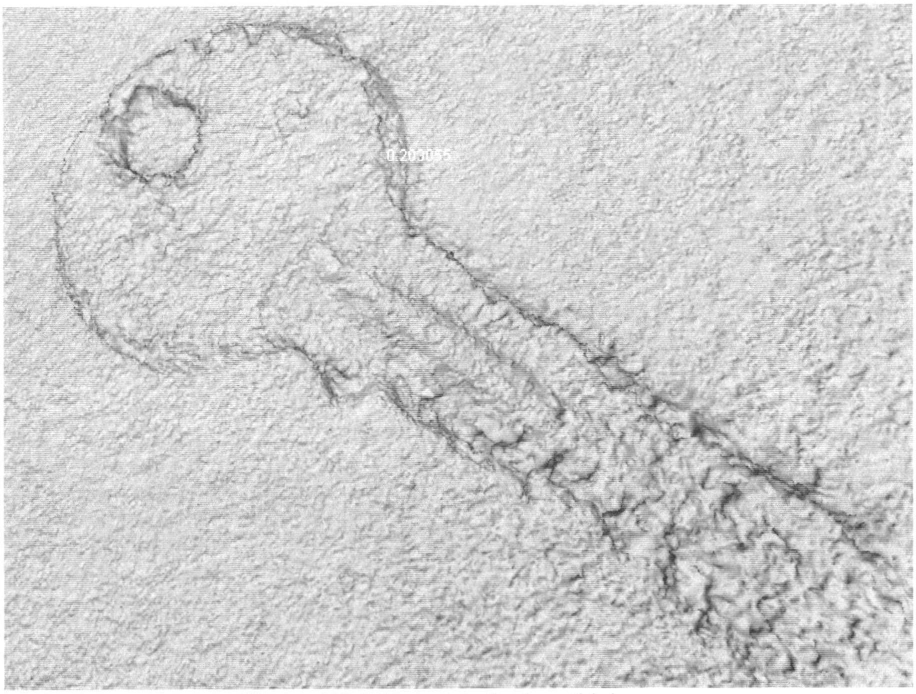

Figure 12. Measuring the key thickness.

The height of the key measured by MeshLab results in 0.203 cm while the real length of the same distance is 0.19 cm, so we have a measurement error of only 0.013 cm as illustrated in Figure 12.

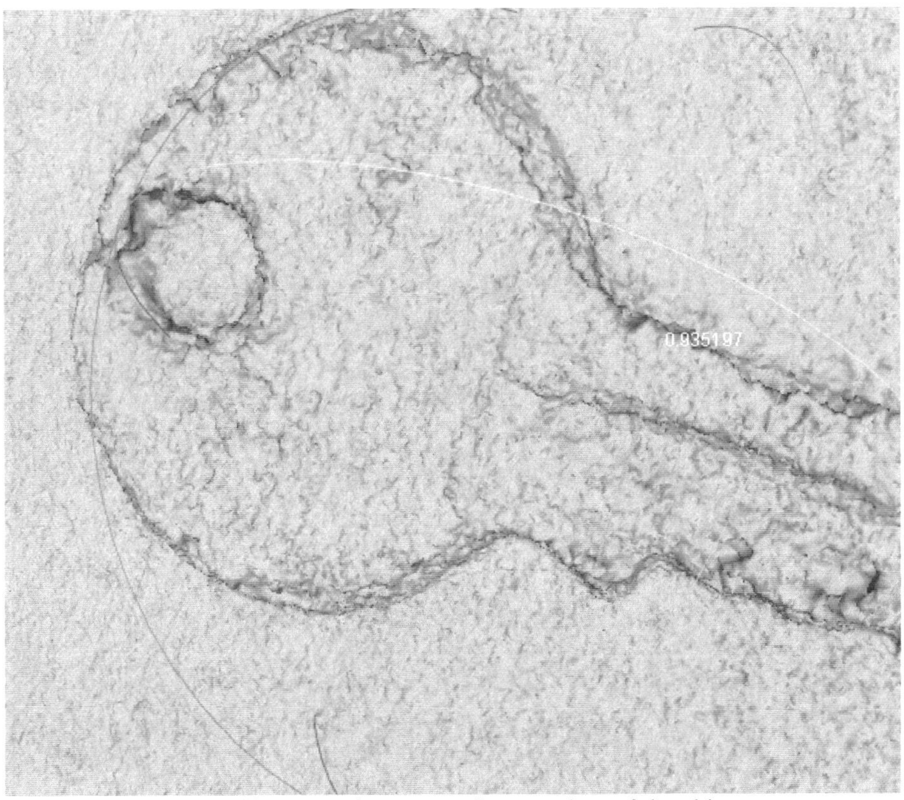

Figure 13. Measuring two ordinary points of the object.

The distance of the two key points as in the picture above measured by MeshLab results in 0.935 cm while the real length of the same distance is 1 cm, so we have a measurement error of only 0.065 cm as shown in Figure 13.

Conclusions

Changing the background from white in a suitable one brought the possibility of processing all Meshroom nodes to generate the reconstructed 3D object. In this paper the focused was on small objects possibly with complex shapes.

The reconstructed object with a set of 31 photos resulted in a very good quality, visually very clear, complemented by content and clear contours. Having a known size element of the object, the reconstructed object is scaled. The dimensions of the small object are measured in MeshLab software and a comparison is done with the real ones. Experimental results show that dimensions of the object are obtained with high accuracy in a range of errors between 0.013 cm to 0.065 cm.

Future work

Experiments will be realized for bodyparts in order to clone them like prothesis, tooth etc. Since in some cases it is difficult to change the background, our suggestion is to change the color of the source light which needs to be experimented.

References

[1] www.meshroom.com
[2] https://support.pix4d.com/hc/en-us/articles/202559369-Reprojection-error
[3] https://github.com/alicevision/meshroom/issues/1075
[4] https://meshroom-manual.readthedocs.io/en/latest/feature- documentation/gui/2dviewer.html
[5] https://en.m.wikipedia.org/wiki/Triangle_mesh
[6] meshroom-manual-readthedocs-io-en-latest
[7] Stark E, Haffner O and Kučera E 2022 Low-Cost Method for 3D Body Measurement Based on Photogrammetry Using Smartphone *Electronics* **11**(7) p 1048 Available from: 10.3390/electronics11071048
[8] Asadpour A 2021 Documenting Historic Tileworks Using Smartphone-based Photogrammetry *Mersin Photogrammetry Journal* . Available from: 10.53093/mephoj.899432
[9] https://www.google.com/url?sa=t&source=web&rct=j&url=https://hal.archives- ouvertes.fr/hal-03351139&ved=2ahUKEwitm4bAwd_2AhX7iv0HHSTWBQEQFnoECBQQAQ&usg=AOvVaw0EPjLKB6gEaXHamKR9gOO3
[10] Berio F and Bayle Y 2020 Scyland 3D: Processing 3D landmarks. Journal of Open Source Software **5**(46) p 1262 doi: 10.21105/joss.01262
[11] Lee W and Yu K 2009 Bundle Block Adjustment with 3D Natural Cubic Splines *Sensors* **9**(12) pp 9629-9665 Available from: 10.3390/s91209629
[12] www.meshlab.net

Acknowledgments. This work was supported by the National Agency for Scientific Research and Innovation under the Contract no. 831.

Evaluation of a panoramic dental radiography properties by using Canny method

C M Stăncioi[1], **M Munteanu**[2], **M Abrudean**[1], **V Mureșan**[1] **and O Bunta**[3]

[1]Automation department, Technical University of Cluj-Napoca, Cluj-Napoca, Romania
[1]Automation department, Technical University of Cluj-Napoca, Cluj-Napoca, Romania
[3]Department of Orthodontics, University of Medicine and Pharmacy, Cluj-Napoca, Romania

E-mail: stancioi_cristina@yahoo.com

Abstract. The main goal of this paper is to analyse the panoramic dental radiography by applying the Canny technique which is customized for this type of images and to obtain an image much more clear to interpret after processing it. From the resulted image then can be extracted all the necessary information for giving a diagnostic. The method can be extended and an application can be designed in such a way that a dental specialist can use it to better examine the characteristics of a patient's intraoral configuration. The simulations on different images hard to read by the human eye were done using the Matlab work environment. Image-processing techniques represent a very important part in this process and the best method which was successful will be detailed in this paper. Some image properties can be identified automatically even before the original image has been processed, but after the processing step, all the information and also the hidden details can be observed by a specialist. As a conclusion, each step performed and described in this article will aid in the analysis of panoramic dental radiography characteristics.

1. Introduction
Understanding the process of image perception by the human eye is very important for the development of techniques for evaluating the quality of an image processing system or algorithm. Visual information, contained by an image, it represents a spatial distribution of a certain sizes, such as the luminance of the objects that make it up image. The information perceived by the human eye can be defined by attributes such as the brightness, colour, or edges of an object.

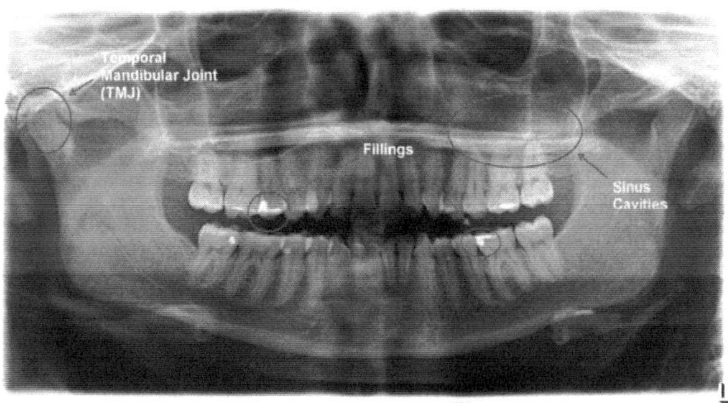

Figure 1. Dentalpanoramic X-ray [1].

The general image enhancement term refers to a broad class operations, which aim to increase the detectability of image components. This detectability depends primarily on a person's visual perception human observer and therefore represents a subjective appreciation of the image [2].

Improving the quality of an image is done without assuming any model degradation or take into account any information related to the image original. Paradoxically, but also an original (ungraded) image can be improved. In general, improvement refers to the emphasis on certain features of the image, such as edges, contours, or contrast. The process of enhancement does not increase the amount of information contained in an image.

Segmentation is the division of the image into areas of interest, according to certain criteria. Each pixel will be assigned a value, 0 or 1, representing its belonging to a certain area or region of interest. Usually, segmentation seeks to extract, identify, or recognize a particular object from an image [3].

In an image, pixel level variations represent changes in the physical or geometric properties of the objects that make up the scene. A large number of cases, these variations in intensity correspond to the boundaries (contours) of regions determined by objects in an image.

The principle of these methods is to define the contour points as those pixels of the image for which there are abrupt changes in the grey level. The measurement of these variations is done by gradient type derivative operators [4].

2. Methods

In this paper, the segmentation of the panoramic dental radiography will be presented, by designing and applying an algorithm based on Canny method, because of the complexity of such an image. This method is part of the segmentation by detection of the edges and contour construction.

Segmentation methods by edge detection are based on the location of light transitions or color that represents the edge of the object and the subsequent construction of closed contours that describes the object.

An optimal method of edge detection is the method proposed by Canny [5]. This method contains several steps:

- smoothing the image;
- building the module and direction map gradient;
- non-maximum suppression;
- frying using two threshold levels with hysteresis along the contour.

Image smoothing is performed with a filter with large neighbourhoods composed according to a two-dimensional Gaussian law [6]:

$$g(x,y,k) = \frac{1}{2\pi k} \cdot e^{\frac{-(x^2+y^2)}{2k}} \quad (1)$$

By varying the coefficient k an optimal answer for filter that was applied. In the given application a5x5 mask with value k = 1.4:

$$f(x,y) = \frac{1}{159} \begin{matrix} 2 & 4 & 5 & 4 & 2 \\ 4 & 9 & 12 & 9 & 4 \\ 5 & 12 & 15 & 12 & 5 \\ 4 & 9 & 12 & 9 & 4 \\ 2 & 4 & 5 & 4 & 2 \end{matrix} \quad (2)$$

After smoothing, the intensity map of gradient according to the relation is built:

$$|G(x,y)| = \sqrt{G_x^2 + G_y^2} \quad (3)$$

Where the gradients G_x and G_y are calculated for each pixel by convolution with a compound mask:

$$G_x = \begin{bmatrix} 15 & 35 & 0 & -35 & -15 \\ 69 & 155 & 0 & -155 & -69 \\ 114 & 255 & 0 & -255 & -114 \\ 69 & 155 & 0 & -155 & -69 \\ 15 & 35 & 0 & -35 & -15 \end{bmatrix} \quad (4)$$

$$G_y = \begin{bmatrix} 15 & 69 & 114 & 69 & 15 \\ 35 & 155 & 255 & 155 & 35 \\ 0 & 0 & 0 & 0 & 0 \\ -35 & -155 & -255 & -155 & -35 \\ -15 & -69 & -114 & -69 & -15 \end{bmatrix} \quad (5)$$

After that, the direction map of gradient for each point in the image is calculated with relation:

$$\alpha(x, y) = \tan^{-1}\frac{G_x}{G_y} \quad (6)$$

The next step is non-maximum removal. For each pixel is made the analysis of the neighbouring pixels in the perpendicular direction with respect to the contour and this pixel is retained only if the modulus of the gradient is greater than the gradient modulus of the neighbouring pixels, otherwise the value which is assigned to the pixel is zero. This procedure allows to obtain the contour with the width of one pixel [7].

Finally, hysteresis is applied. For this, two selection thresholds of the contour points are used:

- a high threshold for selection a secure contour point that has the maximum value in the gradient intensity map;
- a low threshold that selects the points in the vicinity of the high gradient point.

This technique aims to obtain continuous edges even when the edge gradient value is low.

Following the analysis that was performed, it was found that the contour of each cavity is reconstructed with sufficiently high precision. There are some unwanted contours that are detected on the outside of the cavity. Considering that they do not overlap with the outer contour of the cavity, these contours can be ignored, being filtered later by the procedure for reconstructing the shape of the cavity.

3. Results and conclusions

Good results were obtained in the case of detection contour with the method based on Canny technique. This algorithm requires only four iterations for each image, but can be improved by enlarging the memory resources used. Besides this, Canny segmentation is preferable, because the contour of the most important elements is obtained and there is no need post-processing as in the case of segmentation by thresholding.

Figure 2 represents the original image in which is hard to observe all the details necessary in order to establish a proper diagnostic. So that, in Figure 3, after processing the image, it can be seen the obtained resulted image, much more clear. After this, by using the obtained image, a specialist can observe and draw a diagnostic for every patient.

In conclusion, we can say that image processing and analysis has managed to gain ground as due to the spectacular achievements of electronic technology and computer science. The growth of computing power available in computer processing units will probably transform in the near future the processing and analysis of images and development of special applications with reliable solutions of wide industrial consumption.

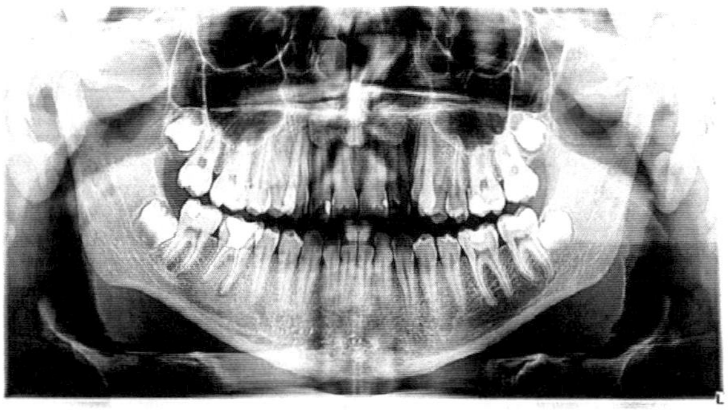

Figure 2. Original image.

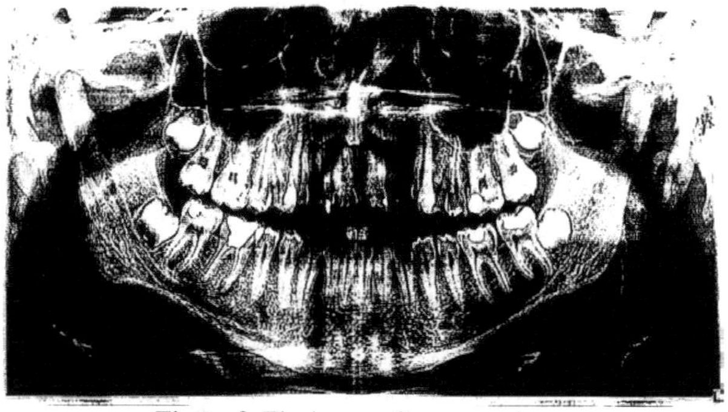

Figure 3. The image after processing.

References

[1] Pritchard B, Akbarian Tefaghi F and Makdissi J 2020 Anatomy in panoramic image interpretation, *British Dental Journal* **228** 229 10.1038/s41415-020-1324-1

[2] Bin Ghazali A B 2020 Interpreting panoramic radiographs for beginners *BDJ Student* **27** 49 10.1038/s41406-020-0144-8

[3] Choi Y J, Young Kim Y H, Han S-S, Jung U-W, Lee C, Lee A and Jeon K J 2020 Alveolar bone height according to the anatomical relationship between the maxillarymolar and sinus *Journal of Periodontal & Implant Science* **50**(1) pp 38-47 10.5051/jpis.2020.50.1.38

[4] de Oliveira Capote T S, de Almeida Gonçalves M, Gonçalves A and Gonçalves M 2015 Panoramic Radiography—Diagnosis of Relevant Structures That Might Compromise Oral and GeneralHealth of the Patient *Emerging Trends in Oral Health Sciences and Dentistry* (InTech: London, UK)

[5] Kamburoglu K, Kolsuz E, Murat S, Yüksel S and Ozen T 2012 Proximal caries detection accuracy using intraoral bitewing radiography, extraoral bitewing radiography and panoramic radiography *Dentomaxillofacial Radiol* **41**(6) pp 450-459

[6] Tikhe S V, Naik A M, Bhide S D, Saravanan T and Kaliyamurthie K P 2016 Algorithm to Identify Enamel Caries and Interproximal Caries Using Dental Digital Radiographs *IEEE 6th International Conference on Advanced Computing (IACC)* pp 226-228

[7] Dinç I and Marc L. Pusey 2015 Emerging Trends in Image Processing *Computer Vision and Pattern Recognition*

IOP Conf. Series: Materials Science and Engineering 1254 (2022) 012041 doi:10.1088/1757-899X/1254/1/012041

Study on the implementation of the Industry 4.0 concept in medical device manufacturing

M Ciorap[1], D Andrițoi[2], R Ciorap[2] and C Dumitraș[1]

[1]Faculty of Machine Manufacturing and Industrial Management, Gheorghe Asachi Technical University of Iasi

[2]Biomedical Sciences Department, Faculty of Medical Bioengineering, Grigore T. Popa University of Medicine and Pharmacy Iasi, Romania

E-mail: mariana.ciorap@academic.tuiasi.ro; doru.andritoi@umfiasi.ro

Abstract. The importance of Industry 4.0 in healthcare is obvious. Medical device manufacturing is a highly regulated field making it challenging to introduce new technologies into the manufacturing process. Computer-aided design and manufacturing technology is used to develop custom medical devices. Respiratory diseases are due to the inability of the lungs to cope with the gas exchange at rest and effort. The causes can be multiple, from stenosis of the respiratory tract to the consequences of infections with various pathogens. Spirometry is one of the main methods used in this respect. This paper aims to develop a spirometer using a Venturi tube connected to the differential pressure sensor type MPX10DP. The Venturi tube was designed using FUSION 360 and 3D printed. The pressure sensor is connected to an amplifier made with the INA118 circuit, and the ARDUINO UNO microcontroller takes the signal. The program running on the computer calculates the forced vital capacity (FVC) and forced expiratory volume in one second (FEV1).

1. Introduction

Medical device manufacturing is a highly regulated field, making it challenging to introduce new technologies into the manufacturing process. In addition, extra precautions must be taken as medical devices will be in direct contact with the patient. Therefore, many ethical considerations will need to be addressed in the research, development, and production of medical devices [1].

Market demands drove the fourth industrial revolution to reduce product implementation time, optimize production and minimize human errors in production. The importance of Industry 4.0 in healthcare is obvious. Digitization in medical device production and healthcare delivery[2], modern 5G communication technologies, additive manufacturing [3], and artificial intelligence open up new possibilities in this field.

From Industry 4.0 was derived the new term Healthcare 4.0 [4]. Some examples of Healthcare 4.0 are eHealth, telemedicine, mobile health (mHealth), and pervasive health [5].

The use of special software in creating, designing, and verifying the latest product from the CAD/CAM/CAE (Computer-Aided Design/Computer-Aided Manufacturing/Computer-Aided Engineering) allows the development of a 3D model to prepare the product for manufacturing [6]. One of the core outcomes of the advances in digital manufacturing is 3D printing technology [7]. 3D printing can be operated and controlled remotely using cloud technology and does not require a

IOP Conf. Series: Materials Science and Engineering 1254 (2022) 012041 doi:10.1088/1757-899X/1254/1/012041

large workforce. It also doesn't need industrial infrastructure to manufacture a product. Computer-aided design and manufacturing technology manufacture implants, instruments, and custom medical devices [8].

When a person had symptoms of an illness, they requested medical advice from their family doctor, and it took time to access the consultation. In the Healthcare 4.0 concept, the patient can have real-time access to their medical data and use wearable devices [8] to keep track of their health.

Digital medical data recording and processing systems such as EHRs (Electronic Healthcare Records) enable personalized treatment [9]. In addition, 3D printing solutions are available for customized production of pharmaceutical dosage forms, drug combinations, and drug delivery forms tailored to each patient's specific needs [10-11].

Respiratory diseases are due to the inability of the lungs to cope with the gas exchange at rest and effort. The causes can be multiple, from stenosis of the respiratory tract to the consequences of infections with various pathogens. To effectively diagnose and monitor respiratory conditions, the doctor needs a medical history, a physical examination, and two tests that measure lung function. Spirometry is one of the main methods used in this respect and has the advantage that it can also be used at the patient's home, allowing the treatment to be optimized from a distance [12].

The spirometer is one of the basic equipment for respiratory function tests. It is a very effective and widespread diagnostic tool, non-invasive and well standardized. It is commonly used to diagnose and evaluate lung function in people with restrictive or obstructive airway diseases. Simple spirometry can easily rule out lung diseases like asthma, bronchitis, and emphysema. In addition, the spirometer is often used to investigate the cause of reported breathing difficulties, to assess the effects of contaminants on lung function, the impact of bronchodilators, and the progress of treatment of the disease. The information obtained by using the spirometer is used to assess lung function.

2. System design

In the current medical context, when there are many patients with lung diseases, the medical investigation called spirometry has become more and more requested. Almost all modern spirometers are connected to a computer that receives a signal from the transducer and converts it into numeric values and graphics. The software also provides a basic interpretation of spirometry and an assessment of the quality of the test performed. Another significant advantage of modern spirometers is the portability associated with good measurement accuracy.

To make the spirometer, we used a Venturi tube connected to the differential pressure sensor type MPX10DP. The Venturi tube was designed using FUSION 360. The pressure sensor is connected to an amplifier made with the INA118 circuit, and the ARDUINO UNO microcontroller takes the signal. The program running on the computer calculates the forced vital capacity (FVC) and forced expiratory volume in one second (FEV1).

We will describe the steps in obtaining a spirometer model using Fusion 360 made by AUTODESK.

Step 1. Sketch generations step–constraints introductions – (Figure 1)

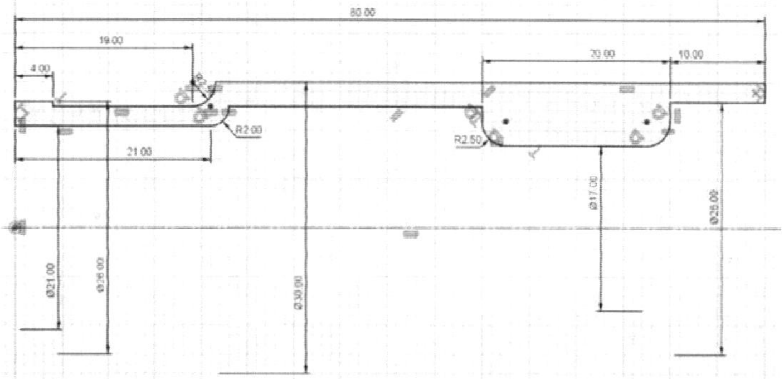

Figure 1.

IOP Conf. Series: Materials Science and Engineering 1254 (2022) 012041 doi:10.1088/1757-899X/1254/1/012041

Step 2. Making the spirometer body - using the command **Revolve**

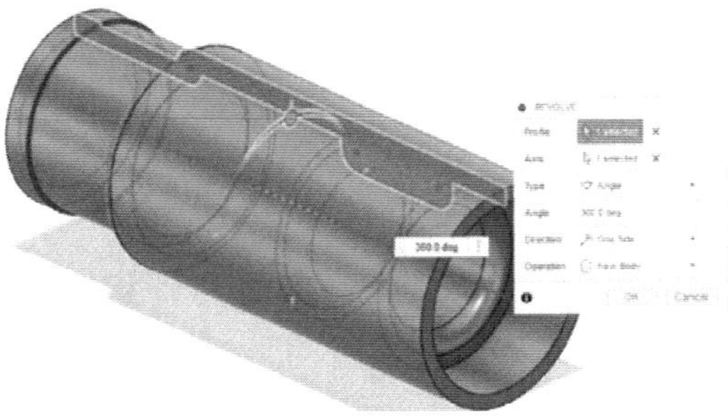

Figure 2.

Step 3. Making the spirometer support element - we used the command **Extrude**

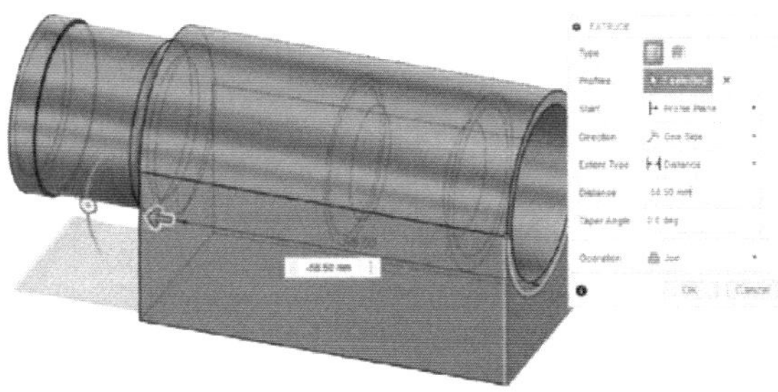

Figure 3.

Step 4. To make the first cavity for the sensor foot, a sketch should be constructed in a plane 3.5 mm from the plane XoY.

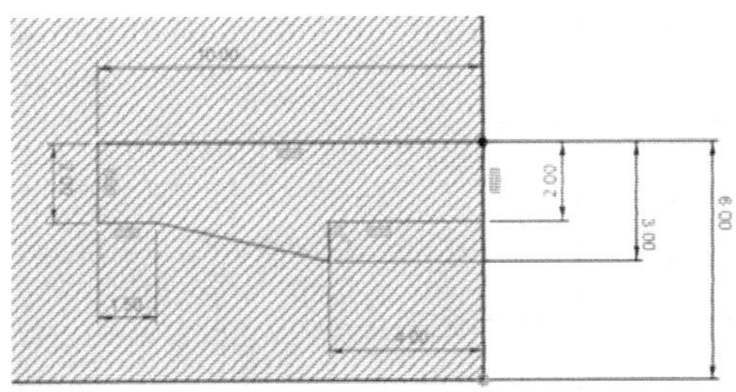

Figure 4.

Step 5. Making the first cavity with the control **Revolve,** with the option **cut.**

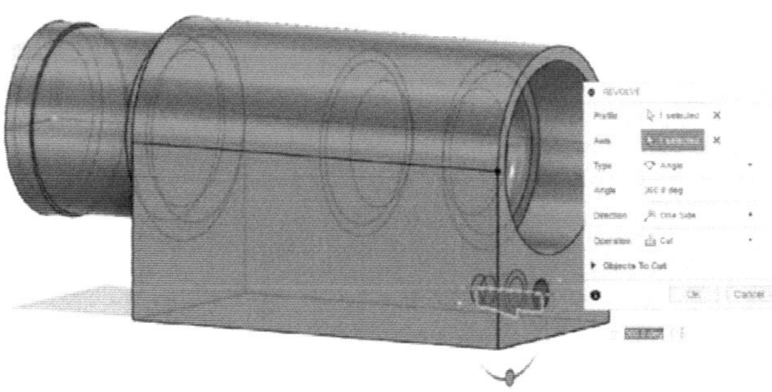

Figure 5.

IOP Conf. Series: Materials Science and Engineering 1254 (2022) 012041 doi:10.1088/1757-899X/1254/1/012041

Step 6. Repeat for the second cavity.

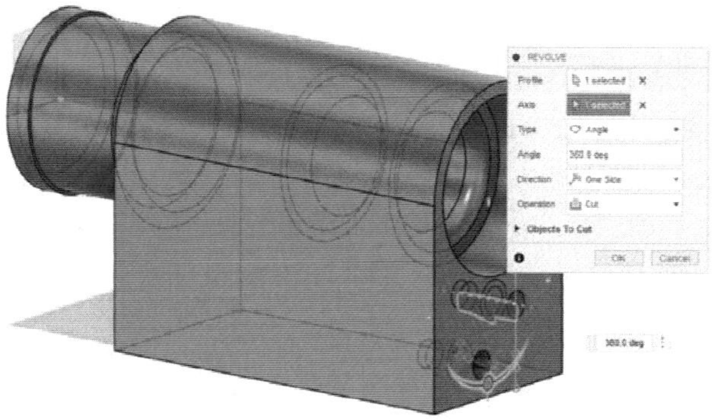

Figure 6.

Step 7. Define the plane and centre point of the first hole inside the spirometer.

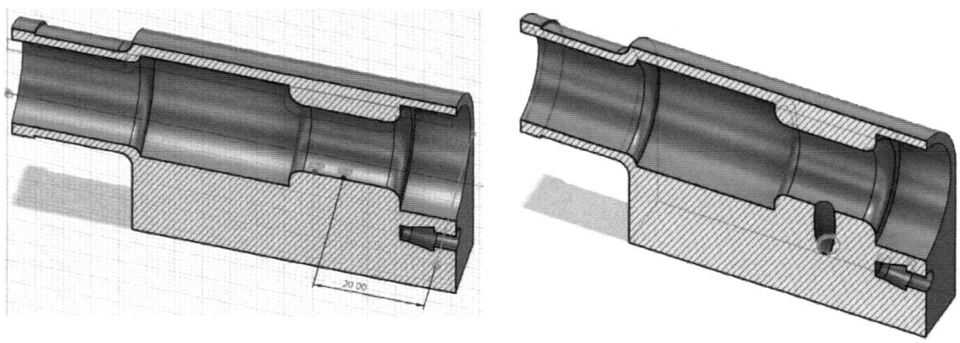

Figure 7.

3. Results

The patient should be seated during this test. In some cases, a small clip will be attached to the nose to prevent air from escaping in other ways than those intended to be evaluated. It is strongly inhaled (forced) during recording to fill the lungs with air completely. The lips must be effectively positioned around the part attached to the equipment to prevent accidental escape of air;

He then exhaled hard and fast, completely emptying his lungs. This test may be repeated several times to ensure the accuracy of the results. In some cases, it may be repeated a quarter of an hour after the administration of a bronchodilator. The spirometry test can take an average of half an hour. The spirometry test is elementary and very safe; some people may feel dizzy or tired for a short time immediately after performing this test.

Based on these records, numerical and graphical results are obtained, which allow the specialist to interpret the respiratory capacities and volumes.

The pressure variation measured by the sensor during a normal breath is shown in Figure 8.

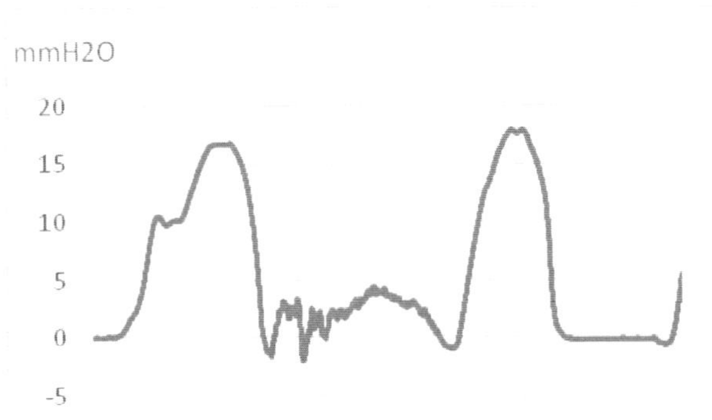

Figure 8.

4. Conclusion

The portable spirometer can measure lung volumes and capacities using low-cost materials, and the results obtained are within the range of accuracy suitable for such measurements.

References

[1] Lepasepp T K and Hurst W 2021 A systematic literature review of industry 4.0 technologies within medical device manufacturing *Future Internet* **13**(10)

[2] Pomazan V M, Petcu L, Sintea S R and Ciorap R 2009 Active data transportation and processing for chronic diseases remote monitoring *2009 International Conference on Signal Processing Systems* (IEEE Computer Soc, 2009) p 853

[3] Zarrabeitia-Bilbao E, Alvarez-Meaza I, Rio-Belver R M and Garechana-Anacabe G 2019 Additive manufacturing technologies for biomedical engineering applications: Research trends and scientific impact *Prof Inform* **28**(2)

[4] Bongomin O, Yemane A, Kembabazi B, Malanda C, Mwape M C, Mpofu N S and Tigalana D 2020 The hype and disruptive technologies of industry 4.0 in major industrial sectors: A state of the art *Journal of Engineering* **2020** 8090521

[5] Mustapha I, Khan N, Qureshi M I, Harasis A A and Van N T 2021 Impact of industry 4.0 on healthcare: A systematic literature review (SLR) from the last decade *International Journal of Interactive Mobile Technologies (iJIM)* **15**(18) p 116

[6] Carvalho N, Chaim O, Cazarini E and Gerolamo M 2018 Manufacturing in the fourth industrial revolution: A positive prospect in sustainable manufacturing *Procedia Manufacturing* **21** p 671

[7] Popov V V, Kudryavtseva E V, Katiyar N K, Shishkin A, Stepanov S I and Goel S 2022 Industry 4.0 and digitalisation in healthcare *Materials* **15**(6) p 2140

[8] Chandra M, Kumar K, Thakur P, Chattopadhyaya S, Alam F and Kumar S 2022 Digital technologies, healthcare and Covid-19: insights from developing and emerging nations *Health Technol-Ger* **12**(2) p 547-568

[9] Condry M W, Quan X I and Fang M 2020 Digital health: innovation, opportunity and challenges *IEEE Ind Elec* p 3408

[10] Awad A, Trenfield S J, Pollard T D, Ong J J, Elbadawi M, McCoubrey L E, Goyanes A, Gaisford S and Basit A W 2021 Connected healthcare: Improving patient care using digital health technologies *Adv Drug Deliver Rev* **178**

[11] Elbadawi M, McCoubrey L E, Gavins F K H, Ong J J, Goyanes A, Gaisford S and Basit A W 2021 Harnessing artificial intelligence for the next generation of 3D printed medicines *Adv Drug Deliver Rev* **175**

[12] Ciorap R, Corciova C, Ciorap M and Zaharia D 2011 *Int Symp Adv Top* 143

Material energy balance at articular cartilage: Bio-tribology

P TOMAR
IGDTUW/GGSIPU, India

Abstract. The synergy of heterogeneous macromolecules at the cartilage-cartilage tribological interface prevents friction under quasi-static mechanical loading. Viscoelastic rheology of soft biological membrane materials, hydration lubrication, and biomechanical diffusion integrate boundary lubrication at the superficial zone. Synchronization of mechanical efficiency is viable in alignment with mechanical work, energy expenditure, and reducing oxidative stress of environmental load in urban areas. Carbon nanoparticle's evolution from anthropogenic activities inversely influence the quality of fuel oxidation. Anisotropic fibrous honeycomb structure panel is included for trapping random environmental carbon nanoparticles/particulate matter for favourable environmental indicators.

1. Introduction

Bio-tribology is useful in daily life for synchronization of friction from third bodies heterogeneous environment at materials energy interface [1-3]. The surface, interfaces, interphase, and biomolecular surface chemistry environment define a boundary for a materials energy balance useful for integrity of biological systems in rationalization of bio-tribology [4]. The deviation of molecular level friction, boundary lubrication, nanotechnology, and mixed lubrication from macroscopic Amontons' law of friction have been broadly presented by researchers in engineering application as a function of state variables [5-7]. Adsorption of a few layers of macromolecular coat over articulating surfaces is termed boundary lubrication for reducing ultra-low friction. Proteoglycan 4 is a boundary lubricant biomechanically diffused across the amphiphilic superficial zone from the cellular domain. Stick-slip friction is evolved as a function of boundary lubricant scarcity, degradation of extracellular matrix, static mechanical loadings, and environmental indicators. Amino and carboxyl functional groups biomacromolecule of lubricin of ultra-fluidity, hyaluronic acid act as a viscous biological ingredient of synovial fluid for impact resistance in hydrodynamic lubrication, and cell membrane act interfacial biological material [8-11]. The tribological performance of soft biological interface is a function of real area of contact, state variables, normal stress, gender diversity, environment and degeneration for ruling out of chronic friction.

2. Bioengineering

Collagen, elastin, and laminins integrate extracellular matrix, provide biological stiffness, and explain the morphological transformation against aging/degeneration/environmental factors. Biomechanical, soft biological materials, diffusion, physiological, and anatomical factors (Fig 1) are responsible for resolving the complexity of friction at articular cartilage. Elevated friction stress at quasi-static mechanical loading is in scarcity of biomacromolecules or degradation of biological interfacial material due to stick-slip shearing mechanics. Healthy articulating joints have a coefficient of friction relatively lesser than lubrication of mechanical sliding surfaces in a boundary lubrication regime [12-14]. Mechanical indicators at the mating domain create cell membrane tension for biophysical diffusion across an amphiphilic surface. Viscoelastic rheology of synovial lubricant is a natural boon for equitable pumping of lubricant at TRIBO interface from the synovial cavity. Static mechanical loading arises from anthropometry of body mass index inversely influence the performance of synovial lubrication. Material energy balance as a function of rational fuel oxidation has been included for synchronization of mechanical efficiency. The coefficient of friction is low in a biological system under normalcy and generate additional shear stresses under chronic conditions. Synovial fluid at the native joint is a biomechanical lubricant, reducing friction, lowering wear of cartilage biological material, and providing impact resistance in dynamic conditions. Lubricin shares biological properties with mucin family members, binds and regulates bio-tribology, is found at superficial zone/synovial fluid, and plays a significant role in reducing cartilage adhesion.

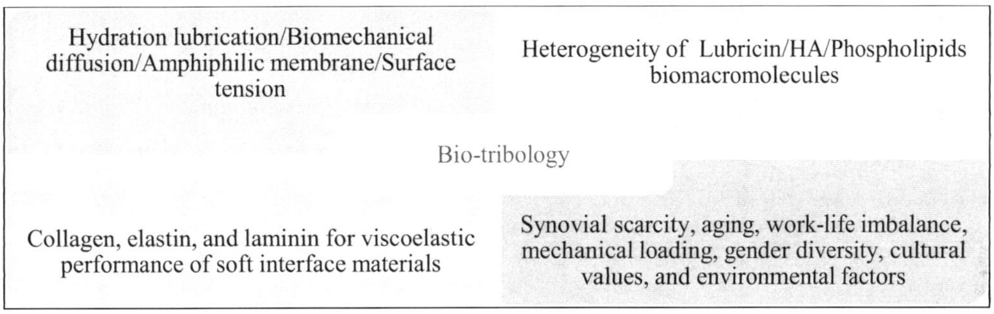

Fig 1. Fundamental indicators for rationalization of bio-tribology: *Friction*, *Wear*, and *Adsorption*

'Biomimetics' or Bio-inspired science of mimicry nature for innovation at mechanics and mechanism interface from the biomechanical phenomenon. Osteoarthritis is a degenerative disease of high prevalence for no disease-modifying therapy prevails: predicted degradation of extracellular matrix or loss of cartilage biological material [15]. Bioscience at the surface, interface, and interphase has evolved for the study of tissue functions and relatively a little attention has been given to reversibility of bio-tribology performance in repairing, manufacturing of new tissue, cell adhesion, and bio-inspired scaffolds [16]. Bio-tribology is a rapidly emerging field of study for quality of life inspired by a natural tribological systems incorporating materials, mechanics, biology, mechanical surface and interface, physics, chemistry, and technology in general [17]. Bioengineering plays a key role in researching joint morphology disorder, abnormal growth of the femoral head and acetabulum, genetic risk factors based on diversity, and mechanical environment for regulation of growth or maturation processes [18]. TRIBO performance of boundary lubrication provides information on synovial lubrication viz. sliding surfaces are separated by a thin molecular level of a biological coating.

3. Energy

Clean water and air are energy-transportation molecules in a conscious umbrella of '*Nature*' for life on the planet. Conservation of mass and energy principle is included for transformation from starved synovial lubrication to rational bio-tribology by lowering the static mechanical load, enhancement of mechanical efficiency, and rational fuel oxidation. Enriched carbohydrate meal is preferred by a mass due to the bioavailability of pulses and grains, cultural values, and personal impression for modulation of energy ingredients in the requirement of negative energy balance. Energy molecules (Table 1) are tabulated in the prediction of synovial diffusion at the interface, nutritional values, micro energy reservoir of triglycerides, chemical oxidation, respiratory quotient, mechanical work, energy expenditure at rest, and mechanical efficiency. Triglycerides formation from fatty acids, cardiovascular pathology, positive energy balance, and work-life imbalance inversely influence TRIBO performance from material energy imbalance in the reciprocity of sustainable development goals. Energy transportation from nutrition values of negative energy balance, high protein low carbohydrate meal, and environmental factors explored in concurrence for a sustainable pattern of life. Biomechanical endurance at articulating cartilage is fundamentally summarized for the reversibility of lubrication performance from the static mechanical loadings, a transformation of moderate risk domain into a rational domain, clean air/water, and validation of mass and energy principle. Bio-tribology: a material energy balance at articular cartilage is a complex biomolecular mechanism for optimization of "Third bodies" carbon nanoparticles present in a local environment, transported by nature consciousness, and trapped by anisotropic fibrous panel driven mechanically.

Table 1. Biochemical and biomechanical energy transportation mechanism

Molecules	Chemical Reaction	V_{CO_2}/V_{O_2}	Energy
Carbohydrate	$C_6H_{12}O_6 + 6O_2 \xrightarrow{\text{Oxidation}} 6CO_2 + 6H_2O$	1.0	4 kcal/g
Fat	$RCOOH + xO_2 \xrightarrow{\text{Oxidation}} yCO_2 + zH_2O$	~ 0.7	9 kcal/g
Protein	$R'CH(NH_2)COOH + xO_2 \xrightarrow{\text{Oxidation}} yCO_2 + zH_2O$	~ 0.8 - 0.9	4 kcal/g

4. Applied thermodynamics

Applied thermodynamics explores the transformation of thermal energy into mechanical work useful for solving applications of physics, chemistry, science, technology, and biomechanics. The first law of thermodynamics ($Q = \Delta U + W$) applicable for metabolism is a conversion principle of fuel energy into heat transfer, mechanical work, and static mechanical load. Applied thermodynamics expression of heat and mass modulation based on second law thermodynamics is fundamentally included in bio-energetics. Biochemical oxidation of fuel is treated as a thermodynamic heat engine for validation of mass and energy principles. Biochemical conversion of fuel energy into triglyceride/fat/residual biomolecules is termed static mechanical load as a function of body mass index. The quantitative mechanical work done due to muscle contraction in biomechanics by fuel energy expenditure $(Q_e + Q_i)$ above resting is a function of the bio-tribology domain. The metabolic energy involved at rest (q_0) for maintaining vital function is borrowed from basal metabolism rate as per the requirement of applied thermodynamics. Mechanical efficiency (η_{mech}) is a ratio of mechanical work to fuel energy expenditure-a performance indicator for assessment of bio-tribology.

$$\eta_{mec} = \left[\frac{W_{mech}}{Q_e+Q_i-q_0}\right] \times 100 \quad \text{(a)}$$

An elementary expression of reducing chronic friction evolved from work-life imbalance, mechanical loading, environmental load, and carbohydrate enriched fuel intake is included from a biomechanical perspective. The sustainable development goals under good health and well-being is fundamentally reinforced for the achievement of sustainability.

5. Friction

Supramolecular interaction is viable at articular cartilage for reducing dynamic friction by the heterogeneity of biomacromolecules over soft biological materials. Nature has evolved biomechanisms to adhere to any kind of surface, interphase, and interphases in a "*hydrophilic*" with "*hydrophobic*" spectrum. Bio-tribology is '*amphiphilic*' by an amalgamation of hydration lubrication with the heterogeneity of *in situ* biomacromolecules. Biological lubrication regimes at the mating interface are trifurcated by locomotion speeds; boundary lubrication of few-layers of biomacromolecules coat; mixed lubrication ahead of boundary lubrication attained with increasing order of locomotion speed; hydrodynamic lubrication prevails ahead of mixed lubrication by separation of surfaces by the thick synovial fluid film. Friction in academic terminology (Table 2) is classified as atomic friction, nanotribology, microscopic, and macroscopic for effective solutions of bio-tribology. The bio-molecular mechanisms for explaining ultra-low coefficient of friction (~0.001), sustaining normal stress (~100 atm) in healthy articular cartilage, and bio-tribology cannot biomimicry by bio-inspired surface/artificial synovial lubrication. The coefficient of friction at articular cartilage deviates from "*Coulomb's friction model*" and obeys a nonlinear increasing trend of coefficient of friction with sliding speeds.

Table 2. Fundamental expression of TRIBO interface as a function of state variables

Lubrication	Expression	Domain
Molecular	Hydration lubrication, H_20 dipole-dipole interaction, energy dissipation, biomechanical diffusion, and supramolecular adhesion	<1 nm
Boundary	Few layers of lubricin coat over superficial zone, heterogeneity, cartilage protection, and synergistic molecular expression	<100 nm
Mixed	Real area of contact, apparent area of contact, shear strength of asperity, load/aging dependence, and mixed lubrication models	Intermediate domain
Stick-slip	Bio-lubricant digestion, degradation of extracellular matrix, mechanical loading, sticking/sliping interfaces, and chronic friction	-do-
HD	Separation of biological interface by thick layers of synovial lubricant, dynamic state variable, and viscoelastic performance	~µm

Conclusion

Hydration lubrication, hydrogen bonds, supramolecular interaction, and heterogeneity with biomacromolecules are valuable for resolving the illusion of ultra-low coefficient of friction in healthy articular cartilage. Membrane surface tension reduction, biomechanical softening of interface material, disruption of the superficial zone, viscoelastic rheology, and synovial digestion evolve interfacial shear stress at the cartilage-cartilage interface. Biochemistry, biomechanical, and physiological boundaries at the biological interface influence bio-tribological performance under a quasi-static environment.

Conflict of Interests

Author primary interest is achievement of academic performance indicators profoundly none competing interest arises from residual indicators

Author contribution

Author during political lockdowns of SARS-CoV-2 pandemic outbreak virtually modulated at bioengineering spectrum for achievement of academic performance indicators. SDGs have been promoted from academic frontier for rationality of good health and well-being. Author perspective or personal view point on biomechanics is presented as per the requirement of academic indicators.

Funding resources

Academic content is prepared by self-funding for academic investment in aligned with sustainable pattern of academic journey

Acknowledgments

Author may like to acknowledge Department for Promotion of Industry and Internal Trade, Govt. of India with reference to registration of a few academic reports useful before basic structuring of perspective. Department of Applied Mechanics, IIT Delhi is acknowledged in aligned with a course of study innovative for re-searching at biomechanical spectrum.

References

[1] Siddaiah A and Menezes P L 2016 Advances in Bio-inspired Tribology for Engineering Applications *J. Bio- Tribo-Corros.* **2** p 23
[2] Zin Z and Dowson D 2013 Bio-friction *Friction* **1** pp 100-113
[3] Stachowiak G W 2017 How tribology has been helping us to advance and to survive *Friction* **5** pp 233–247
[4] Somorjai G A and Li Y 2011 Impact of surface chemistry *Proc. Natl. Acad. Sci. U.S.A.* **108** pp 917-924.

[5] Otsuki M and Matsukawa H 2013 Systematic Breakdown of Amontons' Law of Friction for an Elastic Object Locally Obeying Amontons' Law *Sci Rep* **3 p** 1586
[6] Gao J, Luedtke W D, Gourdon D, Ruths M, Israelachvili J N, and Landman U 2014 Frictional Forces and Amontons' Law: From the Molecular to the Macroscopic Scale *J. Phys. Chem. B* **108** pp 3410-3425
[7] Tomar P 2018 Efficient Friction Models in Extrusion Processes—A Review *Adv. Sci. Eng. Med.* **10** pp 230-233
[8] Bayer I S 2018 Advances in Tribology of Lubricin and Lubricin-Like Synthetic Polymer Nanostructures *Lubricants* **6** p 30
[9] Stachowiak G W, Batchelor A W and Griffiths L J 1994 Friction and wear changes in synovial joints *Wear* **171** pp 135-142
[10] Jay G D and Waller K A 2014 The biology of lubricin: Near frictionless joint motion *Matrix Biol.* **39** pp 17-24
[11] de Boer G N, Raske N, Soltanahmadi S, Dowson D, Bryant M G and Hewson R W 2020 A porohyperelastic lubrication model for articular cartilage in the natural synovial joint *Tribol. Int.* **149** p 105760
[12] Bali R and Shukla A K 2001 Rheological effects of synovial fluid on nutritional transport *Tribol. Lett.* **9** pp 233–239
[13] Crockett R 2009 Boundary Lubrication in Natural Articular Joints *Tribol. Lett.* **35** pp 77–84
[14] Tomar P 2021 Biological lubrication at articulating cartilage in moderate risk domain: PRG4/HA starved diffusion *23rd Int. Conf. on Wear of Mat.* (WOM2021/Elsevier)
[15] Shi Y, Hu X, Cheng J, *et al.* 2019 A small molecule promotes cartilage extracellular matrix generation and inhibits osteoarthritis development *Nat. Commun.* **10** p 1914
[16] Neu C P, Komvopoulos K and Reddi A H 2008 The Interface of Functional Bio-tribology and Regenerative Medicine in Synovial Joints *Tissue Eng. Part B Rev.* **14** pp 235-247
[17] Zhou Z R and Jin Z M 2015 Bio-tribology: Recent progresses and future perspectives *Biosurface Bio-tribology* **1** pp 3-24
[18] Williams G M, Klisch S M and Sah R L 2008 Bioengineering Cartilage Growth, Maturation, and Form *Pediatr. Res.* **63** pp 527-534

AUTHOR INDEX

Institute of Physics
Dirac House, Temple Back
Bristol BS1 6BE UK

ISSN: 1757-8981
ISBN 978-1-7138-6154-6

CURRAN ASSOCIATES INC.
proceedings.com